Mathematical Foundation of the Boundary Integro-Differential Equation Method

Houde Han · Dongsheng Yin

Mathematical Foundation of the Boundary Integro-Differential Equation Method

清華大學出版社
Tsinghua University Press

Springer

Houde Han
Department of Mathematical Sciences
Tsinghua University
Beijing, China

Dongsheng Yin
Department of Mathematical Sciences
Tsinghua University
Beijing, China

ISBN 978-981-95-1087-0 ISBN 978-981-95-1088-7 (eBook)
https://doi.org/10.1007/978-981-95-1088-7

Jointly published with Tsinghua University Press
The print edition is not for sale in China (Mainland). Customers from China (Mainland) please order the print book from: Tsinghua University Press.
ISBN of the Co-publisher's edition: 978-730-26-6473-4

This Springer imprint is published by the registered company Springer Nature Singapore Pte Ltd.
The registered company address is: 152 Beach Road, #21-01/04 Gateway East, Singapore 189721, Singapore

Preface

This book is focused on the mathematical foundation of the boundary integro-differential equation method. It is well known that the boundary integral equation method (or boundary element method) has become one effective numerical computational method for solving the boundary value problems of partial differential equations, that have been formulated as boundary integral equations including the hypersingular boundary integral equations. The hypersingular boundary integral operators are derived from the derivative of double layer potential corresponding to the given problem. From the view of science and engineering computing, the appearance of the hypersingular boundary integral operators has caused new difficulties in the boundary integral equation method. One way to resolve the difficulties is to find the computing method for the hypersingular boundary integral operators. In this book, we prefer to resolve the difficulties from a different perspective. The hypersingular boundary integral operators are regularized as boundary integro-differential operators with only weak singular boundary integral operators. It means that each of the hypersingular boundary integral operators is equivalent to a corresponding boundary integro-differential operator with only a weak singular boundary integral operator. The hypersingular boundary integral operators no longer appear; they are replaced by the corresponding integro-differential operators. Therefore, the difficulties in computing hypersingular integral operators have completely disappeared. It is the motivation for the title of this book.

The book also pays attention to the boundary integral equations of the first kind and the boundary integro-differential equations of the first kind. They occur in the acoustic scattering theory, as pointed out by Colton and Kress in their book (2013) "Traditionally, the use of integral equations of the first kind for studying boundary-value problems in acoustic scattering theory has been neglected due to the lack of a Riesz-Fredholm theory for equations of the first kind and the fact that integral equations of the first kind are improperly posed." But in Chap. 5 of this book boundary integral equations of the first kind and boundary integro-differential equations of the first kind are used to study the boundary value problems of the Helmholtz equation, because the Fredholm alternative theorems are established for the corresponding boundary integral equations of the first kind and boundary integro-differential equations of the first kind.

We also discuss in the last two chapters, the application of the boundary integro-differential equation method to nonlinear problems and the coercive and symmetrical coupling method of finite element method and boundary element method based on the boundary integro-differential operators.

This book contains nine chapters, as listed below.

Chapter 1: Distributions

Chapter 2: Fundamental Solutions of Linear Differential Operators

Chapter 3: Boundary Value Problems of the Laplace Equation

Chapter 4: Boundary Value Problems of Modified Helmholtz Equation

Chapter 5: Boundary Value Problems of Helmholtz Equation

Chapter 6: Boundary Value Problems of the Navier Equations

Chapter 7: Boundary Value Problems of the Stokes Equations

Chapter 8: Some Nonlinear Problems

Chapter 9: Coercive and Symmetrical Coupling Methods of Finite Element Method and Boundary Element Method

Due to the limited knowledge of the authors, errors are inevitable. We would be most grateful to learn of any errors in the book and any suggestions for the revision of a future printing.

We are very grateful for the support and encouragement from our colleagues and friends during the preparing this book. Professor Hermann Brunner carefully read the whole book and gave helpful suggestions for revision. This book has benefited from the works of many other researchers, including our co-authors Weijun Tang, Zhi Guan, Bin He, Chongqing Yu, Jeng-Tzong Chen, Ying-Te Lee, and Wenjun Ying.

Beijing, China

Houde Han
Dongsheng Yin

Introduction

Potential theory originates from physics, especially from gravitational and electrostatic attraction. The gravity force and electrostatic force could be represented by the functions called gravitational potential and electrostatic potential, which satisfy the Laplace equation in a vacuum or the Poisson equation in other cases.

In the nineteenth century potential theory formed as a branch of mathematics. Based on the potential theory a strong method was created to study the principal boundary value problems of linear partial differential equations (LPDE) in modeling a wide range of physical phenomena including gravitational and electrostatic problems. The special parameter-dependent integrals were introduced by the fundamental solution of the corresponding linear partial differential equation (equations) with constant coefficients, such as the single layer potential, the double layer potential, and the volume potential. The properties of those potentials were studied, which play an important role in potential theory. By the potentials and their properties, the principal boundary value problems of LPDE can be reduced to boundary integral equations. This idea can be traced back to the work by Green (1828), Gauss (1839, 1840), then a series of research results appeared on this subject, such as Kellogg (1953), Günter (1967), Muskhelishvili (1967), Mikhlin (2014), and Kupradze (1967). The main goal of these contributions is to study the existence, regularity, and uniqueness of the solution of the principal boundary value problems appearing in broad fields of mathematical physics rather than their numerical solutions.

After the 1960s, this framework was applied to the numerical solution of partial differential equations with the widespread use of computers in engineering and science. The early works by Hess & Smith (1962, 1964), Jaswon (1963), Symm (1963), Hsiao & Wendland (1977), Brebbia (1978), and many others, which can be found in the review papers by Cheng & Cheng (2005). This method was called the boundary integral equation method (Jaswon 1963; Symm 1963; Cruse 1972) or the boundary element method (Brebbia, 1978).

Today the boundary integral equation method has become one of the most effective and important numerical method for the boundary value problems of partial differential equations. The key point of the boundary integral equation method is to reduce the boundary value problems to the equivalent boundary integral equations by the potential theory. It is well known that there exist two main procedures for this boundary reduction. One of them, the direct procedure, is based on the solution of the given boundary value problem which can be represented by Green's formula. The second one, the indirect procedure, is established on the introduction of the new auxiliary function on the boundary. The solution to the given problem is represented by the (single or double) layer potential with the auxiliary function as the density function, which is unknown. In this book, we focus on the indirect procedure; however, the results we obtain also can be applied to the direct procedure. After the boundary reduction procedure, the first and second kind boundary integral equations with singular boundary integral operator and hypersingular boundary integral operator (Hadamard's finite part integral operator) are obtained. From the view of science and engineering computing, the remarkable advantage of the boundary integral equation method is to reduce the dimensions of the given problem by one dimension. Thus, the computing scale of the given problem is greatly reduced. However the appearance of the singular boundary integral operators, especially the hypersingular boundary integral operators has brought a new difficulty in the boundary integral method. One way to overcome the difficulty is to study the computing method of the hypersingular integral operator (Tanaka et al., 1994; Gray et al., 2004). In this book, we emphasize to overcome this difficulty from a different perspective. Namely the new expressions of the hypersingular boundary integral operators arising in the boundary integral equations are given, in which the hypersingular boundary integrals are regularized as the boundary integro-differential operators with only weak singular boundary integral operators. When the hypersingular boundary integral operator appears in the boundary integral equations, we replace it with its new expression (boundary integro-differential operator). Therefore, we obtain the equivalent boundary integro-differential equations. From the perspective of numerical computing, the hypersingular boundary integral operators no longer appear, and hence the difficulty of computing the hypersingular boundary integral operators disappeared completely. This is one of the characteristics of this book, also the motivation for the book title.

Our attention will focus on the mathematical foundation of the boundary integro-differential equation method in this book. It is well known that the fundamental solutions of LPDE play a key role in the boundary reduction process, furthermore, the distribution plays a definite role in the definition and finding the fundamental solutions (Kythe, 1996; Ortner & Wagner, 2015). Naturally "Distribution" and "Fundamental Solution of Linear Differential Operators" become the first two chapters of this book. In the following chapters, the principal boundary value problems of Laplace equation, Helmholtz equation, Navier equations, Stokes equations, etc., are discussed. The applications of the boundary integro-differential equation method to some nonlinear problems are studied in Chap. 8 including the heat radiation problem, variational inequalities (Laplace equation with unilateral boundary conditions, Signorini problems in linear elasticity), and Steklov eigenvalue problems. Also based on the boundary integro-differential equation method, the coercive and symmetric coupling method of finite element method and boundary element method is given in Chap. 9.

In addition to the previously quoted books, the interested readers can also refer to other books related to this subject, such as the books by the following authors, Jaswon and Symm (1977), Kupradze (2012), Brebbia and Walker (2016), Colton and Kress (1998, 2013), Brebbia et al. (2012), Chen and Zhou (1992), Nédélec (2001), Yu (2002), Linkov (2002), Hsiao and Wendland (2008), Katsikadelis (2016).

References

Brebbia C. A. (1978). *The boundary element method for engineers*. Pentech Press Ltd, Plymouth.

Brebbia C. A., Telles J. C. F. & Wrobel L. C. (2012). *Boundary element techniques: theory and applications in engineering*. Springer Science & Business Media.

Brebbia C. A. & Walker S. (2016). *Boundary element techniques in engineering*. Elsevier.

Chen G. & Zhou J. (1992). *Boundary element methods* (Vol. 92). Academic press, London.

Cheng A. H. D. & Cheng D. T. (2005). Heritage and early history of the boundary element method. *Engineering analysis with boundary elements*, *29*(3), 268–302.

Colton D. & Kress R. (1998). *Inverse acoustic and electromagnetic scattering theory* (Vol. 93). Springer.

Colton D. & Kress R. (2013). *Integral equation methods in scattering theory*. SIAM.

Cruse T. (1972). Application of the boundary-integral equation solution method in solid mechanics. *Variational Methods in Engineering. Department of Civil Engineering, Southampton University, England*, 9–1.

Gauss C. (1839). Allgemeine lehrsätze in beziehung auf die im verkelirten verhdltrnisse des. *Quadrats der Entfernung wirkenden Anziehungs-und Abstossungskrdfte Gauss Werke*, 5.

Gauss C. (1840). *Atlas des erdmagnetismus, werke 12*. Göttingen.

Gray L. J., Glaeser J. & Kaplan T. (2004). Direct evaluation of hypersingular galerkin surface integrals. *SIAM Journal on Scientific Computing*, *25*(5), 1534–1556.

Green G. (1828). *An essay on the application of mathematical analysis to the theories of electricity and magnetism*. Notingham, UK

Günter N. M. (1967). *Potential theory and its applications to basic problems of mathematical physics*. Ungar, New York.

Hess J. L. & Smith A. (1964). Calculation of nonlifting potential flow about arbitrary three-dimensional bodies. *Journal of ship research*, *8*(04), 22–44.

Hsiao G. C. & Wendland W. L. (1977). A finite element method for some integral equations of the first kind. *Journal of Mathematical Analysis and Applications*, *58*(3), 449–481.

Hsiao G. C. & Wendland W. L. (2008). *Boundary integral equations*. Springer.

Jaswon M. (1963). Integral equation methods in potential theory. I. *Proceedings of the Royal Society of London. Series A. Mathematical and Physical Sciences*, *275*(1360), 23–32.

Jaswon M. A. & Symm G. T. (1977). *Integral equation methods in potential theory and elastostatics* (Vol. 2). Academic Press.

Katsikadelis J. T. (2016). *The boundary element method for engineers and scientists: theory and applications*. Academic Press.

Kellogg O. D. (1953). *Foundations of potential theory* (Vol. 31). Courier Corporation.

Kupradze V. D. (1967). *Potential methods in elasticity theory* (Tech. Rep.).

Kupradze V. D. (2012). *Three-dimensional problems of elasticity and thermoelasticity*. Elsevier.

Kythe P. (1996). *Fundamental solutions for differential operators and applications*. Springer Science & Business Media.

Linkov A. M. (2002). *Boundary integral equations in elasticity theory* (Vol. 99). Springer Science & Business Media.

Mikhlin S. G. (2014). *Integral equations: and their applications to certain problems in mechanics, mathematical physics and technology*. Elsevier.

Muskhelishvili N. I. (1967). *Some basic problems of the mathematical theory of elasticity* (Vol. 15). Noordhoff, Groningen.

Nédélec J. C. (2001). *Acoustic and electromagnetic equations: integral representations for harmonic problems* (Vol. 144). Springer.

Ortner N. & Wagner P. (2015). *Fundamental solutions of linear partial differential operators*. Springer.

Symm G. (1963). Integral equation methods in potential theory. II. *Proceedings of the Royal Society of London. Series A. Mathematical and Physical Sciences*, *275*(1360), 33–46.

Tanaka M., Sladek V. & Sladek J. (1994). Regularization techniques applied to boundary element methods.

Yu D. H. (2002). *Natural boundary integral method and its applications* (Vol. 539). Springer Science & Business Media.

Contents

About the Authors

Houde Han is a Professor at Tsinghua University who has devoted himself to the teaching and scientific research of computational mathematics. He has achieved significant breakthroughs in the study of numerical solutions of partial differential equations, demonstrating a series of creative research results.

In particular, he has made important contributions to the numerical solutions of partial differential equations on unbounded domains, and the coupling method of the finite element method and boundary element method. Additionally, he has contributed to the numerical solutions of boundary integro-differential equations and variational inequalities problems, as well as the numerical solutions of singular perturbation problems, ill-posed problems, and infinite element methods.

As recognition for his outstanding contributions, Han has been awarded several accolades. He was awarded the second prize of the National Natural Science Award in 2008, the Pieter Hemker Prize in 2008, the Second Prize of the Beijing Science and Technology Progress Award in 2002, the State Education Commission Division First Prize of Technology Progress Award in 1995, the Second Prize in 1988, and the National Science Conference Award in 1978.

Dongsheng Yin is an Associate Professor at Tsinghua University whose research interest is mainly focused on high-frequency waves, partial differential equations on unbounded domains, and fractional differential equations.

Chapter 1
Distributions

The establishment and perfection of distribution (generalized function) theory is an important mathematical achievement in the 20th century, which has a long development process. The δ-function (usually called Dirac's delta function) was introduced by Dirac (1927) for describing a unit point mass (or charge). The δ-function was defined by

$$\delta(x) = \begin{cases} 0, & \text{if } x \neq 0 \\ \infty, & \text{otherwise} \end{cases} \quad \text{and} \quad \int_{-\infty}^{\infty} \delta(x)\,\mathrm{d}x = 1.$$

It has the following properties

$$\delta(-x) = \delta(x), \qquad x\delta(x) = 0, \quad \int_{-\infty}^{+\infty} f(x)\delta(x)\,\mathrm{d}x = f(0)$$

for any continuous function $f(x)$.

Before Dirac, Heaviside (1893) introduced Heaviside function $H(x)$, which is defined by

$$H(x) = \begin{cases} 1, & x > 0, \\ \frac{1}{2}, & x = 0, \\ 0, & x < 0. \end{cases}$$

Heaviside function $H(x)$ is a discontinuous function, the derivative of which was defined as δ-function, namely

$$\frac{\mathrm{d}H(x)}{\mathrm{d}x} = \delta(x).$$

These definitions of δ-function have contradictions with the classical mathematical analysis, which is not as a genuine mathematical function for a long time. However, the δ-function was successfully used in physics. For example, it eventually let

H. Han and D. Yin, *Mathematical Foundation of the Boundary Integro-Differential Equation Method*, https://doi.org/10.1007/978-981-95-1088-7_1

Dirac develop the theories of matrix mechanics and wave mechanics. The δ-function becomes a prominent tool in quantum mechanical research. The goal of mathematicians is not to exclude such improper functions. Mathematicians should try to understand the contradictions and define such kind improper functions mathematically. The generalized functions were invented in 1936 by Sobolev (1936). These ideas were developed by Schwartz in an extended form in the 1940s. The term "distribution" was introduced by Schwartz in his monograph (Schwartz, 1966), which by analogy with a distribution of electrical charge.

The key point in distribution theory is to reinterpret functions as linear functionals on a given space of test functions. The space of smooth functions with compact supports is the basic space of test functions used in distribution theory, leading to the standard distributions. It is said that "Sobolev invented distributions, but it was Schwartz who created the theory of distributions" (Lützen, 2012; Kythe, 1996).

1.1 Space of Test Functions

Let $\mathbb{R}^n$ be the n-dimensional Euclidean space. Let $\boldsymbol{\alpha} = (\alpha_1, \alpha_2, \ldots, \alpha_n)$ be a multi-index of dimension n with $\alpha_1, \alpha_2, \ldots, \alpha_n$ be non-negative integers and

$$|\boldsymbol{\alpha}| := \sum_{j=1}^{n} \alpha_j = n. \tag{1.1.1}$$

The $\boldsymbol{\alpha}$-differential operator is defined by

$$D^{\alpha} = \frac{\partial^{|\alpha|}}{\partial x_1^{\alpha_1} \partial x_2^{\alpha_2} \cdots \partial x_n^{\alpha_n}}, \qquad \forall \boldsymbol{x} = (x_1, x_2, \ldots, x_n) \in \mathbb{R}^n. \tag{1.1.2}$$

Definition 1.1.1 For a real-valued function φ defined on $\mathbb{R}^n$, the support of φ is defined as the smallest closed set containing all points not mapped to zero. Namely,

$$\mathrm{Supp}\{\varphi\} := \overline{\{\boldsymbol{x} \in \mathbb{R}^n : \varphi(\boldsymbol{x}) \neq 0\}}.$$

If $\mathrm{Supp}\{\varphi\}$ is bounded, then the function φ is said to be *compactly supported* or one may say the function φ has compact support.

Let $C_0^{\infty}(\mathbb{R}^n)$ denote the space of all infinitely differentiable real-valued functions on $\mathbb{R}^n$ with compact support. It can be equipped with a topology by defining the limit of a sequence of elements of $C_0^{\infty}(\mathbb{R}^n)$.

Definition 1.1.2 A sequence $\{\varphi_m, m = 1, 2, \ldots\}$, where $\varphi_m \in C_0^{\infty}(\mathbb{R}^n)$, converges to φ if the following two conditions hold:

1. There exists a bounded domain $\bar{\Omega} \subset \mathbb{R}^n$ such that

$$\mathrm{Supp}\{\varphi_m\} \subset \bar{\Omega}, \qquad \text{for } m = 1, 2, \ldots;$$

2. $D^\alpha \varphi_m \to D^\alpha \varphi$ uniformly as $m \to \infty$ for all multi-indices $\boldsymbol{\alpha}$.

It is not difficult to see that $C_0^\infty(\mathbb{R}^n)$ is closed (or is complete) with respect to this definition of convergence (Kanwal, 2004).

Definition 1.1.3 Let $\mathcal{D}(\mathbb{R}^n)$ denote the space $C_0^\infty(\mathbb{R}^n)$ with the convergence defined by Definition 1.1.2. It is called the *basic space of test functions*.

1.2 Definition of Distributions and Their Operations

A linear functional $\ell(\varphi)$ on the space of test functions $\mathcal{D}(\mathbb{R}^n)$ is an operator (or a rule) by which we assign to every $\varphi \in \mathcal{D}(\mathbb{R}^n)$ a real number denoted by $\ell(\varphi)$ such that

$$\ell(c_1\varphi_1 + c_2\varphi_2) = c_1\ell(\varphi_1) + c_2\ell(\varphi_2), \quad \forall\, \varphi_1,\, \varphi_2 \in \mathcal{D}(\mathbb{R}^n), \quad c_1,\, c_2 \in \mathbb{R}. \tag{1.2.1}$$

Definition 1.2.1 A linear functional $\ell(\varphi)$ on $\mathcal{D}(\mathbb{R}^n)$ is continuous if and only if

$$\lim_{m\to\infty} \ell(\varphi_m) = \ell(\varphi), \quad \text{when} \quad \lim_{m\to\infty} \varphi_m = \varphi,$$

in the sense of the convergence defined by Definition 1.1.2.

We are now ready to define the distributions.

Definition 1.2.2 A continuous linear functional on the space of test functions $\mathcal{D}(\mathbb{R}^n)$ is called a distribution. Let $\mathcal{D}'(\mathbb{R}^n)$ denote all distributions on $\mathcal{D}(\mathbb{R}^n)$, which is called the dual space of $\mathcal{D}(\mathbb{R}^n)$.

Example 1.2.1 (*Regular distributions*) For a given locally integrable function $f(\boldsymbol{x})$ on $\mathbb{R}^n$ and test function $\varphi \in \mathcal{D}(\mathbb{R}^n)$, we have

$$f(\varphi) = \langle f, \varphi \rangle = \int_{\mathbb{R}^n} f(\boldsymbol{x})\varphi(\boldsymbol{x})\,\mathrm{d}\boldsymbol{x}, \qquad \forall\, \varphi \in \mathcal{D}(\mathbb{R}^n), \tag{1.2.2}$$

which is a linear functional on $\mathcal{D}(\mathbb{R}^n)$, and

$$|\langle f, \varphi \rangle| \leqslant \max_{\boldsymbol{x}\in \mathrm{Supp}\{\varphi\}} |\varphi(\boldsymbol{x})| \int_{\mathrm{Supp}\{\varphi\}} |f(\boldsymbol{x})|\,\mathrm{d}\boldsymbol{x} < +\infty.$$

If the sequence $\{\varphi_m\}$ converges to zero in $\mathcal{D}(\mathbb{R}^n)$ as $m \to \infty$, then

$$\lim_{m\to\infty} \langle f, \varphi_m \rangle = 0.$$

It means $f(\varphi)$ is a continuous linear functional on $\mathcal{D}(\mathbb{R}^n)$, which is a distribution and denoted by f. The distributions defined by (1.2.2) are *regular* distributions.

Example 1.2.2 (*δ-function*) The δ-function is defined by

$$\langle \delta, \varphi \rangle = \varphi(0), \qquad \forall \varphi \in \mathcal{D}(\mathbb{R}^n). \tag{1.2.3}$$

Obviously this functional is linear. If $\lim_{m\to\infty} \varphi_m = 0$ in $\mathcal{D}(\mathbb{R}^n)$, then $\lim_{m\to\infty} \varphi_m(0) = 0$. Therefore, $\lim_{m\to\infty} \langle \delta, \varphi_m \rangle = \lim_{m\to\infty} \varphi_m(0)$. Continuity of this functional follows directly. The δ-function is a distribution, but it is not locally integrable. Instead, it is a singular distribution.

Example 1.2.3 The function $\dfrac{1}{x}$ is not locally integrable on $\mathbb{R}^1$. The integral $\displaystyle\int_{-\infty}^{+\infty} \frac{1}{x}\varphi(x)\,\mathrm{d}x$ is not convergent for $\varphi \in \mathcal{D}(\mathbb{R}^1)$ and $\varphi(0) \neq 0$. Therefore $\dfrac{1}{x}$ is not a distribution. However, we can define the Cauchy principle value of this integral by

$$\text{P.V.} \int_{-\infty}^{+\infty} \frac{\varphi(x)}{x}\,\mathrm{d}x = \lim_{\epsilon\to+0} \int_{|x|\geqslant\epsilon} \frac{\varphi(x)}{x}\,\mathrm{d}x.$$

Let $[-A, A]$ be the support of the test function $\varphi \in \mathcal{D}(\mathbb{R}^1)$. There is a continuous function ψ such that $\varphi(x) = \varphi(0) + x\psi(x)$. Therefore,

$$\begin{aligned}\text{P.V.} \int_{-\infty}^{+\infty} \frac{\varphi(x)}{x}\,\mathrm{d}x &= \lim_{\epsilon\to+0} \int_{A\geqslant|x|\geqslant\epsilon} \left(\frac{\varphi(0)}{x} + \psi(x)\right)\mathrm{d}x \\ &= \int_{-A}^{A} \psi(x)\,\mathrm{d}x := \left\langle \mathrm{P}\left(\frac{1}{x}\right), \varphi \right\rangle.\end{aligned}$$

The functional $\left\langle \mathrm{P}\left(\frac{1}{x}\right), \varphi \right\rangle$ is linear. We have

$$\left\langle \mathrm{P}\left(\frac{1}{x}\right), \varphi \right\rangle = \int_{-A}^{A} \psi(x)\,\mathrm{d}x = 2A\psi(\xi) = 2A\frac{\varphi(\xi) - \varphi(0)}{\xi}$$

for some $\xi \in [-A, A]$. As a result, we get

$$\left|\left\langle \mathrm{P}\left(\frac{1}{x}\right), \varphi \right\rangle\right| \leqslant 2A \max_{x\in\mathbb{R}^1} |\varphi'(x)|.$$

The functional $\left\langle \mathrm{P}\left(\frac{1}{x}\right), \varphi \right\rangle$ is continuous and $\mathrm{P}\left(\frac{1}{x}\right)$ is a distribution.

Example 1.2.4 The function $\dfrac{1}{|\boldsymbol{x}|^2}$ is not locally integrable on $\mathbb{R}^2$ and the integral

$$\int_{\mathbb{R}^2} \frac{\varphi(\boldsymbol{x})}{|\boldsymbol{x}|^2}\,\mathrm{d}\boldsymbol{x}$$

is divergent for $\varphi(\boldsymbol{x}) \in \mathcal{D}(\mathbb{R}^2)$ with $\varphi(0) \neq 0$. Therefore, $\dfrac{1}{|\boldsymbol{x}|^2}$ is not a distribution. Similarly, we can define the Cauchy principle value of the divergent integral by

$$\begin{aligned}\text{P.V.} \int_{\mathbb{R}^2} \frac{\varphi(\boldsymbol{x})}{|\boldsymbol{x}|^2} \mathrm{d}\boldsymbol{x} &= \int_{|\boldsymbol{x}|\leqslant 1} \frac{\varphi(\boldsymbol{x}) - \varphi(0)}{|\boldsymbol{x}|^2} \mathrm{d}\boldsymbol{x} + \int_{|\boldsymbol{x}|>1} \frac{\varphi(\boldsymbol{x})}{|\boldsymbol{x}|^2} \mathrm{d}\boldsymbol{x} \\ &= \left\langle \mathrm{P}\left(\frac{1}{|\boldsymbol{x}|^2}\right), \varphi \right\rangle, \qquad \forall \varphi \in \mathcal{D}(\mathbb{R}^2).\end{aligned}$$

The functional above is linear and continuous. Therefore, $\mathrm{P}\left(\dfrac{1}{|\boldsymbol{x}|^2}\right)$ is a distribution, namely, $\mathrm{P}\left(\dfrac{1}{|\boldsymbol{x}|^2}\right) \in \mathcal{D}'(\mathbb{R}^2)$.

Definition 1.2.3 (*Derivatives of Distributions*) For a given distribution f, its derivative $\dfrac{\partial f}{\partial x_i}$ is defined by

$$\left\langle \frac{\partial f}{\partial x_i}, \varphi \right\rangle = -\left\langle f, \frac{\partial \varphi}{\partial x_i} \right\rangle, \qquad \forall \varphi \in \mathcal{D}(\mathbb{R}^n). \tag{1.2.4}$$

In fact for any $\varphi \in \mathcal{D}(\mathbb{R}^n)$, $\dfrac{\partial \varphi(\boldsymbol{x})}{\partial x_i}$ is also a test function in $\mathcal{D}(\mathbb{R}^n)$. It is straightforward to show that the functional on $\mathcal{D}(\mathbb{R}^n)$ defined by (1.2.4) yields the following theorem (Kanwal, 2004).

Theorem 1.2.1 *1. For any distribution $f \in \mathcal{D}'(\mathbb{R}^n)$, which is infinitely differentiable, and*

$$\langle D^\alpha f, \varphi \rangle = (-1)^{|\alpha|} \langle f, D^\alpha \varphi \rangle, \qquad \forall \varphi \in \mathcal{D}(\mathbb{R}^n).$$

2. The equalities $\dfrac{\partial f}{\partial x_i} = 0$ $(i = 1, 2, \ldots, n)$ hold if and only if the distribution f is a constant.

3. The generalized derivative agrees with the classical derivative whenever the latter exists.

Product of a Distribution and a Function

In general, it may not be possible to define the product of two distributions, even for the product of a distribution and a locally integrable function. But the product of a distribution $f \in \mathcal{D}'(\mathbb{R}^n)$ and an infinitely differentiable $\psi \in C^\infty(\mathbb{R}^n)$ can be defined by

$$\langle \psi f, \varphi \rangle = \langle f, \psi \varphi \rangle, \qquad \forall \varphi \in \mathcal{D}(\mathbb{R}^n), \tag{1.2.5}$$

because the function $\psi\varphi \in \mathcal{D}(\mathbb{R}^n)$.

Example 1.2.5 For a constant c and a distribution $f \in \mathcal{D}'(\mathbb{R}^n)$, we have $cf \in \mathcal{D}'(\mathbb{R}^n)$, which is defined by

$$\langle cf, \varphi\rangle = c\,\langle f, \varphi\rangle\,, \qquad \forall\,\varphi \in \mathcal{D}(\mathbb{R}^n). \tag{1.2.6}$$

Example 1.2.6 For the δ-function and $\psi \in C^\infty(\mathbb{R}^n)$, we know that $\psi\delta \in \mathcal{D}'(\mathbb{R}^n)$ and

$$\langle \psi\delta, \varphi\rangle = \langle \delta, \psi\varphi\rangle = \psi(0)\varphi(0) = \langle \psi(0)\delta, \varphi\rangle\,.$$

That is, we get

$$\psi\delta = \psi(0)\delta. \tag{1.2.7}$$

For a distribution $f \in \mathcal{D}'(\mathbb{R}^n)$ and a non-singular affine transformation $\boldsymbol{y} = \boldsymbol{A}\boldsymbol{x} - \boldsymbol{b} : \mathbb{R}^n \to \mathbb{R}^n$. Then $f(\boldsymbol{A}\boldsymbol{x} - \boldsymbol{b})$ is a distribution defined by

$$\langle f(\boldsymbol{A}\boldsymbol{x} + \boldsymbol{b}), \varphi\rangle = \big\langle f, \varphi(\boldsymbol{A}^{-1}\boldsymbol{y} + \boldsymbol{b})\big\rangle\,, \qquad \forall\,\varphi \in \mathcal{D}(\mathbb{R}^n). \tag{1.2.8}$$

Example 1.2.7 For the transformation $\boldsymbol{y} = -\boldsymbol{x}$, $f(-\boldsymbol{x})$ is a distribution provided that $f \in \mathcal{D}'(\mathbb{R}^n)$ as

$$\langle f(-\boldsymbol{x}), \varphi(\boldsymbol{x})\rangle = \langle f(\boldsymbol{y}), \varphi(-\boldsymbol{y})\rangle\,, \qquad \forall\,\varphi \in \mathcal{D}(\mathbb{R}^n). \tag{1.2.9}$$

Example 1.2.8 For $\boldsymbol{y} = \lambda\boldsymbol{x}$ with real number $\lambda \neq 0$, $f(\lambda\boldsymbol{x})$ is also a distribution provided that $f \in \mathcal{D}'(\mathbb{R}^n)$ as

$$\langle f(\lambda\boldsymbol{x}), \varphi(\boldsymbol{x})\rangle = \big\langle f(\boldsymbol{y}), \varphi(\lambda^{-1}\boldsymbol{y})\big\rangle\,, \qquad \forall\,\varphi \in \mathcal{D}(\mathbb{R}^n). \tag{1.2.10}$$

If $f \in \mathcal{D}'(\mathbb{R}^n)$ satisfies

$$f(\lambda\boldsymbol{x}) = \lambda^k f(\boldsymbol{x}),$$

then f is called a k-order homogeneous distribution.

Example 1.2.9 For $\boldsymbol{y} = \boldsymbol{x} - \boldsymbol{b}$, $f(\boldsymbol{x} - \boldsymbol{b})$ is a distribution provided that $f \in \mathcal{D}'(\mathbb{R}^n)$ as

$$\langle f(\boldsymbol{x} - \boldsymbol{b}), \varphi(\boldsymbol{x})\rangle = \langle f(\boldsymbol{y}), \varphi(\boldsymbol{y} + \boldsymbol{b})\rangle\,, \qquad \forall\,\varphi \in \mathcal{D}(\mathbb{R}^n). \tag{1.2.11}$$

Therefore, $\delta(\boldsymbol{x} - \boldsymbol{\xi})$ is a distribution and

$$\langle \delta(\boldsymbol{x} - \boldsymbol{\xi}), \varphi(\boldsymbol{x})\rangle = \varphi(\boldsymbol{\xi}), \qquad \forall\,\varphi \in \mathcal{D}(\mathbb{R}^n). \tag{1.2.12}$$

Convergence of Distribution

A given sequence of distributions $f_j \in \mathcal{D}'(\mathbb{R}^n)$, $j = 1,\ 2,\ \ldots$, is said to converge to $f \in \mathcal{D}'(\mathbb{R}^n)$ if

$$\lim_{j\to\infty} \left\langle f_j, \varphi \right\rangle = \langle f, \varphi \rangle, \qquad \forall \varphi \in \mathcal{D}(\mathbb{R}^n). \tag{1.2.13}$$

The distribution $f \in \mathcal{D}'(\mathbb{R}^n)$ is called the limit of the sequence $\left\{f_j,\ j = 1, 2, \ldots\right\}$ and we write

$$\lim_{j\to\infty} f_j = f, \qquad \text{in } \mathcal{D}'(\mathbb{R}^n). \tag{1.2.14}$$

We have the following theorem (Kanwal, 2004).

Theorem 1.2.2 *If the distribution sequence $\{f_j,\ j = 1, 2, \ldots\}$ converges to distribution f, then the distribution sequence $\{D^\alpha f_j,\ j = 1, 2, \ldots\}$ converges to distribution $D^\alpha f$.*

Proof By definition, we have

$$\begin{aligned}
\lim_{j\to\infty} \left\langle D^\alpha f_j, \varphi \right\rangle &= \lim_{j\to\infty} (-1)^\alpha \left\langle f_j, D^\alpha \varphi \right\rangle \\
&= (-1)^\alpha \left\langle f, D^\alpha \varphi \right\rangle \\
&= \left\langle D^\alpha f, \varphi \right\rangle, \qquad \forall \varphi \in \mathcal{D}(\mathbb{R}^n).
\end{aligned}$$

□

Now we consider the distribution f_h, which involves a parametric variable $h \in \mathbb{R}^1$.

Definition 1.2.4 For given f_h, distributions involving parametric variable $h \in \mathbb{R}^1$, and a distribution $f \in \mathcal{D}'(\mathbb{R}^n)$, if

$$\lim_{h\to 0} \langle f_h, \varphi \rangle = \langle f, \varphi \rangle, \qquad \forall \varphi \in \mathcal{D}(\mathbb{R}^n), \tag{1.2.15}$$

then f is called the limit of f_h, when $h \to 0$. We denote it by

$$\lim_{h\to 0} f_h = f, \qquad \text{in } \mathcal{D}'(\mathbb{R}^n). \tag{1.2.16}$$

Any distribution can be approximated by the classical functions, especially for the δ-function. The approximation of δ-function has been widely applied in physics.

Definition 1.2.5 A function sequence $\{S_j,\ j = 1, 2, \ldots\}$ is called a delta-convergence sequence if

$$\lim_{j\to\infty} \int_{\mathbb{R}^n} S_j(\boldsymbol{x})\varphi(\boldsymbol{x})\, \mathrm{d}\boldsymbol{x} = \varphi(0), \qquad \forall \varphi \in \mathcal{D}(\mathbb{R}^n). \tag{1.2.17}$$

Then we can say the limit of a delta-convergence sequence $\{S_j,\ j = 1, 2, \ldots\}$ is the δ-function, i.e.,

$$\lim_{j\to\infty} S_j = \delta, \qquad \text{in } \mathcal{D}'(\mathbb{R}^n). \tag{1.2.18}$$

Example 1.2.10 (Kanwal, 2004)

$$\lim_{j\to\infty} \frac{1}{\pi}\frac{j}{1+j^2x^2} = \delta, \qquad \text{in } \mathcal{D}'(\mathbb{R}^1). \tag{1.2.19}$$

$$\lim_{j\to\infty} \left(\frac{j}{\pi}\right)^{1/2} \mathrm{e}^{-jx^2} = \delta, \qquad \text{in } \mathcal{D}'(\mathbb{R}^1). \tag{1.2.20}$$

Support of Distribution

By the definition of distributions, one may have realized that we can not talk about properties of distributions at a point $\boldsymbol{x} \in \mathbb{R}^n$. We can only consider their action on the test function in $\mathcal{D}(\mathbb{R}^n)$.

Definition 1.2.6 For a distribution $f \in \mathcal{D}'(\mathbb{R}^n)$, if there exists an open domain $\Omega \subset \mathbb{R}^n$ such that

$$\langle f, \varphi\rangle = 0, \qquad \forall \varphi \in \mathcal{D}(\mathbb{R}^n) \ \text{ and } \ \mathrm{Supp}\{\varphi\} \subset \Omega, \tag{1.2.21}$$

then we say that the distribution f vanishes in the domain Ω.

Definition 1.2.7 A point $\boldsymbol{x}$ is called an essential point of a distribution f if there does not exist a neighborhood of $\boldsymbol{x}$ in which $f(\boldsymbol{x})$ vanishes.

Definition 1.2.8 The closure of the set of all essential points of f is defined as support of f and denoted as $\mathrm{Supp}\{f\}$, the same notation as for classical functions. It is clear that the support of δ and each of its derivatives is a single point $\boldsymbol{x} = 0$. For $a \in C^\infty(\mathbb{R}^n)$ and $f \in \mathcal{D}'(\mathbb{R}^n)$, we have

$$\mathrm{Supp}\{af\} \subset \mathrm{Supp}\{a\} \cap \mathrm{Supp}\{f\}. \tag{1.2.22}$$

1.3 Direct Products and Convolution of Distributions

Direct Products of Distributions

Let $\boldsymbol{x} = (x_1, x_2, \ldots, x_m)$ and $\boldsymbol{y} = (y_1, y_2, \ldots, y_n)$ denote the points of Euclidean spaces $\mathbb{R}^m$ and $\mathbb{R}^n$, respectively. Then $(\boldsymbol{x}, \boldsymbol{y}) = (x_1, x_2, \ldots, x_m; y_1, y_2, \ldots, y_n)$ is a point of Cartesian product $\mathbb{R}^{m+n}$. Suppose that $f(\boldsymbol{x})$ and $g(\boldsymbol{y})$ are locally integrable functions in $\mathbb{R}^m$ and $\mathbb{R}^n$, then $f(\boldsymbol{x})g(\boldsymbol{y})$ is also a locally integrable function in $\mathbb{R}^{m+n}$. It defines the regular distribution function $fg \in \mathcal{D}'(\mathbb{R}^{m+n})$,

$$\langle fg, \varphi\rangle = \int_{\mathbb{R}^m} f(\boldsymbol{x}) \int_{\mathbb{R}^n} g(\boldsymbol{y})\varphi(\boldsymbol{x}, \boldsymbol{y})\,\mathrm{d}\boldsymbol{y}\,\mathrm{d}\boldsymbol{x}$$

$$= \langle f, \langle g, \varphi \rangle \rangle, \qquad \forall \varphi \in \mathcal{D}(\mathbb{R}^{m+n}).$$

Given two distributions $f \in \mathcal{D}'(\mathbb{R}^m)$ and $g \in \mathcal{D}'(\mathbb{R}^n)$, is it possible to define the direct product of two distributions f and g ? Before answering this question, we recall the following lemma.

Lemma 1.3.1 (Kanwal, 2004) *For any $g \in \mathcal{D}'(\mathbb{R}^n)$ and $\varphi \in \mathcal{D}(\mathbb{R}^{m+n})$, we have*

$$\psi = \langle g, \varphi \rangle \in \mathcal{D}(\mathbb{R}^m)$$

and

$$D^\alpha \psi = \left\langle g, D_x^\alpha \varphi \right\rangle, \tag{1.3.1}$$

for all multi-index α, where D_x^α means differentiation with respect to $\boldsymbol{x} = (x_1, x_2, \ldots, x_m)$.

By the results in Lemma 1.3.1, we obtain that $\langle f, \langle g, \varphi \rangle \rangle$ is a continuous linear functional on $\mathcal{D}(\mathbb{R}^{m+n})$.

Definition 1.3.1 For any two distributions $f \in \mathcal{D}'(\mathbb{R}^m)$ and $g \in \mathcal{D}'(\mathbb{R}^n)$, the direct product $f \otimes g$ of f and g is defined by

$$\langle f \otimes g, \varphi \rangle = \langle f, \langle g, \varphi \rangle \rangle, \qquad \forall \varphi \in \mathcal{D}(\mathbb{R}^{m+n}). \tag{1.3.2}$$

Some properties of the direct products are listed as follows (Kanwal, 2004):

1. Commutativity.

$$f \otimes g = g \otimes f. \tag{1.3.3}$$

2. Continuity. If the sequence $\{f_j,\ j = 1, 2, \ldots\}$ converges to f in $\mathcal{D}'(\mathbb{R}^m)$, then

$$\lim_{j \to \infty} \{f_j \otimes g\} = f \otimes g, \qquad \text{in } \mathcal{D}'(\mathbb{R}^{m+n}). \tag{1.3.4}$$

3. Associativity. For $f \in \mathcal{D}'(\mathbb{R}^m)$, $g \in \mathcal{D}'(\mathbb{R}^n)$ and $h \in \mathcal{D}'(\mathbb{R}^p)$, we have

$$f \otimes \left[g \otimes h\right] = \left[f \otimes g\right] \otimes h. \tag{1.3.5}$$

4. Support.

$$\text{Supp}\{f \otimes g\} = \text{Supp}\{f\} \times \text{Supp}\{g\}. \tag{1.3.6}$$

5. Differentiation.

$$D_x^\alpha \left[f \otimes g\right] = D_x^\alpha f \otimes g. \tag{1.3.7}$$

6. Multiplication. Given a function $a \in C^\infty(\mathbb{R}^m)$, we have

$$a\left[f \otimes g\right] = \left[af\right] \otimes g. \tag{1.3.8}$$

7. Translation.

$$(f \otimes g)(\boldsymbol{x}+\boldsymbol{h}, \boldsymbol{y}) = f(\boldsymbol{x}+\boldsymbol{h}) \otimes g(\boldsymbol{y}). \tag{1.3.9}$$

Convolution of Distributions

For two locally integrable functions f and g in $\mathbb{R}^n$, the support of one of which is bounded, their convolution is given by

$$(f * g) = \int_{\mathbb{R}^n} f(\boldsymbol{y}) g(\boldsymbol{x}-\boldsymbol{y})\, \mathrm{d}\boldsymbol{y}, \tag{1.3.10}$$

and

$$(f * g) = \int_{\mathbb{R}^n} f(\boldsymbol{y}) g(\boldsymbol{x}-\boldsymbol{y})\, \mathrm{d}\boldsymbol{y} = \int_{\mathbb{R}^n} g(\boldsymbol{y}) f(\boldsymbol{x}-\boldsymbol{y})\, \mathrm{d}\boldsymbol{y}. \tag{1.3.11}$$

The convolution $(f * g)$ is locally integrable in $\mathbb{R}^n$ and a regular distribution in the sense that

$$\begin{aligned}\langle f * g, \varphi \rangle &= \int_{\mathbb{R}^n} (f * g)(\boldsymbol{z}) \varphi(\boldsymbol{z})\, \mathrm{d}\boldsymbol{z} \\ &= \int_{\mathbb{R}^n} \Big[\int_{\mathbb{R}^n} g(\boldsymbol{y}) f(\boldsymbol{z}-\boldsymbol{y})\, \mathrm{d}\boldsymbol{y} \Big] \varphi(\boldsymbol{z})\, \mathrm{d}\boldsymbol{z} \\ &= \int_{\mathbb{R}^n} \int_{\mathbb{R}^n} f(\boldsymbol{x}) g(\boldsymbol{y}) \varphi(\boldsymbol{x}+\boldsymbol{y})\, \mathrm{d}\boldsymbol{x} \mathrm{d}\boldsymbol{y} \\ &= \langle f \otimes g, \varphi \rangle, \qquad \forall \varphi \in \mathcal{D}(\mathbb{R}^n).\end{aligned} \tag{1.3.12}$$

Is it possible to define the convolution of two distributions f and g by the Eq. (1.3.12)? If $\langle f \otimes g, \varphi \rangle$ is a continuous linear functional on the test function space $\mathcal{D}(\mathbb{R}^n)$, the answer is yes. Given two distributions f and g in $\mathcal{D}'(\mathbb{R}^n)$, we know that their direct product $f \otimes g \in \mathcal{D}'(\mathbb{R}^{2n})$. But for the test function $\varphi \in \mathcal{D}(\mathbb{R}^n)$, in general $\varphi(\boldsymbol{x}+\boldsymbol{y})$ does not have compact support in $\mathbb{R}^{2n}$ (Its support can be the infinite strip that lies between $\boldsymbol{x}+\boldsymbol{y} = A$ and $\boldsymbol{x}+\boldsymbol{y} = -A$, where the constant A depends on the support of the test function $\varphi(\boldsymbol{x})$). In order to ensure $\langle f \otimes g, \varphi \rangle$ is a continuous linear functional on $\mathcal{D}(\mathbb{R}^n)$, we have to make certain restrictions on the distribution f or g. We assume the support of one of the two distributions f and g is bounded. In this case $\langle f \otimes g, \varphi \rangle$ is a continuous linear functional on $\mathcal{D}(\mathbb{R}^n)$ (Kanwal, 2004).

Definition 1.3.2 Suppose that the support of one of the two distributions f and g in $\mathcal{D}'(\mathbb{R}^n)$ is bounded. Their convolution $(f * g) \in \mathcal{D}'(\mathbb{R}^n)$ is defined by

$$\langle f * g, \varphi \rangle = \langle f \otimes g, \varphi \rangle, \qquad \forall \varphi \in \mathcal{D}(\mathbb{R}^n). \tag{1.3.13}$$

Some properties of the convolution of distributions are listed as follows (Kanwal, 2004):

1. Commutativity.

$$f * g = g * f. \tag{1.3.14}$$

2. Associativity.

$$(f * g) * h = f * (g * h), \tag{1.3.15}$$

If the supports of two of these three distributions are bounded.

3. Differentiation.

If the convolution $f * g$ exists, then $(D^\alpha f) * g$ and $f * (D^\alpha g)$ exist with

$$(D^\alpha f) * g = D^\alpha (f * g) = f * (D^\alpha g). \tag{1.3.16}$$

4. Continuity. In some cases the convolution is a continuous operator. See the theorem below.

Theorem 1.3.1 *Suppose that the sequence of distributions $f_j \to f$ as $j \to \infty$. We have $f_j * g \to f * g$ under the following conditions:*

1. All distributions f_j are concentrated on the same bounded set;

2. The distribution g is concentrated on a bounded set.

1.4 Tempered Distributions and Fourier Transform

In order to define the Fourier transform of distributions, we recall the definition of the Fourier transform for classical functions. For an integrable function $f(\boldsymbol{x})$ on $\mathbb{R}^n$, the Fourier transform of $f(\boldsymbol{x})$ is defined by

$$\hat{f}(\boldsymbol{\xi}) = \mathfrak{F}\big[f(\boldsymbol{x})\big] = \int_{\mathbb{R}^n} \mathrm{e}^{-\mathrm{i}\boldsymbol{x}\cdot\boldsymbol{\xi}} f(\boldsymbol{x})\,\mathrm{d}\boldsymbol{x}. \tag{1.4.1}$$

However, the function $\mathrm{e}^{-\mathrm{i}\boldsymbol{x}\cdot\boldsymbol{\xi}}$ is not a test function in $\mathcal{D}(\mathbb{R}^n)$. Therefore the action of a distribution f on $\mathrm{e}^{-\mathrm{i}\boldsymbol{x}\cdot\boldsymbol{\xi}}$ is not defined. The formula (1.4.1) can not be used to define the Fourier transform of a distribution directly. For two functions $f(\boldsymbol{x})$ and $\varphi(\boldsymbol{x})$, which are square-integrable on $\mathbb{R}^n$, we have Parseval's formula

$$\int_{\mathbb{R}^n} \hat{f}(\boldsymbol{x})\varphi(\boldsymbol{x})\,\mathrm{d}\boldsymbol{x} = \int_{\mathbb{R}^n} f(\boldsymbol{x})\hat{\varphi}(\boldsymbol{x})\,\mathrm{d}\boldsymbol{x}.$$

Especially, for $\varphi \in \mathcal{D}(\mathbb{R}^n)$, we have

$$\left\langle \hat{f}, \varphi \right\rangle = \left\langle f, \hat{\varphi} \right\rangle, \qquad \forall \varphi \in \mathcal{D}(\mathbb{R}^n). \tag{1.4.2}$$

If we define the Fourier transform of distributions using formula (1.4.2), once again we may get into trouble since $\hat{\varphi}$ may not be a test function in $\mathcal{D}(\mathbb{R}^n)$ even though φ is. This means it is not the right way to define the Fourier transform for distributions.

In the next, we will first introduce $\mathcal{S}'(\mathbb{R}^n)$, a new class of distributions, which is a subset of $\mathcal{D}'(\mathbb{R}^n)$, then define the Fourier transform of distributions on $\mathcal{S}'(\mathbb{R}^n)$.

Test Function Space of Rapid Decay

Definition 1.4.1 Let $\mathcal{S}(\mathbb{R}^n)$ denote the set of all test functions of rapid decay $\varphi(\boldsymbol{x})$ that have the following two properties:

1. $\varphi(\boldsymbol{x})$ is infinitely differentiable, $\varphi \in C^\infty(\mathbb{R}^n)$;
2. $\varphi(\boldsymbol{x})$ as well as its derivatives of all orders vanishes at infinity faster than reciprocal of any polynomial.

The second property could be expressed by the following inequality

$$|\boldsymbol{x}^{\boldsymbol{p}} D^{\alpha}\varphi(\boldsymbol{x})| \leqslant C_{\boldsymbol{p},\alpha}, \tag{1.4.3}$$

where $\boldsymbol{p}$ and $\boldsymbol{\alpha}$ are two multi-index of dimension n and $C_{\boldsymbol{p},\alpha}$ is a constant depending on $\boldsymbol{p}$, $\boldsymbol{\alpha}$ and $\varphi(\boldsymbol{x})$.

Definition 1.4.2 (Convergence in $\mathcal{S}(\mathbb{R}^n)$) A sequence $\{\varphi_j,\ j = 1, 2, \ldots\} \subset \mathcal{S}(\mathbb{R}^n)$ is said to converge to φ if and only if the functions φ_j and all their derivatives converge to φ and corresponding derivatives of φ uniformly with respect to $\boldsymbol{x}$ in every bounded domain $\Omega \subset \mathbb{R}^n$.

This definition means that the constant $C_{\boldsymbol{p},\alpha}$ can be chosen independent of j such that

$$|\boldsymbol{x}^{\boldsymbol{p}}\big(D^{\alpha}\varphi_j(\boldsymbol{x}) - D^{\alpha}\varphi(\boldsymbol{x})\big)| \leqslant C_{\boldsymbol{p},\alpha}$$

for all value of j. It is straightforward to show that $\varphi \in \mathcal{S}(\mathbb{R}^n)$. Therefore $\mathcal{S}(\mathbb{R}^n)$ is closed with respect to this convergence. Obviously $\mathcal{D}(\mathbb{R}^n)$ is a subspace of $\mathcal{S}(\mathbb{R}^n)$ and $\mathcal{D}(\mathbb{R}^n)$ is dense in $\mathcal{S}(\mathbb{R}^n)$ (Kanwal, 2004).

Tempered Distributions (Distribution of Slow Growth)

We are now ready to define the tempered distribution space, which is a subspace of $\mathcal{D}'(\mathbb{R}^n)$.

Definition 1.4.3 A continuous linear functional $\ell(\varphi)$ on the space $\mathcal{S}(\mathbb{R}^n)$ of test functions is called a tempered distribution or a distribution of slow growth.

It is obvious that a tempered distribution $\ell(\varphi)$ has the following properties:
1.

$$\langle \ell, c_1\varphi_1 + c_2\varphi_2\rangle = c_1 \langle \ell, \varphi_1\rangle + c_2 \langle \ell, \varphi_2\rangle, \qquad \forall\, \varphi_1,\ \varphi_2 \in \mathcal{S}(\mathbb{R}^n);$$

2.

$$\lim_{j\to\infty} \langle \ell, \varphi_j\rangle = 0, \quad \text{for every } \varphi_j \to 0, \text{ as } j \to \infty \text{ in } \mathcal{S}(\mathbb{R}^n).$$

Let $\mathcal{S}'(\mathbb{R}^n)$ denote the set of all tempered distributions. The definitions of $\mathcal{D}(\mathbb{R}^n)$ and $\mathcal{S}(\mathbb{R}^n)$ imply that a sequence $\{\varphi_j,\ j = 1, 2, \ldots\}$ converging to φ in the sense of $\mathcal{D}(\mathbb{R}^n)$ also converges to φ in the sense of $\mathcal{S}(\mathbb{R}^n)$. Therefore, every continuous linear functional on $\mathcal{S}(\mathbb{R}^n)$ is also a continuous linear functional on $\mathcal{D}(\mathbb{R}^n)$. This means $\mathcal{S}'(\mathbb{R}^n) \subset \mathcal{D}'(\mathbb{R}^n)$, $\mathcal{S}'(\mathbb{R}^n)$ is a subspace of $\mathcal{D}'(\mathbb{R}^n)$.

The Fourier Transform

At first, we consider the Fourier transform of the test function $\varphi \in \mathcal{S}(\mathbb{R}^n)$,

$$\hat{\varphi}(\boldsymbol{\xi}) = \mathfrak{F}(\varphi(\boldsymbol{x})) = \int_{\mathbb{R}^n} \mathrm{e}^{-\mathrm{i}\boldsymbol{x}\cdot\boldsymbol{\xi}}\varphi(\boldsymbol{x})\,\mathrm{d}\boldsymbol{x}, \tag{1.4.4}$$

the inverse Fourier transform takes

$$\varphi(\boldsymbol{x}) = \mathfrak{F}^{-1}\big[\hat{\varphi}(\boldsymbol{\xi})\big] = \frac{1}{(2\pi)^n}\int_{\mathbb{R}^n} \mathrm{e}^{\mathrm{i}\boldsymbol{\xi}\cdot\boldsymbol{x}}\hat{\varphi}(\boldsymbol{\xi})\,\mathrm{d}\boldsymbol{\xi}. \tag{1.4.5}$$

For the function $\varphi \in \mathcal{S}(\mathbb{R}^n)$, we have

Theorem 1.4.1 (Kanwal, 2004) *1. If φ is in $\mathcal{S}(\mathbb{R}^n)$, then $\hat{\varphi}(\boldsymbol{\xi})$ exists and $\hat{\varphi} \in \mathcal{S}(\mathbb{R}^n)$, and*

$$\mathfrak{F}\big[\mathfrak{F}(\varphi)\big] = (2\pi)^n\varphi(-\boldsymbol{x}); \tag{1.4.6}$$

2. The Fourier transform and its inverse is continuous, linear one to one mappings of $\mathcal{S}(\mathbb{R}^n)$ onto itself.

We are now ready to define the Fourier transform of tempered distributions. For any $\varphi \in \mathcal{S}(\mathbb{R}^n)$, we have known that $\hat{\varphi} \in \mathcal{S}(\mathbb{R}^n)$, and whenever $\varphi_j \to 0$, then $\hat{\varphi}_j \to 0$ as $j \to \infty$. It is clear that for any tempered distribution $f \in \mathcal{S}'(\mathbb{R}^n)$, $\langle f, \hat{\varphi}\rangle$ is defined as a tempered distribution, which is called the Fourier transform of f.

Definition 1.4.4 For a tempered distribution $f \in \mathcal{S}'(\mathbb{R}^n)$, its Fourier transform $\hat{f}$ is defined by Parseval's formula

$$\big\langle \hat{f}, \varphi\big\rangle = \langle f, \hat{\varphi}\rangle, \qquad \forall\, \varphi \in \mathcal{S}(\mathbb{R}^n). \tag{1.4.7}$$

Suppose a sequence $\{f_j,\ j=1,2,\ldots\}\subset\mathcal{S}'(\mathbb{R}^n)$ converges to $f\in\mathcal{S}'(\mathbb{R}^n)$, then

$$\begin{aligned}\lim_{j\to\infty}\left\langle \hat{f}_j,\varphi\right\rangle &= \lim_{j\to\infty}\left\langle f_j,\hat{\varphi}\right\rangle\\ &= \left\langle f,\hat{\varphi}\right\rangle = \left\langle \hat{f},\varphi\right\rangle.\end{aligned}$$

It means that the Fourier transform is continuous, linear one to one mapping of $\mathcal{S}'(\mathbb{R}^n)$ onto itself. The equality (1.4.6) indicates

$$\mathfrak{F}\big[\mathfrak{F}(f)\big] = (2\pi)^n f(-\boldsymbol{x}),\qquad \forall f\in\mathcal{S}'(\mathbb{R}^n). \tag{1.4.8}$$

Example 1.4.1 The Fourier transform of δ. It is clear that $\delta\in\mathcal{S}'(\mathbb{R}^n)$. We have

$$\begin{aligned}\left\langle\hat{\delta},\varphi\right\rangle &= \left\langle\delta,\hat{\varphi}\right\rangle\\ &= \left\langle\delta,\int_{\mathbb{R}^n}\varphi(\boldsymbol{\xi})\mathrm{e}^{-\mathrm{i}\boldsymbol{x}\cdot\boldsymbol{\xi}}\,\mathrm{d}\boldsymbol{\xi}\right\rangle\\ &= \int_{\mathbb{R}^n}\varphi(\boldsymbol{\xi})\,\mathrm{d}\boldsymbol{\xi}\\ &= \langle 1,\varphi\rangle,\qquad \forall\varphi\in\mathcal{S}(\mathbb{R}^n).\end{aligned}$$

Therefore, we get

$$\hat{\delta} = 1. \tag{1.4.9}$$

Example 1.4.2 The Fourier transform of 1. 1 is a tempered distribution. We have

$$\hat{1} = \mathfrak{F}\big[\hat{\delta}\big] = (2\pi)^n\delta(-\boldsymbol{x}) = (2\pi)^n\delta(\boldsymbol{x}). \tag{1.4.10}$$

Example 1.4.3 The Fourier transform of $D^{\alpha}\delta$ with multi-index $\boldsymbol{\alpha}=(\alpha_1,\alpha_2,\ldots,\alpha_n)$ of dimension n.

$$\begin{aligned}\left\langle\mathfrak{F}\big[D^{\alpha}\delta\big],\varphi\right\rangle &= \left\langle D^{\alpha}\delta,\hat{\varphi}\right\rangle\\ &= (-1)^{|\alpha|}\left\langle\delta,D^{\alpha}\hat{\varphi}\right\rangle\\ &= (-1)^{|\alpha|}\left\langle\delta,D^{\alpha}_{\xi}\int_{\mathbb{R}^n}\mathrm{e}^{-\mathrm{i}\boldsymbol{x}\cdot\boldsymbol{\xi}}\varphi(\boldsymbol{x})\,\mathrm{d}\boldsymbol{x}\right\rangle\\ &= (-1)^{|\alpha|}\left\langle\delta(\boldsymbol{\xi}),\prod_{j=1}^{n}\int_{\mathbb{R}^n}(-\mathrm{i}x_j)^{\alpha_j}\mathrm{e}^{-\mathrm{i}\boldsymbol{x}\cdot\boldsymbol{\xi}}\varphi(\boldsymbol{x})\,\mathrm{d}\boldsymbol{x}\right\rangle\\ &= (-1)^{|\alpha|}\prod_{j=1}^{n}\int_{\mathbb{R}^n}(-\mathrm{i}x_j)^{\alpha_j}\varphi(\boldsymbol{x})\,\mathrm{d}\boldsymbol{x}\\ &= \left\langle\prod_{j=1}^{n}(\mathrm{i}\xi_j)^{\alpha_j},\varphi(\boldsymbol{\xi})\right\rangle,\qquad\forall\varphi\in\mathcal{S}(\mathbb{R}^n).\end{aligned}$$

Therefore,

$$\mathfrak{F}\left[D^{\alpha}\delta\right](\boldsymbol{\xi})=\prod_{j=1}^{n}(\mathrm{i}\xi_j)^{\alpha_j}. \tag{1.4.11}$$

The main materials in this chapter are from the books by Kanwal (2004) and by Qi and Wu (1999).

References

Dirac, P. A. M. (1927). The physical interpretation of the quantum dynamics. *Proceedings of the Royal Society of London. Series A, Containing Papers of a Mathematical and Physical Character, 113*(765), 621–641.

Kanwal, R. P. (2004). *Generalized functions: Theory and applications.* Springer Science & Business Media.

Kythe, P. (1996). *Fundamental solutions for differential operators and applications.* Springer Science & Business Media.

Lützen, J. (2012). *The prehistory of the theory of distributions* (Vol. 7). Springer Science & Business Media.

Qi, M., & Wu, F. (1999). *Generalized functions and mathematical physical equations (in chinese).* Beijing: Higher Education Press.

Schwartz, L. (1966). *Théorie des distributions* (Vol. 2). Hermann, Paris.

Sobolev, S. (1936). Méthode nouvelle à résoudre le problème de cauchy pour les équations linéaires hyperboliques normales. *Matematiceskij sbornik, 43*(1), 39–72.

Chapter 2
Fundamental Solutions of Linear Differential Operators

After introducing the theory of distributions in Chap. 1, we are now ready to discuss the fundamental solutions of linear differential operators with constant coefficients, which are the foundation of the potential theory and the boundary integro-differential method.

2.1 Definition of Fundamental Solution

Let $\boldsymbol{\alpha} = (\alpha_1, \alpha_2, \ldots, \alpha_n)^{\mathrm{T}}$ be a multi-index of dimension n, namely $\alpha_1, \alpha_2, \ldots, \alpha_n$ are non-negative integers and

$$|\boldsymbol{\alpha}| = \sum_{j=1}^{n} \alpha_j. \tag{2.1.1}$$

The $\boldsymbol{\alpha}$-differential operator is defined by

$$D^{\alpha} = \frac{\partial^{|\alpha|}}{\partial x_1^{\alpha_1} \partial x_2^{\alpha_2} \ldots \partial x_n^{\alpha_n}}, \qquad \forall \boldsymbol{x} = (x_1, x_2, \ldots, x_n)^{\mathrm{T}} \in \mathbb{R}^n. \tag{2.1.2}$$

The mth-order linear differential operator $L(D)$ is defined by

$$L(D) = \sum_{|\alpha| \leqslant m} a_{\alpha} D^{\alpha}, \tag{2.1.3}$$

where $a_{\alpha} \in \mathbb{C}$ are given constants and $\mathbb{C}$ is the complex number space.

Corresponding to the differential operator $L(D)$, we obtain the mth-order linear differential equation

H. Han and D. Yin, *Mathematical Foundation of the Boundary Integro-Differential Equation Method*, https://doi.org/10.1007/978-981-95-1088-7_2

$$L(D)u(\boldsymbol{x}) = f(\boldsymbol{x}), \qquad \forall \boldsymbol{x} \in \mathbb{R}^n, \tag{2.1.4}$$

with given function $f : \mathbb{R}^n \to \mathbb{C}$, and the unknown function $u : \mathbb{R}^n \to \mathbb{C}$. When $f(\boldsymbol{x}) \equiv 0$, the Eq. (2.1.4) is reduced to its homogeneous equation

$$L(D)u(\boldsymbol{x}) = 0, \qquad \forall \boldsymbol{x} \in \mathbb{R}^n. \tag{2.1.5}$$

Next the linear differential matrix operator is defined by

$$\mathcal{L}_k(D) = (L_{ij}(D))_{k\times k}, \tag{2.1.6}$$

with

$$L_{ij}(D) = \sum_{|\alpha|\leqslant m_{ij}} D^{\alpha},$$

and $k \geqslant 2$ be an integer. The system of linear differential equations corresponding to the matrix operator $\mathcal{L}_k(D)$ is given by

$$\mathcal{L}_k(D)\boldsymbol{u}(\boldsymbol{x}) = \boldsymbol{f}(\boldsymbol{x}), \qquad \forall \boldsymbol{x} \in \mathbb{R}^n, \tag{2.1.7}$$

with a given vector-valued function $\boldsymbol{f}(\boldsymbol{x}) = (f_1(\boldsymbol{x}),\ f_2(\boldsymbol{x}),\ \ldots,\ f_k(\boldsymbol{x}))^{\mathrm{T}}$ and the unknown function vector-valued function $\boldsymbol{u} = (u_1(\boldsymbol{x}),\ u_2(\boldsymbol{x}),\ \ldots,\ u_k(\boldsymbol{x}))^{\mathrm{T}}$. When $\boldsymbol{f}(\boldsymbol{x}) = 0$, the system (2.1.7) is reduced to its homogeneous counterpart.

We also consider the linear differential operator dependent on time $t \in \mathbb{R}^+$, namely the transient operator (Kythe, 1996). Let

$$L\left(D, \frac{\partial}{\partial t}\right) = \frac{\partial^j}{\partial t^j} + L(D), \tag{2.1.8}$$

with $j \geqslant 1$ be an integer and the linear differential operator $L(D)$ is defined by (2.1.3). It is straightforward to obtain the evolution equation corresponding to the transient operator $L(D, \frac{\partial}{\partial t})$

$$\frac{\partial^j u(\boldsymbol{x}, t)}{\partial t^j} + L(D)u(\boldsymbol{x}, t) = f(\boldsymbol{x}, t), \qquad \forall \boldsymbol{x} \in \mathbb{R}^n,\ t \in \mathbb{R}^+. \tag{2.1.9}$$

and its homogeneous one

$$\frac{\partial^j u(\boldsymbol{x}, t)}{\partial t^j} + L(D)u(\boldsymbol{x}, t) = 0, \qquad \forall \boldsymbol{x} \in \mathbb{R}^n,\ t \in \mathbb{R}^+. \tag{2.1.10}$$

Similarly we have the linear differential matrix operator dependent on time $t \in \mathbb{R}^+$ and corresponding evolution systems of differential equations. Namely,

$$\mathcal{L}_k\left(D, \frac{\partial}{\partial t}\right) = \frac{\partial^j}{\partial t^j}\boldsymbol{I}_k + \mathcal{L}(D), \tag{2.1.11}$$

and

$$\mathcal{L}_k\left(D, \frac{\partial}{\partial t}\right)\boldsymbol{u}(\boldsymbol{x}, t) = \frac{\partial^j \boldsymbol{u}(\boldsymbol{x}, t)}{\partial t^j} + \mathcal{L}(D)\boldsymbol{u}(\boldsymbol{x}, t) = \boldsymbol{f}(\boldsymbol{x}, t), \quad \forall \boldsymbol{x} \in \mathbb{R}^n,\ t \in \mathbb{R}^+. \tag{2.1.12}$$

Here, $\boldsymbol{I}_k$ denotes the kth-order identity matrix with $k \geqslant 2$, $\boldsymbol{f}(\boldsymbol{x}, t) = (f_1(\boldsymbol{x}, t), f_2(\boldsymbol{x}, t), \ldots, f_k(\boldsymbol{x}, t))^{\mathrm{T}}$ is given and the unknown function $\boldsymbol{u}(\boldsymbol{x}, t) = (u_1(\boldsymbol{x}, t), u_2(\boldsymbol{x}, t), \ldots, u_k(\boldsymbol{x}, t))^{\mathrm{T}}$. When $\boldsymbol{f}(\boldsymbol{x}, t) = 0$, the nonhomogeneous system (2.1.12) reduces to the homogeneous one, too.

Definition 2.1.1 A distribution $E(\boldsymbol{x})$ that satisfies the equation

$$L(D)E(\boldsymbol{x}) = \delta(\boldsymbol{x}), \qquad \forall \boldsymbol{x} \in \mathbb{R}^n, \tag{2.1.13}$$

in the distribution sense is called a *fundamental solution*of the differential operator $L(D)$.

Definition 2.1.2 A kth-order distribution matrix $\mathbb{E}_k(\boldsymbol{x})$, which satisfies the following system of differential equations

$$\mathcal{L}_k(D)\mathbb{E}_k(\boldsymbol{x}) = \delta(\boldsymbol{x})\boldsymbol{I}_k, \qquad \forall \boldsymbol{x} \in \mathbb{R}^n, \tag{2.1.14}$$

in the sense of distribution, is called a *fundamental solution*of the matrix differential operator $\mathcal{L}_k(D)$.

A fundamental solution of the transient operator can be defined similarly.

Definition 2.1.3 A distribution $E(\boldsymbol{x}, t)$ that satisfies the equation

$$L\left(D, \frac{\partial}{\partial t}\right)E(\boldsymbol{x}, t) = \delta(\boldsymbol{x})\delta(t), \qquad \forall \boldsymbol{x} \in \mathbb{R}^n,\ t \in \mathbb{R}^+, \tag{2.1.15}$$

is call a *fundamental solution* of the transient operator $L\left(D, \frac{\partial}{\partial t}\right)$.

Definition 2.1.4 A kth-order distribution matrix $\mathbb{E}_k(\boldsymbol{x}, t)$, which satisfies the following system of differential equations

$$\mathcal{L}_k\left(D, \frac{\partial}{\partial t}\right)\mathbb{E}_k(\boldsymbol{x}, t) = \delta(\boldsymbol{x})\delta(t)\boldsymbol{I}_k, \qquad \forall \boldsymbol{x} \in \mathbb{R}^n, t \in \mathbb{R}^+,$$

is called a *fundamental solution* of the matrix transient differential operator $\mathcal{L}_k(D, t)$.

In views of the definitions (2.1.1)–(2.1.4), the fundamental solution $E(\boldsymbol{x})$ $(\mathbb{E}_k(\boldsymbol{x}), E(\boldsymbol{x}, t), \mathbb{E}_k(\boldsymbol{x}, t))$ of operator $L(D)\left(\mathcal{L}_k(D), L\left(D, \frac{\partial}{\partial t}\right), \mathcal{L}_k\left(D, \frac{\partial}{\partial t}\right)\right)$ is unique up to a solution of the corresponding homogeneous equation or system $L(D)w(\boldsymbol{x}) = 0$ $\left(\mathcal{L}_k(D)\boldsymbol{w}_k(\boldsymbol{x}) = 0,\ L\left(D, \frac{\partial}{\partial t}\right) w(\boldsymbol{x}, t) = 0,\ \mathcal{L}_k\left(D, \frac{\partial}{\partial t}\right) \boldsymbol{w}_k(\boldsymbol{x}, t) = 0\right)$.

In the coming sections of this chapter, the fundamental solutions of some important differential operators will be given.

2.2 Elliptic Operators

In this section, we will derive the fundamental solutions of the Laplace, Helmholtz and biharmonic operators, which are the classical elliptic differential operators.

2.2.1 *Laplace Operator*

We consider the Laplace operator Δ given by

$$\Delta = \frac{\partial^2}{\partial x_1^2} + \frac{\partial^2}{\partial x_2^2} + \cdots + \frac{\partial^2}{\partial x_n^2}, \qquad \text{in } \mathbb{R}^n. \tag{2.2.1}$$

Let $L(D) = -\Delta$, and $E_n(\boldsymbol{x})$ denote the fundamental solution $L(D)$ in $\mathbb{R}^n$, which satisfies

$$-\Delta E_n(\boldsymbol{x}) = \delta(\boldsymbol{x}), \qquad \forall \boldsymbol{x} \in \mathbb{R}^n. \tag{2.2.2}$$

Applying the Fourier transform to Eq. (2.2.2), we obtain

$$|\xi|^2 \mathfrak{F}[E_n](\boldsymbol{\xi}) = 1, \tag{2.2.3}$$

with $\mathfrak{F}[E_n](\boldsymbol{\xi})$ be the Fourier transform of $E_n(\boldsymbol{x})$. Now we will find the solution of the transformed equation (2.2.3) in the dual space $\mathcal{D}'$ of the fundamental space $\mathcal{D}$ of test functions, which are compactly supported and infinitely differentiable in $\mathbb{R}^n$.

In the case $n = 2$, unfortunately $\frac{1}{|\xi|^2}$ is not a distribution, since it is not locally integrable in $\mathbb{R}^2$. But the distribution $P\left(\frac{1}{|\boldsymbol{\xi}|^2}\right)$ defined in Chap. 1 Example 1.2.4 is a solution of (2.2.3) for $n = 2$, since for any $\varphi \in \mathcal{D}$,

$$\left\langle |\xi|^2 P\left(\frac{1}{|\xi|^2}\right), \varphi \right\rangle = \left\langle P\left(\frac{1}{|\xi|^2}\right), |\xi|^2\varphi \right\rangle$$
$$= \int_{|\xi|<1} \frac{|\xi|^2\varphi(\xi) - 0}{|\xi|^2}\,\mathrm{d}\xi + \int_{|\xi|>1} \frac{|\xi|^2\varphi(\xi)}{|\xi|^2}\,\mathrm{d}\xi$$
$$= \int_{\mathbb{R}^2} \varphi(\xi)\,\mathrm{d}\xi = \langle 1, \varphi \rangle.$$

Namely,

$$\mathfrak{F}[E_2](\xi) = P\left(\frac{1}{|\xi|^2}\right),$$

the inverse Fourier transform of $P\left(\frac{1}{|\xi|^2}\right)$ is given by

$$E_2^*(\xi) = \mathfrak{F}^{-1}\left[P\left(\frac{1}{|\xi|^2}\right)\right] = \frac{1}{4\pi^2}\mathfrak{F}\left[P\left(\frac{1}{|\xi|^2}\right)\right]$$
$$= -\frac{1}{2\pi}\ln|\boldsymbol{x}| - \frac{C_0}{2\pi},$$

with a constant C_0. The last equality is from the result in Example A.1.5 (Kythe, 1996). A constant is always a solution of the Laplace equation. We choose the fundamental solution of the Laplace operator in $\mathbb{R}^2$ by

$$E_2(\boldsymbol{x}) = -\frac{1}{2\pi}\ln|\boldsymbol{x}|. \tag{2.2.4}$$

In the case that $n \geqslant 3$, the function $\frac{1}{|\xi|^2}$ is locally integrable in $\mathbb{R}^n$ and hence a distribution on $\mathbb{R}^n$. The solution of (2.2.3) is given by

$$\mathfrak{F}[E_n] = \frac{1}{|\xi|^2}. \tag{2.2.5}$$

By the inverse Fourier transform, the fundamental solution of the Laplace operator is given by

$$E_n(\boldsymbol{x}) = \frac{1}{(n-2)S_n(1)}|\boldsymbol{x}|^{2-n}, \qquad n \geqslant 3. \tag{2.2.6}$$

Here, $S_n(1) = 2\pi^{\frac{n}{2}} / \Gamma\left(\frac{n}{2}\right)$ is the surface area of the unit ball in $\mathbb{R}^n$.

Specially, when $n = 3$,

$$E_3(\boldsymbol{x}) = \frac{1}{4\pi}\frac{1}{|\boldsymbol{x}|}. \tag{2.2.7}$$

2.2.2 Helmholtz Operator

In this subsection, we derive the fundamental solution of the Helmholtz operator $L = -(\Delta + \kappa^2)$ in $\mathbb{R}^n$ with κ be a non-zero complex number by a different approach. Since the fundamental solution $E_n \in \mathcal{D}'$ satisfies

$$-(\Delta + \kappa^2)E_n(\boldsymbol{x}) = \delta(\boldsymbol{x}), \qquad \text{in } \mathbb{R}^n. \tag{2.2.8}$$

By the symmetry of the Helmholtz operator and the distribution $\delta(\boldsymbol{x})$, we look for a radial distribution solution $E_n(\boldsymbol{x}) = E_n(|\boldsymbol{x}|)$ of (2.2.8). Then $E_n(r)$ with $r = |\boldsymbol{x}|$ satisfies the following ordinary equation

$$\frac{\mathrm{d}}{\mathrm{d}r}\left(r^{n-1}\frac{\mathrm{d}E_n(r)}{\mathrm{d}r}\right) + \kappa^2 r^{n-1}E_n(r) = 0, \qquad r > 0. \tag{2.2.9}$$

Let

$$E_n(r) = r^{1-\frac{n}{2}}W_n(r). \tag{2.2.10}$$

Substituting (2.2.10) into (2.2.9) yields

$$r^2W_n''(r) + rW_n'(r) + \left[\kappa^2r^2 - \left(\frac{n}{2} - 1\right)^2\right]W_n(r) = 0, \tag{2.2.11}$$

which is the Bessel equation of order $\dfrac{n}{2} - 1$ with parameter κ. The general solution of the Bessel equation (2.2.11) can be given by Hankel functions

$$W_n(r) = c_1H^{(1)}_{\frac{n}{2}-1}(\kappa r) + c_2H^{(2)}_{\frac{n}{2}-1}(\kappa r), \qquad n \geqslant 2,$$

and

$$E_n(r) = c_1r^{1-\frac{n}{2}}H^{(1)}_{\frac{n}{2}-1}(\kappa r) + c_2r^{1-\frac{n}{2}}H^{(2)}_{\frac{n}{2}-1}(\kappa r), \qquad n \geqslant 2$$

with $H^{(1)}_\nu(z)$ and $H^{(2)}_\nu(z)$ be the first kind and second kind Hankel functions, respectively.

In the case that $\kappa^2 \not\subset [0, \infty)$ and $\kappa \in \mathbb{C}$ has positive imaginary, the function $H^{(2)}_{\frac{n}{2}-1}(\kappa r)$ becomes exponentially large as $r \to +\infty$, and $H^{(1)}_{\frac{n}{2}-1}(\kappa r)$ is exponentially going to zero as $r \to +\infty$. To get a bounded function, we take and $c_2 = 0$ and obtain

$$E_n(r) = c_1r^{1-\frac{n}{2}}H^{(1)}_{\frac{n}{2}-1}(\kappa r), \qquad n \geqslant 2, \tag{2.2.12}$$

which vanishes at $r = +\infty$.

The function $E_n(\boldsymbol{x}) = E_n(|\boldsymbol{x}|)$ given by (2.2.12) is a distribution on $\mathbb{R}^n$. In the next, we need to choose a suitable constant c_1 such that $E_n(\boldsymbol{x})$ is a solution of equation (2.2.8) in the sense of distribution.

For any $\varphi \in \mathcal{D}$, we consider

$$\begin{aligned}\big\langle -(\Delta+\kappa^2)E_n, \varphi\big\rangle &= -\int_{\mathbb{R}^n} \big((\Delta+\kappa^2)E_n(\boldsymbol{x})\big)\,\varphi(\boldsymbol{x})\,\mathrm{d}\boldsymbol{x} \\ &= -\lim_{\epsilon\to+0}\int_{|\boldsymbol{x}|\leqslant\epsilon} \big((\Delta+\kappa^2)E_n(\boldsymbol{x})\big)\,\varphi(\boldsymbol{x})\,\mathrm{d}\boldsymbol{x}. \end{aligned} \tag{2.2.13}$$

On the other hand, by the divergence theorem, we have

$$\begin{gathered} -\int_{|\boldsymbol{x}|\leqslant\epsilon} \big((\Delta+\kappa^2)E_n\big)\,\varphi\,\mathrm{d}\boldsymbol{x} = \\ \int_{|\boldsymbol{x}|\leqslant\epsilon} \nabla E_n\cdot\nabla\varphi\,\mathrm{d}\boldsymbol{x} - \kappa^2\int_{|\boldsymbol{x}|\leqslant\epsilon} E_n\varphi\,\mathrm{d}\boldsymbol{x} - \int_{|\boldsymbol{x}|=\epsilon}\frac{\partial E_n}{\partial r}\varphi\,\mathrm{d}s. \end{gathered} \tag{2.2.14}$$

By an asymptotic expansion of $H_\nu^{(1)}(z)$ for small z, we have

$$H_\nu^{(1)}(z) \sim \begin{cases} \mathrm{i}\dfrac{2}{\pi}\ln z, & \nu = 0, \\ -\dfrac{\mathrm{i}\Gamma(\nu)}{\pi}\left(\dfrac{2}{z}\right)^\nu, & \nu > 0. \end{cases}$$

Then we can obtain the dominant behavior of $E_n(r)$ and $\dfrac{\mathrm{d}}{\mathrm{d}r}E_n(r)$, when r is small, i.e.,

$$E_n(r) \sim \begin{cases} c_1\dfrac{2\mathrm{i}}{\pi}\ln r\,, & n = 2, \\ -c_1\dfrac{\mathrm{i}\,\Gamma\left(\frac{n}{2}-1\right)}{\pi}\left(\dfrac{2}{\kappa}\right)^{\frac{n}{2}-1} r^{2-n}, & n \geqslant 3\,, \end{cases} \tag{2.2.15}$$

and

$$\frac{\mathrm{d}}{\mathrm{d}r}E_n(r) \sim \begin{cases} c_1\dfrac{2\mathrm{i}}{\pi}\dfrac{1}{r}, & n = 2, \\ c_1\dfrac{\mathrm{i}\,(n-2)\Gamma\left(\frac{n}{2}-1\right)}{\pi}\left(\dfrac{2}{\kappa}\right)^{\frac{n}{2}-1} r^{1-n}, & n \geqslant 3\,. \end{cases} \tag{2.2.16}$$

Using the asymptotic expansions (2.2.15)–(2.2.16), a direct computation shows that

$$\lim_{\epsilon\to+0}\int_{|\boldsymbol{x}|\leqslant\epsilon} \nabla E_n\cdot\nabla\varphi\,\mathrm{d}\boldsymbol{x} = 0, \tag{2.2.17}$$

$$\lim_{\epsilon\to+0}\int_{|\boldsymbol{x}|\leqslant\epsilon} E_n\varphi\,\mathrm{d}\boldsymbol{x} = 0, \tag{2.2.18}$$

and

$$\lim_{\epsilon\to+0}\int_{|x|\leqslant\epsilon}\frac{\partial E_n}{\partial r}\varphi\,\mathrm{d}x=\begin{cases}c_1 4\,\mathrm{i}\,\varphi(0), & n=2,\\ c_1\dfrac{(n-2)\mathrm{i}\,\Gamma\left(\frac{n}{2}-1\right)S_n(1)}{\pi}\left(\dfrac{2}{\kappa}\right)^{\frac{n}{2}-1}\varphi(0), & n\geqslant 3,\end{cases} \tag{2.2.19}$$

for any $\varphi\in\mathcal{D}$. Taking

$$c_1=\begin{cases}\dfrac{\mathrm{i}}{4}, & n=2,\\ \dfrac{\mathrm{i}\,\pi}{(n-2)\Gamma\left(\frac{n}{2}-1\right)S_n(1)}\left(\dfrac{\kappa}{2}\right)^{\frac{n}{2}-1}=\dfrac{\mathrm{i}}{4}\left(\dfrac{\kappa}{2\pi}\right)^{\frac{n}{2}-1}, & n\geqslant 3.\end{cases}$$

Namely,

$$c_1=\frac{\mathrm{i}}{4}\left(\frac{\kappa}{2\pi}\right)^{\frac{n}{2}-1},\qquad n\geqslant 2. \tag{2.2.20}$$

Then the distribution $E_n(x)$ with the constant c_1 given by (2.2.20) is a solution of equation (2.2.8). Therefore for $\kappa^2\notin[0,+\infty)$ and $\kappa\in\mathbb{C}$ has a positive imaginary, the fundamental solution of the Helmholtz operator is given by

$$E_n(x)=\frac{\mathrm{i}}{4}\left(\frac{\kappa}{2\pi}\right)^{\frac{n}{2}-1}|x|^{1-\frac{n}{2}}H^{(1)}_{\frac{n}{2}-1}(\kappa|x|),\qquad n\geqslant 2. \tag{2.2.21}$$

Especially, we have

$$\begin{cases}E_2(x)=\dfrac{\mathrm{i}}{4}H_0^{(1)}(\kappa|x|),\\ E_3(x)=\dfrac{\mathrm{i}\kappa}{4\pi}\left(\dfrac{\pi}{2\kappa|x|}\right)^{1/2}H^{(1)}_{1/2}(\kappa|x|)=\dfrac{1}{4\pi}\dfrac{\mathrm{e}^{\mathrm{i}\kappa|x|}}{|x|}.\end{cases} \tag{2.2.22}$$

For the case $\kappa^2\in(0,+\infty)$ and $\kappa>0$, which denotes the wave number, let

$$E_n(x)=c_2|x|^{1-\frac{n}{2}}H^{(2)}_{\frac{n}{2}-1}(\kappa|x|)$$

be the fundamental solution of the Helmholtz operator with constant c_2 to be determined. Repeating the discussion above, we can obtain

$$c_2=-\frac{\mathrm{i}}{4}\left(\frac{\kappa}{2\pi}\right)^{\frac{n}{2}-1},\qquad n\geqslant 2,$$

and arrive at

$$E_n(\boldsymbol{x}) = -\frac{\mathrm{i}}{4}\left(\frac{\kappa}{2\pi}\right)^{\frac{n}{2}-1} |\boldsymbol{x}|^{1-\frac{n}{2}} H^{(2)}_{\frac{n}{2}-1}(\kappa|\boldsymbol{x}|), \quad n \geqslant 2. \tag{2.2.23}$$

For $n = 2,\ 3$, we have

$$\begin{cases} E_2(\boldsymbol{x}) = -\dfrac{\mathrm{i}}{4} H^{(2)}_0(\kappa|\boldsymbol{x}|), \\ E_3(\boldsymbol{x}) = -\dfrac{\mathrm{i}}{4}\dfrac{\kappa}{\pi}\left(\dfrac{\pi}{2\kappa|\boldsymbol{x}|}\right) H^{(2)}_{1/2}(\kappa|\boldsymbol{x}|) = \dfrac{1}{4\pi}\dfrac{\mathrm{e}^{-\mathrm{i}k|\boldsymbol{x}|}}{|\boldsymbol{x}|}. \end{cases} \tag{2.2.24}$$

The fundamental solution for the operator $L = -\Delta + \kappa^2$ with κ be a positive real number is also given by (2.2.21), in which κ should be replaced with $\mathrm{i}\kappa$. For $n = 2,\ 3$, we have

$$\begin{cases} E_2(\boldsymbol{x}) = \dfrac{\mathrm{i}}{4} H^{(1)}_0(\mathrm{i}\kappa|\boldsymbol{x}|) = \dfrac{1}{2\pi} K_0(\kappa|\boldsymbol{x}|), \\ E_3(\boldsymbol{x}) = \dfrac{1}{4\pi}\dfrac{\mathrm{e}^{-\kappa|\boldsymbol{x}|}}{|\boldsymbol{x}|}, \end{cases} \tag{2.2.25}$$

where $K_0(z)$ is the modified Bessel function of the zeroth order.

2.2.3 *Biharmonic Operator*

The fundamental solution of the Biharmonic operator $L = \Delta^2$ in $\mathbb{R}^n$ is a distribution solution of the following equation

$$-\Delta^2 E_n(\boldsymbol{x}) = \delta(\boldsymbol{x}), \qquad \text{in } \mathbb{R}^n, \tag{2.2.26}$$

which can be rewritten as

$$-\Delta(\Delta E_n(\boldsymbol{x})) = \delta(\boldsymbol{x}) \qquad \text{in } \mathbb{R}^n.$$

It means that $\Delta E_n(\boldsymbol{x})$ is the fundamental solution of the Laplace operator. Thus we have

$$\Delta E_n(\boldsymbol{x}) = \begin{cases} -\dfrac{1}{2\pi}\ln|\boldsymbol{x}|, & \text{in } \mathbb{R}^2, \\ \dfrac{1}{(n-2)S_n(1)}|\boldsymbol{x}|^{2-n}, & \text{in } \mathbb{R}^n, \quad n \geqslant 3. \end{cases} \tag{2.2.27}$$

Using the symmetry of the Laplace operator and the functions on the right hand side of the equation (2.2.27), we consider the radial solution $E_n(\boldsymbol{x}) = E_n(|\boldsymbol{x}|) := E_n(r)$, then the function $E_n(r)$ satisfies the following ordinary differential equation:

$$\frac{d^2E_n(r)}{dr^2}+\frac{n-1}{r}\frac{dE_n(r)}{dr}=\begin{cases}-\dfrac{1}{2\pi}\ln r, & n=2,\\[2mm] \dfrac{1}{(n-2)S_n(1)}r^{2-n}, & n\geqslant 3.\end{cases} \tag{2.2.28}$$

For $n=2,\ 3$, we obtain the fundamental solutions of the biharmonic operator Δ^2 directly,

$$\begin{cases}E_2(\boldsymbol{x})=-\dfrac{|\boldsymbol{x}|^2}{8\pi}(\ln|\boldsymbol{x}|-1),\\[2mm] E_3(\boldsymbol{x})=\dfrac{|\boldsymbol{x}|}{8\pi}.\end{cases} \tag{2.2.29}$$

By the way, we present the fundamental solution of the triple-harmonic operator $E_n(\boldsymbol{x})$ that satisfies

$$-\Delta^3E_n(\boldsymbol{x})=\delta(\boldsymbol{x}),\qquad \forall\boldsymbol{x}\in\mathbb{R}^n. \tag{2.2.30}$$

It is given by

$$E_n(\boldsymbol{x})=\begin{cases}-\dfrac{|\boldsymbol{x}|^4\ln|\boldsymbol{x}|}{128\,\pi}, & n=2,\\[2mm] \dfrac{|\boldsymbol{x}|^3}{96\pi}, & n=3.\end{cases} \tag{2.2.31}$$

2.3 Transient Operator

The fundamental solutions of three important transient operators, heat conduction operator, Schrödinger operator and wave operator are derived in this section.

2.3.1 Heat Conduction Operator

We consider the heat conduction operator

$$L\left(D,\frac{\partial}{\partial t}\right)=\frac{\partial}{\partial t}-a\,\Delta \tag{2.3.1}$$

with $a>0$ be a constant. Physically, the operator often arises in case a is the thermal diffusivity. The fundamental solution $E_n(\boldsymbol{x},t)$ of the heat conduction operator is a distribution solution of the following equation

$$\frac{\partial E_n(\boldsymbol{x},t)}{\partial t}-a\,\Delta E_n(\boldsymbol{x},t)=\delta(\boldsymbol{x})\delta(t),\qquad \forall\boldsymbol{x}\in\mathbb{R}^n,\ t\in\mathbb{R}^+. \tag{2.3.2}$$

Applying the Fourier transform $\mathfrak{F}_x$ to both sides of equation (2.3.2) and using notation $\hat{E}_n(\boldsymbol{\xi}, t) = \mathfrak{F}_x[E_n](\boldsymbol{\xi}, t)$, it yields

$$\frac{\partial \hat{E}_n(\boldsymbol{\xi}, t)}{\partial t} + a\,|\boldsymbol{\xi}|^2 \hat{E}_n(\boldsymbol{\xi}, t) = \delta(t), \tag{2.3.3}$$

which has the solution

$$\hat{E}_n(\boldsymbol{\xi}, t) = H(t)\,\mathrm{e}^{-a|\boldsymbol{\xi}|^2 t},$$

with $H(t)$ is the Heaviside unit step function. Obviously, $\hat{E}_n(\boldsymbol{\xi}, t) \in \mathcal{D}'$. Applying the inverse Fourier transform $\mathfrak{F}_{\xi}^{-1}$ and the result given in Example A.1.3 (Kythe, 1996) leads to

$$\begin{aligned} E_n(\boldsymbol{x}, t) &= \mathfrak{F}_{\xi}^{-1}[\hat{E}_n(\boldsymbol{\xi}, t)](\boldsymbol{x}, t) \\ &= \frac{H(t)}{(2\pi)^n} \int_{\mathbb{R}^n} \mathrm{e}^{-a|\boldsymbol{\xi}|^2 t - \mathrm{i}\boldsymbol{\xi}\cdot\boldsymbol{x}}\,\mathrm{d}\boldsymbol{\xi} \\ &= \frac{H(t)}{(4\pi a t)^{\frac{n}{2}}} \mathrm{e}^{-\frac{|\boldsymbol{x}|^2}{4at}}. \end{aligned}$$

Namely,

$$E_n(\boldsymbol{x}, t) = \begin{cases} \dfrac{1}{(4\pi a t)^{\frac{n}{2}}} \mathrm{e}^{-\frac{|\boldsymbol{x}|^2}{4at}}, & t \geqslant 0, \\ 0, & t < 0, \end{cases} \tag{2.3.4}$$

is the fundamental solution of the heat conduction operator.

2.3.2 Schrödinger Operator

The Schrödinger equation is given by

$$\mathrm{i}\hbar \frac{\partial u(\boldsymbol{x}, t)}{\partial t} + \frac{\hbar^2}{2m} \Delta u(\boldsymbol{x}, t) = f(\boldsymbol{x}, t), \qquad \forall \boldsymbol{x} \in \mathbb{R}^n,\ t \in \mathbb{R}^+. \tag{2.3.5}$$

with $m > 0$ be the mass of the quantum particle and $\hbar > 0$ be the Planck's constant. The differential operator

$$L\left(D, \frac{\partial}{\partial t}\right) = -\mathrm{i}\hbar \frac{\partial}{\partial t} - \frac{\hbar^2}{2m} \Delta. \tag{2.3.6}$$

is the Schrödinger operator. The fundamental solution of the Schrödinger operator is a distributional solution of the following equation

$$-\mathrm{i}\hbar\frac{\partial}{\partial t}E_n(\boldsymbol{x},t)-\frac{\hbar^2}{2m}\,\triangle E_n(\boldsymbol{x},t)=\delta(\boldsymbol{x})\delta(t). \tag{2.3.7}$$

Let

$$E_n(\boldsymbol{x},t)=\frac{\mathrm{i}}{\hbar}E_n^*(\boldsymbol{x},t),$$

then $E_n^*(\boldsymbol{x},t)$ satisfies

$$\frac{\partial E_n^*(\boldsymbol{x},t)}{\partial t}-\mathrm{i}\frac{\hbar}{2m}\,\triangle E_n^*(\boldsymbol{x},t)=\delta(\boldsymbol{x})\delta(t),$$

which is the Eq. (2.3.2) with $a=\mathrm{i}\dfrac{\hbar}{2m}$. By the fundamental solution (2.3.4), we obtain $E_n^*(\boldsymbol{x},t)$ and

$$E_n(\boldsymbol{x},t)=\frac{\mathrm{i}}{\hbar}E_n^*(\boldsymbol{x},t)=\frac{H(t)}{\hbar}\mathrm{e}^{-\mathrm{i}\pi(n-2)/4}\left(\frac{2\pi\hbar t}{m}\right)^{-\frac{n}{2}}\mathrm{e}^{\mathrm{i}\frac{m|\boldsymbol{x}|^2}{2\hbar t}}. \tag{2.3.8}$$

Especially for the scaled Schrödinger operator $-\left(\mathrm{i}\dfrac{\partial}{\partial t}+\triangle\right)$, the fundamental solution can be obtained from (2.3.8). Namely,

$$E_n(\boldsymbol{x},t)=H(t)\,\mathrm{e}^{-\pi(n-2)/4}(4\pi t)^{-\frac{n}{2}}\mathrm{e}^{\mathrm{i}\frac{|\boldsymbol{x}|^2}{4t}}. \tag{2.3.9}$$

2.3.3 Wave Operator

We consider the wave equation

$$L\left(D,\frac{\partial}{\partial t}\right)u\equiv\frac{\partial^2}{\partial t^2}u-c^2\,\triangle u=f, \tag{2.3.10}$$

which is a hyperbolic partial differential equation. It typically concerns wave propagation. The unknown scalar function $u(\boldsymbol{x},t)$ describes the mechanical displacement of a wave. The constant c is identified as the propagation speed of the wave. The fundamental solution $E_n(\boldsymbol{x},t)$ of the wave operator $L\left(D,\dfrac{\partial}{\partial t}\right)=\dfrac{\partial^2}{\partial t^2}-c^2\,\triangle$ is a distribution solution of the equation

$$\frac{\partial^2}{\partial t^2}E_n(\boldsymbol{x},t)-c^2\,\triangle E_n(\boldsymbol{x},t)=\delta(\boldsymbol{x})\delta(t),\qquad\forall\boldsymbol{x}\in\mathbb{R}^n,\ t\in\mathbb{R}^+. \tag{2.3.11}$$

Applying the Fourier transform with respect to $\boldsymbol{x}\in\mathbb{R}^n$ on both sides of equation (2.3.11) yields

$$\frac{\mathrm{d}^2}{\mathrm{d}t^2}\hat{E}_n(\boldsymbol{\xi}, t) + c^2|\boldsymbol{\xi}|^2\hat{E}_n(\boldsymbol{\xi}, t) = \delta(t) \tag{2.3.12}$$

with $\hat{E}_n(\boldsymbol{\xi}, t) = \mathfrak{F}_x[E_n(\boldsymbol{x}, t)](\boldsymbol{\xi}, t)$. It is straightforward to check that

$$\hat{E}_n(\boldsymbol{\xi}, t) = H(t)\frac{\sin(c|\boldsymbol{\xi}|t)}{c|\boldsymbol{\xi}|} \tag{2.3.13}$$

is a distribution solution of equation (2.3.12). Therefore

$$E_n(\boldsymbol{x}, t) = H(t)\mathfrak{F}_\xi^{-1}\left[\frac{\sin(c|\boldsymbol{\xi}|t)}{c|\boldsymbol{\xi}|}\right](\boldsymbol{x}, t). \tag{2.3.14}$$

Using the results in Examples A.1.7, A.1.8 and A.1.4 (Kythe, 1996) for $n = 1,\ 2,\ 3$ from formula (2.3.14), we obtain

$$E_1(\boldsymbol{x}, t) = \frac{1}{2c}H(ct - |\boldsymbol{x}|), \tag{2.3.15}$$

$$E_2(\boldsymbol{x}, t) = \frac{H(ct - |\boldsymbol{x}|)}{2\pi c\sqrt{c^2t^2 - |\boldsymbol{x}|^2}}, \tag{2.3.16}$$

$$E_3(\boldsymbol{x}, t) = \frac{H(t)}{2c}\delta(c^2t^2 - |\boldsymbol{x}|^2), \tag{2.3.17}$$

which are the fundamental solutions of the wave operator for $n = 1,\ 2,\ 3$.

The main materials in Sects. 2.2–2.3 are from the book by Kythe (1996).

2.4 Matrix Operator

In this section, we discuss the fundamental solutions of four matrix operators, which correspond to the steady-state Navier equations, harmonic elastic waves, steady-state Stokes equations and Oseen equations (linear Navier–Stokes equations).

The Navier equations are the basic partial differential equations in the theory of linear elasticity (Sokolniko, 1983). Let $\boldsymbol{u} = (u_1, u_2, \ldots, u_n)^{\mathrm{T}}$ $(n = 2,\ 3)$ denote the displacement vector the elastic medium, $\boldsymbol{\epsilon} = (\epsilon_{i,j})_{n\times n}$ and $\boldsymbol{\sigma} = (\sigma_{i,j})_{n\times n}$ denote the strain and stress tensors, respectively. The strain-displacement equations are given by

$$\epsilon_{ij}(\boldsymbol{u}) = \frac{1}{2}\left(\frac{\partial u_i}{\partial x_j} + \frac{\partial u_j}{\partial x_i}\right), \qquad 1 \leqslant i,\ j \leqslant n. \tag{2.4.1}$$

Namely,

$$\boldsymbol{\epsilon} = \frac{1}{2}\left[\nabla\boldsymbol{u} + (\nabla\boldsymbol{u})^{\mathrm{T}}\right].$$

The strain and stress tensors satisfy Hooke's law,

$$\sigma_{ij} = \lambda\Big(\sum_{k=1}^{n} \epsilon_{kk}\Big)\delta_{ij} + 2\mu\epsilon_{ij}, \qquad 1 \leqslant i,\ j \leqslant n, \tag{2.4.2}$$

where δ_{ij} denotes the Kronecker symbol, the positive constants λ and μ are the Lamé constants of the isotropic elastic medium. The Lamé constants can be written in terms of Young's modulus E and Poisson's ratio ν as follows

$$\lambda = \frac{E\nu}{(1+\nu)(1-2\nu)}, \qquad \mu = \frac{E}{2(1+\nu)}. \tag{2.4.3}$$

The Navier equations for motion of a linear elastic medium are given (Sokolniko, 1983):

$$\rho\frac{\partial^2 \boldsymbol{u}}{\partial t^2} = \nabla\cdot\boldsymbol{\sigma} + \boldsymbol{f}, \tag{2.4.4}$$

with the constant $\rho > 0$, the density of the elastic medium, and the given body force $\boldsymbol{f} = (f_1, f_2, \dots, f_n)^{\mathrm{T}}$.

Combining the equations (2.4.1)–(2.4.2) and (2.4.4), we obtain the Navier equations for motion of an isotropic linear elastic medium in terms of displacement,

$$\rho\frac{\partial^2 u_i}{\partial t^2} = \mu\,\Delta u_i + (\lambda+\mu)\frac{\partial}{\partial x_i}\Big(\sum_{j=1}^{n}\frac{\partial u_j}{\partial x_j}\Big) + f_i, \quad 1 \leqslant i \leqslant n.$$

Its vector form reads

$$\rho\frac{\partial^2 \boldsymbol{u}}{\partial t^2} = \mu\,\Delta\boldsymbol{u} + (\lambda+\mu)\nabla(\nabla\cdot\boldsymbol{u}) + \boldsymbol{f}. \tag{2.4.5}$$

When $\boldsymbol{f} = 0$, we obtain the homogeneous Navier equations,

$$\rho\frac{\partial^2 \boldsymbol{u}}{\partial t^2} = \mu\,\Delta\boldsymbol{u} + (\lambda+\mu)\nabla(\nabla\cdot\boldsymbol{u}). \tag{2.4.6}$$

Furthermore, we consider the harmonic solutions of equations (2.4.6). Let

$$\boldsymbol{u}(\boldsymbol{x}, t) = \mathrm{e}^{\mathrm{i}\omega t}\boldsymbol{u}(\boldsymbol{x}), \qquad \omega \in \mathbb{R}^+. \tag{2.4.7}$$

Substituting (2.4.7) into (2.4.6) leads to

$$-\mu\,\Delta\boldsymbol{u}(\boldsymbol{x}) + (\lambda+\mu)\nabla\big(\nabla\cdot\boldsymbol{u}(\boldsymbol{x})\big) - \rho\omega^2\boldsymbol{u}(\boldsymbol{x}) = 0, \tag{2.4.8}$$

which are the harmonic elastic wave equations, also called the harmonic Navier equations.

In the steady-state case, the nonhomogeneous equations (2.4.5) are reduced to

$$-\mu\,\Delta\boldsymbol{u}-(\lambda+\mu)\nabla(\nabla\cdot\boldsymbol{u})=\boldsymbol{f}. \tag{2.4.9}$$

2.4.1 Steady-State Navier Operator

The steady-state Navier equations (2.4.9) can be rewritten as

$$-\mathcal{L}_n(D)\boldsymbol{u}=\boldsymbol{f}, \tag{2.4.10}$$

with

$$\begin{cases}\mathcal{L}_n(D)=(L_{ij}(D))_{m\times n},\\[2mm] L_{ij}(D)=\mu\,\delta_{ij}\,\Delta+(\lambda+\mu)\dfrac{\partial^2}{\partial x_i\partial x_j}.\end{cases} \tag{2.4.11}$$

We now derive the fundamental solution $\mathbb{E}_n(\boldsymbol{x})$ of the steady-state Navier operator (2.4.11), which is an nth-order distribution matrix satisfying the differential equations

$$-\mathcal{L}_n(D)\mathbb{E}_n(\boldsymbol{x})=\delta(\boldsymbol{x})\boldsymbol{I}_n,\qquad \forall\boldsymbol{x}\in\mathbb{R}^n$$

in the distributional sense.

At first we consider the case $n=2$,

$$\mathcal{L}_2(D)=\begin{pmatrix}\mu\,\Delta+(\lambda+\mu)\dfrac{\partial^2}{\partial x_1^2} & (\lambda+\mu)\dfrac{\partial^2}{\partial x_1\partial x_2}\\[3mm] (\lambda+\mu)\dfrac{\partial^2}{\partial x_1\partial x_2} & \mu\,\Delta+(\lambda+\mu)\dfrac{\partial^2}{\partial x_2^2}\end{pmatrix}.$$

Let $\mathcal{L}_2^*(D)$ denote the adjoint matrix of $\mathcal{L}_2(D)$. It has the form

$$\mathcal{L}_2^*(D)=\begin{pmatrix}\mu\,\Delta+(\lambda+\mu)\dfrac{\partial^2}{\partial x_2^2} & -(\lambda+\mu)\dfrac{\partial^2}{\partial x_1\partial x_2}\\[3mm] -(\lambda+\mu)\dfrac{\partial^2}{\partial x_1\partial x_2} & \mu\,\Delta+(\lambda+\mu)\dfrac{\partial^2}{\partial x_1^2}\end{pmatrix}:=\begin{pmatrix}L_{1,1}^*(D) & L_{1,2}^*(D)\\ L_{2,1}^*(D) & L_{2,2}^*(D)\end{pmatrix}.$$

A direct computation shows

$$\mathcal{L}_2(D)\mathcal{L}_2^*(D)=\begin{pmatrix}\mu(\lambda+2\mu)\,\Delta^2 & 0\\ 0 & \mu(\lambda+2\mu)\,\Delta^2\end{pmatrix}. \tag{2.4.12}$$

Let $\mathbb{G}_2(\boldsymbol{x})$ denote the fundamental solution of the matrix operator $\mathcal{L}_2(D)\mathcal{L}_2^*(D)$. By the identity (2.4.12), one may get an expression of $\mathbb{G}_2(\boldsymbol{x})$

$$\mathbb{G}_2(\boldsymbol{x}) = \begin{pmatrix} G_2(\boldsymbol{x}) & 0 \\ 0 & G_2(\boldsymbol{x}) \end{pmatrix}, \tag{2.4.13}$$

with the distribution $G_2(\boldsymbol{x})$ satisfying

$$-\mu(\lambda + 2\mu)\,\Delta^2 G_2(\boldsymbol{x}) = \delta(\boldsymbol{x}), \qquad \forall\, \boldsymbol{x} \in \mathbb{R}^2. \tag{2.4.14}$$

It is obvious that $\mu(\lambda + 2\mu)G_2(\boldsymbol{x})$ is the fundamental solution of the biharmonic operator. By (2.2.29), we obtain

$$G_2(\boldsymbol{x}) = -\frac{1}{8\pi\mu(\lambda + 2\mu)}(|\boldsymbol{x}|^2 \ln|\boldsymbol{x}| - |\boldsymbol{x}|^2), \tag{2.4.15}$$

and

$$-\mathcal{L}_2(D)\mathcal{L}_2^*(D)\mathbb{G}_2(\boldsymbol{x}) = \delta(\boldsymbol{x})\boldsymbol{I}_2, \qquad \forall\, \boldsymbol{x} \in \mathbb{R}^2. \tag{2.4.16}$$

Let

$$\mathbb{E}_2^*(\boldsymbol{x}) = \mathcal{L}_2^*(D)\mathbb{G}_2(\boldsymbol{x}), \tag{2.4.17}$$

which is a fundamental solution of the steady-state Navier operator $-\mathcal{L}_2(D)$.

Combining (2.4.13), (2.4.15), (2.4.17) yields

$$\mathbb{E}_2^*(\boldsymbol{x}) = \big(E_{ij}^*(\boldsymbol{x})\big)_{2\times 2} \tag{2.4.18}$$

with

$$E_{ij}^*(\boldsymbol{x}) = L_{ij}^*(D)G_2(\boldsymbol{x}), \qquad 1 \leqslant i,\ j \leqslant 2.$$

Furthermore, substituting (2.4.15) into (2.4.18) leads to

$$\begin{aligned}
L_{11}^*(D)G_2(\boldsymbol{x}) &= \mu\,\Delta G + (\lambda + \mu)\frac{\partial^2 G}{\partial x_2^2} = (\lambda + 2\mu)\,\Delta G - (\lambda + \mu)\frac{\partial^2 G}{\partial x_1^2} \\
&= \frac{\Delta + 3\mu}{4\pi\mu(\Delta + 2\mu)}\ln\frac{1}{|\boldsymbol{x}|} + \frac{\lambda + \mu}{4\pi\mu(\lambda + 2\mu)}\frac{x_1^2}{|\boldsymbol{x}|^2} - \frac{\lambda + \mu}{8\pi\mu(\lambda + 2\mu)}, \\
L_{12}^*(D)G_2(\boldsymbol{x}) &= L_{2,1}^*(D)G_2(\boldsymbol{x}) = -\frac{\partial^2 G}{\partial x_1 \partial x_2} = \frac{\lambda + \mu}{4\pi\mu(\lambda + 2\mu)}\frac{x_1 x_2}{|\boldsymbol{x}|^2}, \\
L_{22}^*(D)G_2(\boldsymbol{x}) &= \mu\,\Delta G + (\lambda + \mu)\frac{\partial^2 G}{\partial x_1^2} = (\lambda + 2\mu)\,\Delta G - (\lambda + \mu)\frac{\partial^2 G}{\partial x_2^2} \\
&= \frac{\Delta + 3\mu}{4\pi\mu(\Delta + 2\mu)}\ln\frac{1}{|\boldsymbol{x}|} + \frac{\lambda + \mu}{4\pi\mu(\lambda + 2\mu)}\frac{x_2^2}{|\boldsymbol{x}|^2} - \frac{\lambda + \mu}{8\pi\mu(\lambda + 2\mu)}.
\end{aligned}$$

Let

$$\begin{cases} \mathbb{E}_2(\boldsymbol{x}) = \big(E_{i,j}(\boldsymbol{x})\big)_{2\times 2} \\ E_{ij}(\boldsymbol{x}) = \dfrac{\lambda+3\mu}{4\pi\mu(\lambda+2\mu)}\delta_{ij}\ln\dfrac{1}{|\boldsymbol{x}|} + \dfrac{\lambda+\mu}{4\pi\mu(\lambda+2\mu)}\dfrac{x_i x_j}{|\boldsymbol{x}|^2}, \qquad 1 \leqslant i,\ j \leqslant 2, \end{cases} \tag{2.4.19}$$

Then we have

$$\mathbb{E}_2^*(\boldsymbol{x}) = \mathbb{E}_2(\boldsymbol{x}) - \frac{\lambda+\mu}{8\pi\mu(\lambda+2\mu)}\boldsymbol{I}_2.$$

Obviously $\mathbb{E}_2(\boldsymbol{x})$ also is a fundamental solution of the steady-state Navier Operator $-\mathcal{L}_2(D)$, which is the fundamental solution of the steady-state Navier operator in the literature.

Next we consider the case $n = 3$ with

$$\mathcal{L}_3(D) = \begin{pmatrix} \mu\,\Delta + (\lambda+\mu)\dfrac{\partial^2}{\partial x_1^2} & (\lambda+\mu)\dfrac{\partial^2}{\partial x_1\partial x_2} & (\lambda+\mu)\dfrac{\partial^2}{\partial x_1\partial x_3} \\ (\lambda+\mu)\dfrac{\partial^2}{\partial x_1\partial x_2} & \mu\,\Delta + (\lambda+\mu)\dfrac{\partial^2}{\partial x_2^2} & (\lambda+\mu)\dfrac{\partial^2}{\partial x_2\partial x_3} \\ (\lambda+\mu)\dfrac{\partial^2}{\partial x_1\partial x_3} & (\lambda+\mu)\dfrac{\partial^2}{\partial x_2\partial x_3} & \mu\,\Delta + (\lambda+\mu)\dfrac{\partial^2}{\partial x_3^2} \end{pmatrix}. \tag{2.4.20}$$

Let

$$\begin{cases} \mathcal{L}_3^*(D) = \big(L_{ij}^*(D)\big)_{3\times 3} \\ L_{ij}^*(D) = \mu(\lambda+2\mu)\,\Delta^2\delta_{ij} - \mu(\lambda+\mu)\,\Delta\dfrac{\partial^2}{\partial x_i\partial x_j}, \qquad 1 \leqslant i,\ j \leqslant 3, \end{cases}$$

be the adjoint matrix of $\mathcal{L}_3(D)$. A direct computation shows that

$$\sum_{\ell=1}^{3} L_{i\ell}(D)L_{\ell j}^*(D) = \mu^2(\lambda+2\mu)\,\Delta^3\delta_{ij}, \qquad 1 \leqslant i,\ j \leqslant 3,$$

that is,

$$-\,\mathcal{L}_3(D)\mathcal{L}_3(D)^* = -\mu^2(\lambda+2\mu)\,\Delta^3\boldsymbol{I}_3.$$

By the fundamental solution of the tri-harmonic operator Δ^3 given by (2.2.31), then the fundamental solution of the operator $-\mu^2(\lambda+2\mu)\,\Delta^3$ reads

$$G_3(\boldsymbol{x}) = \frac{|\boldsymbol{x}|^3}{96\pi\mu^2(\lambda+2\mu)}, \tag{2.4.21}$$

which satisfies

$$-\mu^2(\lambda+2\mu)\,\Delta^3 G_3(\boldsymbol{x}) = \delta(\boldsymbol{x}), \qquad \forall \boldsymbol{x} \in \mathbb{R}^3.$$

Let

$$\mathbb{G}_3(\boldsymbol{x}) = G_3(\boldsymbol{x})\boldsymbol{I}_3.$$

Then we have

$$-\mathcal{L}_3(D)\mathcal{L}_3^*(D)\mathbb{G}_3(\boldsymbol{x}) = \delta(\boldsymbol{x})\boldsymbol{I}_3, \qquad \forall \boldsymbol{x} \in \mathbb{R}^3. \tag{2.4.22}$$

Let

$$\mathbb{E}_3(\boldsymbol{x}) = \mathcal{L}_3^*(D)\mathbb{G}_3(\boldsymbol{x}) := \left(E_{ij}^{(3)}(\boldsymbol{x})\right)_{3\times 3}. \tag{2.4.23}$$

It can be verified that $\mathbb{E}_3(\boldsymbol{x})$ satisfies the equations

$$-\mathcal{L}_3(D)\mathbb{E}_3(\boldsymbol{x}) = \delta(\boldsymbol{x})\boldsymbol{I}_3, \qquad \forall \boldsymbol{x} \in \mathbb{R}^3. \tag{2.4.24}$$

So $\mathbb{E}_3(\boldsymbol{x})$ is a fundamental solution of the Navier operator $\mathcal{L}_3(D)$. By the equality (2.4.23), we obtain

$$E_{ij}^{(3)}(\boldsymbol{x}) = \mu(\lambda+2\mu)\,\Delta^2 G_3(\boldsymbol{x})\delta_{ij} - \mu(\lambda+\mu)\,\Delta\frac{\partial^2 G_3(\boldsymbol{x})}{\partial x_i \partial x_j}, \qquad 1 \leqslant i,\ j \leqslant 3. \tag{2.4.25}$$

Substituting (2.4.21) into (2.4.25) yields

$$E_{ij}^{(3)}(\boldsymbol{x}) = \frac{\lambda+3\mu}{8\pi\mu(\lambda+2\mu)}\frac{1}{|\boldsymbol{x}|}\delta_{ij} + \frac{\lambda+\mu}{8\pi\mu(\lambda+2\mu)}\frac{x_i x_j}{|\boldsymbol{x}|^3}, \qquad 1 \leqslant i,\ j \leqslant 3. \tag{2.4.26}$$

The fundamental solution $\mathbb{E}_3(\boldsymbol{x})$ of the three dimensional steady-state Navier operator is given by (2.4.23) and (2.4.26).

2.4.2 *Harmonic Navier Operator*

In this section, we derive the fundamental solution of the harmonic Navier operator by the approach used in Sect. 2.4.1.

We consider the harmonic elastic wave equations (2.4.8). Let

$$\kappa_1 = \sqrt{\frac{\rho}{\mu}}\omega, \qquad \kappa_2 = \sqrt{\frac{\rho}{\lambda+2\mu}}\omega. \tag{2.4.27}$$

Then the Eq. (2.4.8) can be rewritten as

$$-\mathcal{L}_n(D)\boldsymbol{u} = 0, \tag{2.4.28}$$

with

$$\mathcal{L}_n(D) = \left(\mu(\Delta + \kappa_1^2)\delta_{i,j} + (\lambda + \mu)\frac{\partial^2}{\partial x_i \partial x_j}\right)_{n\times n}. \tag{2.4.29}$$

At first we discuss the case $n = 2$, the adjoint matrix operator $\mathcal{L}_2^*(D)$ of $\mathcal{L}_2(D)$ is given by

$$\mathcal{L}_2^*(D) = \left((\lambda + 2\mu)(\Delta + \kappa_2^2)\delta_{i,j} - (\lambda + \mu)\frac{\partial^2}{\partial x_i \partial x_j}\right)_{2\times 2}. \tag{2.4.30}$$

It is straightforward to check that

$$\mathcal{L}_2(D)\mathcal{L}_2^*(D) = \mu(\lambda + 2\mu)(\Delta + \kappa_1^2)(\Delta + \kappa_2^2)\boldsymbol{I}_2. \tag{2.4.31}$$

To find the fundamental solution of the matrix operator $-\mathcal{L}_2(D)\mathcal{L}_2^*(D)$, we only need to get the fundamental solution of the operator $-\mu(\lambda + 2\mu)(\Delta + \kappa_1^2)(\Delta + \kappa_2^2)$, namely a distribution $G_2(\boldsymbol{x})$ such that

$$-\mu(\lambda + 2\mu)(\Delta + \kappa_1^2)(\Delta + \kappa_2^2)G_2(\boldsymbol{x}) = \delta(\boldsymbol{x}), \qquad \forall \boldsymbol{x} \in \mathbb{R}^2. \tag{2.4.32}$$

Let

$$W_2(\boldsymbol{x}) = (\lambda + 2\mu)(\Delta + \kappa_2^2)G_2(\boldsymbol{x}). \tag{2.4.33}$$

It satisfies

$$-\mu(\Delta + \kappa_1^2)W_2(\boldsymbol{x}) = \delta(\boldsymbol{x}), \qquad \forall \boldsymbol{x} \in \mathbb{R}^2. \tag{2.4.34}$$

By (2.2.25), we have

$$W_2(\boldsymbol{x}) = \frac{1}{2\pi\mu}K_0(\kappa_1|\boldsymbol{x}|). \tag{2.4.35}$$

Substituting (2.4.35) into (2.4.33) gives us

$$(\lambda + 2\mu)(\Delta + \kappa_2^2)G_2(\boldsymbol{x}) = \frac{1}{2\pi\mu}K_0(\kappa_1|\boldsymbol{x}|). \tag{2.4.36}$$

A solution of equation (2.4.36) is given by

$$G_2(\boldsymbol{x}) = \frac{1}{2\pi\mu(\lambda + \mu)\kappa_1^2}\left\{K_0(\kappa_2|\boldsymbol{x}|) - K_0(\kappa_1|\boldsymbol{x}|)\right\},$$

which is the fundamental solution of the operator $-\mu(\lambda + 2\mu)(\Delta + \kappa_1^2)(\Delta + \kappa_2^2)$. Therefore it yields the fundamental solution of the matrix operator $-\mathcal{L}_2(D)\mathcal{L}_2^*(D)$, which is given by $G_2(\boldsymbol{x})\boldsymbol{I}_2$. Furthermore, the fundamental solution of the harmonic

Navier operator $-\mathcal{L}_2(D)$ is obtained by

$$\mathbb{E}_2(\boldsymbol{x}) = \mathcal{L}_2^*(D)\big(G_2(\boldsymbol{x})\boldsymbol{I}_2\big). \tag{2.4.37}$$

After direct computation, one may rewrite Eq. (2.4.37) as

$$\mathbb{E}_2(\boldsymbol{x}) = \big(E_{ij}(\boldsymbol{x})\big)_{2\times 2}, \tag{2.4.38}$$

with

$$E_{ij}(\boldsymbol{x}) = \frac{1}{2\pi\mu}K_0(\kappa_1|\boldsymbol{x}|)\delta_{ij} - (\lambda+\mu)\frac{\partial^2}{\partial x_i \partial x_j}G_2(\boldsymbol{x}), \quad 1 \leqslant i,\ j \leqslant 2.$$

We now discuss the case $n = 3$. The adjoint matrix operator $\mathcal{L}_3^*(D)$ of $\mathcal{L}_3(D)$ is given by

$$\mathcal{L}_3^*(D) = \left(\mu(\lambda+2\mu)(\Delta+\kappa_1^2)(\Delta+\kappa_2^2)\delta_{ij} - \mu(\lambda+\mu)(\Delta+\kappa_1)\frac{\partial^2}{\partial x_i \partial x_j}\right)_{3\times 3}.$$

We have

$$\begin{aligned} \mathcal{L}_3(D)\mathcal{L}_3^*(D) &= \left(\mu(\lambda+2\mu)(\Delta+\kappa_1^2)^2(\Delta+\kappa_2^2)\delta_{ij}\right)_{3\times 3} \\ &= \mu(\lambda+2\mu)(\Delta+\kappa_1^2)^2(\Delta+\kappa_2^2)\boldsymbol{I}_3. \end{aligned} \tag{2.4.39}$$

Let $\mathbb{G}_3(\boldsymbol{x})$ be the fundamental solution of the operator $-\mathcal{L}_3(D)\mathcal{L}_3^*(D)$, which satisfies

$$-\mathcal{L}_3(D)\mathcal{L}_3^*(D)\mathbb{G}_3(\boldsymbol{x}) = \delta(\boldsymbol{x})\boldsymbol{I}_3, \qquad \forall \boldsymbol{x} \in \mathbb{R}^3. \tag{2.4.40}$$

We see that $\mathbb{G}_3(\boldsymbol{x})$ can be written as

$$\mathbb{G}_3(\boldsymbol{x}) = G_3(\boldsymbol{x})\boldsymbol{I}_3,$$

provided that $G_3(\boldsymbol{x})$ is the fundamental solution of the operator $-\mu(\lambda+2\mu)(\Delta+\kappa_1^2)^2(\Delta+\kappa_2^2)$, i.e.,

$$-\mu(\lambda+2\mu)(\Delta+\kappa_1^2)^2(\Delta+\kappa_2^2)G_3(\boldsymbol{x}) = \delta(\boldsymbol{x}), \qquad \forall \boldsymbol{x} \in \mathbb{R}^3.$$

Let

$$W_3(\boldsymbol{x}) = \mu(\lambda+2\mu)(\Delta+\kappa_1^2)(\Delta+\kappa_2^2)G_3(\boldsymbol{x}). \tag{2.4.41}$$

Then $W_3(\boldsymbol{x})$ satisfies

$$-(\Delta+\kappa_1^2)W_3(\boldsymbol{x}) = \delta(\boldsymbol{x}), \qquad \forall \boldsymbol{x} \in \mathbb{R}^3.$$

By (2.2.25), we have

$$W_3(\boldsymbol{x}) = \frac{1}{4\pi}\frac{\mathrm{e}^{-\kappa_1|\boldsymbol{x}|}}{|\boldsymbol{x}|}. \tag{2.4.42}$$

Furthermore, let

$$U_3(\boldsymbol{x}) = (\Delta + \kappa_1^2)G_3(\boldsymbol{x}). \tag{2.4.43}$$

Combining (2.4.41)–(2.4.43) yields

$$-(\Delta + \kappa_2^2)U_3(\boldsymbol{x}) = -\frac{1}{\mu(\lambda+2\mu)}W_3(\boldsymbol{x}) = -\frac{1}{4\pi\mu(\lambda+2\mu)}\frac{\mathrm{e}^{-\kappa_1|\boldsymbol{x}|}}{|\boldsymbol{x}|}. \tag{2.4.44}$$

A solution of equation (2.4.44) takes

$$U_3(\boldsymbol{x}) = \frac{1}{\mu(\lambda+\mu)(\lambda+2\mu)\kappa_1\kappa_2}\left\{\frac{1}{4\pi}\frac{\mathrm{e}^{-\kappa_2|\boldsymbol{x}|}}{|\boldsymbol{x}|} - \frac{1}{4\pi}\frac{\mathrm{e}^{-\kappa_1|\boldsymbol{x}|}}{|\boldsymbol{x}|}\right\}.$$

From (2.4.43), we can obtain the fundamental solution $G_3(\boldsymbol{x})$ satisfies

$$(\Delta + \kappa_1^2)G_3(\boldsymbol{x}) = U_3(\boldsymbol{x}) = \frac{1}{\mu(\lambda+\mu)(\lambda+2\mu)\kappa_1\kappa_2}\left\{\frac{1}{4\pi}\frac{\mathrm{e}^{-\kappa_2|\boldsymbol{x}|}}{|\boldsymbol{x}|} - \frac{1}{4\pi}\frac{\mathrm{e}^{-\kappa_1|\boldsymbol{x}|}}{|\boldsymbol{x}|}\right\},$$

and

$$\mu(\lambda+2\mu)(\Delta+\kappa_1^2)(\Delta+\kappa_2^2)G_3(\boldsymbol{x}) = W_3(\boldsymbol{x}) = \frac{1}{4\pi}\frac{\mathrm{e}^{-\kappa_1|\boldsymbol{x}|}}{|\boldsymbol{x}|}.$$

By Eq. (2.4.40), the fundamental solution of the harmonic Navier operator $\mathcal{L}_3(D)$ is given by

$$\begin{aligned}
\mathbb{E}_3(\boldsymbol{x}) &= \mathcal{L}_3^*(D)\mathbb{G}_3(\boldsymbol{x}) \\
&= \left(\mu(\lambda+2\mu)(\Delta+\kappa_1^2)(\Delta+\kappa_2^2)G_3(\boldsymbol{x})\delta_{i,j} - \mu(\lambda+\mu)(\Delta+\kappa_1^2)\frac{\partial^2 G_3(\boldsymbol{x})}{\partial x_i \partial x_j}\right) \\
&= \left(W_3(\boldsymbol{x})\delta_{i,j} - \mu(\lambda+\mu)\frac{\partial^2 U_3(\boldsymbol{x})}{\partial x_i\partial x_j}\right)_{3\times 3} \\
&= \left[\frac{1}{4\pi}\frac{\mathrm{e}^{-\kappa_1^2|\boldsymbol{x}|}}{|\boldsymbol{x}|}\delta_{i,j} - \frac{1}{(\lambda+2\mu)\kappa_1\kappa_2}\frac{\partial^2}{\partial x_i\partial x_j}\left(\frac{1}{4\pi}\frac{\mathrm{e}^{-\kappa_2|\boldsymbol{x}|}}{|\boldsymbol{x}|} - \frac{1}{4\pi}\frac{\mathrm{e}^{-\kappa_1|\boldsymbol{x}|}}{|\boldsymbol{x}|}\right)\right]_{3\times 3}.
\end{aligned} \tag{2.4.45}$$

In the remaining two subsections, we will discuss the fundamental solutions of the steady-state Stokes operator and the steady-state Oseen operator. First, we consider the incompressible Navier–Stokes equations in velocity-pressure formulation (Happel & Brenner, 1983):

$$\frac{\partial \boldsymbol{u}}{\partial t} + (\boldsymbol{u} \cdot \nabla)\boldsymbol{u} - \nu \Delta \boldsymbol{u} + \nabla p = \boldsymbol{f}, \tag{2.4.46}$$

$$\nabla \cdot \boldsymbol{u} = 0, \tag{2.4.47}$$

where $\boldsymbol{u} = (u_1, u_2, \ldots, u_n)^{\mathrm{T}}$ and p are the velocity and the pressure variables of the flow, $\nabla = \left(\dfrac{\partial}{\partial x_1}, \dfrac{\partial}{\partial x_2}, \ldots, \dfrac{\partial}{\partial x_n}\right)^{\mathrm{T}}$ is the gradient operator, constant $\nu > 0$ is the kinematic viscosity coefficient, $\boldsymbol{f} = (f_1, f_2, \ldots, f_n)^{\mathrm{T}}$ is the given body force. The Navier–Stokes equations (2.4.46)–(2.4.47) are the basic governing equations for the incompressible viscous fluid, which are nonlinear partial differential equations. After linearization of equation (2.4.46), we obtain the Oseen equations (linear Navier–Stokes equations),

$$\frac{\partial \boldsymbol{u}}{\partial t} + (\boldsymbol{v} \cdot \nabla)\boldsymbol{u} - \nu \Delta \boldsymbol{u} + \nabla p = \boldsymbol{f}, \tag{2.4.48}$$

$$\nabla \cdot \boldsymbol{u} = 0, \tag{2.4.49}$$

where $\boldsymbol{v} = (v_1, v_2, \ldots, v_n)^{\mathrm{T}}$ is a known constant vector.

For the low Reynolds number case, the nonlinear term in Eq. (2.4.46) can be ignored, resulting in the Stokes equations

$$\frac{\partial \boldsymbol{u}}{\partial t} - \nu \Delta \boldsymbol{u} + \nabla p = \boldsymbol{f}, \tag{2.4.50}$$

$$\nabla \cdot \boldsymbol{u} = 0. \tag{2.4.51}$$

The steady-state Oseen equations and steady-state Stokes equations respectively read

$$(\boldsymbol{v} \cdot \nabla)\boldsymbol{u} - \nu \Delta \boldsymbol{u} + \nabla p = \boldsymbol{f}, \tag{2.4.52}$$

$$\nabla \cdot \boldsymbol{u} = 0, \tag{2.4.53}$$

and

$$- \nu \Delta \boldsymbol{u} + \nabla p = \boldsymbol{f}, \tag{2.4.54}$$

$$\nabla \cdot \boldsymbol{u} = 0. \tag{2.4.55}$$

2.4.3 Steady-State Stokes Operator

An nth-order distribution matrix $\mathbb{E}(\boldsymbol{x}) = (E_{ij}(\boldsymbol{x}))_{n\times n} \equiv (\mathbb{E}_1(\boldsymbol{x}), \mathbb{E}_2(\boldsymbol{x}), \ldots, \mathbb{E}_n(\boldsymbol{x}))$ and an nth-order distribution vector $\mathbb{P}(\boldsymbol{x}) = (P_1(\boldsymbol{x}), P_2(\boldsymbol{x}), \ldots, P_n(\boldsymbol{x}))$, which satisfies the Stokes equations

$$- \nu \Delta \mathbb{E}(\boldsymbol{x}) + \nabla \mathbb{P}(\boldsymbol{x}) = \delta(\boldsymbol{x})\boldsymbol{I}_n, \qquad \forall \boldsymbol{x} \in \mathbb{R}^n, \tag{2.4.56}$$

$$\nabla \cdot \mathbb{E}(\boldsymbol{x}) = 0, \qquad \forall \boldsymbol{x} \in \mathbb{R}^n, \tag{2.4.57}$$

in the distributional sense, are called a fundamental solution of the Stokes operator.

Applying the divergence operator $\nabla\cdot$ to the Eq. (2.4.56) and taking into account the incompressibility condition (2.4.57), we get

$$\triangle \mathbb{P}(\boldsymbol{x}) = \nabla \cdot \delta(\boldsymbol{x}) \boldsymbol{I}_n, \qquad \forall \boldsymbol{x} \in \mathbb{R}^n,$$

i.e.,

$$\triangle P_j(\boldsymbol{x}) = \frac{\partial}{\partial x_j} \delta(\boldsymbol{x}), \qquad \forall \boldsymbol{x} \in \mathbb{R}^n, \ 1 \leqslant j \leqslant n. \tag{2.4.58}$$

Let $V_n(\boldsymbol{x})$ denote the fundamental solution of the Laplace equation in $\mathbb{R}^n$. By (2.2.4) and (2.2.7), we know

$$V_n(\boldsymbol{x}) = \begin{cases} -\dfrac{1}{2\pi} \ln |\boldsymbol{x}|, & n = 2, \\ \dfrac{1}{4\pi} \dfrac{1}{|\boldsymbol{x}|}, & n = 3. \end{cases} \tag{2.4.59}$$

Since

$$- \triangle V_n(\boldsymbol{x}) = \delta(\boldsymbol{x}), \qquad \forall \boldsymbol{x} \in \mathbb{R}^n,$$

a solution of equation (2.4.58) is obtained

$$P_j(\boldsymbol{x}) = -\frac{\partial V_n(\boldsymbol{x})}{\partial x_j}, \qquad 1 \leqslant j \leqslant n. \tag{2.4.60}$$

Furthermore, we have

$$\begin{aligned} &\mathbb{P}(\boldsymbol{x}) = (P_1(\boldsymbol{x}), \ P_2(\boldsymbol{x}), \ \ldots, \ P_n(\boldsymbol{x})) = -\nabla^{\mathrm{T}} V_n(\boldsymbol{x}), \\ &\nabla \mathbb{P}(\boldsymbol{x}) = -\nabla \nabla^{\mathrm{T}} V_n(\boldsymbol{x}). \end{aligned} \tag{2.4.61}$$

Then the momentum equations (2.4.56) can be rewritten as

$$- \nu \triangle \mathbb{E}(\boldsymbol{x}) = \delta(\boldsymbol{x}) \boldsymbol{I}_n + \nabla \nabla^{\mathrm{T}} V_n(\boldsymbol{x}),$$

i.e.,

$$- \nu \triangle E_{ij}(\boldsymbol{x}) = \delta(\boldsymbol{x}) \delta_{ij} + \frac{\partial^2}{\partial x_i \partial x_j} V_n(\boldsymbol{x}), \qquad 1 \leqslant i, \ j \leqslant n. \tag{2.4.62}$$

In the case that $i \neq j$, we obtain

$$- \nu \triangle E_{ij}(\boldsymbol{x}) = \frac{\partial^2}{\partial x_i \partial x_j} V_n(\boldsymbol{x}), \quad 1 \leqslant i \neq j \leqslant n,$$

and

$$\nu\,\Delta^2 E_{ij}(\boldsymbol{x}) = \frac{\partial^2}{\partial x_i \partial x_j}\delta(\boldsymbol{x}), \quad 1 \leqslant i \neq j \leqslant n. \tag{2.4.63}$$

Let $W_n(\boldsymbol{x})$ denote the fundamental solution of the biharmonic operator. By (2.2.29), we have

$$W_n(\boldsymbol{x}) = \begin{cases} -\dfrac{|\boldsymbol{x}|^2}{8\pi}(\ln|\boldsymbol{x}| - 1), & n = 2, \\ \dfrac{|\boldsymbol{x}|}{8\pi}, & n = 3. \end{cases}$$

Since

$$-\,\Delta^2 W_n(\boldsymbol{x}) = \delta(\boldsymbol{x}), \qquad \forall \boldsymbol{x} \in \mathbb{R}^n,$$

we simply obtain a solution of equation (2.4.63),

$$E_{ij}(\boldsymbol{x}) = -\frac{1}{\nu}\frac{\partial^2 W_n(\boldsymbol{x})}{\partial x_i \partial x_j}, \quad 1 \leqslant i \neq j \leqslant n. \tag{2.4.64}$$

In the case that $i = j$, it is straightforward to obtain the solution of equation (2.4.62). It is given by

$$E_{ii}(\boldsymbol{x}) = \frac{1}{\nu}V_n(\boldsymbol{x}) - \frac{1}{\nu}\frac{\partial^2 W_n(\boldsymbol{x})}{\partial x_i^2}, \quad 1 \leqslant i \leqslant n. \tag{2.4.65}$$

We rewrite (2.4.64) and (2.4.65) uniformly as

$$E_{ij}(\boldsymbol{x}) = \frac{1}{\nu}V_n(\boldsymbol{x})\delta_{ij} - \frac{1}{\nu}\frac{\partial^2 W_n(\boldsymbol{x})}{\partial x_i \partial x_j}, \quad 1 \leqslant i,\ j \leqslant n. \tag{2.4.66}$$

Let $\{\mathbb{E}_j(\boldsymbol{x})\}_{j=1}^n$ be the column vectors of $\mathbb{E}(\boldsymbol{x}) = (E_{ij}(\boldsymbol{x}))_{n\times n}$. Direct computation shows that

$$\nabla \cdot \mathbb{E}_i(\boldsymbol{x}) = \frac{1}{\nu}\frac{\partial V_n(\boldsymbol{x})}{\partial x_i} - \frac{1}{\nu}\frac{\partial}{\partial x_i}\big(\triangle W_n(\boldsymbol{x})\big) = 0.$$

This indicates

$$\nabla \cdot \mathbb{E}(\boldsymbol{x}) = \big(\nabla \cdot \mathbb{E}_1(\boldsymbol{x}),\ \nabla \cdot \mathbb{E}_2(\boldsymbol{x}),\ \ldots,\ \nabla \cdot \mathbb{E}_n(\boldsymbol{x})\big) = 0.$$

The distribution matrix $\mathbb{E}(\boldsymbol{x}) = (E_{i,j})_{n\times n}$ with $E_{i,j}$ be given by (2.4.66) and the distribution vector $\mathbb{P}(\boldsymbol{x}) = (P_1(\boldsymbol{x}), P_2(\boldsymbol{x}), \ldots, P_n(\boldsymbol{x}))$ with P_j given by (2.4.60) are the distribution solution of equations (2.4.56)–(2.4.57). In other words, $\{\mathbb{E}(\boldsymbol{x}), \mathbb{P}(\boldsymbol{x})\}$ is a fundamental solution of the steady-state Stokes operator.

Furthermore a direct computation we obtain

$$\mathbb{E}(\boldsymbol{x}) = \frac{1}{4(n-1)\pi\nu}\left\{\left(\gamma_n(\boldsymbol{x}) + \frac{3-n}{2}\right)\mathbb{I} + \frac{\boldsymbol{x}\boldsymbol{x}^{\mathrm{T}}}{|\boldsymbol{x}|^n}\right\},$$

$$\mathbb{P}(\boldsymbol{x}) = \frac{1}{2(n-1)\pi}\Big(-\nabla\gamma_n(\boldsymbol{x})\Big)^{\mathrm{T}} = -\frac{1}{2(n-1)\pi}\frac{\boldsymbol{x}^{\mathrm{T}}}{|\boldsymbol{x}|^n},$$

with

$$\gamma_n(\boldsymbol{x}) = \begin{cases} -\ln|\boldsymbol{x}|, & n = 2, \\ \dfrac{1}{|\boldsymbol{x}|}, & n = 3. \end{cases}$$

Let

$$\mathbb{E}_n(\boldsymbol{x}) = \frac{1}{4(n-1)\pi\nu}\left\{\gamma_n(\boldsymbol{x})\mathbb{I} + \frac{\boldsymbol{x}\boldsymbol{x}^{\mathrm{T}}}{|\boldsymbol{x}|^n}\right\}, \tag{2.4.67}$$

$$\mathbb{P}_n(\boldsymbol{x}) = \frac{1}{2(n-1)\pi}\Big(-\nabla\gamma_n(\boldsymbol{x})\Big)^{\mathrm{T}} = -\frac{1}{2(n-1)\pi}\frac{\boldsymbol{x}^{\mathrm{T}}}{|\boldsymbol{x}|^n}. \tag{2.4.68}$$

Obviously $\{\mathbb{E}_n(\boldsymbol{x}), \mathbb{P}_n(\boldsymbol{x})\}$ is also a fundamental solution of the steady-state Stokes equations, which is used in literature.

2.4.4 Steady-State Oseen Operator

We now derive the fundamental solution of the steady-state Oseen operator. For simplicity in explanation, suppose that the known constant vector $\boldsymbol{v} = (U, 0, \ldots, 0)^{\mathrm{T}} \in \mathbb{R}^n$. Then the steady-state Oseen equations can be rewritten as

$$-\nu\Delta\boldsymbol{u} + U\frac{\partial}{\partial x_1}\boldsymbol{u} + \nabla p = \boldsymbol{f}, \tag{2.4.69}$$

$$\nabla\cdot\boldsymbol{u} = 0. \tag{2.4.70}$$

An nth-order distribution matrix $\mathbb{E}(\boldsymbol{x}) = (E_{ij}(\boldsymbol{x}))_{n\times n} = \big(\mathbb{E}_1(\boldsymbol{x}), \mathbb{E}_2(\boldsymbol{x}), \ldots, \mathbb{E}_n(\boldsymbol{x})\big)$ and nth-order distribution vector $\Pi(\boldsymbol{x}) = (\Pi_1(\boldsymbol{x}), \Pi_2(\boldsymbol{x}), \ldots, \Pi_n(\boldsymbol{x}))^{\mathrm{T}}$, which satisfies the steady-state Oseen equations

$$-\nu\Delta\mathbb{E}(\boldsymbol{x}) + U\frac{\partial}{\partial x_1}\mathbb{E}(\boldsymbol{x}) + \nabla\Pi(\boldsymbol{x})^{\mathrm{T}} = \delta(\boldsymbol{x})\boldsymbol{I}_n, \quad \forall\boldsymbol{x}\in\mathbb{R}^n, \tag{2.4.71}$$

$$\nabla\cdot\mathbb{E}(\boldsymbol{x}) = 0, \quad \forall\boldsymbol{x}\in\mathbb{R}^n, \tag{2.4.72}$$

in the sense of distribution, are called a fundamental solution of the steady-state Oseen operator.

Taking divergence of Eq. (2.4.71) leads to

$$\Delta \Pi(\boldsymbol{x})^{\mathrm{T}} = \nabla\big(\delta(\boldsymbol{x})\boldsymbol{I}_n\big), \qquad \forall \boldsymbol{x} \in \mathbb{R}^n,$$

namely,

$$\Delta \Pi_j(\boldsymbol{x}) = \frac{\partial}{\partial x_j}\delta(\boldsymbol{x}), \qquad \forall \boldsymbol{x} \in \mathbb{R}^n. \tag{2.4.73}$$

A solution of (2.4.73) is given by

$$\Pi_j(\boldsymbol{x}) = -\frac{\partial V_n(\boldsymbol{x})}{\partial x_j}, \qquad 1 \leqslant j \leqslant n$$

with $V_n(\boldsymbol{x})$ be the fundamental solution of the Laplace operator in $\mathbb{R}^n$. That is,

$$\Pi(\boldsymbol{x}) = -\nabla V_n(\boldsymbol{x}) \tag{2.4.74}$$

and

$$\nabla \Pi(\boldsymbol{x})^{\mathrm{T}} = -\nabla\nabla^{\mathrm{T}} V_n(\boldsymbol{x}). \tag{2.4.75}$$

Equation (2.4.71) can be rewritten as

$$-\nu\,\Delta \mathbb{E} + U\frac{\partial}{\partial x_1}\mathbb{E} = \delta(\boldsymbol{x})\boldsymbol{I}_n + \nabla\nabla^{\mathrm{T}} V_n(\boldsymbol{x}), \qquad \forall \boldsymbol{x} \in \mathbb{R}^n,$$

i.e.,

$$-\nu\,\Delta E_{ij}(\boldsymbol{x}) + U\frac{\partial}{\partial x_1}E_{ij}(\boldsymbol{x}) = \delta(\boldsymbol{x})\delta_{ij} + \frac{\partial^2}{\partial x_i \partial x_j}V_n(\boldsymbol{x}), \qquad 1 \leqslant i,\ j \leqslant n. \tag{2.4.76}$$

Before finding the solution $\{\, E_{ij}(\boldsymbol{x}),\ 1 \leqslant i,\ j \leqslant n \,\}$ of Eq. (2.4.76), we introduce two fundamental solutions. Let $C_n(\boldsymbol{x})$ be the fundamental solution of the conduction-diffusion operator, which satisfies

$$-\nu\,\Delta C_n(\boldsymbol{x}) + U\frac{\partial}{\partial x_1}C_n(\boldsymbol{x}) = \delta(\boldsymbol{x}), \qquad \forall \boldsymbol{x} \in \mathbb{R}^n. \tag{2.4.77}$$

Let $D_n(\boldsymbol{x})$ be the fundamental solution of the following fourth-order elliptic operator,

$$-\nu\,\Delta^2 D_n(\boldsymbol{x}) + U\frac{\partial}{\partial x_1}\Delta D_n(\boldsymbol{x}) = \delta(\boldsymbol{x}), \qquad \forall \boldsymbol{x} \in \mathbb{R}^n. \tag{2.4.78}$$

For finding the solution of equation (2.4.77), we introduce a new unknown $G_n(\boldsymbol{x})$ by

$$C_n(\boldsymbol{x}) = \mathrm{e}^{\lambda x_1} G_n(\boldsymbol{x}), \tag{2.4.79}$$

with

$$\lambda = \frac{U}{2\nu}.$$

Substituting (2.4.79) into Eq. (2.4.77) yields

$$-\Delta\big(\nu\, G_n(\boldsymbol{x})\big) + \lambda^2\big(\nu\, G_n(\boldsymbol{x})\big) = \delta(\boldsymbol{x}), \qquad \forall \boldsymbol{x} \in \mathbb{R}^n.$$

It means that $\nu G_n(\boldsymbol{x})$ is the fundamental solution of the Helmholtz operator with $i\lambda$, which is given by (2.2.25). Therefore, we have

$$C_n(\boldsymbol{x}) = \begin{cases} \dfrac{\mathrm{e}^{\lambda x_1}}{2\pi\nu} K_0(\lambda|\boldsymbol{x}|), & n = 2, \\[2ex] \dfrac{\mathrm{e}^{\lambda x_1}}{4\pi\nu}\dfrac{\mathrm{e}^{-\lambda|\boldsymbol{x}|}}{|\boldsymbol{x}|}, & n = 3. \end{cases} \tag{2.4.80}$$

Next we discuss Eq. (2.4.78). It is obvious that

$$\Delta D_n(\boldsymbol{x}) = C_n(\boldsymbol{x}), \qquad \forall \boldsymbol{x} \in \mathbb{R}^n. \tag{2.4.81}$$

Differentiating both sides of Eq. (2.4.81) with respect to x_1 yields

$$\Delta \frac{\partial}{\partial x_1} D_n(\boldsymbol{x}) = \frac{\partial}{\partial x_1} C_n(\boldsymbol{x}), \qquad \forall \boldsymbol{x} \in \mathbb{R}^n. \tag{2.4.82}$$

It is straightforward to check that

$$\frac{\partial}{\partial x_1} D_n(\boldsymbol{x}) = \frac{1}{U}\big(\nu C_n(\boldsymbol{x}) - V_n(\boldsymbol{x})\big) \tag{2.4.83}$$

is a solution of (2.4.82) with $C_n(\boldsymbol{x})$ and $V_n(\boldsymbol{x})$ given by (2.4.79) and (2.4.59). Therefore, $D_n(\boldsymbol{x})$ is obtained from (2.4.83),

$$D_n(\boldsymbol{x}) = \frac{1}{U}\int_{-\infty}^{x_1} \big(\nu C_n(\xi, x_2, \ldots, x_n) - V_n(\xi, x_2, \ldots, x_n)\big)\, \mathrm{d}\boldsymbol{\xi}.$$

After $C_n(\boldsymbol{x})$ and $D_n(\boldsymbol{x})$ are obtained, the solution $\big(E_{ij}(\boldsymbol{x})\big)_{n\times n}$ of Eq. (2.4.76) can be written out

$$E_{ij}(\boldsymbol{x}) = C_n(\boldsymbol{x})\delta_{ij} - \frac{\partial^2 D_n(\boldsymbol{x})}{\partial x_i \partial x_j}, \quad 1 \leqslant i,\ j \leqslant n. \tag{2.4.84}$$

It is straightforward to check

$$\nabla \cdot \mathbb{E}(\boldsymbol{x}) = \big(\nabla \cdot \mathbb{E}_1(\boldsymbol{x}),\ \nabla \cdot \mathbb{E}_2(\boldsymbol{x}),\ \ldots,\ \nabla \cdot \mathbb{E}_n(\boldsymbol{x})\big) = \nabla\big(C_n(\boldsymbol{x}) - \Delta D_n(\boldsymbol{x})\big) = 0.$$

The fundamental solution $(\mathbb{E}(\boldsymbol{x}), \Pi(\boldsymbol{x}))$ of the steady-state Oseen operator is given by (2.4.84) and (2.4.74).

In the practical applications of the fundamental solutions, a parameter $\boldsymbol{y} \in \mathbb{R}^n$ is always introduced in the given fundamental solutions of the steady-state operators and matrix operators. For example, for the fundamental solution of the Laplace operator, we have

$$E_n(\boldsymbol{x}, \boldsymbol{y}) = \begin{cases} -\dfrac{1}{2\pi} \ln |\boldsymbol{x} - \boldsymbol{y}|, & n = 2, \\ \dfrac{1}{(n-2)S_n(1)} |\boldsymbol{x} - \boldsymbol{y}|^{2-n}, & n \geqslant 3. \end{cases} \tag{2.4.85}$$

which is a distribution solution of the equation

$$-\Delta_x E_n(\boldsymbol{x}, \boldsymbol{y}) = \delta(\boldsymbol{x} - \boldsymbol{y}), \qquad \forall \boldsymbol{x}, \boldsymbol{y} \in \mathbb{R}^n. \tag{2.4.86}$$

The fundamental solutions of the transient operators and matrix operators, parameter $\boldsymbol{y} \in \mathbb{R}^n$ and $\tau \in \mathbb{R}$ are always introduced in practical applications. For instance, the fundamental solution of the heat conduction operator is given

$$E_n(\boldsymbol{x}, \boldsymbol{y}; t, \tau) = \begin{cases} \dfrac{1}{[4\pi a(t-\tau)]^{\frac{n}{2}}} \mathrm{e}^{-\frac{|\boldsymbol{x}-\boldsymbol{y}|^2}{4a(t-\tau)}}, & t \geqslant \tau, \\ 0, & t < \tau. \end{cases}$$

which is a distribution solution of the equation

$$\frac{\partial E_n(\boldsymbol{x}, \boldsymbol{y}; t, \tau)}{\partial t} - a\,\Delta_x E_n(\boldsymbol{x}, \boldsymbol{y}; t, \tau) = \delta(\boldsymbol{x} - \boldsymbol{y})\delta(t - \tau), \qquad \forall \boldsymbol{x}, \boldsymbol{y} \in \mathbb{R}^n,\ t \in \mathbb{R}^+.$$

In this chapter, we restrict ourselves only to discussing the fundamental solutions of some typical linear constant coefficients differential operators. For more about the fundamental solutions of differential operators, we refer the readers to the books by Kythe (1996) and Ortner and Wagner (2015).

In fact, the fundamental solution of a given linear partial differential operator (matrix operator), is a special solution of the corresponding linear partial differential equation (equations) with the given δ-function on its right hand side. The fundamental solution can be used to construct the potential integrals including the surface potential integrals (single layer potential integrals and double layer potential integrals) and volume potential integrals of the corresponding equation (equations). Furthermore, the solution of the given linear partial differential equation (equations) can be expressed as a sum of different potential integrals. Fundamental solutions have played an important role in the development of the theory of partial differential equations, which provided an established method to study the existence, and regularity of the solution of the boundary (or initial boundary) value problem of PDE since the time to use Green's functions. In the sixties of the last century, the boundary integral equations, provided by the potential theory, were used to obtain numerical solutions to various problems in engineering computation. Those works formed the well-known boundary element method (Brebbia, 1978).

In the coming chapters, we discuss the properties of the potential integrals and the boundary integro-differential equations for the principal boundary value problems of various partial differential equations.

References

Brebbia, C. A. (1978). *The boundary element method for engineers*. Wiley.

Happel, J., & Brenner, H. (1983). *Low reynolds number hydrodynamics: With special applications to particulate media* (Vol. 1). Springer Science & Business Media.

Kythe, P. (1996). *Fundamental solutions for differential operators and applications*. Springer Science & Business Media.

Ortner, N., & Wagner, P. (2015). *Fundamental solutions of linear partial differential operators*. Springer.

Sokolnikoff, I. S. (1983). *Mathematical theory of elasticity*. Krieger Pub Co.

Chapter 3
Boundary Value Problems of the Laplace Equation

In this chapter, we study the boundary integro-differential equations of the boundary value problems of the Laplace equation. First, we recall some results on the boundary value problems of the Laplace equation. Then we discuss the potential theory corresponding to the Laplace equation. Finally, we reduce the boundary value problems of the Laplace equation to boundary integro-differential equations (or boundary integral equations).

Let $\Omega \subset \mathbb{R}^n$ $(n = 2, 3)$ be an open bounded domain with boundary Γ and $\Omega_e = \mathbb{R}^n \setminus \bar{\Omega}$ be the exterior domain, the complement of $\bar{\Omega}$.

3.1 Function Spaces

In this section, a brief review of function spaces is given (Grisvard, 2011).

3.1.1 Continuous and Continuously Differential Function Spaces

Definition 3.1.1 (i) Let $C^0(\bar{\Omega})\big(C^0(\bar{\Omega}_e)\big)$ denote the space of all continuous and bounded functions u defined on $\bar{\Omega}$ $(\bar{\Omega}_e)$ with norm

$$\|u\|_{C^0(\bar{\Omega})} = \max_{x \in \bar{\Omega}} |u(\boldsymbol{x})|, \tag{3.1.1}$$

$$\|u\|_{C^0(\bar{\Omega}_e)} = \sup_{x \in \bar{\Omega}_e} \{|u(\boldsymbol{x})|\}. \tag{3.1.2}$$

Let $C^{(m)}(\bar{\Omega})(C^{(m)}(\bar{\Omega}_e))$ denote the space of all functions u defined on $\bar{\Omega}$ $(\bar{\Omega}_e)$, u and its derivatives, up to the order m, are continuous and bounded on $\bar{\Omega}(\bar{\Omega}_e)$, with norm

H. Han and D. Yin, *Mathematical Foundation of the Boundary Integro-Differential Equation Method*, https://doi.org/10.1007/978-981-95-1088-7_3

$$\|u\|_{C^m(\bar{\Omega})} = \sum_{|\alpha|\leqslant m} \max_{x\in\bar{\Omega}} |D^\alpha u(x)|, \tag{3.1.3}$$

$$\|u\|_{C^m(\bar{\Omega}_e)} = \sum_{|\alpha|\leqslant m} \sup_{x\in\bar{\Omega}_e} \{|D^\alpha u(x)|\}. \tag{3.1.4}$$

(ii) Let $C^0(\Omega)\,\big(C^0(\Omega_e)\big)$ denote the space of all continuous functions on $\Omega\,(\Omega_e)$, $C^m(\Omega)\,(C^m(\Omega_e))$ denote the space of all continuously differentiable functions with the order m on $\Omega\,(\Omega_e)$.

(iii) Let $C_0^m(\Omega)\,\big(C_0^m(\Omega_e)\big)$ denote the subspace of $C^m(\Omega)\,(C^m(\Omega_e))$, in which the functions in $C^m(\Omega)\,(C^m(\Omega_e))$ with compact support in $\Omega\,(\Omega_e)$ with nonnegative integer m.

Specially for $m = 0$, the following notations are also used.

$$C(\overline{\Omega}) := C^0(\overline{\Omega}), \qquad C(\overline{\Omega}_e) := C^0(\overline{\Omega}_e),$$
$$C(\Omega) := C^0(\Omega), \qquad C(\Omega_e) := C^0(\Omega_e).$$

3.1.2 *Hölder Spaces*

Definition 3.1.2 Let m be a non-negative integer and β be a real number with $0 < \beta \leqslant 1$. Let $C^{m,\beta}(\bar{\Omega})$ $(C^{m,\beta}(\bar{\Omega}_e))$ denote the space of all functions u defined on $\bar{\Omega}$ $(\bar{\Omega}_e)$, their derivatives, up to the order m, are continuous and bounded and their derivatives of order m are uniformly Hölder continuous with exponent $0 < \beta \leqslant 1$ with norm

$$\|u\|_{C^{m,\beta}(\bar{\Omega})} = \|u\|_{C^m(\bar{\Omega})} + \sum_{|\alpha|=m} \sup_{\boldsymbol{x},\, \boldsymbol{y}\in\bar{\Omega}} \frac{|D^\alpha u(\boldsymbol{x}) - D^\alpha u(\boldsymbol{y})|}{|\boldsymbol{x}-\boldsymbol{y}|^\beta}, \tag{3.1.5}$$

$$\|u\|_{C^{m,\beta}(\bar{\Omega}_e)} = \|u\|_{C^m(\bar{\Omega}_e)} + \sum_{|\alpha|=m} \sup_{\boldsymbol{x},\, \boldsymbol{y}\in\bar{\Omega}_e} \frac{|D^\alpha u(\boldsymbol{x}) - D^\alpha u(\boldsymbol{y})|}{|\boldsymbol{x}-\boldsymbol{y}|^\beta}. \tag{3.1.6}$$

Let $C^{m,0}(\bar{\Omega}) := C^m(\bar{\Omega})$, $C^{m,0}(\bar{\Omega}_e) := C^m(\bar{\Omega}_e)$, $(m = 0,\ 1,\ \ldots)$. The function spaces $C^{m,\beta}(\bar{\Omega})$ $(C^{m,\beta}(\bar{\Omega}_e))$ with nonnegative integer m and $0 \leqslant \beta \leqslant 1$ are Banach spaces.

We now are ready to define the smoothness of the boundary Γ.

Definition 3.1.3 Suppose that $\Omega \subset \mathbb{R}^n$ is an open bounded domain with boundary Γ. We call that the boundary Γ belongs to the class $C^{m,\beta}$ (m is a nonnegative integer and $0 < \beta \leqslant 1$) if for every $\boldsymbol{x} \in \Gamma$ there exists a neighborhood V of $\boldsymbol{x}$ in $\mathbb{R}^n$ and a new orthogonal coordinates $\{\, y_1,\ y_2,\ \ldots,\ y_n \,\}$ such that

1. V is a hypercube in the new coordinates:

$$V = \{ (y_1, y_2, \ldots, y_n) \mid -a_j \leqslant y_j \leqslant a_j, \ 1 \leqslant j \leqslant n \}.$$

2. There exists a function φ defined on

$$V' = \{ (y_1, y_2, \ldots, y_{n-1}) \mid -a_j \leqslant y_j \leqslant a_j, \ 1 \leqslant j \leqslant n-1 \}$$

and $\varphi \in C^{m,\beta}(V')$ so that

$$|\varphi(\mathbf{y}')| \leqslant \frac{a_n}{2} \quad \text{for every } \mathbf{y}' = (y_1, y_2, \ldots, y_{n-1}) \in V',$$

$$\Omega \cap V = \{ \mathbf{y} = (\mathbf{y}', y_n) \in V \mid y_n < \varphi(\mathbf{y}') \},$$

$$\Gamma \cap V = \{ \mathbf{y} = (\mathbf{y}', y_n) \in V \mid y_n = \varphi(\mathbf{y}') \}.$$

Especially if $\Gamma \in C := C^{0,0}$, the boundary Γ is said to be continuous; if $\Gamma \in C^{0,1}$, the boundary Γ is said to be Lipschitz continuous; and if $\Gamma \in C^m := C^{m,0}$, the boundary Γ is said to be continuously differentiable of order m.

Furthermore space $C^{m,\alpha}(\Gamma)$, the Hölder function space on Γ, can be defined (e.g., Colton & Kress, 2013 for $m = 0, 1$) provided that $\Gamma \in C^m$.

Definition 3.1.4 For the case $\Gamma \in C^{m,1}$ $(m \geqslant 0)$, the unit outward normal $\boldsymbol{\nu}$ on Γ of Ω is defined by

$$\boldsymbol{\nu}(\mathbf{y}', \varphi(\mathbf{y}')) = \frac{(-D_1\varphi(\mathbf{y}'), \ldots, -D_{n-1}\varphi(\mathbf{y}'), 1)}{\sqrt{1 + D_1\varphi(\mathbf{y}')^2 + \cdots + D_{n-1}\varphi(\mathbf{y}')^2}}, \qquad \text{for } \mathbf{y}' \in V'. \tag{3.1.7}$$

This vector field can be extended to the whole domain V by defining it independently of y_n. Furthermore, by a unity partition, one may define an L^∞ vector field $\boldsymbol{\nu}$ in a neighborhood of $\bar{\Omega}$ with $\boldsymbol{\nu}$ be the unit outward normal a.e. on Γ. When $m \geqslant 1$, the vector field $\boldsymbol{\nu}$ belongs to the class $C^{m-1,1}(\Gamma)$ (Grisvard, 2011).

3.1.3 The Spaces $L_p(\Omega)$

Definition 3.1.5 Let $L_p(\Omega)$ denote the space of all equivalence classes of the measurable functions on Ω whose powers of order p $(p \geqslant 1)$ are integrable, with norm

$$\|u\|_{L_p(\Omega)} := \left\{ \int_\Omega |u(\mathbf{x})|^p \, \mathrm{d}\mathbf{x} \right\}^{1/p}, \qquad \text{for } 1 \leqslant p < \infty. \tag{3.1.8}$$

Furthermore, $L_\infty(\Omega)$ is the space of functions u which are measurable and bounded almost with the norm

$$\|u\|_{L_\infty(\Omega)} := \operatorname*{ess\,sup}_{x\in\Omega}\{\,|u(x)|\,\}. \tag{3.1.9}$$

The spaces $L_p(\Omega)$ $(1 \leqslant p < \infty)$ are Banach spaces with respect to norm $\|\cdot\|_{L_p(\Omega)}$.

For $u \in L_p(\Omega)$ and $v \in L_q(\Omega)$ with adjoint parameters p and q which satisfy

$$\frac{1}{p} + \frac{1}{q} = 1,$$

the Hölder's inequality holds

$$\int_\Omega |u(x)v(x)|\,\mathrm{d}x \leqslant \|u\|_{L_p(\Omega)}\|v\|_{L_q(\Omega)}, \qquad \forall\, u \in L_p(\Omega),\ v \in L_q(\Omega). \tag{3.1.10}$$

Defining the duality pairing

$$\langle u, v\rangle_\Omega := \int_\Omega u(x)v(x)\,\mathrm{d}x,$$

we have

$$\|v\|_{L_q(\Omega)} = \sup_{0\neq u\in L_p(\Omega)} \frac{\langle u, v\rangle_\Omega}{\|u\|_{L_p(\Omega)}} \quad \text{for } 1 \leqslant p < \infty,\ \frac{1}{p} + \frac{1}{q} = 1.$$

Then $L_q(\Omega)$ is the dual space of $L_p(\Omega)$ $(1 \leqslant p < \infty)$, $\dfrac{1}{p} + \dfrac{1}{q} = 1$.

For the special case $p = 2$, the space $L_2(\Omega)$ is the space of all square integrable functions. For any $u,\ v \in L_2(\Omega)$, we can define the inner product

$$\langle u, v\rangle_{L_2(\Omega)} := \int_\Omega u(x)v(x)\,\mathrm{d}x$$

and

$$\langle u, u\rangle_{L_2(\Omega)} = \|u\|^2_{L_2(\Omega)}, \qquad \forall\, u \in L_2(\Omega).$$

The space $L_2(\Omega)$ is a Hilbert space.

3.1.4 Sobolev Spaces

Definition 3.1.6 Let $W^{s,p}(\Omega)$ denote the space of all distributions u in Ω such that

1. When $s = m$ is a non-negative integer, $D^\alpha u \in L_p(\Omega)$ for $|\alpha| \leqslant m$. We define a Banach norm on $W^{m,p}(\Omega)$ by

$$\|u\|_{W^{m,p}(\Omega)} = \left\{ \sum_{|\alpha| \leqslant m} \int_\Omega |D^\alpha u|^p \, \mathrm{d}\boldsymbol{x} \right\}^{1/p};$$

2. When s is non-negative but not an integer, assuming $s = m + \sigma$ with non-negative integer m and $0 < \sigma < 1$, $u \in W^{m,p}(\Omega)$ and

$$\int_\Omega \int_\Omega \frac{|D^\alpha u(\boldsymbol{x}) - D^\alpha u(\boldsymbol{y})|^p}{|\boldsymbol{x} - \boldsymbol{y}|^{n+\sigma p}} \, \mathrm{d}\boldsymbol{x} \, \mathrm{d}\boldsymbol{y} < +\infty$$

for $|\alpha| = m$. We define a Banach norm on $W^{s,p}(\Omega)$ by

$$\|u\|_{W^{s,p}(\Omega)} = \left\{ \|u\|^p_{W^{m,p}(\Omega)} + \sum_{|\alpha|=m} \int_\Omega \int_\Omega \frac{|D^\alpha u(\boldsymbol{x}) - D^\alpha u(\boldsymbol{y})|^p}{|\boldsymbol{x} - \boldsymbol{y}|^{n+\sigma p}} \, \mathrm{d}\boldsymbol{x} \, \mathrm{d}\boldsymbol{y} \right\}^{1/p}.$$

Definition 3.1.7 For $s > 0$, let $W_0^{s,p}(\Omega)$ denote the closure of $\mathcal{D}(\Omega)$ in $W^{s,p}(\Omega)$.

Definition 3.1.8 For $s < 0$, let $W_0^{s,p}(\Omega)$ denote the dual space of $W_0^{-s,q}(\Omega)$ with $\frac{1}{p} + \frac{1}{q} = 1$.

In the special case $p = 2$, the common notation $H^s(\Omega)$ instead of $W^{s,2}(\Omega)$ and $H_0^s(\Omega)$ instead of $W_0^{s,2}(\Omega)$ are used. They are Hilbert spaces.

We now recall the main properties of Sobolev spaces without proofs (Adams & Fournier, 2003; Gilbarg & Trudinger, 2001).

Theorem 3.1.1 (Density) *Let Ω be an open domain of $\mathbb{R}^n$ with a continuous boundary. Then $C^\infty(\bar{\Omega})$ is dense in $W^{s,p}(\Omega)$ for all $s > 0$ and $\mathcal{D}(\Omega)$ is dense in $W_0^{s,p}(\Omega)$.*

Theorem 3.1.2 (Imbedding Theorem of Sobolev) *Let Ω be a bounded domain in $\mathbb{R}^n$ with Lipschitz boundary Γ. Then*

1. In case $sp < n$, the space $W^{s,p}(\Omega)$ is continuously imbedded in L_{q^*}, $q^* = \frac{np}{n - sp}$, and compactly imbedded in $L_q(\Omega)$ for any $q < q^*$.

2. In case $0 \leqslant m < s - \frac{n}{p} < m + 1$, the space $W^{s,p}(\Omega)$ is continuously imbedded in $C^{m,\alpha}(\bar{\Omega})$, $\alpha = s - \frac{n}{p} - m$, and compact in $C^{m,\beta}(\bar{\Omega})$ for any $0 \leqslant \beta < \alpha$.

Theorem 3.1.3 (Trace Theorem) *Let Ω be a bounded domain in $\mathbb{R}^n$ with a $C^{k,1}$ boundary Γ. Assume that $s - \frac{1}{p}$ is not an integer and $s \leqslant k + 1$, $s - \frac{1}{p} = \ell + \sigma$ with $0 < \sigma < 1$, ℓ is a nonnegative integer. Then the mapping*

$$u \mapsto \left\{ \gamma u := u\Big|_\Gamma, \gamma\frac{\partial u}{\partial \boldsymbol{\nu}} := \frac{\partial u}{\partial \boldsymbol{\nu}}\Big|_\Gamma, \ldots, \gamma\frac{\partial^\ell u}{\partial \boldsymbol{\nu}^\ell} := \frac{\partial^\ell u}{\partial \boldsymbol{\nu}^\ell}\Big|_\Gamma \right\},$$

which is defined for $u \in C^{k,1}(\bar{\Omega})$*, has a unique continuous extension as an operator from* $W^{s,p}(\Omega)$ *onto* $\prod_{j=0}^{\ell} W^{s-j-1/p,p}(\Gamma)$.

3.2 The Dirichlet and Neumann Problems of the Laplace Equation

The Laplace equation is perhaps the most important equation in partial differential equations, which has many applications in science and engineering. We consider the following Dirichlet and Neumann boundary value problems of the Laplace equation.

The interior Dirichlet problem reads

$$\begin{cases} -\Delta u(\boldsymbol{x}) = 0, & \forall \boldsymbol{x} \in \Omega, \quad (3.2.1a) \\ u(\boldsymbol{x}) = g(\boldsymbol{x}), & \forall \boldsymbol{x} \in \Gamma, \quad (3.2.1b) \end{cases}$$

with given function g on Γ.

There are two exterior Dirichlet problems, the first one reading

$$\begin{cases} -\Delta u(\boldsymbol{x}) = 0, & \forall \boldsymbol{x} \in \Omega_e, \quad (3.2.2a) \\ u(\boldsymbol{x}) = g(\boldsymbol{x}), & \forall \boldsymbol{x} \in \Gamma, \quad (3.2.2b) \\ u(\boldsymbol{x}) = C + \mathcal{O}(|\boldsymbol{x}|^{1-n}), & \text{as } |\boldsymbol{x}| \to \infty, \quad (3.2.2c) \end{cases}$$

with given function g on Γ and unknown constant C, and the second one taking

$$\begin{cases} -\Delta u(\boldsymbol{x}) = 0, & \forall \boldsymbol{x} \in \Omega_e, \quad (3.2.3a) \\ u(\boldsymbol{x}) = g(\boldsymbol{x}), & \forall \boldsymbol{x} \in \Gamma, \quad (3.2.3b) \\ u(\boldsymbol{x}) = \mathcal{O}(|\boldsymbol{x}|^{1-n}) + \begin{cases} C \ln |\boldsymbol{x}|, & n = 2, \\ C|\boldsymbol{x}|^{2-n}, & n \geqslant 3 \end{cases} & \text{as } |\boldsymbol{x}| \to \infty, \quad (3.2.3c) \end{cases}$$

with given function g on Γ and unknown constant C.

The interior Neumann problem reads

$$\begin{cases} -\Delta u(\boldsymbol{x}) = 0, & \forall \boldsymbol{x} \in \Omega, \quad (3.2.4a) \\ \left.\dfrac{\partial u}{\partial \boldsymbol{\nu}}\right|_{\Gamma} = h(\boldsymbol{x}), & \forall \boldsymbol{x} \in \Gamma, \quad (3.2.4b) \end{cases}$$

with the given function h on Γ and $\left.\frac{\partial u}{\partial \boldsymbol{\nu}}\right|_{\Gamma}$ denotes the derivative of u in the outward normal direction on Γ of Ω. Furthermore, we consider two exterior Neumann problems, the first reading

$$\begin{cases} -\Delta u(\boldsymbol{x}) = 0, & \forall \boldsymbol{x} \in \Omega_e, & (3.2.5a) \\ \left.\dfrac{\partial u}{\partial \boldsymbol{\nu}}\right|_{\Gamma} = h(\boldsymbol{x}), & \forall \boldsymbol{x} \in \Gamma, & (3.2.5b) \\ u(\boldsymbol{x}) = \mathcal{O}(|\boldsymbol{x}|^{1-n}), & \text{as } |\boldsymbol{x}| \to \infty, & (3.2.5c) \end{cases}$$

and the second one taking

$$-\begin{cases} \Delta u(\boldsymbol{x}) = 0, & \forall \boldsymbol{x} \in \Omega_e, & (3.2.6a) \\ \left.\dfrac{\partial u}{\partial \boldsymbol{\nu}}\right|_{\Gamma} = h(\boldsymbol{x}), & \forall \boldsymbol{x} \in \Gamma, & (3.2.6b) \\ u(\boldsymbol{x}) = \mathcal{O}(|\boldsymbol{x}|^{1-n}) + \begin{cases} C \ln |\boldsymbol{x}|, & n = 2, \\ C|\boldsymbol{x}|^{2-n}, & n \geqslant 3 \end{cases} & \text{as } |\boldsymbol{x}| \to \infty, & (3.2.6c) \end{cases}$$

with given function h on Γ and unknown constant C.

3.2.1 Classical Solutions

At first we consider the classical solutions of those six Dirichlet and Neumann problems of the Laplace equation.

Definition 3.2.1 1. If a function $u \in C^0(\bar{\Omega}) \cap C^2(\Omega)$ satisfies the Laplace equation in Ω and the Dirichlet boundary condition $u\big|_{\Gamma} = g\big|_{\Gamma}$ for a given function $g \in C^0(\Gamma)$, then u is called a classical solution to the interior Dirichlet problem (3.2.1).

2. If a function $u \in C^0(\bar{\Omega}_e) \cap C^2(\Omega_e)$ satisfies the Laplace equation in Ω_e and the Dirichlet boundary condition $u\big|_{\Gamma} = g$ on Γ with given function $g \in C(\Gamma)$. Furthermore,

(a) If u satisfies the boundary condition (3.2.2c), then u is called a classical solution to the exterior Dirichlet problem (3.2.2);

(b) If u satisfies the boundary condition (3.2.3c), then u is called a classical solution to the exterior Dirichlet problem (3.2.3).

To define the classical solution to the Neumann problems, we introduce the space $C_\nu(\bar{\Omega})$ and $C_\nu(\bar{\Omega}_e)$ (Folland, 1995). $C_\nu(\bar{\Omega})$ is defined to be the space of function $u \in C(\bar{\Omega}) \cap C^1(\Omega)$ and the limit

$$\frac{\partial u(\boldsymbol{x})}{\partial \boldsymbol{\nu}^-} = \lim_{t<0,\, t\to 0} \boldsymbol{\nu}(\boldsymbol{x}) \cdot \nabla u(\boldsymbol{x} + t\,\boldsymbol{\nu}(\boldsymbol{x})), \tag{3.2.7}$$

exists for each $\boldsymbol{x} \in \Gamma$, the convergence being uniform on Γ (therefore $\dfrac{\partial u(\boldsymbol{x})}{\partial \boldsymbol{\nu}^-}$ is continuous function on Γ). Similarly, $C_\nu(\overline{\Omega}_e)$ is defined to be the space of functions $u \in C(\bar{\Omega}_e) \cap C^1(\Omega_e)$ and the limit

$$\frac{\partial u(\boldsymbol{x})}{\partial \boldsymbol{\nu}^+} = \lim_{t>0,\, t\to 0} \boldsymbol{\nu}(\boldsymbol{x}) \cdot \nabla u(\boldsymbol{x} + t\,\boldsymbol{\nu}(\boldsymbol{x})), \tag{3.2.8}$$

exists for each $\boldsymbol{x} \in \Gamma$, the convergence being uniform on Γ. Obviously if $u \in C^1(\bar{\Omega})$, then we have

$$\frac{\partial u(\boldsymbol{x})}{\partial \boldsymbol{\nu}^-} = \left.\frac{\partial u(\boldsymbol{x})}{\partial \boldsymbol{\nu}}\right|_{\Gamma-0},$$

and if $u \in C^1(\bar{\Omega}_e)$, then we have

$$\frac{\partial u(\boldsymbol{x})}{\partial \boldsymbol{\nu}^+} = \left.\frac{\partial u(\boldsymbol{x})}{\partial \boldsymbol{\nu}}\right|_{\Gamma+0}.$$

Definition 3.2.2 1. If a function $u \in C_{\nu}(\bar{\Omega}) \cap C^2(\Omega)$ satisfies the Laplace equation in Ω and the Neumann boundary condition

$$\left.\frac{\partial u(\boldsymbol{x})}{\partial \boldsymbol{\nu}^-}\right|_{\Gamma} = h(\boldsymbol{x})$$

with a given continuous function h on Γ, then u is called a classical solution to the interior Neumann problem (3.2.4);

2. Suppose a function $u \in C_{\nu}(\bar{\Omega}_e) \cap C^2(\Omega_e)$ satisfies the Laplace equation in Ω_e and the Neumann boundary condition

$$\left.\frac{\partial u(\boldsymbol{x})}{\partial \boldsymbol{\nu}^+}\right|_{\Gamma} = h(\boldsymbol{x})$$

with a given continuous function h on Γ.

(a) If u satisfies the condition (3.2.5c), then u is called a classical solution to the exterior Neumann boundary value problem (3.2.5);

(b) If u satisfies the condition (3.2.6c), then u is called a classical solution to the exterior Neumann boundary value problem (3.2.6).

In the next, we present the main results on the classical solutions of the Dirichlet and Neumann boundary value problems of the Laplace equation for the case that the boundary Γ is smooth, for example, $\Gamma \in C^2$.

Theorem 3.2.1 *Suppose that the bounded domain Ω is simply connected and its boundary $\Gamma \in C^2$.*

1. For any given function $g \in C(\Gamma)$, the interior Dirichlet problem (3.2.1) has a unique classical solution (Folland, 1995, p. 138).

2. For any given function $g \in C(\Gamma)$, the exterior Dirichlet problem (3.2.2) has a unique classical solution (for $n = 2$, see Folland, 1995, p. 138; for $n = 3$, see Kondratév & Oleinik, 1987).

3. For any given function $g \in C(\Gamma)$, the exterior Dirichlet problem (3.2.3) ($n = 3$) has a unique classical solution (Folland, 1995, p. 138).

4. For any given function $h \in C(\Gamma)$, the interior Neumann problem (3.2.4) has a classical solution if and only if $\int_\Gamma h(\boldsymbol{x})\,\mathrm{d}s_x = 0$. The solution is unique modulo functions, which is constant on $\bar{\Omega}$ (Folland, 1995, p. 138).

5. For any given function $h \in C(\Gamma)$, the exterior Neumann problem (3.2.5) has a classical solution if and only if $\int_\Gamma h(\boldsymbol{x})\,\mathrm{d}s_x = 0$, and the solution is unique (Folland, 1995, p. 138).

6. For any given function $h \in C(\Gamma)$, the exterior Neumann problem (3.2.6) has a unique classical solution.

The solvability of the exterior Dirichlet problem (3.2.3) for the case $n = 2$ will be discussed in the coming subsection.

Remark 3.2.1 The condition $\Gamma \in C^2$ in Theorem 3.2.1 can be reduced. For instance, suppose that the boundary Γ is continuous and the boundary points are all regular. The interior Dirichlet problem (3.2.1) has a unique classical solution (Gilbarg & Trudinger, 2001).

3.2.2 Generalized Solutions and Variational Problems

In this subsection, we discuss the generalized solutions of the Dirichlet and Neumann problems of the Laplace equation. For the interior problem and given $g \in H^{1/2}(\Gamma)$, let

$$H_g^1(\Omega) = \left\{ v \;\middle|\; v \in H^1(\Omega) \text{ and } v\big|_\Gamma = g \right\}, \tag{3.2.9}$$

which is a subset of $H^1(\Omega)$.

Definition 3.2.3 1. Given function $g \in H^{1/2}(\Gamma)$, find $u \in H_g^1(\Omega)$ such that

$$a(u, v) = 0, \qquad \forall\, v \in H_0^1(\Omega) \tag{3.2.10}$$

with

$$a(u, v) = \int_\Omega \nabla u \cdot \nabla v \,\mathrm{d}\boldsymbol{x}, \qquad \text{for } u, v \in H^1(\Omega); \tag{3.2.11}$$

then u is called a generalized solution of the interior Dirichlet problem (3.2.1).

2. Given function $h \in H^{-1/2}(\Gamma)$, find $u \in H^1(\Omega)$ such that

$$a(u, v) = \int_\Gamma h v \,\mathrm{d}s, \qquad \forall\, v \in H^1(\Omega); \tag{3.2.12}$$

then u is called a generalized solution of the interior Neumann problem (3.2.4).

For discussing the equivalent variational problems of the exterior problems (3.2.2), (3.2.3), (3.2.5) and (3.2.6), introduce the following two spaces (Hsiao & Wendland, 2008):

$$H^1_{\mathrm{loc}}(\Omega_e) = \left\{ v \middle| v \in H^1(G), \text{ for any bounded domain } G \subset \Omega_e \right\},$$
$$H^1_{\mathrm{comp}}(\Omega_e) = \left\{ v \middle| v \in H^1_{\mathrm{loc}}(\Omega_e) \text{ and the support of } v \text{ is compact} \right\}.$$

Obviously

$$H^1_{\mathrm{comp}}(\Omega_e) \subset H^1(\Omega_e) \subset H^1_{\mathrm{loc}}(\Omega_e).$$

Furthermore let

$$H^1_{\mathrm{loc},g}(\Omega_e) = \left\{ v \middle| v \in H^1_{\mathrm{loc}}(\Omega_e) \text{ and } v\Big|_\Gamma = g \right\},$$
$$H^1_{\mathrm{loc},0}(\Omega_e) = \left\{ v \middle| v \in H^1_{\mathrm{loc}}(\Omega_e) \text{ and } v\Big|_\Gamma = 0 \right\}.$$

Definition 3.2.4 Given functions $g \in H^{1/2}(\Gamma)$ and $h \in H^{-1/2}(\Gamma)$.

1.

$$\begin{cases} \textit{Find } u \in H^1_{\mathrm{loc},g}(\Omega_e) \textit{ such that} \\ a_e(u, v) = 0, \quad \forall v \in H^1_{\mathrm{comp},0}(\Omega_e). \\ u(\boldsymbol{x}) = C + \mathcal{O}\Big(|\boldsymbol{x}|^{1-n}\Big), \quad \text{as } |\boldsymbol{x}| \to \infty, \end{cases} \tag{3.2.13}$$

with

$$a_e(u, v) = \int_{\Omega_e} \nabla u \cdot \nabla v \mathrm{d}\boldsymbol{x}, \quad \text{for } u \in H^1_{\mathrm{loc}}(\Omega_e), \ v \in H^1_{\mathrm{comp}}(\Omega_e). \tag{3.2.14}$$

Then u is called a generalized solution of the exterior Dirichlet problem (3.2.2).

2.

$$\begin{cases} \textit{Find } u \in H^1_{\mathrm{loc},g}(\Omega_e) \textit{ such that} \\ a_e(u, v) = 0, \quad \forall v \in H^1_{\mathrm{comp},0}(\Omega_e). \\ u(\boldsymbol{x}) = \begin{Bmatrix} C \ln |\boldsymbol{x}|, & n = 2, \\ C|\boldsymbol{x}|^{2-n}, & n \geqslant 3 \end{Bmatrix} + \mathcal{O}(|\boldsymbol{x}|^{1-n}), \quad \text{as } |\boldsymbol{x}| \to \infty, \end{cases} \tag{3.2.15}$$

then u is called a generalized solution of the exterior Dirichlet problem (3.2.3).

3.

$$\begin{cases} \textit{Find } u \in H^1_{\mathrm{loc},g}(\Omega_e) \textit{ such that} \\ a_e(u, v) = -\int_\Gamma h v \mathrm{d}s, \quad \forall v \in H^1_{\mathrm{comp}}(\Omega_e). \\ u(\boldsymbol{x}) = \mathcal{O}(|\boldsymbol{x}|^{1-n}), \quad \text{as } |\boldsymbol{x}| \to \infty, \end{cases} \tag{3.2.16}$$

then u is called a generalized solution of the exterior Dirichlet problem (3.2.5).

4.

$$
\begin{cases}
\textit{Find } u \in H^1_{\mathrm{loc},g}(\Omega_e) \textit{ such that} \\
a_e(u, v) = -\int_\Gamma h v \mathrm{d}s, \quad \forall v \in H^1_{\mathrm{comp}}(\Omega_e). \\
u(x) = \begin{Bmatrix} C \ln |x|, & n = 2, \\ C|x|^{2-n}, & n \geqslant 3 \end{Bmatrix} + \mathcal{O}(|x|^{1-n}), \quad \text{as } |x| \to \infty,
\end{cases}
\tag{3.2.17}
$$

then u is called a generalized solution of the exterior Dirichlet problem (3.2.6).

3.3 Single Layer and Double Layer Potentials

Given functions $\varphi(\boldsymbol{x})$ and $\psi(\boldsymbol{x})$ on Γ, consider the single layer potential

$$
V_I(\boldsymbol{x}) = \int_\Gamma E_n(\boldsymbol{x}, \boldsymbol{y})\varphi(\boldsymbol{y})\, \mathrm{d}s_y, \qquad \forall \boldsymbol{x} \in \mathbb{R}^n, \tag{3.3.1}
$$

with the given density function $\varphi(\boldsymbol{x})$ on Γ; the double layer potential

$$
V_{II}(\boldsymbol{x}) = \int_\Gamma \frac{\partial}{\partial \boldsymbol{\nu}_y} E_n(\boldsymbol{x}, \boldsymbol{y})\psi(\boldsymbol{y})\, \mathrm{d}\boldsymbol{y}, \qquad \forall \boldsymbol{x} \in \mathbb{R}^n, \tag{3.3.2}
$$

with the given density function $\psi(\boldsymbol{y})$ on Γ and $\boldsymbol{\nu}_y$ denoting the unit outward normal vector on Γ of Ω at point $\boldsymbol{y} \in \Gamma$, which is defined by (3.1.7). $E_n(\boldsymbol{x}, \boldsymbol{y})$ is given by (2.2.4)–(2.2.6), the fundamental solution of the Laplace equation, namely

$$
E_n(\boldsymbol{x}, \boldsymbol{y}) = \begin{cases} -\dfrac{1}{2\pi} \ln |\boldsymbol{x} - \boldsymbol{y}|, & n = 2, \\ \dfrac{1}{(n-2)S_n(1)} |\boldsymbol{x} - \boldsymbol{y}|^{2-n}, & n \geqslant 3. \end{cases} \tag{3.3.3}
$$

For studying single and double layer potentials, we recall some results of weakly singular integral operators on the boundary Γ of the bounded domain $\Omega \in \mathbb{R}^n$ (Folland, 1995; Colton & Kress, 2013).

3.3.1 *Weakly Singular Integral Operators on* Γ

We consider the integral operator T_k defined by

$$
(T_k\varphi)(\boldsymbol{x}) = \int_\Gamma K(\boldsymbol{x}, \boldsymbol{y})\varphi(\boldsymbol{y})\, \mathrm{d}s_y, \tag{3.3.4}
$$

with a given kernel $K(\boldsymbol{x}, \boldsymbol{y})$.

If a kernel $K(\boldsymbol{x}, \boldsymbol{y})$ is defined and continuous for all $\boldsymbol{x}, \boldsymbol{y} \in \Gamma, \boldsymbol{x} \neq \boldsymbol{y}$ and there exist two constants $M > 0$ and $\alpha \in [0, n-1)$ such that for all $\boldsymbol{x}, \boldsymbol{y} \in \Gamma$ and $\boldsymbol{x} \neq \boldsymbol{y}$, the following estimate holds

$$|K(\boldsymbol{x}, \boldsymbol{y})| \leqslant M|\boldsymbol{x} - \boldsymbol{y}|^{-\alpha}, \qquad \forall \boldsymbol{x}, \boldsymbol{y} \in \Gamma, \tag{3.3.5}$$

then $K(\boldsymbol{x}, \boldsymbol{y})$ is called a weakly singular kernel of order α.

Suppose $K(\boldsymbol{x}, \boldsymbol{y})$ is a weakly singular kernel of order α. The corresponding integral operator T_k is called a weakly singular integral operator of order α.

At first we recall a lemma on the boundary Γ (Folland, 1995).

Lemma 3.3.1 *Suppose that $\Gamma \in C^2$, then there exists a positive constant M such that for all $\boldsymbol{x}, \boldsymbol{y} \in \Gamma$,*

$$|(\boldsymbol{x} - \boldsymbol{y}) \cdot \boldsymbol{\nu}_y| \leqslant M|\boldsymbol{x} - \boldsymbol{y}|^2. \tag{3.3.6}$$

Furthermore, we have

Theorem 3.3.1 (Folland, 1995) *Suppose that the kernel $K(\boldsymbol{x}, \boldsymbol{y})$ is a weakly singular kernel of order $\alpha \in [0, n-1)$. Then the weakly singular integral operator T_k is bounded on $L_p(\Gamma)$ for $1 \leqslant p \leqslant \infty$. Namely there exists a constant $C > 0$ depending only on $\alpha \in [0, n-1)$ and the boundary Γ such that*

$$\|T_k\varphi\|_{L_p(\Gamma)} \leqslant C\|\varphi\|_{L_p(\Gamma)}. \tag{3.3.7}$$

Theorem 3.3.2 (Colton & Kress, 2013) *1. The weakly singular integral operator T_k is a compact operator on $C(\Gamma)$.*

2. Suppose that the kernel $K(\boldsymbol{x}, \boldsymbol{y})$ is defined and continuous for all $\boldsymbol{x} \in \bar{\Omega}$, $\boldsymbol{y} \in \Gamma$, $\boldsymbol{x} \neq \boldsymbol{y}$ and there exist two constants $M > 0$ and $\alpha \in [0, n-1)$ such that the estimate (3.3.5) holds for $\boldsymbol{x} \in \bar{\Omega}$, $\boldsymbol{y} \in \Gamma$. Moreover, assume there exists a constant $M > 0$ and a positive integer m such that

$$|K(\boldsymbol{x}_1, \boldsymbol{y}) - K(\boldsymbol{x}_2, \boldsymbol{y})| \leqslant M \sum_{j=1}^{m} |\boldsymbol{x}_1 - \boldsymbol{y}|^{-\alpha-j} |\boldsymbol{x}_1 - \boldsymbol{x}_2|^j \tag{3.3.8}$$

for all $\boldsymbol{x}_1, \boldsymbol{x}_2 \in \Omega$, $\boldsymbol{y} \in \Gamma$ with $2|\boldsymbol{x}_1 - \boldsymbol{x}_2| \leqslant |\boldsymbol{x}_1 - \boldsymbol{y}|$. Then the potential $u(\boldsymbol{x})$ defined on $\bar{\Omega}$ by

$$u(\boldsymbol{x}) = \int_\Gamma K(\boldsymbol{x}, \boldsymbol{y})\varphi(\boldsymbol{y})\, \mathrm{d}s_y, \qquad \forall \boldsymbol{x} \in \bar{\Omega} \tag{3.3.9}$$

with density $\varphi \in C(\Gamma)$ belonging to the Hölder space $C^{0,\beta}(\bar{\Omega})$ for all $\beta \in (0, \beta^*)$ with

$$\beta^* = \begin{cases} (n-1) - \alpha, & (n-2) < \alpha < (n-1), \\ 1, & 0 \leqslant \alpha \leqslant (n-2), \end{cases}$$

and

$$\|u\|_{C^{0,\beta}(\bar{\Omega})} \leqslant C_\beta \|\varphi\|_{C(\Gamma)} \tag{3.3.10}$$

with constant C_β independent of density φ.

3.3.2 Double Layer Potentials

We now discuss the properties of the double layer potential (3.3.2) with given density $\psi \in C(\Gamma)$, namely

$$V_{II}(\boldsymbol{x}) = \int_\Gamma \frac{\partial}{\partial \boldsymbol{\nu}_y} E_n(\boldsymbol{x}, \boldsymbol{y})\psi(\boldsymbol{y})\, \mathrm{d}s_y := \int_\Gamma K_{II}(\boldsymbol{x}, \boldsymbol{y})\psi(\boldsymbol{y})\, \mathrm{d}s_y, \qquad \boldsymbol{x} \in \mathbb{R}^n, \tag{3.3.11}$$

with the kernel

$$K_{II}(\boldsymbol{x}, \boldsymbol{y}) = \frac{\partial}{\partial \boldsymbol{\nu}_y} E_n(\boldsymbol{x}, \boldsymbol{y}), \qquad \forall \boldsymbol{x} \in \mathbb{R}^n,\ \boldsymbol{y} \in \Gamma. \tag{3.3.12}$$

We note that

$$K_{II}(\boldsymbol{x}, \boldsymbol{y}) = \frac{\partial}{\partial \boldsymbol{\nu}_y} E_n(\boldsymbol{x}, \boldsymbol{y}) = \frac{(\boldsymbol{x} - \boldsymbol{y}) \cdot \boldsymbol{\nu}_y}{S_n(1)|\boldsymbol{x} - \boldsymbol{y}|^n}, \qquad \text{for } \boldsymbol{x} \in \mathbb{R}^n,\ \boldsymbol{y} \in \Gamma.$$

For the case $\boldsymbol{x} \in \mathbb{R}^n \setminus \Gamma$, $\boldsymbol{y} \in \Gamma$, $K_{II}(\boldsymbol{x}, \boldsymbol{y})$ is continuous in $\boldsymbol{y}$ and harmonic in $\boldsymbol{x}$ and it is $\mathcal{O}(|\boldsymbol{x}|^{1-n})$ as $|\boldsymbol{x}| \to \infty$. Here, $V_{II}(\boldsymbol{x})$ is harmonic on $\mathbb{R}^n \setminus \Gamma$ and $V_{II}(\boldsymbol{x}) = \mathcal{O}(|\boldsymbol{x}|^{1-n})$ as $|\boldsymbol{x}| \to \infty$. For the case $\boldsymbol{x} \in \Gamma$, from the estimate (3.3.6), we know that

$$|K_{II}(\boldsymbol{x}, \boldsymbol{y})| = \left| \frac{(\boldsymbol{x} - \boldsymbol{y}) \cdot \boldsymbol{\nu}_y}{S_n(1)|\boldsymbol{x} - \boldsymbol{y}|^n} \right| \leqslant C|\boldsymbol{x} - \boldsymbol{y}|^{-(n-2)}, \qquad \forall \boldsymbol{x},\ \boldsymbol{y} \in \Gamma.$$

Therefore $K_{II}(\boldsymbol{x}, \boldsymbol{y})$ is a weakly singular kernel of order $(n-2)$, by the result given in Theorem 3.3.2. The double layer potential defined in (3.3.2), i.e.,

$$V_{II}(\boldsymbol{x}) = \int_\Gamma K_{II}(\boldsymbol{x}, \boldsymbol{y})\psi(\boldsymbol{y})\, \mathrm{d}s_y := (\mathbb{K}_{II}\psi)(\boldsymbol{x}), \qquad \text{for } \boldsymbol{x} \in \Gamma,$$

introduces a compact operator $\mathbb{K}_{II}$ on $C(\Gamma)$. It means that $V_{II}(\boldsymbol{x})$ is defined on $\mathbb{R}^n$, but $V_{II}(\boldsymbol{x})$ is not continuous on $\mathbb{R}^n$ except $\psi = 0$ on Γ. Generally $V_{II}(\boldsymbol{x})$ has the discontinuity of the first kind on Γ. To show this phenomenon clearly, we consider the simplest case, when $\psi = 1$ on Γ.

Lemma 3.3.2 (Folland, 1995, p. 125)

$$\int_\Gamma \frac{\partial}{\partial \boldsymbol{\nu}_y} E_n(\boldsymbol{x},\boldsymbol{y})\,\mathrm{d}s_y = \begin{cases} 0, & \textit{if } \boldsymbol{x}\in\Omega_e, \\ -1, & \textit{if } \boldsymbol{x}\in\Omega, \\ -1/2, & \textit{if } \boldsymbol{x}\in\Gamma. \end{cases} \tag{3.3.13}$$

Proof 1. For $\boldsymbol{x}\in\Omega_e$, $E_n(\boldsymbol{x},\boldsymbol{y})$ is a harmonic function of $\boldsymbol{y}$ on $\bar{\Omega}$, namely $\triangle_y E_n(\boldsymbol{x},\boldsymbol{y}) = 0, \forall\, \boldsymbol{y}\in\bar{\Omega}$. This leads to

$$\int_\Gamma \frac{\partial}{\partial \boldsymbol{\nu}_y} E_n(\boldsymbol{x},\boldsymbol{y})\,\mathrm{d}s_y = 0, \qquad \forall \boldsymbol{x}\in\Omega_e.$$

2. For $\boldsymbol{x}\in\Omega$, we choose a sufficiently small constant $\epsilon > 0$ such that $\bar{B}_\epsilon := \bar{B}_\epsilon(\boldsymbol{x}) \subset \Omega$ denote the ball centered at $\boldsymbol{x}$ with radius $\epsilon > 0$. Then $E_n(\boldsymbol{x},\boldsymbol{y})$ is a harmonic function of $\boldsymbol{y}$ on $\bar{\Omega}\setminus B_\epsilon$, i.e.,

$$\triangle_y E_n(\boldsymbol{x},\boldsymbol{y}) = 0, \qquad \forall\, \boldsymbol{y}\in\bar{\Omega}\setminus B_\epsilon.$$

We get

$$0 = \int_\Gamma \frac{\partial}{\partial \boldsymbol{\nu}_y} E_n(\boldsymbol{x},\boldsymbol{y})\,\mathrm{d}s_y + \frac{\epsilon^{1-n}}{S_n(1)}\int_{\partial B_\epsilon} \mathrm{d}s = \int_\Gamma \frac{\partial}{\partial \boldsymbol{\nu}_y} E_n(\boldsymbol{x},\boldsymbol{y})\,\mathrm{d}s_y + 1.$$

That is,

$$\int_\Gamma \frac{\partial}{\partial \boldsymbol{\nu}_y} E_n(\boldsymbol{x},\boldsymbol{y})\,\mathrm{d}s_y = -1, \qquad \forall \boldsymbol{x}\in\Omega.$$

3. For the case that $\boldsymbol{x}\in\Gamma$, once again we let $\bar{B}_\epsilon := \bar{B}_\epsilon(\boldsymbol{x})$, $\Gamma_\epsilon := \Gamma\cap\bar{B}_\epsilon$, $B_\epsilon^* := \Omega\cap B_\epsilon$ and $\Gamma_\epsilon^{**} := \partial B_\epsilon^*\setminus\Gamma_\epsilon$. On the domain $\bar{\Omega}_\epsilon^* = \bar{\Omega}\setminus B_\epsilon^*$, $E_n(\boldsymbol{x},\boldsymbol{y})$ is a harmonic function in $\boldsymbol{y}$, i.e.,

$$\triangle_y E_n(\boldsymbol{x},\boldsymbol{y}) = 0, \qquad \forall\, \boldsymbol{y}\in\bar{\Omega}_\epsilon^*.$$

We arrive at

$$0 = \int_{\partial\bar{\Omega}_\epsilon^*} \frac{\partial}{\partial \boldsymbol{\nu}_y} E_n(\boldsymbol{x},\boldsymbol{y})\,\mathrm{d}s_y.$$

Note that the boundary of domain $\bar{\Omega}_\epsilon^*$ is given by $\partial(\bar{\Omega}_\epsilon^*) = (\Gamma\setminus\Gamma_\epsilon)\cup(\Gamma_\epsilon^{**})$. We have

$$0 = \int_{\Gamma\setminus\Gamma_\epsilon} \frac{\partial}{\partial \boldsymbol{\nu}_y} E_n(\boldsymbol{x},\boldsymbol{y})\,\mathrm{d}s_y + \int_{\Gamma_\epsilon^{**}} \frac{\partial}{\partial \boldsymbol{\nu}_y} E_n(\boldsymbol{x},\boldsymbol{y})\,\mathrm{d}s_y.$$

Let $\epsilon\to 0$ in the equality above. It is straightforward to get

$$0 = \int_\Gamma \frac{\partial}{\partial \boldsymbol{\nu}_y} E_n(\boldsymbol{x},\boldsymbol{y})\,\mathrm{d}s_y + \frac{1}{2}.$$

That is,

$$\int_{\Gamma} \frac{\partial}{\partial \boldsymbol{\nu}_{\boldsymbol{y}}} E_n(\boldsymbol{x}, \boldsymbol{y})\, \mathrm{d}s_{\boldsymbol{y}} = -\frac{1}{2}, \qquad \forall\, \boldsymbol{x} \in \Gamma.$$

□

Lemma 3.3.3 (Folland, 1995, p. 126) *There exists a positive constant C such that*

$$\int_{\Gamma} \left| \frac{\partial}{\partial \boldsymbol{\nu}_{\boldsymbol{y}}} E_n(\boldsymbol{x}, \boldsymbol{y}) \right| \mathrm{d}s_{\boldsymbol{y}} \leqslant C, \qquad \forall\, \boldsymbol{x} \in \mathbb{R}^n. \tag{3.3.14}$$

Proof Let $\operatorname{dist}(\boldsymbol{x}, \Gamma)$ denote the distance from $\boldsymbol{x} \in \mathbb{R}^n$ to the nearest point of Γ. Then we can find a constant $\delta > 0$ such that ① $\delta < 1/(2M)$ where $M > 0$ is the constant in Lemma 3.3.1; ② if $\operatorname{dist}(\boldsymbol{x}, \Gamma) < \delta/2$, there exists a unique point $\boldsymbol{x}_0 \in \Gamma$ so that $\boldsymbol{x} = \boldsymbol{x}_0 + t\boldsymbol{\nu}(\boldsymbol{x}_0)$ with $t \in (-\delta/2, \delta/2)$. Note $\boldsymbol{x}_0$ is the closest point in Γ to $\boldsymbol{x}$ by the Lagrange multiplier principle.

For the case that $\operatorname{dist}(\boldsymbol{x}, \Gamma) \geqslant \delta/2$, obviously

$$\left| \frac{\partial}{\partial \boldsymbol{\nu}_{\boldsymbol{y}}} E_n(\boldsymbol{x}, \boldsymbol{y}) \right| \leqslant c_1 \delta^{1-n}, \qquad \forall\, \boldsymbol{y} \in \Gamma,$$

therefore

$$\int_{\Gamma} \left| \frac{\partial}{\partial \boldsymbol{\nu}_{\boldsymbol{y}}} E_n(\boldsymbol{x}, \boldsymbol{y}) \right| \mathrm{d}s_{\boldsymbol{y}} \leqslant c_1 \delta^{1-n} \int_{\Gamma} \mathrm{d}s_{\boldsymbol{y}} = c_2.$$

For the case $\operatorname{dist}(\boldsymbol{x}, \Gamma) \leqslant \delta/2$, let $\boldsymbol{x}_0$ be the unique point of Γ such that $\boldsymbol{x} = \boldsymbol{x}_0 + t\boldsymbol{\nu}(\boldsymbol{x}_0)$ with $|t| < \delta/2$ and let

$$\Gamma_\delta(\boldsymbol{x}_0) = \{\, \boldsymbol{y} \in \Gamma \quad \text{and} \quad |\boldsymbol{x}_0 - \boldsymbol{y}| < \delta \,\}.$$

In the case $\boldsymbol{y} \in \Gamma \setminus \Gamma_\delta$, we have

$$|\boldsymbol{x} - \boldsymbol{y}| \geqslant |\boldsymbol{x}_0 - \boldsymbol{y}| - |\boldsymbol{x} - \boldsymbol{x}_0| \geqslant \delta - \frac{1}{2}\delta = \frac{1}{2}\delta.$$

Hence

$$\left| \frac{\partial}{\partial \boldsymbol{\nu}_{\boldsymbol{y}}} E_n(\boldsymbol{x}, \boldsymbol{y}) \right| \leqslant c_1 \delta^{1-n}$$

and

$$\int_{\Gamma \setminus \Gamma_\delta} \left| \frac{\partial}{\partial \boldsymbol{\nu}_{\boldsymbol{y}}} E_n(\boldsymbol{x}, \boldsymbol{y}) \right| \mathrm{d}s_{\boldsymbol{y}} \leqslant c_2.$$

In the case $\boldsymbol{y} \in \Gamma_\delta$, we have

$$\begin{aligned}\left|\frac{\partial}{\partial \boldsymbol{\nu}_y} E_n(\boldsymbol{x}, \boldsymbol{y})\right| &\leqslant \frac{|(\boldsymbol{x}-\boldsymbol{y}) \cdot \boldsymbol{\nu}_y|}{S_n(1)|\boldsymbol{x}-\boldsymbol{y}|^n} \\ &\leqslant \frac{|(\boldsymbol{x}-\boldsymbol{x}_0) \cdot \boldsymbol{\nu}_y| + |(\boldsymbol{x}_0-\boldsymbol{y}) \cdot \boldsymbol{\nu}_y|}{S_n(1)|\boldsymbol{x}-\boldsymbol{y}|^n} \\ &\leqslant \frac{|\boldsymbol{x}-\boldsymbol{x}_0| + M|\boldsymbol{x}_0-\boldsymbol{y}|^2}{S_n(1)|\boldsymbol{x}-\boldsymbol{y}|^n}.\end{aligned}$$

The last inequality is from (3.3.6). Furthermore, we have

$$|\boldsymbol{x}-\boldsymbol{y}|^2 = |\boldsymbol{x}-\boldsymbol{x}_0|^2 + |\boldsymbol{x}_0-\boldsymbol{y}|^2 + 2(\boldsymbol{x}-\boldsymbol{x}_0) \cdot (\boldsymbol{x}_0-\boldsymbol{y})$$

and

$$\begin{aligned}|2(\boldsymbol{x}-\boldsymbol{x}_0) \cdot (\boldsymbol{x}_0-\boldsymbol{y})| &= 2|t\boldsymbol{\nu}(\boldsymbol{x}_0) \cdot (\boldsymbol{x}_0-\boldsymbol{y})| \\ &\leqslant \delta|\boldsymbol{\nu}(\boldsymbol{x}_0) \cdot (\boldsymbol{x}_0-\boldsymbol{y})| \\ &\leqslant \delta M|\boldsymbol{x}_0-\boldsymbol{y}|^2 \\ &\leqslant \frac{1}{2}|\boldsymbol{x}_0-\boldsymbol{y}|^2.\end{aligned}$$

Therefore,

$$|\boldsymbol{x}-\boldsymbol{y}|^2 \geqslant \frac{1}{2}\left(|\boldsymbol{x}-\boldsymbol{x}_0|^2 + |\boldsymbol{x}_0-\boldsymbol{y}|^2\right)$$

and

$$\begin{aligned}\left|\frac{\partial}{\partial \boldsymbol{\nu}_y} E_n(\boldsymbol{x}, \boldsymbol{y})\right| &\leqslant c_3 \frac{|\boldsymbol{x}-\boldsymbol{x}_0| + c|\boldsymbol{x}_0-\boldsymbol{y}|^2}{\left(|\boldsymbol{x}-\boldsymbol{x}_0|^2 + |\boldsymbol{x}_0-\boldsymbol{y}|^2\right)^{n/2}} \\ &\leqslant \frac{c_3|\boldsymbol{x}-\boldsymbol{x}_0|}{\left(|\boldsymbol{x}-\boldsymbol{x}_0|^2 + |\boldsymbol{x}_0-\boldsymbol{y}|^2\right)^{n/2}} + \frac{c_3 c}{|\boldsymbol{x}_0-\boldsymbol{y}|^{n-2}}.\end{aligned}$$

Setting $r = |\boldsymbol{x}-\boldsymbol{y}|$ and $a = |\boldsymbol{x}-\boldsymbol{x}_0| \geqslant 0$, integrating in polar coordinates and making the substitution $r = as$, we arrive at

$$\begin{aligned}\int_{\Gamma_\delta}\left|\frac{\partial}{\partial \boldsymbol{\nu}_y} E_n(\boldsymbol{x}, \boldsymbol{y})\right| \mathrm{d}s_y &\leqslant c_4 \int_0^\delta \left[\frac{a}{(a^2+r^2)^{n/2}} + \frac{1}{r^{n-2}}\right] r^{n-2}\,\mathrm{d}r \\ &\leqslant c_4 \int_0^\infty \frac{s^{n-2}}{(1+s^2)^{n/2}}\,\mathrm{d}s + c_4\delta.\end{aligned}$$

The last integral is convergent. Thus the proof is completed. □

Lemma 3.3.4 *Suppose that the density function $\psi \in C(\Gamma)$, and $\psi(\boldsymbol{x}_0) = 0$ at a point $\boldsymbol{x}_0 \in \Gamma$. Then the double layer potential $V_{I\!I}(\boldsymbol{x})$ is continuous at $\boldsymbol{x} = \boldsymbol{x}_0 \in \Gamma$.*

Proof For the point $\boldsymbol{x}_0 \in \Gamma$, let $\Gamma_\eta = \{\boldsymbol{x} \in \Gamma : |\boldsymbol{x} - \boldsymbol{x}_0| < \eta\}$ with $\eta \in (0, 1]$ be a small positive constant. Then for $\boldsymbol{x} \in \mathbb{R}^n$, we have

$$\begin{aligned} |V_{I\!I}(\boldsymbol{x}) - V_{I\!I}(\boldsymbol{x}_0)| \leqslant & \int_{\Gamma_\eta} \left(\left| \frac{\partial}{\partial \boldsymbol{\nu}_y} E_n(\boldsymbol{x}, \boldsymbol{y}) \right| + \left| \frac{\partial}{\partial \boldsymbol{\nu}_y} E_n(\boldsymbol{x}_0, \boldsymbol{y}) \right| \right) |\psi(\boldsymbol{y})| \, \mathrm{d}s_y + \\ & \int_{\Gamma \setminus \Gamma_\eta} \left| \frac{\partial}{\partial \boldsymbol{\nu}_y} E_n(\boldsymbol{x}, \boldsymbol{y}) - \frac{\partial}{\partial \boldsymbol{\nu}_y} E_n(\boldsymbol{x}_0, \boldsymbol{y}) \right| |\psi(\boldsymbol{y})| \, \mathrm{d}s_y \\ := & \; I + I\!I. \end{aligned} \tag{3.3.15}$$

The estimate (3.3.14) yields

$$I \leqslant 2\, c \max_{y \in \Gamma_\eta} |\psi(\boldsymbol{y})|.$$

Furthermore, suppose that $|\boldsymbol{x}_0 - \boldsymbol{x}| \leqslant \frac{1}{2}\eta$. For $\boldsymbol{y} \in \Gamma \setminus \Gamma_\eta$, we have

$$|\boldsymbol{x}_0 - \boldsymbol{y}| \geqslant \eta, \qquad |\boldsymbol{x} - \boldsymbol{y}| \geqslant \frac{1}{2}\eta.$$

We note that

$$\begin{aligned} & \left| \frac{\partial}{\partial \boldsymbol{\nu}_y} E_n(\boldsymbol{x}, \boldsymbol{y}) - \frac{\partial}{\partial \boldsymbol{\nu}_y} E_n(\boldsymbol{x}_0, \boldsymbol{y}) \right| \\ = & \frac{1}{S_n(1)} \left| \frac{(\boldsymbol{x} - \boldsymbol{y}) \cdot \boldsymbol{\nu}_y}{|\boldsymbol{x} - \boldsymbol{y}|^n} - \frac{(\boldsymbol{x}_0 - \boldsymbol{y}) \cdot \boldsymbol{\nu}_y}{|\boldsymbol{x}_0 - \boldsymbol{y}|^n} \right| \\ = & \frac{1}{S_n(1)} \left| \frac{(\boldsymbol{x} - \boldsymbol{x}_0) \cdot \boldsymbol{\nu}_y}{|\boldsymbol{x} - \boldsymbol{y}|^n} + (\boldsymbol{x}_0 - \boldsymbol{y}) \cdot \boldsymbol{\nu}_y \left(\frac{1}{|\boldsymbol{x} - \boldsymbol{y}|^n} - \frac{1}{|\boldsymbol{x}_0 - \boldsymbol{y}|^n} \right) \right| \\ \leqslant & \, c_5 \eta^{-n-1} |\boldsymbol{x} - \boldsymbol{x}_0|, \end{aligned}$$

where the constant c_5 is independent of η and $\boldsymbol{x}$. Furthermore $I\!I \leqslant C_6 \eta^{-n-1} |\boldsymbol{x} - \boldsymbol{x}_0|$. Given $\epsilon > 0$, we first take $\eta \in (0, 1]$ such that

$$2\, c \max_{y \in \Gamma_\eta} |\psi(\boldsymbol{y})| < \frac{1}{2}\epsilon;$$

and then choose

$$\delta = \min\{1, \frac{\epsilon\, \eta^{n+1}}{2\, c_6}\}.$$

Therefore, when $|\boldsymbol{x} - \boldsymbol{x}_0| < \delta$, we have

$$|V_{I\!I}(\boldsymbol{x}) - V_{I\!I}(\boldsymbol{x}_0)| < \epsilon.$$

That is,

$$\lim_{x\to x_0} V_{II}(x) = V_{II}(x_0).$$

□

Lemma 3.3.5 *Suppose that the density function $\psi \in C(\Gamma)$, then the restriction of double layer potential $V_{II}(x)$ in Ω has a continuous extension to $\bar{\Omega}$, and the restriction of $V_{II}(x)$ in Ω_e has a continuous extension to $\bar{\Omega}_e$. Furthermore, we have*

$$\lim_{\substack{x'\mapsto x\in\Gamma\\ x'\in\Omega}} V_{II}(x') = -\frac{1}{2}\psi(x) + \int_\Gamma \frac{\partial}{\partial \nu_y} E_n(x, y)\psi(y)\,\mathrm{d}s_y,$$

$$\lim_{\substack{x'\mapsto x\in\Gamma\\ x'\in\Omega_e}} V_{II}(x') = \frac{1}{2}\psi(x) + \int_\Gamma \frac{\partial}{\partial \nu_y} E_n(x, y)\psi(y)\,\mathrm{d}s_y.$$

Proof For a point $x \in \Gamma$ and $x' \in \mathbb{R}^n$, we have

$$\begin{aligned} V_{II}(x') &= \int_\Gamma \frac{\partial}{\partial \nu_y} E_n(x', y)\psi(y)\,\mathrm{d}s_y \\ &= \psi(x)\int_\Gamma \frac{\partial}{\partial \nu_y} E_n(x', y)\,\mathrm{d}s_y + \int_\Gamma \frac{\partial}{\partial \nu_y} E_n(x', y)\big(\psi(y) - \psi(x)\big)\,\mathrm{d}s_y \\ &:= I_1(x') + I_2(x'). \end{aligned}$$

By the identities (3.3.13), we get

$$I_1(x') = \begin{cases} -\psi(x), & \forall x' \in \Omega, \\ -\dfrac{1}{2}\psi(x), & \forall x' \in \Gamma, \\ 0, & \forall x' \in \Omega_e. \end{cases} \tag{3.3.16}$$

By Lemma 3.3.4, we know that $I_2(x')$ is continuous at $x' = x$ and

$$\begin{aligned} \lim_{x'\to x\in\Gamma} I_2(x') &= \int_\Gamma \frac{\partial}{\partial \nu_y} E_n(x, y)\big(\psi(y) - \psi(x)\big)\,\mathrm{d}s_y \\ &= \int_\Gamma \frac{\partial}{\partial \nu_y} E_n(x, y)\psi(y)\,\mathrm{d}s_y + \frac{1}{2}\psi(x). \end{aligned} \tag{3.3.17}$$

□

Combining the results above, we arrive at

Theorem 3.3.3 *Suppose that the boundary $\Gamma \in C^2$ and the density function $\psi \in C(\Gamma)$, then*

1. *The double layer potential $V_{II}(\boldsymbol{x})$ is well defined for all $\boldsymbol{x} \in \mathbb{R}^n$.*
2. *$V_{II}(\boldsymbol{x})$ is harmonic on $\mathbb{R}^n \setminus \Gamma$ and*

$$V_{II}(\boldsymbol{x}) = \mathcal{O}(|\boldsymbol{x}|^{1-n}) \qquad as \qquad \boldsymbol{x} \to \infty. \tag{3.3.18}$$

3. *Restriction of $V_{II}(\boldsymbol{x})$ in Ω (Ω_e) has a continuous extension to $\bar{\Omega}$ ($\bar{\Omega}_e$) and*

$$V_{II}(\boldsymbol{x})\Big|_{\Gamma-0} = \lim_{\substack{\boldsymbol{x}'\mapsto \boldsymbol{x}\in\Gamma \\ \boldsymbol{x}'\in\Omega}} V_{II}(\boldsymbol{x}') = -\frac{1}{2}\psi(\boldsymbol{x}) + \int_\Gamma \frac{\partial}{\partial \boldsymbol{\nu}_{\boldsymbol{y}}} E_n(\boldsymbol{x}, \boldsymbol{y})\psi(\boldsymbol{y})\, \mathrm{d}s_{\boldsymbol{y}}, \tag{3.3.19}$$

$$V_{II}(\boldsymbol{x})\Big|_{\Gamma+0} = \lim_{\substack{\boldsymbol{x}'\mapsto \boldsymbol{x}\in\Gamma \\ \boldsymbol{x}'\in\Omega_e}} V_{II}(\boldsymbol{x}') = \frac{1}{2}\psi(\boldsymbol{x}) + \int_\Gamma \frac{\partial}{\partial \boldsymbol{\nu}_{\boldsymbol{y}}} E_n(\boldsymbol{x}, \boldsymbol{y})\psi(\boldsymbol{y})\, \mathrm{d}s_{\boldsymbol{y}}. \tag{3.3.20}$$

Let

$$V_{II}^-(\boldsymbol{x}) = \begin{cases} V_{II}(\boldsymbol{x}), & \forall \boldsymbol{x} \in \Omega, \\ -\frac{1}{2}\psi(\boldsymbol{x}) + \int_\Gamma \frac{\partial}{\partial \boldsymbol{\nu}_{\boldsymbol{y}}} E_n(\boldsymbol{x}, \boldsymbol{y})\psi(\boldsymbol{y})\, \mathrm{d}s_{\boldsymbol{y}}, & \forall \boldsymbol{x} \in \Gamma; \end{cases}$$

and

$$V_{II}^+(\boldsymbol{x}) = \begin{cases} V_{II}(\boldsymbol{x}), & \forall \boldsymbol{x} \in \Omega_e, \\ \frac{1}{2}\psi(\boldsymbol{x}) + \int_\Gamma \frac{\partial}{\partial \boldsymbol{\nu}_{\boldsymbol{y}}} E_n(\boldsymbol{x}, \boldsymbol{y})\psi(\boldsymbol{y})\, \mathrm{d}s_{\boldsymbol{y}}, & \forall \boldsymbol{x} \in \Gamma. \end{cases}$$

$V_{II}^-(\boldsymbol{x})$ is a continuous function on $\bar{\Omega}$ and $V_{II}^+(\boldsymbol{x})$ is a continuous on $\bar{\Omega}_e$, and

$$\left[V_{II}(\boldsymbol{x})\right]_\Gamma := V_{II}(\boldsymbol{x})\Big|_{\Gamma+0} - V_{II}(\boldsymbol{x})\Big|_{\Gamma-0} = \psi(\boldsymbol{x}), \qquad \forall \boldsymbol{x} \in \Gamma. \tag{3.3.21}$$

3.3.3 Single Layer Potentials

Consider the single layer potential (3.3.1) with given density function $\varphi \in C(\Gamma)$. Let

$$K_I(\boldsymbol{x}, \boldsymbol{y}) = E_n(\boldsymbol{x}, \boldsymbol{y}),$$

then

$$V_I(\boldsymbol{x}) = \int_\Gamma E_n(\boldsymbol{x}, \boldsymbol{y})\varphi(\boldsymbol{y})\, \mathrm{d}s_{\boldsymbol{y}} := \int_\Gamma K_I(\boldsymbol{x}, \boldsymbol{y})\varphi(\boldsymbol{y})\, \mathrm{d}s_{\boldsymbol{y}} = (\mathbb{K}_I\varphi)(\boldsymbol{x}). \tag{3.3.22}$$

By (3.3.3) and (3.3.16), we know that

$$|K_I(\boldsymbol{x},\boldsymbol{y})| \leqslant C \begin{cases} \big|\ln|\boldsymbol{x}-\boldsymbol{y}|\big|, & n=2,\ \boldsymbol{x}\in\mathbb{R}^n,\ \boldsymbol{y}\in\Gamma, \\ |\boldsymbol{x}-\boldsymbol{y}|^{2-n}, & n\geqslant 3,\ \boldsymbol{x}\in\mathbb{R}^n,\ \boldsymbol{y}\in\Gamma. \end{cases} \tag{3.3.23}$$

Obviously, $K_I(\boldsymbol{x},\boldsymbol{y})$ is a weakly singular kernel and integral operator K_I is a compact operator on $C(\Gamma)$. It means that $V_I(\boldsymbol{x})$ is well defined on Γ. In the case $\boldsymbol{x}\in\mathbb{R}^n\setminus\Gamma$, $\boldsymbol{y}\in\Gamma$, $K_I(\boldsymbol{x},\boldsymbol{y})$ is continuous in $\boldsymbol{y}$ and harmonic in $\boldsymbol{x}$. Therefore, $V_I(\boldsymbol{x})$ is harmonic on $\mathbb{R}^n\setminus\Gamma$. For single layer potential $V_I(\boldsymbol{x})$, we have

Theorem 3.3.4 *Suppose that $\Gamma\in C^2$ and $\varphi\in C(\Gamma)$, then*

1. $V_I(\boldsymbol{x})$ is well defined on $\mathbb{R}^n$, $V_I(\boldsymbol{x})$ is continuous on $\mathbb{R}^n$ and $V_I(\boldsymbol{x})$ is harmonic on $\mathbb{R}^n\setminus\Gamma$.

2.

$$V_I(\boldsymbol{x}) = \begin{cases} \mathcal{O}(|\ln|\boldsymbol{x}||), & \text{as } \boldsymbol{x}\to\infty,\ n=2, \\ \mathcal{O}(|\boldsymbol{x}|^{2-n}), & \text{as } \boldsymbol{x}\to\infty,\ n>2. \end{cases} \tag{3.3.24}$$

3. If and only if $\int_\Gamma \varphi(\boldsymbol{y})\,\mathrm{d}s_y = 0$, then

$$V_I(\boldsymbol{x}) = \mathcal{O}(|\boldsymbol{x}|^{1-n}) \ \text{as } \boldsymbol{x}\to\infty. \tag{3.3.25}$$

Proof 1. We only need to prove that $V_I(\boldsymbol{x})$ is continuous on Γ. Note that the estimate (3.3.23) holds for all $\boldsymbol{x}\in\mathbb{R}^n$, $\boldsymbol{y}\in\Gamma$. The single layer potential $V_I(\boldsymbol{x})$ as a parameter-variable integral is uniformly convergent at $B_\delta(\boldsymbol{x}) = \{\boldsymbol{z}\in\mathbb{R}^n \mid |\boldsymbol{x}-\boldsymbol{z}|\leqslant\delta\}$ with $\boldsymbol{x}\in\Gamma$. It follows that $V_I(\boldsymbol{x})$ is continuous at $\boldsymbol{x}\in\Gamma$.

2. The conclusion follows directly from estimate (3.3.23).

3. Let

$$\varphi(\boldsymbol{x}) = \varphi^*(\boldsymbol{x}) + \frac{1}{\mathrm{meas}\{\Gamma\}}\int_\Gamma \varphi(\boldsymbol{x})\,\mathrm{d}s_x := \varphi^*(\boldsymbol{x}) + c^*.$$

We know that

$$\int_\Gamma \varphi^*(\boldsymbol{x})\,\mathrm{d}s_x = 0.$$

Therefore,

$$\begin{aligned} V_I(\boldsymbol{x}) &= \int_\Gamma E_n(\boldsymbol{x},\boldsymbol{y})\varphi^*(\boldsymbol{y})\,\mathrm{d}s_y + c^*\int_\Gamma E_n(\boldsymbol{x},\boldsymbol{y})\,\mathrm{d}s_y \\ &= \int_\Gamma \Big[E_n(\boldsymbol{x},\boldsymbol{y}) - E_n(\boldsymbol{x},\boldsymbol{y}_0)\Big]\varphi^*(\boldsymbol{y})\,\mathrm{d}s_y + c^*\int_\Gamma E_n(\boldsymbol{x},\boldsymbol{y})\,\mathrm{d}s_y \end{aligned}$$

with $\boldsymbol{y}_0 \in \Omega$ be a fixed point. Since

$$E_n(\boldsymbol{x}, \boldsymbol{y}) - E_n(\boldsymbol{x}, \boldsymbol{y}_0) = \mathcal{O}(|\boldsymbol{x}|^{1-n}), \qquad \text{as } |\boldsymbol{x}| \to +\infty,$$

uniformly for $\boldsymbol{y} \in \Gamma$, the third conclusion simply follows. □

3.3.4 The Derivatives of Single Layer Potentials

If using single layer potentials to represent the solutions of Neumann problems of the Laplace equation, we need to extend the first derivatives of single layer potential $V_I(\boldsymbol{x})$ onto the boundary Γ from Ω (Ω_e). It is our main goal in this subsection.

The counter examples given by Günter (1967) show that for a single layer potential $V_I(\boldsymbol{x})$ with merely continuous density function φ, the gradient of $V_I(\boldsymbol{x})$ can not extend to the boundary Γ continuously from Ω (or Ω_e) in general.

Suppose that the density function $\varphi \in C^1(\Gamma)$. For the unit outward vector $\boldsymbol{\nu}_y = \boldsymbol{\nu}(\boldsymbol{y}) = (\nu_1(\boldsymbol{y}), \nu_2(\boldsymbol{y}), \ldots, \nu_n(\boldsymbol{y}))^{\mathrm{T}}$ on Γ, let

$$\boldsymbol{\tau}_y^{i,j} = \boldsymbol{\tau}^{i,j}(\boldsymbol{y}) = \left(\tau_1^{i,j}(\boldsymbol{y}), \tau_2^{i,j}(\boldsymbol{y}), \cdots, \tau_n^{i,j}(\boldsymbol{y})\right)$$

with

$$\tau_k^{i,j}(\boldsymbol{y}) = -\delta_{i,k}\nu_j(\boldsymbol{y}) + \delta_{j,k}\nu_i(\boldsymbol{y}), \qquad 1 \leqslant i,\ j \leqslant n,\ 1 \leqslant k \leqslant n.$$

Obviously, the vectors $\{\boldsymbol{\tau}^{i,j}(\boldsymbol{y}), 1 \leqslant i, j \leqslant n\}$ are orthogonal to $\boldsymbol{\nu}(\boldsymbol{y})$, which are tangential vectors of Γ. We now recall the Stokes formula on Γ

$$\int_\Gamma \frac{\partial}{\partial \boldsymbol{\tau}_y^{i,j}} u(\boldsymbol{y})\, \mathrm{d}s_y = 0, \qquad \forall\, u \in C^1(\Gamma). \tag{3.3.26}$$

Theorem 3.3.5 *The first derivatives of single layer potential $V_I(\boldsymbol{x})$ with density function $\varphi \in C^1(\Gamma)$ can be continuously extended from Ω (Ω_e) to the boundary Γ, namely*

$$\left.\begin{aligned} \lim_{\substack{z \to \boldsymbol{x} \in \Gamma \\ z \in \Omega}} \operatorname{grad}_z V_I(z) &= \int_\Gamma \operatorname{grad}_x E_n(\boldsymbol{x}, \boldsymbol{y})\varphi(\boldsymbol{y})\, \mathrm{d}s_y + \frac{1}{2}\boldsymbol{\nu}(\boldsymbol{x})\varphi(\boldsymbol{x}), \\ \lim_{\substack{z \to \boldsymbol{x} \in \Gamma \\ z \in \Omega_e}} \operatorname{grad}_z V_I(z) &= \int_\Gamma \operatorname{grad}_x E_n(\boldsymbol{x}, \boldsymbol{y})\varphi(\boldsymbol{y})\, \mathrm{d}s_y - \frac{1}{2}\boldsymbol{\nu}(\boldsymbol{x})\varphi(\boldsymbol{x}), \end{aligned}\right\} \tag{3.3.27}$$

where the integral exists as a Hardamard finite part integral.

Proof For $z \in \mathbb{R}^n \setminus \Gamma$, we consider the ith-component of $\mathrm{grad}_z V_I(z)$,

$$\begin{aligned}
\frac{\partial V_I(z)}{\partial z_i} &= \int_\Gamma \frac{\partial}{\partial z_i} E_n(z, y)\varphi(y)\,\mathrm{d}s_y \\
&= -\int_\Gamma \varphi(y)\frac{\partial}{\partial y_i} E_n(z, y)\,\mathrm{d}s_y \\
&= -\int_\Gamma \varphi(y)\left[\sum_{k=1}^{n} \nu_k(y)^2\right]\frac{\partial}{\partial y_i} E_n(z, y)\,\mathrm{d}s_y \\
&= -\int_\Gamma \varphi(y)\nu_i(y)\left[\sum_{k=1}^{n} \nu_k(y)\frac{\partial}{\partial y_k}\right] E_n(z, y)\,\mathrm{d}s_y + \\
&\quad \int_\Gamma \varphi(y)\left\{-\left[\sum_{k=1}^{n} \nu_k(y)^2\right]\frac{\partial}{\partial y_i} + \sum_{k=1}^{n}\left[\nu_i(y)\nu_k(y)\frac{\partial}{\partial y_k}\right]\right\} E_n(z, y)\,\mathrm{d}s_y \\
&= -\int_\Gamma \nu_i(y)\varphi(y)\frac{\partial}{\partial \boldsymbol{\nu}_y} E_n(z, y)\,\mathrm{d}s_y + \\
&\quad \int_\Gamma \varphi(y)\left\{\sum_{k=1}^{n} \nu_k(y)\left[-\nu_k(y)\frac{\partial}{\partial y_i} + \nu_i(y)\frac{\partial}{\partial y_k}\right]\right\} E_n(z, y)\,\mathrm{d}s_y \\
&= -\int_\Gamma \nu_i(y)\varphi(y)\frac{\partial}{\partial \boldsymbol{\nu}_y} E_n(z, y)\,\mathrm{d}s_y + \int_\Gamma \varphi(y)\sum_{k=1}^{n} \nu_k(y)\frac{\partial E_n(z, y)}{\partial \boldsymbol{\tau}_y^{k,i}}\,\mathrm{d}s_y \\
&= -\int_\Gamma \nu_i(y)\varphi(y)\frac{\partial}{\partial \boldsymbol{\nu}_y} E_n(z, y)\,\mathrm{d}s_y - \int_\Gamma \sum_{k=1}^{n} \frac{\partial}{\partial \boldsymbol{\tau}_y^{k,i}}\left[\nu_k(y)\varphi(y)\right] E_n(z, y)\,\mathrm{d}s_y \\
&:= U_1(z) + U_2(z).
\end{aligned}$$

It is clear that $U_1(z)$ is a double layer potential with density function $-\nu_i\varphi \in C^1(\Gamma)$, which can be extended from Ω (Ω_e) to the boundary Γ, and $U_2(z)$ is a single layer potential with density function $\left[-\sum_{k=1}^{n} \frac{\partial}{\partial \boldsymbol{\tau}_y^{k,i}}\left(\nu_k(y)\varphi(y)\right)\right] \in C(\Gamma)$, which is continuous on $\mathbb{R}^n$. Therefore, we obtain

$$\begin{aligned}
\lim_{\substack{z\to x\in\Gamma \\ z\in\Omega}} \frac{V_I(z)}{\partial z_i} &= -\int_\Gamma \nu_i(y)\varphi(y)\frac{\partial}{\partial \boldsymbol{\nu}_y} E_n(x, y)\,\mathrm{d}s_y - \\
&\quad \int_\Gamma \sum_{k=1}^{n} \frac{\partial}{\partial \boldsymbol{\tau}_y^{k,i}}\left(\nu_k(y)\varphi(y)\right) E_n(x, y)\,\mathrm{d}s_y + \frac{1}{2}\nu_i(x)\varphi(x) \\
&= \int_\Gamma \frac{\partial}{\partial x_i} E_n(x, y)\varphi(y)\,\mathrm{d}s_y + \frac{1}{2}\nu_i(x)\varphi(x), \qquad \forall x \in \Gamma
\end{aligned}$$

and

$$\lim_{\substack{z\to x\in\Gamma\\ z\in\Omega_e}} \frac{V_I(z)}{\partial z_i} = \int_\Gamma \frac{\partial}{\partial x_i} E_n(\boldsymbol{x},\boldsymbol{y})\varphi(\boldsymbol{y})\,\mathrm{d}s_y - \frac{1}{2}\nu_i(\boldsymbol{x})\varphi(\boldsymbol{x}), \qquad \forall\,\boldsymbol{x}\in\Gamma.$$

The equalities (3.3.27) simply follow. □

Now for $\forall\,\boldsymbol{x}\in\Gamma$, we have

$$\left.\begin{aligned}
\left.\frac{\partial V_I(\boldsymbol{x})}{\partial \boldsymbol{\nu}_x}\right|_{\Gamma-0} &= \lim_{\substack{z\to x\in\Gamma\\ z\in\Omega}} \boldsymbol{\nu}_x\cdot \operatorname{grad}_z V_I(z) = \int_\Gamma \frac{\partial}{\partial \boldsymbol{\nu}_x} E_n(\boldsymbol{x},\boldsymbol{y})\varphi(\boldsymbol{y})\,\mathrm{d}s_y + \frac{1}{2}\varphi(\boldsymbol{x}),\\
\left.\frac{\partial V_I(\boldsymbol{x})}{\partial \boldsymbol{\nu}_x}\right|_{\Gamma+0} &= \lim_{\substack{z\to x\in\Gamma\\ z\in\Omega_e}} \boldsymbol{\nu}_x\cdot \operatorname{grad}_z V_I(z) = \int_\Gamma \frac{\partial}{\partial \boldsymbol{\nu}_x} E_n(\boldsymbol{x},\boldsymbol{y})\varphi(\boldsymbol{y})\,\mathrm{d}s_y - \frac{1}{2}\varphi(\boldsymbol{x}),
\end{aligned}\right\} \tag{3.3.28}$$

and

$$\left[\frac{\partial V_I(\boldsymbol{x})}{\partial \boldsymbol{\nu}_x}\right]_\Gamma = \left.\frac{\partial V_I(\boldsymbol{x})}{\partial \boldsymbol{\nu}_x}\right|_{\Gamma+0} - \left.\frac{\partial V_I(\boldsymbol{x})}{\partial \boldsymbol{\nu}_x}\right|_{\Gamma-0} = -\varphi(\boldsymbol{x}), \qquad \forall\,\boldsymbol{x}\in\Gamma.$$

Note that the integral on the right hand side of (3.3.28) is a weakly singular integral. In fact, we have

$$\int_\Gamma \frac{\partial}{\partial \boldsymbol{\nu}_x} E_n(\boldsymbol{x},\boldsymbol{y})\varphi(\boldsymbol{y})\,\mathrm{d}s_y = \int_\Gamma K^*_{I\!I}(\boldsymbol{x},\boldsymbol{y})\varphi(\boldsymbol{y})\,\mathrm{d}s_y = (\mathbb{K}^*_{I\!I}\varphi)(\boldsymbol{x}), \qquad \forall\,\boldsymbol{x}\in\Gamma, \tag{3.3.29}$$

with

$$K^*_{I\!I}(\boldsymbol{x},\boldsymbol{y}) = \frac{\partial}{\partial \boldsymbol{\nu}_x} E_n(\boldsymbol{x},\boldsymbol{y}) = K_{I\!I}(\boldsymbol{y},\boldsymbol{x}) = \frac{(\boldsymbol{y}-\boldsymbol{x})\cdot\boldsymbol{\nu}(\boldsymbol{x})}{S_n(1)|\boldsymbol{x}-\boldsymbol{y}|^n} \tag{3.3.30}$$

and

$$|K_{I\!I}(\boldsymbol{x},\boldsymbol{y})^*| \leqslant c\,|\boldsymbol{x}-\boldsymbol{y}|^{-(n-2)}, \qquad \forall\,\boldsymbol{x},\ \boldsymbol{y}\in\Gamma. \tag{3.3.31}$$

Obviously the integral operator $\mathbb{K}^*_{I\!I}$ is compact on $C(\Gamma)$. For the tangential derivatives of $V_I(\boldsymbol{x})$ on Γ, from (3.3.20), immediately we obtain

$$\left.\frac{\partial V_I(\boldsymbol{x})}{\partial \boldsymbol{\tau}_x^{k,j}}\right|_{\Gamma-0} = \lim_{\substack{z\to x\in\Gamma\\ z\in\Omega}} \boldsymbol{\tau}(\boldsymbol{x})^{k,j}\cdot \operatorname{grad}_z V_I(z) = \frac{\partial}{\partial \boldsymbol{\tau}_x^{i,i}} \int_\Gamma E_n(\boldsymbol{x},\boldsymbol{y})\,\varphi(\boldsymbol{y})\,\mathrm{d}s_y,$$

$$\left.\frac{\partial V_I(\boldsymbol{x})}{\partial \boldsymbol{\tau}_x^{k,j}}\right|_{\Gamma+0} = \lim_{\substack{z\to x\in\Gamma\\ z\in\Omega_e}} \boldsymbol{\tau}(\boldsymbol{x})^{k,j}\cdot \operatorname{grad}_z V_I(z) = \frac{\partial}{\partial \boldsymbol{\tau}_x^{i,i}} \int_\Gamma E_n(\boldsymbol{x},\boldsymbol{y})\,\varphi(\boldsymbol{y})\,\mathrm{d}s_y.$$

Namely the tangential derivatives of $V_I(\boldsymbol{x})$ on Γ are continuous and

$$\begin{aligned}
\left.\frac{\partial V_I(\boldsymbol{x})}{\partial \boldsymbol{\tau}_x^{k,j}}\right|_{\Gamma} &= \lim_{\substack{z\to x\in\Gamma\\ z\in\mathbb{R}^n\setminus\Gamma}} \boldsymbol{\tau}(\boldsymbol{x})^{k,j}\cdot \operatorname{grad}_z V_I(z)\\
&= \frac{\partial}{\partial \boldsymbol{\tau}_x^{i,i}} \int_\Gamma E_n(\boldsymbol{x},\boldsymbol{y})\,\varphi(\boldsymbol{y})\,\mathrm{d}s_y, \qquad 1\leqslant k,\ j\leqslant n.
\end{aligned} \tag{3.3.32}$$

Remark 3.3.1 Suppose that the density function $\varphi \in C(\Gamma)$, the equalities (3.3.28) are still valid in the following sense

$$\left.\begin{aligned}\left.\frac{V_I(\boldsymbol{x})}{\partial \boldsymbol{\nu}_x^-}\right|_\Gamma &= \lim_{\substack{h\to 0\\ h<0}} \frac{\partial V_I(\boldsymbol{x}+h\boldsymbol{\nu}(\boldsymbol{x}))}{\partial \boldsymbol{\nu}_x} = \int_\Gamma \frac{\partial}{\partial \boldsymbol{\nu}_x} E_n(\boldsymbol{x},\boldsymbol{y})\varphi(\boldsymbol{y})\,\mathrm{d}s_y + \frac{1}{2}\varphi(\boldsymbol{x}), \quad \forall \boldsymbol{x}\in\Gamma,\\ \left.\frac{V_I(\boldsymbol{x})}{\partial \boldsymbol{\nu}_x^+}\right|_\Gamma &= \lim_{\substack{h\to 0\\ h>0}} \frac{\partial V_I(\boldsymbol{x}+h\boldsymbol{\nu}(\boldsymbol{x}))}{\partial \boldsymbol{\nu}_x} = \int_\Gamma \frac{\partial}{\partial \boldsymbol{\nu}_x} E_n(\boldsymbol{x},\boldsymbol{y})\varphi(\boldsymbol{y})\,\mathrm{d}s_y - \frac{1}{2}\varphi(\boldsymbol{x}), \quad \forall \boldsymbol{x}\in\Gamma.\end{aligned}\right\} \tag{3.3.33}$$

3.3.5 *The Derivatives of Double Layer Potentials*

In this subsection we discuss the extension of the first derivatives of double layer potential $V_{II}(\boldsymbol{x})$ from Ω (Ω_e) to the boundary continuously. Suppose that the density function $\psi \in C^2(\Gamma)$.

Theorem 3.3.6 *The first derivatives of double layer potential $V_{II}(\boldsymbol{x})$ with density function $\psi \in C^2(\Gamma)$ can be continuously extended from Ω (Ω_e)to the boundary Γ, namely*

$$\left.\begin{aligned}\lim_{\substack{z\to x\in\Gamma\\ z\in\Omega}} \frac{\partial V_{II}(z)}{\partial z_i} &= \frac{\partial}{\partial x_i}\int_\Gamma \frac{\partial}{\partial \boldsymbol{\nu}_y} E_n(\boldsymbol{x},\boldsymbol{y})\psi(\boldsymbol{y})\,\mathrm{d}s_y - \frac{1}{2}\nu_i(\boldsymbol{x})\sum_{k=1}^n \frac{\partial}{\partial \boldsymbol{\tau}_x^{k,i}}\psi(\boldsymbol{x}),\\ \lim_{\substack{z\to x\in\Gamma\\ z\in\Omega_e}} \frac{\partial V_{II}(z)}{\partial z_i} &= \frac{\partial}{\partial x_i}\int_\Gamma \frac{\partial}{\partial \boldsymbol{\nu}_y} E_n(\boldsymbol{x},\boldsymbol{y})\psi(\boldsymbol{y})\,\mathrm{d}s_y + \frac{1}{2}\nu_i(\boldsymbol{x})\sum_{k=1}^n \frac{\partial}{\partial \boldsymbol{\tau}_x^{k,i}}\psi(\boldsymbol{x}).\end{aligned}\right\} \tag{3.3.34}$$

The integrals in (3.3.34) exist as a Hardamard finite part integral for any $\boldsymbol{x} \in \Gamma$. Furthermore, we have

$$\begin{aligned}\left.\frac{\partial V_{II}(\boldsymbol{x})}{\partial \boldsymbol{\nu}_x}\right|_\Gamma &= \lim_{\substack{z\to x\in\Gamma\\ z\in\mathbb{R}^n\setminus\Gamma}} \boldsymbol{\nu}(\boldsymbol{x})\cdot \operatorname{grad}_z V_{II}(z)\\ &= \sum_{1\leqslant k<j\leqslant n} \frac{\partial}{\partial \boldsymbol{\tau}_x^{k,j}} \int_\Gamma E_n(\boldsymbol{x},\boldsymbol{y})\left(\frac{\partial}{\partial \boldsymbol{\tau}_y^{k,j}}\psi(\boldsymbol{y})\right)\mathrm{d}s_y\\ &= \frac{\partial}{\partial \boldsymbol{\nu}_x}\int_\Gamma \frac{\partial}{\partial \boldsymbol{\nu}_y} E_n(\boldsymbol{x},\boldsymbol{y})\psi(\boldsymbol{y})\,\mathrm{d}s_y.\end{aligned} \tag{3.3.35}$$

The last integral also exists as a Hardamard finite part integral.

Proof At first we have noted that the fundamental solution $E_n(\boldsymbol{x}, \boldsymbol{y})$ satisfies

$$\frac{\partial^2 E_n(\boldsymbol{x}, \boldsymbol{y})}{\partial x_i \partial y_k} = \frac{\partial^2 E_n(\boldsymbol{x}, \boldsymbol{y})}{\partial x_k \partial y_i} = -\frac{\partial^2 E_n(\boldsymbol{x}, \boldsymbol{y})}{\partial y_i \partial y_k}. \tag{3.3.36}$$

For $\boldsymbol{z} \in \mathbb{R}^n \setminus \Gamma$, $\boldsymbol{y} \in \Gamma$, we have

$$\begin{aligned}
\frac{\partial}{\partial z_i}\frac{\partial}{\partial \boldsymbol{\nu}_{\boldsymbol{y}}} E_n(\boldsymbol{z}, \boldsymbol{y}) &= \frac{\partial}{\partial z_i}\sum_{k=1}^{n} \nu_k(\boldsymbol{y})\frac{\partial}{\partial y_k} E_n(\boldsymbol{z}, \boldsymbol{y}) \\
&= \sum_{k=1}^{n} \frac{\partial}{\partial z_k}\nu_k(\boldsymbol{y})\frac{\partial}{\partial y_i} E_n(\boldsymbol{z}, \boldsymbol{y}) \\
&= \sum_{k=1}^{n} \frac{\partial}{\partial z_k}\left[\nu_k(\boldsymbol{y})\frac{\partial}{\partial y_i} - \nu_i(\boldsymbol{y})\frac{\partial}{\partial y_k}\right] E_n(\boldsymbol{z}, \boldsymbol{y}) + \\
&\quad \sum_{k=1}^{n} \frac{\partial}{\partial z_k}\Big(\nu_i(\boldsymbol{y})\frac{\partial}{\partial y_k}\Big) E_n(\boldsymbol{z}, \boldsymbol{y}) \\
&= \sum_{k=1}^{n} \frac{\partial}{\partial z_k}\frac{\partial}{\partial \boldsymbol{\tau}_{\boldsymbol{y}}^{k,i}} E_n(\boldsymbol{z}, \boldsymbol{y}).
\end{aligned}$$

Therefore, we get

$$\begin{aligned}
\frac{\partial V_{II}(\boldsymbol{z})}{\partial z_i} &= \sum_{k=1}^{n} \frac{\partial}{\partial z_k}\int_\Gamma \left(\frac{\partial}{\partial \boldsymbol{\tau}_{\boldsymbol{y}}^{k,i}} E_n(\boldsymbol{z}, \boldsymbol{y})\right)\psi(\boldsymbol{y})\,\mathrm{d}s_{\boldsymbol{y}} \\
&= -\sum_{k=1}^{n} \frac{\partial}{\partial z_k}\int_\Gamma E_n(\boldsymbol{z}, \boldsymbol{y})\left(\frac{\partial}{\partial \boldsymbol{\tau}_{\boldsymbol{y}}^{k,i}}\psi(\boldsymbol{y})\right)\mathrm{d}s_{\boldsymbol{y}}, \qquad \forall\, \boldsymbol{z} \in \mathbb{R}^n \setminus \Gamma.
\end{aligned}$$

By Theorem 3.3.5, we arrive at

$$\begin{aligned}
\lim_{\substack{\boldsymbol{z}\to \boldsymbol{x}\in\Gamma \\ \boldsymbol{z}\in\Omega}} \frac{\partial V_{II}(\boldsymbol{z})}{\partial z_i} &= -\sum_{k=1}^{n} \frac{\partial}{\partial x_k}\int_\Gamma E_n(\boldsymbol{x}, \boldsymbol{y})\left(\frac{\partial}{\partial \boldsymbol{\tau}_{\boldsymbol{y}}^{k,i}}\psi(\boldsymbol{y})\right)\mathrm{d}s_{\boldsymbol{y}} - \\
&\quad \frac{1}{2}\nu_i(\boldsymbol{x})\sum_{k=1}^{n}\left(\frac{\partial}{\partial \boldsymbol{\tau}_{\boldsymbol{x}}^{k,i}}\psi(\boldsymbol{x})\right) \\
&= \sum_{k=1}^{n} \frac{\partial}{\partial x_k}\int_\Gamma \left[\nu_k(\boldsymbol{y})\frac{\partial}{\partial y_i} - \nu_i(\boldsymbol{y})\frac{\partial}{\partial y_k}\right] E_n(\boldsymbol{x}, \boldsymbol{y})\psi(\boldsymbol{y})\,\mathrm{d}s_{\boldsymbol{y}} - \\
&\quad \frac{1}{2}\nu_i(\boldsymbol{x})\sum_{k=1}^{n}\left(\frac{\partial}{\partial \boldsymbol{\tau}_{\boldsymbol{x}}^{k,i}}\psi(\boldsymbol{x})\right) \\
&= \frac{\partial}{\partial x_i}\int_\Gamma \frac{\partial}{\partial \boldsymbol{\nu}_{\boldsymbol{y}}} E_n(\boldsymbol{x}, \boldsymbol{y})\psi(\boldsymbol{y})\,\mathrm{d}s_{\boldsymbol{y}} - \frac{1}{2}\nu_i(\boldsymbol{x})\sum_{k=1}^{n}\left(\frac{\partial}{\partial \boldsymbol{\tau}_{\boldsymbol{x}}^{k,i}}\psi(\boldsymbol{x})\right).
\end{aligned}$$

Similarly, we obtain

$$\lim_{\substack{z\to x\in\Gamma\\ z\in\Omega_e}}\frac{\partial V_{II}(z)}{\partial z_i}=\frac{\partial}{\partial x_i}\int_\Gamma\frac{\partial}{\partial \boldsymbol{\nu}_y}E_n(\boldsymbol{x},\boldsymbol{y})\psi(\boldsymbol{y})\,\mathrm{d}s_y+\frac{1}{2}\nu_i(\boldsymbol{x})\sum_{k=1}^n\left(\frac{\partial}{\partial\boldsymbol{\tau}_x^{k,i}}\psi(\boldsymbol{x})\right).$$

Moreover, for $\boldsymbol{x},\ \boldsymbol{y}\in\Gamma$, $z\in\mathbb{R}^n\setminus\Gamma$, we have

$$\boldsymbol{\nu}(\boldsymbol{x})\cdot\mathrm{grad}_z V_{II}(z)=\sum_{j=1}^n\nu_j(\boldsymbol{x})\frac{\partial}{\partial z_j}\int_\Gamma\frac{\partial}{\partial\boldsymbol{\nu}_y}E_n(z,\boldsymbol{y})\psi(\boldsymbol{y})\,\mathrm{d}s_y,$$

and

$$\begin{aligned}
&\sum_{j=1}^n\nu_j(\boldsymbol{x})\frac{\partial}{\partial z_j}\left(\sum_{k=1}^n\nu_k(\boldsymbol{y})\frac{\partial}{\partial y_k}\right)E_n(z,\boldsymbol{y})\\
&=\sum_{k,\,j=1}^n\left(\nu_j(\boldsymbol{x})\frac{\partial}{\partial z_k}\right)\left(\nu_k(\boldsymbol{y})\frac{\partial}{\partial y_j}\right)E_n(z,\boldsymbol{y})\\
&=\frac{1}{2}\sum_{k,\,j=1}^n\left(\nu_j(\boldsymbol{x})\frac{\partial}{\partial z_k}-\nu_k(\boldsymbol{x})\frac{\partial}{\partial z_j}\right)\left(\nu_k(\boldsymbol{x})\frac{\partial}{\partial y_j}-\nu_j(\boldsymbol{x})\frac{\partial}{\partial y_k}\right)E_n(z,\boldsymbol{y})+\\
&\quad\sum_{k,\,j=1}^n\left(\nu_j(\boldsymbol{x})\frac{\partial}{\partial z_k}\right)\left(\nu_k(\boldsymbol{x})\frac{\partial}{\partial z_j}\right)E_n(z,\boldsymbol{y})\\
&=\frac{1}{2}\sum_{k,\,j=1}^n\left(\nu_j(\boldsymbol{x})\frac{\partial}{\partial z_k}-\nu_k(\boldsymbol{x})\frac{\partial}{\partial z_j}\right)\left(\nu_k(\boldsymbol{x})\frac{\partial}{\partial y_j}-\nu_j(\boldsymbol{x})\frac{\partial}{\partial y_k}\right)E_n(z,\boldsymbol{y})\\
&=-\sum_{1\leqslant k<j\leqslant n}\left(\nu_j(\boldsymbol{x})\frac{\partial}{\partial z_k}-\nu_k(\boldsymbol{x})\frac{\partial}{\partial z_j}\right)\left(\nu_k(\boldsymbol{x})\frac{\partial}{\partial y_j}-\nu_j(\boldsymbol{x})\frac{\partial}{\partial y_k}\right)E_n(z,\boldsymbol{y}).
\end{aligned}$$

This yields

$$\begin{aligned}
\boldsymbol{\nu}(\boldsymbol{x})\cdot\mathrm{grad}_z V_{II}(z)&=-\sum_{1\leqslant k<j\leqslant n}\frac{\partial}{\partial\boldsymbol{\tau}_x^{k,j}}\int_\Gamma\left(\frac{\partial}{\partial\boldsymbol{\tau}_y^{k,j}}E_n(z,\boldsymbol{y})\right)\psi(\boldsymbol{y})\,\mathrm{d}s_y\\
&=\sum_{1\leqslant k<j\leqslant n}\frac{\partial}{\partial\boldsymbol{\tau}_x^{k,j}}\int_\Gamma E_n(z,\boldsymbol{y})\left(\frac{\partial}{\partial\boldsymbol{\tau}_y^{k,j}}\psi(\boldsymbol{y})\right)\mathrm{d}s_y,\quad\forall\, z\in\mathbb{R}^n\setminus\Gamma.
\end{aligned}$$

By the continuity of the tangential derivatives of the single layer potentials on Γ, the equalities in (3.3.35) are proved directly. □

Remark 3.3.2 1. For the case $n = 2$, let $\boldsymbol{\tau}_x = \boldsymbol{\tau}_x^{1,2} = (-\nu_2(\boldsymbol{x}), \nu_1(\boldsymbol{x}))^{\mathrm{T}}$ and $\boldsymbol{\tau}_y = \boldsymbol{\tau}_y^{1,2} = (-\nu_2(\boldsymbol{y}), \nu_1(\boldsymbol{y}))^{\mathrm{T}}$ for any $\boldsymbol{x}, \boldsymbol{y} \in \Gamma$. Then the formula (3.3.35) ($n = 2$) is reduced to

$$\left.\frac{\partial V_{II}(\boldsymbol{x})}{\partial \boldsymbol{\nu}_x}\right|_{\Gamma} = \frac{\partial}{\partial \boldsymbol{\tau}_x}\int_{\Gamma} E_2(\boldsymbol{x}, \boldsymbol{y})\left(\frac{\partial \psi(\boldsymbol{y})}{\partial \boldsymbol{\tau}_y}\right) \mathrm{d}s_y = \int_{\Gamma} \frac{\partial^2}{\partial \boldsymbol{\nu}_x \partial \boldsymbol{\nu}_y} E_2(\boldsymbol{x}, \boldsymbol{y})\, \psi(\boldsymbol{y})\, \mathrm{d}s_y. \tag{3.3.37}$$

2. For the case $n = 3$, let

$$\begin{aligned}
\boldsymbol{\tau}_{1,x} &= \boldsymbol{\tau}_x^{2,3} = (0, -\nu_3(\boldsymbol{x}), \nu_2(\boldsymbol{x}))^{\mathrm{T}},\\
\boldsymbol{\tau}_{2,x} &= \boldsymbol{\tau}_x^{3,1} = (\nu_3(\boldsymbol{x}), 0, -\nu_1(\boldsymbol{x}))^{\mathrm{T}},\\
\boldsymbol{\tau}_{3,x} &= \boldsymbol{\tau}_x^{1,2} = (-\nu_2(\boldsymbol{x}), \nu_1(\boldsymbol{x}), 0)^{\mathrm{T}},
\end{aligned}$$

and

$$\boldsymbol{\tau}_{1,y} = \boldsymbol{\tau}_y^{2,3}, \quad \boldsymbol{\tau}_{2,y} = \boldsymbol{\tau}_y^{3,1}, \quad \boldsymbol{\tau}_{3,y} = \boldsymbol{\tau}_y^{1,2}.$$

Then the formula (3.3.27) ($n = 3$) is reduced to

$$\left.\frac{\partial V_{II}(\boldsymbol{x})}{\partial \boldsymbol{\nu}_x}\right|_{\Gamma} = \sum_{k=1}^{3} \frac{\partial}{\partial \boldsymbol{\tau}_{k,x}} \int_{\Gamma} E_3(\boldsymbol{x}, \boldsymbol{y}) \frac{\partial}{\partial \boldsymbol{\tau}_{k,y}} \psi(\boldsymbol{y})\, \mathrm{d}s_y = \int_{\Gamma} \frac{\partial^2}{\partial \boldsymbol{\nu}_x \partial \boldsymbol{\nu}_y} E_3(\boldsymbol{x}, \boldsymbol{y})\psi(\boldsymbol{y})\, \mathrm{d}s_y. \tag{3.3.38}$$

The formulas (3.3.37) and (3.3.38) were proven by Maue (1949) and Günter (1967). The idea of the proof in Theorem 3.3.6 is from Han (1988, 1994).

3. By the formula (3.3.35), it is natural that we obtain a hypersingular integral operator on $C^2(\Gamma)$, namely

$$(\mathbb{K}_{III}\psi)(\boldsymbol{x}) = \int_{\Gamma} \frac{\partial^2}{\partial \boldsymbol{\nu}_x \partial \boldsymbol{\nu}_y} E_n(\boldsymbol{x}, \boldsymbol{y})\psi(\boldsymbol{y})\, \mathrm{d}s_y, \tag{3.3.39}$$

which is equivalent to the integro-differential operator

$$(\mathbb{K}_{III}\psi)(\boldsymbol{x}) = \sum_{1 \leqslant k < j \leqslant n} \frac{\partial}{\partial \boldsymbol{\tau}_x^{k,j}} \int_{\Gamma} E_n(\boldsymbol{x}, \boldsymbol{y})\left(\frac{\partial}{\partial \boldsymbol{\tau}_y^{k,j}} \psi(\boldsymbol{y})\right) \mathrm{d}s_y. \tag{3.3.40}$$

3.3.6 The Single and Double Layer Potentials in Sobolev Spaces

In this subsection we study the single layer potential $V_I(\boldsymbol{x})$ with the density function $\varphi \in H^{-1/2}(\Gamma)$ and the double layer potential $V_{II}(\boldsymbol{x})$ with density function $\psi \in H^{1/2}(\Gamma)$.

At first consider the single layer potential $V_I(\boldsymbol{x})$ with density function $\varphi \in C^1(\Gamma)$. We know that $V_I(\boldsymbol{x})$ is the unique solution to the following transmission problem

$$\begin{cases} \Delta v(\boldsymbol{x}) = 0, \quad \text{in } \mathbb{R}^n \setminus \Gamma, & (3.3.41\text{a}) \\ \left[v(\boldsymbol{x})\right]_\Gamma = 0, \quad \forall \boldsymbol{x} \in \Gamma, & (3.3.41\text{b}) \\ \left[\dfrac{v(\boldsymbol{x})}{\partial \boldsymbol{\nu}_x}\right]_\Gamma = -\varphi(\boldsymbol{x}), \quad \forall \boldsymbol{x} \in \Gamma, & (3.3.41\text{c}) \\ v(\boldsymbol{x}) = \left\{ \begin{matrix} C \ln |\boldsymbol{x}|, & n = 2 \\ C|\boldsymbol{x}|^{2-n}, & n \geqslant 3 \end{matrix} \right\} + \mathcal{O}(|\boldsymbol{x}|^{1-n}), \quad \text{as } |\boldsymbol{x}| \to \infty. & (3.3.41\text{d}) \end{cases}$$

The problem (3.3.41) is a special case of the Theorem 5.6.1 in the book by Hsiao and Wendland (2008).

For a given $\varphi \in H^{-1/2}(\Gamma)$, the problem (3.3.41) has a unique solution, which is given by the single layer potential $V_I(\boldsymbol{x})$ with density function $\varphi \in H^{-1/2}(\Gamma)$. This means

Lemma 3.3.6 *For a given density function $\varphi \in H^{-1/2}$, the corresponding single layer potential $V_I \in H^1_{\text{loc}}(\mathbb{R}^n)$ and*

1. $V_I\big|_\Omega \in H^1(\Omega)$ and its traces on Γ, $V_I\big|_{\Gamma-0} \in H^{1/2}(\Gamma)$, $\left.\dfrac{\partial V_I}{\partial \boldsymbol{\nu}_x}\right|_{\Gamma-0} \in H^{-1/2}(\Gamma)$.

$V_I\big|_{\Omega_e} \in H^1_{\text{loc}}(\Omega_e)$ and its traces on Γ, $V_I\big|_{\Gamma+0} \in H^{1/2}(\Gamma)$, $\left.\dfrac{\partial V_I}{\partial \boldsymbol{\nu}_x}\right|_{\Gamma+0} \in H^{-1/2}(\Gamma)$.

2.

$$V_I(\boldsymbol{x})\big|_\Gamma = V_I(\boldsymbol{x})\big|_{\Gamma\pm 0} = \int_\Gamma E_n(\boldsymbol{x}, \boldsymbol{y})\varphi(\boldsymbol{y})\, \mathrm{d}s_y := (\mathbb{K}_I\varphi)(\boldsymbol{x}).$$

The operator $\mathbb{K}_I$ is bounded from $H^{-1/2}(\Gamma)$ to $H^{1/2}(\Gamma)$, namely, there exists a constant $C > 0$ such that

$$\left\| V_I\big|_\Gamma \right\|_{H^{1/2}(\Gamma)} \leqslant C\|\varphi\|_{H^{-1/2}(\Gamma)}. \tag{3.3.42}$$

3. The following two equalities hold in $H^{-1/2}(\Gamma)$:

$$\left.\frac{\partial V_I(\boldsymbol{x})}{\partial \boldsymbol{\nu}_x}\right|_{\Gamma-0} = \int_\Gamma \frac{\partial}{\partial \boldsymbol{\nu}_x} E_n(\boldsymbol{x}, \boldsymbol{y})\varphi(\boldsymbol{y})\, \mathrm{d}s_y + \frac{1}{2}\varphi(\boldsymbol{x}), \quad \forall \boldsymbol{x} \in \Gamma,$$

$$\left.\frac{\partial V_I(\boldsymbol{x})}{\partial \boldsymbol{\nu}_x}\right|_{\Gamma+0} = \int_\Gamma \frac{\partial}{\partial \boldsymbol{\nu}_x} E_n(\boldsymbol{x}, \boldsymbol{y})\varphi(\boldsymbol{y})\, \mathrm{d}s_y - \frac{1}{2}\varphi(\boldsymbol{x}), \quad \forall \boldsymbol{x} \in \Gamma.$$

That is, the equalities (3.3.28) are correct for $\varphi \in H^{-1/2}(\Gamma)$. The operator $\mathbb{K}^*_{II}$ defined by

$$(\mathbb{K}^*_{II}\varphi)(\boldsymbol{x}) := \int_\Gamma \frac{\partial}{\partial \boldsymbol{\nu}_x} E_n(\boldsymbol{x}, \boldsymbol{y})\varphi(\boldsymbol{y})\, \mathrm{d}s_y, \quad \forall \boldsymbol{x} \in \Gamma$$

is bounded from $H^{-1/2}(\Gamma)$ to $H^{-1/2}(\Gamma)$.

4. For the case $n = 3$ *and* $n = 2$ *with* $\int_\Gamma \varphi(\boldsymbol{y})\,\mathrm{d}s_y = 0$*, it is straightforward to obtain*

$$\left|V_I\right|_{H^1(\mathbb{R}^n)} = \left(\int_{\mathbb{R}^n} \left|\nabla V_I(\boldsymbol{x})\right|^2 \mathrm{d}\boldsymbol{x}\right)^{1/2} < +\infty$$

and

$$\left\| \left.\frac{\partial V_I}{\partial \boldsymbol{\nu}_x}\right|_{\Gamma\pm 0} \right\|_{H^{-1/2}(\Gamma)} \leqslant c\left|V_I(\boldsymbol{x})\right|_{H^1(\mathbb{R}^n)}. \tag{3.3.43}$$

Similarly, the double layer potential $V_{II}(\boldsymbol{x})$ with density function $\psi \in H^{1/2}(\Gamma)$ is the unique solution to the following transmission problem:

$$\begin{cases} \Delta v(\boldsymbol{x}) = 0, \quad \text{in } \mathbb{R}^n \setminus \Gamma, & (3.3.44\text{a}) \\ \left[v(\boldsymbol{x})\right]_\Gamma = \psi(\boldsymbol{x}), \quad \forall \boldsymbol{x} \in \Gamma, & (3.3.44\text{b}) \\ \left[\dfrac{\partial v(\boldsymbol{x})}{\partial \boldsymbol{\nu}_x}\right]_\Gamma = 0, \quad \forall \boldsymbol{x} \in \Gamma, & (3.3.44\text{c}) \\ v(\boldsymbol{x}) = \mathcal{O}(|\boldsymbol{x}|^{1-n}), \quad \text{as } |\boldsymbol{x}| \to \infty. & (3.3.44\text{d}) \end{cases}$$

By Theorem 5.6.1 (Hsiao & Wendland, 2008), we have

Lemma 3.3.7 *1.* $V_{II}\Big|_\Omega \in H^1(\Omega)$*, and its traces on* Γ*,*

$$V_{II}\Big|_{\Gamma-0} \in H^{1/2}(\Gamma), \quad \left.\frac{\partial V_{II}}{\partial \boldsymbol{\nu}_x}\right|_{\Gamma-0} \in H^{-1/2}(\Gamma);$$

$V_{II}\Big|_\Omega \in H^1_{\mathrm{loc}}(\Omega_e)$*, and its traces on* Γ*,*

$$V_{II}\Big|_{\Gamma+0} \in H^{1/2}(\Gamma), \quad \left.\frac{\partial V_{II}}{\partial \boldsymbol{\nu}_x}\right|_{\Gamma+0} \in H^{-1/2}(\Gamma).$$

2. The following equalities hold in $H^{1/2}(\Gamma)$ *with*

$$V_{II}(\boldsymbol{x})\Big|_{\Gamma\pm 0} = \pm\frac{1}{2}\psi(\boldsymbol{x}) + \int_\Gamma \frac{\partial}{\partial \boldsymbol{\nu}_y} E_n(\boldsymbol{x}, \boldsymbol{y})\psi(\boldsymbol{y})\,\mathrm{d}s_y.$$

This means equalities (3.3.19)–(3.3.20) hold for $\psi \in H^{1/2}(\Gamma)$*. Furthermore, the operator* $\mathbb{K}_{II}$ *defined by*

$$(\mathbb{K}_{II}\psi)(\boldsymbol{x}) = \int_\Gamma \frac{\partial}{\partial \boldsymbol{\nu}_y} E_n(\boldsymbol{x}, \boldsymbol{y})\psi(\boldsymbol{y})\,\mathrm{d}s_y$$

is bounded from $H^{1/2}(\Gamma)$ *to* $H^{1/2}(\Gamma)$*.*

3.

$$\left.\frac{\partial V_{II}(\boldsymbol{x})}{\partial \boldsymbol{\nu}_x}\right|_{\Gamma} = \left.\frac{\partial V_{II}(\boldsymbol{x})}{\partial \boldsymbol{\nu}_x}\right|_{\Gamma\pm 0} = \sum_{1\leqslant k<j\leqslant n} \frac{\partial}{\partial \boldsymbol{\tau}_x^{k,j}} \int_{\Gamma} E_n(\boldsymbol{x},\boldsymbol{y})\left(\frac{\partial}{\partial \boldsymbol{\tau}_y^{k,j}}\psi(\boldsymbol{y})\right) \mathrm{d}s_y$$
$$= \int_{\Gamma} \frac{\partial^2}{\partial \boldsymbol{\nu}_x \partial \boldsymbol{\nu}_y} E_n(\boldsymbol{x},\boldsymbol{y})\psi(\boldsymbol{y})\, \mathrm{d}s_y,$$

for $\psi \in H^{1/2}(\Gamma)$. *Furthermore, the operator* $\mathbb{K}^*_{III}$ *defined by*

$$(\mathbb{K}^*_{III}\psi)(\boldsymbol{x}) := \sum_{1\leqslant k<j\leqslant n} \frac{\partial}{\partial \boldsymbol{\tau}_x^{k,j}} \int_{\Gamma} E(\boldsymbol{x},\boldsymbol{y})\left(\frac{\partial}{\partial \boldsymbol{\tau}_y^{k,j}}\psi(\boldsymbol{y})\right) \mathrm{d}s_y$$
$$= \int_{\Gamma} \frac{\partial^2}{\partial \boldsymbol{\nu}_x \partial \boldsymbol{\nu}_y} E_n(\boldsymbol{x},\boldsymbol{y})\psi(\boldsymbol{y})\, \mathrm{d}s_y$$

is a bounded operator from $H^{1/2}(\Gamma)$ *to* $H^{-1/2}(\Gamma)$.

We now recall more results of the operators $\mathbb{K}_{II}$ and $\mathbb{K}^*_{II}$ (Folland, 1995, pp. 121–122).

Lemma 3.3.8 *Suppose that* $\Gamma \in C^2$. *We have*
1. $\mathbb{K}_{II}$ *and* $\mathbb{K}^*_{III}$ *are compact from* $L_2(\Gamma)$ *to* $L_2(\Gamma)$.
2. For given $\varphi \in L_2(\Gamma)$ *and* $\frac{1}{2}\varphi \pm (\mathbb{K}_{II}\varphi) \in C(\Gamma)$ *or* $\frac{1}{2}\varphi \pm (\mathbb{K}^*_{II}\varphi) \in C(\Gamma)$, *then* $\varphi \in C(\Gamma)$.

3.4 Boundary Reduction

We are now ready to reduce the boundary value problems of the Laplace equation to the boundary integral (integro-differential) equations.

3.4.1 Boundary Integral (Integro-Differential) Equations of the First Kind

Let

$$H_*^{-1/2}(\Gamma) = \left\{\varphi \mid \varphi \in H^{-1/2}(\Gamma) \quad \text{and} \quad \int_{\Gamma} \varphi(\boldsymbol{x})\, \mathrm{d}s_x = 0\right\} \tag{3.4.1}$$

$$H_*^{1/2}(\Gamma) = \left\{\varphi \mid \varphi \in H^{1/2}(\Gamma) \quad \text{and} \quad \int_{\Gamma} \varphi(\boldsymbol{x})\, \mathrm{d}s_x = 0\right\}. \tag{3.4.2}$$

Obviously, $H_*^{-1/2}(\Gamma)$ is a subspace of $H^{-1/2}(\Gamma)$ and $H_*^{1/2}(\Gamma)$ is a subspace of $H^{1/2}(\Gamma)$. For any $\psi \in H^{1/2}(\Gamma)$, the tangential derivatives of ψ, $\dfrac{\partial \psi}{\partial \boldsymbol{\tau}_{\boldsymbol{x}}^{k,j}} \in H_*^{-1/2}(\Gamma)$ $(1 \leqslant k < j \leqslant n)$ and the semi-norm of ψ

$$|\psi|_{H^{1/2}(\Gamma)} = \left\{ \sum_{1 \leqslant k < j \leqslant n} \left\| \frac{\partial \psi}{\partial \boldsymbol{\tau}_{\boldsymbol{x}}^{k,j}} \right\|_{H^{-1/2}(\Gamma)} \right\} \tag{3.4.3}$$

is an equivalent norm of the space $H_*^{1/2}(\Gamma)$.

Let $V_I(\boldsymbol{x}) + C$ be the solution of the interior Dirichlet problem (3.2.1) on $\bar{\Omega}$ and the exterior Dirichlet problem (3.2.2) on $\bar{\Omega}_e$, the density function $\varphi \in H_*^{-1/2}(\Gamma)$ and constant C are unknown. By the Dirichlet boundary condition on Γ, the unknowns $(\varphi,\ C)$ satisfy:

$$\begin{cases} \text{Find } \varphi \in H_*^{-1/2}(\Gamma) \text{ and constant } C \text{ such that} \\ \displaystyle\int_\Gamma E_n(\boldsymbol{x}, \boldsymbol{y})\varphi(\boldsymbol{y})\, \mathrm{d}s_y + C = g(\boldsymbol{x}), \qquad \forall \boldsymbol{x} \in \Gamma, \text{ for give } g \in H^{1/2}(\Gamma). \end{cases} \tag{3.4.4}$$

Similarly, let $V_I(\boldsymbol{x})$ be the solution to the interior Dirichlet problem (3.2.1) on $\bar{\Omega}$ and the exterior Dirichlet problem (3.2.3), the density function $\varphi \in H^{-1/2}(\Gamma)$ is unknown. By the Dirichlet condition on Γ, we obtain that φ satisfies:

$$\begin{cases} \text{Find } \varphi \in H^{-1/2}(\Gamma) \text{ such that} \\ \displaystyle\int_\Gamma E_n(\boldsymbol{x}, \boldsymbol{y})\varphi(\boldsymbol{y})\, \mathrm{d}s_y = g(\boldsymbol{x}), \qquad \forall \boldsymbol{x} \in \Gamma. \end{cases} \tag{3.4.5}$$

Let the double layer potential $V_{II}(\boldsymbol{x})$ with density function $\psi \in H_*^{1/2}(\Gamma)$ be the solution to the interior Neumann problem (3.2.4) and exterior Neumann problem (3.2.5). By the Neumann boundary condition on Γ, we obtain the hypersingular integral equation:

$$-\int_\Gamma \frac{\partial^2}{\partial \boldsymbol{\nu}_x \partial \boldsymbol{\nu}_y} E_n(\boldsymbol{x}, \boldsymbol{y})\psi(\boldsymbol{y})\, \mathrm{d}s_y = -h(\boldsymbol{x}), \qquad \forall \boldsymbol{x} \in \Gamma, \tag{3.4.6}$$

which is equivalent to the integro-differential equation

$$-\sum_{1 \leqslant k < j \leqslant n} \frac{\partial}{\partial \boldsymbol{\tau}_{\boldsymbol{x}}^{k,j}} \int_\Gamma E_n(\boldsymbol{x}, \boldsymbol{y}) \left(\frac{\partial \psi(\boldsymbol{y})}{\partial \boldsymbol{\tau}_{\boldsymbol{y}}^{k,j}} \right) \mathrm{d}s_y = -h(\boldsymbol{x}), \qquad \forall \boldsymbol{x} \in \Gamma \tag{3.4.7}$$

with $h \in H_*^{-1/2}(\Gamma)$.

The problem (3.4.4) is equivalent to the following variational problem (Lax & Milgram, 1954; Yosida, 2012):

$$\begin{cases} \text{Find } \varphi \in H_*^{-1/2}(\Gamma) \text{ such that} \\ \int_\Gamma \int_\Gamma E_n(\boldsymbol{x}, \boldsymbol{y})\varphi(\boldsymbol{y})\mu(\boldsymbol{x})\,\mathrm{d}s_y \mathrm{d}s_x = \int_\Gamma g(\boldsymbol{x})\mu(\boldsymbol{x})\,\mathrm{d}s_x, \quad \forall\, \mu \in H_*^{-1/2}(\Gamma) \end{cases} \tag{3.4.8}$$

and

$$C = \frac{1}{\text{meas}(\Gamma)} \left\{ \int_\Gamma g(\boldsymbol{x})\,\mathrm{d}s_x - \int_\Gamma \int_\Gamma E_n(\boldsymbol{x}, \boldsymbol{y})\varphi(\boldsymbol{y})\,\mathrm{d}s_y \mathrm{d}s_x \right\}. \tag{3.4.9}$$

The problem (3.4.5) is equivalent to the variational problem:

$$\begin{cases} \text{Find } \varphi \in H^{-1/2}(\Gamma) \text{ such that} \\ \int_\Gamma \int_\Gamma E_n(\boldsymbol{x}, \boldsymbol{y})\varphi(\boldsymbol{y})\mu(\boldsymbol{x})\,\mathrm{d}s_y \mathrm{d}s_x = \int_\Gamma g(\boldsymbol{x})\mu(\boldsymbol{x})\,\mathrm{d}s_x, \quad \forall\, \mu \in H^{-1/2}(\Gamma). \end{cases} \tag{3.4.10}$$

The integro-differential equation (3.4.7) is equivalent to the variational problem:

$$\begin{cases} \text{Find } \psi \in H_*^{1/2}(\Gamma), \text{ such that,} \\ \sum\limits_{1 \leqslant k < j \leqslant n} \int_\Gamma \int_\Gamma E_n(\boldsymbol{x}, \boldsymbol{y}) \left(\dfrac{\partial \psi(\boldsymbol{y})}{\partial \boldsymbol{\tau}_y^{k,j}} \right) \left(\dfrac{\partial \rho(\boldsymbol{x})}{\partial \boldsymbol{\tau}_x^{k,j}} \right) \mathrm{d}s_y \mathrm{d}s_x \\ \quad = -\int_\Gamma h(\boldsymbol{x})\rho(\boldsymbol{x})\,\mathrm{d}s_x, \quad \forall\, \rho \in H_*^{1/2}(\Gamma). \end{cases} \tag{3.4.11}$$

In order to study the solvabilities of the variational problems above. We recall the Lax–Milgram theorem.

Let X be a Hilbert space equipped with inner product $(\cdot, \cdot)_X$ and X' is the dual space of X. Let $A(u, v)$ be a bounded bilinear form on $X \times X$, namely

$$A(u, v) : X \times X \mapsto \mathbb{R}^1.$$

Given $f(v) \in X'$, a bounded linear functional on X, consider the abstract variational problem:

$$\begin{cases} \text{Find } u \in X, \text{ such that} \\ A(u, v) = f(v), \qquad \forall\, v \in X. \end{cases} \tag{3.4.12}$$

Theorem 3.4.1 (Lax–Milgram theorem) *Suppose that*

1. The bilinear form $A(u, v)$ defined on $X \times X$ is bounded, namely there exists a constant $M > 0$ such that

$$|A(u, v)| \leqslant M \|u\|_X \|v\|_X, \qquad \forall\, u, v \in X. \tag{3.4.13}$$

2. The bilinear form $A(u, v)$ is coercive, namely there is a constant $\delta > 0$ such that

$$A(u, u) \geqslant \delta \|u\|_X^2, \qquad \forall u \in X. \tag{3.4.14}$$

Then for any given function $f(v) \in X'$, the variational problem (3.4.12) has a unique solution $u \in X$ and

$$\|u\|_X \leqslant \frac{M}{\delta} \|f\|_{X'}. \tag{3.4.15}$$

We now return to the variational problems (3.4.8), (3.4.10) and (3.4.11). Let

$$A(\varphi, \mu) := \int_\Gamma \int_\Gamma E_n(\boldsymbol{x}, \boldsymbol{y}) \varphi(\boldsymbol{y}) \mu(\boldsymbol{x}) \, \mathrm{d}s_{\boldsymbol{y}} \mathrm{d}s_{\boldsymbol{x}}, \qquad \forall \varphi, \mu \in H^{-1/2}(\Gamma), \tag{3.4.16}$$

$$B(\psi, \rho) := \sum_{1 \leqslant k < j \leqslant n} \int_\Gamma \int_\Gamma E_n(\boldsymbol{x}, \boldsymbol{y}) \left(\frac{\partial \psi(\boldsymbol{y})}{\partial \boldsymbol{\tau}_{\boldsymbol{y}}^{k,j}} \right) \left(\frac{\partial \rho(\boldsymbol{x})}{\partial \boldsymbol{\tau}_{\boldsymbol{x}}^{k,j}} \right) \mathrm{d}s_{\boldsymbol{y}} \mathrm{d}s_{\boldsymbol{x}}, \quad \forall \psi, \rho \in H_*^{1/2}(\Gamma). \tag{3.4.17}$$

For the bilinear forms $A(\varphi, \mu)$ and $B(\psi, \rho)$, we have

Lemma 3.4.1 *1. $A(\varphi, \mu)$ is bounded on $H^{-1/2}(\Gamma) \times H^{-1/2}(\Gamma)$, namely there exists a constant $M > 0$ such that*

$$|A(\varphi, \mu)| \leqslant M \|\varphi\|_{H^{-1/2}(\Gamma)} \|\mu\|_{H^{-1/2}(\Gamma)}, \qquad \forall \varphi, \mu \in H^{-1/2}(\Gamma).$$

2. $A(\varphi, \mu)$ is coercive on $H_*^{-1/2}(\Gamma) \times H_*^{-1/2}(\Gamma)$, i.e., there is a constant $\delta > 0$ such that

$$A(\varphi, \varphi) \geqslant \delta \|\varphi\|_{H^{-1/2}(\Gamma)}^2, \qquad \forall \varphi \in H_*^{-1/2}(\Gamma).$$

Furthermore for the case $n = 3$, $A(\varphi, \mu)$ is coercive on $H^{-1/2}(\Gamma)$. In other words, there is a constant $\delta > 0$ such that

$$A(\varphi, \varphi) \geqslant \delta \, \|\varphi\|_{H^{-1/2}(\Gamma)}^2, \qquad \forall \varphi \in H^{-1/2}(\Gamma).$$

3. $B(\psi, \rho)$ is bounded and coercive on $H_*^{1/2}(\Gamma) \times H_*^{1/2}(\Gamma)$. That is, there exists two constants $\delta > 0$ and $M > 0$ such that

$$|B(\psi, \rho)| \leqslant M \, \|\psi\|_{H^{1/2}(\Gamma)} \|\rho\|_{H^{1/2}(\Gamma)}, \qquad \forall \psi, \rho \in H_*^{1/2}(\Gamma),$$

$$B(\psi, \psi) \geqslant \delta \|\psi\|_{H^{1/2}(\Gamma)}^2, \qquad \forall \psi \in H_*^{1/2}(\Gamma).$$

Proof 1. Let

$$V_I^\varphi(\boldsymbol{x}) = \int_\Gamma E_n(\boldsymbol{x}, \boldsymbol{y}) \varphi(\boldsymbol{y}) \, \mathrm{d}s_{\boldsymbol{y}},$$

for $\varphi \in H^{-1/2}(\Gamma)$. We have

$$|A(\varphi, \mu)| = \left| \int_\Gamma V_I^\varphi(\boldsymbol{x})\mu(\boldsymbol{x})\,\mathrm{d}s_{\boldsymbol{x}} \right| \leqslant \|V_I^\varphi\|_{H^{1/2}(\Gamma)} \|\mu\|_{H^{-1/2}(\Gamma)}$$
$$\leqslant M\|\varphi\|_{H^{-1/2}(\Gamma)} \|\mu\|_{H^{-1/2}(\Gamma)}, \ \forall \varphi, \ \mu \in H^{-1/2}(\Gamma).$$

The last inequality follows from estimate (3.3.42).

2. For $\varphi \in H_*^{-1/2}(\Gamma)$, the single layer potential $V_I^\varphi(\boldsymbol{x})$ satisfies

$$-\Delta V_I^\varphi(\boldsymbol{x}) = 0, \qquad \text{in } \mathbb{R}^n \setminus \Gamma. \tag{3.4.18}$$

Multiplying the Eq. (3.4.18) with $V_I^\varphi(\boldsymbol{x})$ and integrating it over $\mathbb{R}^n$, after integration by parts, we get

$$\int_{\mathbb{R}^n} \left|\nabla V_I^\varphi(\boldsymbol{x})\right|^2 \mathrm{d}\boldsymbol{x} = \int_\Gamma \varphi(\boldsymbol{x}) V_I^\varphi(\boldsymbol{x})\,\mathrm{d}s_{\boldsymbol{x}}$$
$$= \int_\Gamma \int_\Gamma E_n(\boldsymbol{x}, \boldsymbol{y})\varphi(\boldsymbol{y})\varphi(\boldsymbol{x})\,\mathrm{d}s_{\boldsymbol{y}}\mathrm{d}s_{\boldsymbol{x}} = A(\varphi, \varphi).$$

On the other hand, we have

$$\|\varphi\|^2_{H^{-1/2}(\Gamma)} = \left\| \left.\frac{\partial V_I^\varphi}{\partial \boldsymbol{\nu}_{\boldsymbol{x}}}\right|_{\Gamma+0} - \left.\frac{\partial V_I^\varphi}{\partial \boldsymbol{\nu}_{\boldsymbol{x}}}\right|_{\Gamma-0} \right\|_{H^{-1/2}(\Gamma)} \leqslant C \int_{\mathbb{R}^n} \left|\nabla V_I^\varphi(\boldsymbol{x})\right|^2 \mathrm{d}\boldsymbol{x}.$$

The last inequality follows from estimate (3.3.43). Let $\delta = 1/C$. We arrive at

$$A(\varphi, \varphi) \geqslant \delta \|\varphi\|^2_{H^{1/2}(\Gamma)}, \qquad \forall \varphi \in H_*^{-1/2}(\Gamma).$$

The proof of that $A(\varphi, \mu)$ is coercive on $H^{-1/2}(\Gamma)$ with $n = 3$ is similar.

3. Noting that for any $\psi \in H^{1/2}(\Gamma)$, we have $\dfrac{\partial \psi(\boldsymbol{x})}{\partial \boldsymbol{\tau}_{\boldsymbol{x}}^{k,j}} \in H_*^{-1/2}(\Gamma)$ for $1 \leqslant k < j \leqslant n$ and the semi-norm $|\psi|_{H^{1/2}(\Gamma)}$ defined by (3.4.3) is an equivalent norm of the space $H_*^{1/2}(\Gamma)$. That is, there are two constants $\alpha > 0$ and $\beta > 0$ such that

$$\alpha\|\psi\|_{H^{1/2}(\Gamma)} \leqslant |\psi|_{H^{1/2}(\Gamma)} \leqslant \beta\|\psi\|_{H^{1/2}(\Gamma)}, \qquad \forall \psi \in H_*^{1/2}(\Gamma).$$

Then the conclusions on bilinear form $B(\psi, \rho)$ follow directly from (1) and (2). □

Applying the Lax–Milgram theorem the results in Lemma 3.4.1, it is straightforward to get

Theorem 3.4.2 *1. Given $g \in H^{1/2}(\Gamma)$, the variational problem (3.4.8) has a unique solution $\varphi \in H_*^{-1/2}(\Gamma)$ and (φ, C) with C be given by (3.4.9) is the unique solution to the integral equation (3.4.4).*

2. Given $g \in H^{1/2}(\Gamma)$ and $n = 3$, the variational problem (3.4.10) has a unique solution $\varphi \in H^{-1/2}(\Gamma)$. That is, φ is the unique solution to the integral equation (3.4.5) with $n = 3$.

3. Given $h \in H_*^{-1/2}(\Gamma)$, the variational problem (3.4.11) has a unique solution $\psi \in H_*^{1/2}(\Gamma)$ and $\psi + C$ ($C \in \mathbb{R}^1$) are the solutions to the integro-differential equation (3.4.7), which is unique up to an additive constant.

After solving the first kind integral (integro-differential) equations, we immediately obtain the solutions to the interior Dirichlet problem (3.2.1), the exterior Dirichlet problems (3.2.2) and (3.2.3) with $n = 3$ by the corresponding single layer potentials, and the solutions to the interior Neumann problem (3.2.4) and the exterior Neumann problem (3.2.5) by the corresponding double layer potential.

We now discuss the exterior Neumann problem (3.2.6). Suppose that the origin of coordinates is in Ω. Then the function $E_n(0, \boldsymbol{x})$ satisfies

$$\triangle E_n(0, \boldsymbol{x}) = 0, \qquad \forall \boldsymbol{x} \in \Omega_e,$$

$$E_n(0, \boldsymbol{x}) = \begin{cases} -\dfrac{1}{2\pi} \ln |\boldsymbol{x}| & \text{if } n = 2 \\ \dfrac{1}{(n-2)S_n(1)} |\boldsymbol{x}|^{2-n} & \text{if } n \geqslant 3 \end{cases}$$

and

$$\int_\Gamma \frac{\partial E_n(0, \boldsymbol{x})}{\partial \boldsymbol{\nu}_x} \, \mathrm{d}s_x = -1.$$

1. Suppose that the given function $h \in H^{-1/2}(\Gamma)$ and $\int_\Gamma h(\boldsymbol{x}) \, \mathrm{d}s_x = 0$, then the solution $u(\boldsymbol{x})$ to the problem (3.2.6) satisfies

$$u(\boldsymbol{x}) = \mathcal{O}(|\boldsymbol{x}|^{1-n}), \qquad \text{as } |\boldsymbol{x}| \to \infty.$$

In this case, the exterior Neumann problems (3.2.5) and (3.2.6) are the same.

2. Suppose $\int_\Gamma h(\boldsymbol{x}) \, \mathrm{d}s_x \neq 0$. Let

$$u(\boldsymbol{x}) = w(\boldsymbol{x}) + C^* E_n(0, \boldsymbol{x}), \qquad \forall \boldsymbol{x} \in \Omega_e, \tag{3.4.19}$$

with

$$C^* = -\int_\Gamma h(\boldsymbol{x}) \, \mathrm{d}s_x.$$

Then we solve the exterior Neumann problem (3.2.5),

$$\begin{cases} \triangle w(\boldsymbol{x}) = 0, & \text{in } \Omega_e, \\ \left.\dfrac{\partial w}{\partial \boldsymbol{\nu}}\right|_\Gamma = h(\boldsymbol{x}) - C^* \dfrac{\partial E_n(0, \boldsymbol{x})}{\partial \boldsymbol{\nu}} = h^*(\boldsymbol{x}), & \text{on } \Gamma, \\ w(\boldsymbol{x}) = \mathcal{O}(|\boldsymbol{x}|^{1-n}), & \text{as } |\boldsymbol{x}| \to \infty. \end{cases}$$

Note that

$$\int_\Gamma h^*(\boldsymbol{x})\,\mathrm{d}s_x = 0.$$

After solving the problem (3.2.5) with $\left.\dfrac{\partial w(\boldsymbol{x})}{\partial \boldsymbol{\nu}_x}\right|_\Gamma = h^*(\boldsymbol{x})$, the solution $u(\boldsymbol{x})$ given by (3.4.19) is the unique solution to the exterior Neumann problem (3.2.6).

The exterior Dirichlet problem (3.2.3) with $n = 2$ and the integral equation (3.4.5) with $n = 2$ will be discussed in the coming Sect. 3.4.2.

3.4.2 Solvability of First Kind Integral Equation with $n = 2$ and the Degenerate Scale

We now discuss the first kind integral equation (3.4.5) with $n = 2$, which can be rewritten as

$$(2\pi\mathbb{K}_I\varphi)(\boldsymbol{x}) := -\int_\Gamma \ln|\boldsymbol{x} - \boldsymbol{y}|\varphi(\boldsymbol{y})\,\mathrm{d}s_y = f(\boldsymbol{x}), \qquad \forall \boldsymbol{x} \in \Gamma, \tag{3.4.20}$$

with $f = 2\pi g$ be given in $H^{1/2}(\Gamma)$. The solvability of the first kind integral equation (3.4.20) has attracted the attention of many researchers since last century. Petrovsky (2012) has pointed out that when Γ is a circle with radius 1, the integral equation

$$-\int_\Gamma \ln|\boldsymbol{x} - \boldsymbol{y}|\varphi(\boldsymbol{y})\,\mathrm{d}s_y = 1, \qquad \forall \boldsymbol{x} \in \Gamma \tag{3.4.21}$$

does not have a solution. In fact, suppose that $\varphi \in H^{-1/2}(\Gamma)$ is a solution of (3.4.21). Let

$$u_1(\boldsymbol{x}) = -\int_\Gamma \ln|\boldsymbol{x} - \boldsymbol{y}|\varphi(\boldsymbol{y})\,\mathrm{d}s_y, \qquad \forall \boldsymbol{x} \in \Omega.$$

Then we have

$$\begin{cases} \Delta u_1(\boldsymbol{x}) = 0, \qquad \forall \boldsymbol{x} \in \Omega, & (3.4.22\text{a}) \\ u_1(\boldsymbol{x})\Big|_\Gamma = 1. & (3.4.22\text{b}) \end{cases}$$

By the uniqueness of the problem (3.4.22), we know that

$$u_1(\boldsymbol{x}) \equiv 1, \qquad \forall \boldsymbol{x} \in \bar{\Omega}. \tag{3.4.23}$$

But

$$u_1(\boldsymbol{x}_0) = -\int_\Gamma \ln|\boldsymbol{x}_0 - \boldsymbol{y}|\varphi(\boldsymbol{y})\,\mathrm{d}s_y = 0 \tag{3.4.24}$$

with $\boldsymbol{x}_0 \in \Omega$ be the center of the unit circle Γ. The contradiction between (3.4.23) and (3.4.24) shows the above conclusion by Petrovsky in his textbook (Petrovsky, 2012). Hsiao and Wendland (1977) proposed a restriction condition on Γ, namely,

$$\max_{\boldsymbol{x},\, \boldsymbol{y} \in \Gamma} |\boldsymbol{x} - \boldsymbol{y}| < 1, \tag{3.4.25}$$

and they proved that, under the condition (3.4.25) for any given $f \in H^{1/2}(\Gamma)$, the first kind integral equation (3.4.20) has a unique solution $\varphi \in H^{-1/2}(\Gamma)$. But the condition (3.4.25) is only a sufficient condition for the solvability of the first kind integral equation (3.4.20).

For further discussion, we introduce the degenerate scale of the integral equation (3.4.20) and a computable index γ_Γ of Γ.

Definition 3.4.1 Suppose that $\Gamma \in C^2$, which is a simple closed curve in the plane, if

1. The homogeneous integral equation

$$-\int_\Gamma \ln |\boldsymbol{x} - \boldsymbol{y}| \varphi(\boldsymbol{y})\, \mathrm{d}s_y = 0, \qquad \forall \boldsymbol{x} \in \Gamma$$

has a non-trivial solution.

2. The integral equation

$$-\int_\Gamma \ln |\boldsymbol{x} - \boldsymbol{y}| \varphi(\boldsymbol{y})\, \mathrm{d}s_y = 1, \qquad \forall \boldsymbol{x} \in \Gamma$$

does not have any solution, then the curve Γ is called a *degenerate scale* of the first kind integral equation (3.4.20).

The unit circle Γ is a degenerate scale of equation (3.4.20).

Let

$$u_1(\boldsymbol{x}) = -\int_\Gamma \ln |\boldsymbol{x} - \boldsymbol{y}|\, \mathrm{d}s_y, \qquad \forall \boldsymbol{x} \in \mathbb{R}^2$$

and

$$f_1(\boldsymbol{x}) = u_1(\boldsymbol{x})\big|_\Gamma.$$

We consider the following problem of integral equation:

$$\begin{cases} \text{Find } \varphi_\Gamma \in H_*^{-1/2}(\Gamma) \text{ and } \gamma_\Gamma \in \mathbb{R}^1, \text{ such that} \\ -\displaystyle\int_\Gamma \ln |\boldsymbol{x} - \boldsymbol{y}| \varphi_\Gamma(\boldsymbol{y})\, \mathrm{d}s_y + \gamma_\Gamma = f_1(\boldsymbol{x}), \qquad \forall \boldsymbol{x} \in \Gamma. \end{cases} \tag{3.4.26}$$

By the Theorem 3.4.2, we know that the problem (3.4.26) has a unique solution $(\varphi_\Gamma, \gamma_\Gamma) \in H_*^{-1/2}(\Gamma) \times \mathbb{R}^1$ and

$$\gamma_\Gamma = \frac{1}{\text{meas}(\Gamma)}\left\{-\int_\Gamma\int_\Gamma \ln|\boldsymbol{x}-\boldsymbol{y}|\,\mathrm{d}s_x\mathrm{d}s_y + \int_\Gamma\int_\Gamma \ln|\boldsymbol{x}-\boldsymbol{y}|\varphi_\Gamma(\boldsymbol{y})\,\mathrm{d}s_x\mathrm{d}s_y\right\}. \tag{3.4.27}$$

Definition 3.4.2 The real number γ_Γ given by (3.4.27) is called γ_Γ-index of the boundary Γ regarding the first kind integral equation (3.4.20).

Let $\mathbb{N}(2\pi\mathbb{K}_I)$ denote the null space of the operator $2\pi\mathbb{K}_I$.

Theorem 3.4.3 *Suppose that* $\Gamma \in C^2$, *we have*
1. If $\gamma_\Gamma = 0$, *then* $\dim \mathbb{N}(2\pi\mathbb{K}_I) = 1$ *and*

$$\mathbb{N}(2\pi\mathbb{K}_I) = \text{span}\{\varphi_0(\boldsymbol{x})\} \quad \textit{with} \quad \int_\Gamma \varphi_0(\boldsymbol{x})\,\mathrm{d}s_x = 1.$$

The base function $\varphi_0(\boldsymbol{x})$ *will be given in the following proof.*
2. Suppose that $\gamma_\Gamma = 0$. *For any given* $f \in H^{1/2}(\Gamma)$, *the first kind integral equation (3.4.20) has a solution if and only if*

$$\int_\Gamma f(\boldsymbol{x})\varphi_0(\boldsymbol{x})\,\mathrm{d}s_x = 0. \tag{3.4.28}$$

3. If $\gamma_\Gamma \neq 0$, *for any given* $f \in H^{1/2}(\Gamma)$, *the first kind integral equation (3.4.20) has a unique solution* $\varphi \in H^{-1/2}(\Gamma)$.

Proof 1. Since $\gamma_\Gamma = 0$, we have

$$-\int_\Gamma \ln|\boldsymbol{x}-\boldsymbol{y}|\varphi_\Gamma(\boldsymbol{y})\,\mathrm{d}s_y + \int_\Gamma \ln|\boldsymbol{x}-\boldsymbol{y}|\,\mathrm{d}s_y = 0$$

with $\varphi_\Gamma \in H_*^{-1/2}(\Gamma)$, i.e.,

$$-\int_\Gamma \ln|\boldsymbol{x}-\boldsymbol{y}|\left[\frac{1}{\text{meas}(\Gamma)}\Big(\varphi_\Gamma(\boldsymbol{y})-1\Big)\right]\mathrm{d}s_y = 0.$$

Let

$$\varphi_0(\boldsymbol{x}) = \frac{1}{\text{meas}(\Gamma)}\big(1-\varphi_\Gamma(\boldsymbol{x})\big).$$

We have

$$\int_\Gamma \ln|\boldsymbol{x}-\boldsymbol{y}|\varphi_0(\boldsymbol{y})\,\mathrm{d}s_y = 0, \quad \text{with} \quad \int_\Gamma \varphi_0(\boldsymbol{x})\,\mathrm{d}s_x = 1.$$

By the uniqueness of integral equation (3.4.4), we get $\dim \mathbb{N}(2\pi\mathbb{K}_I) = 1$ and

$$\mathbb{N}(2\pi\mathbb{K}_I) = \text{span}\{\varphi_0(\boldsymbol{x})\}. \tag{3.4.29}$$

2. Suppose $\gamma_{\Gamma} = 0$ and φ_f is a solution to (3.4.20) for given $f \in H^{1/2}(\Gamma)$, i.e., $\varphi_f(\boldsymbol{x})$ satisfying

$$-\int_{\Gamma} \ln|\boldsymbol{x}-\boldsymbol{y}|\varphi_f(\boldsymbol{y})\,\mathrm{d}s_y = f(\boldsymbol{x}), \qquad \forall \boldsymbol{x} \in \Gamma. \tag{3.4.30}$$

Multiplying the Eq. (3.4.30) with $\varphi_0(\boldsymbol{x})$ and integrating on Γ yields

$$\int_{\Gamma} f(\boldsymbol{x})\varphi_0(\boldsymbol{x})\,\mathrm{d}s_x = -\int_{\Gamma}\int_{\Gamma} \ln|\boldsymbol{x}-\boldsymbol{y}|\varphi_0(\boldsymbol{x})\varphi_f(\boldsymbol{y})\,\mathrm{d}s_y\mathrm{d}s_x = 0.$$

The last equality follows from $\varphi_0 \in \mathbb{N}(2\pi\mathbb{K}_I)$. On the other hand, suppose $f \in H^{1/2}(\Gamma)$ and $\int_{\Gamma} f(\boldsymbol{x})\varphi_0(\boldsymbol{x})\,\mathrm{d}s_x = 0$. There exists a unique $(\varphi_f, C_f) \in H_*^{-1/2}(\Gamma) \times \mathbb{R}^1$ such that

$$-\int_{\Gamma} \ln|\boldsymbol{x}-\boldsymbol{y}|\varphi_f(\boldsymbol{y})\,\mathrm{d}s_y + C_f = f(\boldsymbol{x}), \qquad \forall \boldsymbol{x} \in \Gamma. \tag{3.4.31}$$

Similarly, multiplying the Eq. (3.4.31) with $\varphi_0(\boldsymbol{x})$ and integrating over Γ yields

$$C_f = \int_{\Gamma} f(\boldsymbol{x})\varphi_0(\boldsymbol{x})\,\mathrm{d}s_x + \int_{\Gamma}\int_{\Gamma} \ln|\boldsymbol{x}-\boldsymbol{y}|\varphi_0(\boldsymbol{x})\varphi_f(\boldsymbol{y})\,\mathrm{d}s_y\mathrm{d}s_x = 0.$$

This implies that $\varphi_f \in H_*^{-1/2}(\Gamma)$ satisfies

$$-\int_{\Gamma} \ln|\boldsymbol{x}-\boldsymbol{y}|\varphi_f(\boldsymbol{y})\,\mathrm{d}s_y = f(\boldsymbol{x}), \qquad \forall \boldsymbol{x} \in \Gamma$$

and the general solution to the first kind integral equation (3.4.20) with given $f \in H^{1/2}(\Gamma)$ and $\int_{\Gamma} f(\boldsymbol{x})\varphi_0(\boldsymbol{x})\,\mathrm{d}s_x = 0$ takes the form

$$\varphi(\boldsymbol{x}) = \varphi_f(\boldsymbol{x}) + C\varphi_0(\boldsymbol{x}), \qquad \forall C \in \mathbb{R}^1. \tag{3.4.32}$$

3. Suppose that $\gamma_{\Gamma} \neq 0$, i.e.

$$\gamma_{\Gamma} = \int_{\Gamma} \ln|\boldsymbol{x}-\boldsymbol{y}|\big(\varphi_{\Gamma}(\boldsymbol{y}) - 1\big)\,\mathrm{d}s_y \neq 0, \qquad \forall \boldsymbol{x} \in \Gamma. \tag{3.4.33}$$

For any given $f \in H^{1/2}(\Gamma)$, there exists unique $(\varphi_f, C_f) \in H_*^{-1/2}(\Gamma) \times \mathbb{R}^1$ such that

$$-\int_{\Gamma} \ln|\boldsymbol{x}-\boldsymbol{y}|\varphi_f(\boldsymbol{y})\,\mathrm{d}s_y + C_f = f(\boldsymbol{x}), \qquad \forall \boldsymbol{x} \in \Gamma. \tag{3.4.34}$$

By (3.4.33), we get

$$C_f = \frac{C_f}{\gamma_\Gamma} \int_\Gamma \ln|\boldsymbol{x} - \boldsymbol{y}|\big(\varphi_\Gamma(\boldsymbol{y}) - 1\big)\,\mathrm{d}s_y, \qquad \forall \boldsymbol{x} \in \Gamma. \tag{3.4.35}$$

Substituting (3.4.35) into (3.4.34) leads to

$$-\int_\Gamma \ln|\boldsymbol{x} - \boldsymbol{y}|\varphi(\boldsymbol{y})\,\mathrm{d}s_y = f(\boldsymbol{x}), \qquad \forall \boldsymbol{x} \in \Gamma$$

with

$$\varphi(\boldsymbol{x}) = \varphi_f(\boldsymbol{x}) + \frac{C_f}{\gamma_\Gamma}(1 - \varphi_\Gamma(\boldsymbol{x})). \tag{3.4.36}$$

Clearly, $\varphi \in H^{-1/2}(\Gamma)$ given in (3.4.36) is a solution to the first kind integral equation (3.4.20). The uniqueness of the first kind integral equation (3.4.20) follows from the fact $\gamma_\Gamma \neq 0$. □

We now return to the exterior Dirichlet problem (3.2.3) with $n = 2$. We consider the following auxiliary exterior Dirichlet problem

$$\begin{cases} -\Delta u = 0, \quad \text{in } \Omega_e, & (3.4.37\text{a}) \\ u\big|_\Gamma = \alpha, & (3.4.37\text{b}) \\ u(\boldsymbol{x}) = \mathcal{O}(|\boldsymbol{x}|^{1-n}), \quad \text{as } |\boldsymbol{x}| \to \infty, & (3.4.37\text{c}) \end{cases}$$

with constant $\alpha \in \mathbb{R}^1$. The existence and uniqueness of the solution to the exterior Dirichlet problem (3.2.2) yield

Lemma 3.4.2 *The problem (3.4.37) has a solution if and only if $\alpha = 0$, and $u(\boldsymbol{x}) = 0$ is the unique solution to the auxiliary problem (3.4.37) with $\alpha = 0$.*

Combining the results in Theorem 3.4.3 and Lemma 3.4.2, we get

Theorem 3.4.4 *1. If $\gamma_\Gamma \neq 0$, then for any given $f \in H^{1/2}(\Gamma)$ the exterior Dirichlet problem (3.2.3) with $n = 2$ has a unique solution.*

2. If $\gamma_\Gamma = 0$, then for given $f \in H^{1/2}(\Gamma)$, the exterior Dirichlet problem (3.2.3) with $n = 2$ has a solution if and only if

$$\int_\Gamma f(\boldsymbol{x})\varphi_0(\boldsymbol{x})\,\mathrm{d}s_x = 0,$$

with $\varphi_0 \in \mathbb{N}(2\pi\mathbb{K}_I)$.

Corollary 3.4.1 *The boundary Γ is a generate scale of the first kind integral equation (3.4.20) if and only if the computable index $\gamma_\Gamma = 0$.*

Example 3.4.1 Suppose that Γ is a circle with radius $R > 0$ and center $\boldsymbol{x}^0 = (x_1^0, x_2^0)$, i.e., $\Gamma = \{\boldsymbol{x} \mid |\boldsymbol{x} - \boldsymbol{x}^0| = R\}$. By the mean value theorem of harmonic functions on the domain $\bar{\Omega} = \{\boldsymbol{x} \mid |\boldsymbol{x} - \boldsymbol{x}^0| \leqslant R\}$, we get

$$\int_\Gamma \int_\Gamma \ln |x - y| \, ds_y ds_x = \text{meas}(\Gamma) \int_\Gamma \ln |x^0 - y| \, ds_y = \left(\text{meas}(\Gamma)\right)^2 \ln R.$$

On the other hand, we have

$$\int_\Gamma \int_\Gamma \ln |x - y| \varphi_\Gamma(x) \, ds_y ds_x = \text{meas}(\Gamma) \int_\Gamma \ln |x^0 - y| \varphi_\Gamma(x) \, ds_y = 0,$$

function $\varphi_\Gamma(x)$ is given by (3.4.26). By the definition of γ_Γ, we obtain

$$\gamma_\Gamma = -2\pi R \ln R.$$

Obviously, when $R = 1$, $\gamma_\Gamma = 0$. That is, the unit circle is a degenerate scale of the equation (3.4.20) given by Petrovsky. There are many works on the degenerate scale problems; for more results on them, see Yan and Sloan (1988), Chen et al. (2005), and Han et al. (2015).

3.4.3 Boundary Integral Equations of the Second Kind

Let the double layer potential $V_{II}(x)$ with density function $\psi \in C(\Gamma)$ be the solution of interior Dirichlet problem (3.2.1) and the exterior Dirichlet problem (3.2.2) for the given function $g \in C(\Gamma)$. By the boundary condition in the interior Dirichlet problem (3.2.1), it yields the second kind Fredholm integral equation to determine the density function $\psi(x)$

$$-\frac{1}{2}\psi(x) + (\mathbb{K}_{II}\psi)(x) = g(x), \qquad \forall x \in \Gamma. \tag{3.4.38}$$

Corresponding to the boundary condition in the exterior Dirichlet problem (3.2.2), we get

$$\frac{1}{2}\psi(x) + (\mathbb{K}_{II}\psi)(x) = g(x), \qquad \forall x \in \Gamma. \tag{3.4.39}$$

Similarly, let the single layer potential $V_I(x)$ with density function $\varphi \in C(\Gamma)$ be the solution to the interior Neumann problem (3.2.4) and the solution to the exterior Neumann problem (3.2.5). The Neumann boundary conditions on Γ for above interior and exterior Neumann problems yield

$$\frac{1}{2}\varphi(x) + (\mathbb{K}_{II}^*\varphi)(x) = h(x), \qquad \forall x \in \Gamma, \tag{3.4.40}$$

$$-\frac{1}{2}\varphi(x) + (\mathbb{K}_{II}^*\varphi)(x) = h(x), \qquad \forall x \in \Gamma. \tag{3.4.41}$$

For studying the solvability of the second kind Fredholm integral equations (3.4.38)–(3.4.41), we recall the Fredholm alternative theorem (Hsiao & Wendland, 2008).

Let X be a Hilbert space with inner product $(\cdot,\cdot)_X$ and $\mathbb{K}$ is a compact linear operator on the Hilbert space X into itself. We consider the second kind operator equation

$$Tu := (I - \mathbb{K})u = g, \qquad \text{in } X, \tag{3.4.42}$$

with given $g \in X$.

Let $\mathbb{K}^* : X \to X$ denote the adjoint operator of $\mathbb{K}$, which is defined by

$$(\mathbb{K}u, v) = (u, \mathbb{K}^* v)_X, \qquad \forall u, v \in X.$$

$\mathbb{K}^*$ is also a compact linear operator on the Hilbert space X. We introduce the adjoint equation

$$T^* v = (I - \mathbb{K}^*)v = g^*, \qquad \text{in } X, \tag{3.4.43}$$

with given $g^* \in X$. Let $\mathbb{R}(T)$ denote the range of T and $\mathbb{N}(T) = \ker(T)$ denote the null space of T.

Theorem 3.4.5 (Fredholm's Alternative Theorem) *For the operator equations (3.4.42) and (3.4.43), the following alternatives holds,*

Either

1. $\mathbb{N}(T) = \mathbb{N}(T^*) = 0, \quad \mathbb{R}(T) = \mathbb{R}(T^*) = X;$

Or

2. $0 < \dim \mathbb{N}(T) = \dim \mathbb{N}(T^*) < \infty$, *in the case the operator equations (3.4.42) and (3.4.43) have solutions if and only if the corresponding right hand sides satisfy the finitely many orthogonality conditions*

$$(g, v)_X = 0, \qquad \forall v \in \mathbb{N}(T^*), \tag{3.4.44}$$

and

$$(g^*, w)_X = 0, \qquad \forall w \in \mathbb{N}(T), \tag{3.4.45}$$

respectively. If condition (3.4.44) holds, then the general solution to the Eq. (3.4.42) is of the following form

$$u = u^* + \sum_{\ell=1}^{m} c_\ell w_\ell^0,$$

where u^* *is a particular solution of (3.4.42) depending continuously on* g*:*

$$\|u^*\|_H \leqslant c\|g\|_X,$$

$\{w_\ell^0,\ \ell = 1, 2, \ldots, m\}$ *are the linearly independent eigensolutions with* $\mathbb{N}(T) = \operatorname{span}\{w_\ell^0,\ \ell = 1, 2, \ldots, m\}$. *Similarly, if (3.4.45) holds, the general solution to the equation (3.4.43) takes the form*

$$v = v^* + \sum_{\ell=1}^{m} c_\ell v_\ell^0,$$

where v^* is a particular solution to the equation (3.4.43) depending continuously on g^*, and $\{ v_\ell^0, \ \ell = 1, 2, \ldots, m \}$ are the linearly independent eigensolutions with $\mathbb{N}(T^*) = \text{span}\{v_\ell^0, \ \ell = 1, 2, \ldots, m\}$.

Now we return to the second kind Fredholm integral equations (3.4.38)–(3.4.41), which can be rewritten as

$$T_-\psi := \psi(\boldsymbol{x}) - 2(\mathbb{K}_{II}\psi)(\boldsymbol{x}) = -2g(\boldsymbol{x}), \qquad \forall\, \boldsymbol{x} \in \Gamma, \tag{3.4.46}$$

$$T_+\psi := \psi(\boldsymbol{x}) + 2(\mathbb{K}_{II}\psi)(\boldsymbol{x}) = 2g(\boldsymbol{x}), \qquad \forall\, \boldsymbol{x} \in \Gamma, \tag{3.4.47}$$

$$T_+^*\varphi := \varphi(\boldsymbol{x}) + 2(\mathbb{K}_{II}^*\varphi)(\boldsymbol{x}) = 2h(\boldsymbol{x}), \qquad \forall\, \boldsymbol{x} \in \Gamma, \tag{3.4.48}$$

$$T_-^*\varphi := \varphi(\boldsymbol{x}) - 2(\mathbb{K}_{II}^*\varphi)(\boldsymbol{x}) = -2h(\boldsymbol{x}), \qquad \forall\, \boldsymbol{x} \in \Gamma. \tag{3.4.49}$$

The operator $2\mathbb{K}_{II}^*$ is the adjoint operator of $2\mathbb{K}_{II}$ on the Hilbert space $L_2(\Gamma)$. Both operators $2\mathbb{K}_{II}$ and $2\mathbb{K}_{II}^*$ are compact (Lemma 3.3.8).

We first discuss the solvability of the Eqs. (3.4.46)–(3.4.49) in the Hilbert space $L_2(\Gamma)$.

Lemma 3.4.3 *For the operators T_-, T_+, T_+^* and T_-^*, we have*

1. $\mathbb{N}(T_-) = \mathbb{N}(T_-^*) = 0$ *and* $\mathbb{R}(T) = \mathbb{R}(T_-^*) = L_2(\Gamma)$*;*

2. $\dim \mathbb{N}(T_+) = \dim \mathbb{N}(T_+^*) = 1$ *and* $\mathbb{N}(T_+) = \text{span}\{1\}$ *and* $\mathbb{N}(T_+^*) = \text{span}\{\varphi_0(\boldsymbol{x})\}$, *where the base function $\varphi_0(\boldsymbol{x})$ will be given in the following proof.*

Proof 1. We only need to prove that $\mathbb{N}(T_-) = 0$. Suppose that $\psi_0 \in \mathbb{N}(T_-)$. Then we have $\psi_0 \in L_2(\Gamma)$ and $\psi_0(\boldsymbol{x}) - 2(\mathbb{K}_{II}\psi_0)(\boldsymbol{x}) = 0$. By Lemma 3.3.8, we know that $\psi_0 \in C(\Gamma)$. Let

$$V_{II}^0(\boldsymbol{x}) = \int_\Gamma \frac{\partial}{\partial \boldsymbol{\nu}_y} E_n(\boldsymbol{x}, \boldsymbol{y})\psi_0(\boldsymbol{y})\mathrm{d}s_y,$$

then on the domain Ω, $V_{II}^0(\boldsymbol{x})$ satisfies:

$$\begin{cases} -\triangle V_{II}^0(\boldsymbol{x}) = 0, & \text{in } \Omega, \\ V_{II}^0(\boldsymbol{x})\Big|_{\Gamma-0} = 0. \end{cases}$$

This implies that $V_{II}^0(\boldsymbol{x}) \equiv 0, \forall\, \boldsymbol{x} \in \bar{\Omega}$; in particular,

$$\frac{\partial V_{II}^0(\boldsymbol{x})}{\partial \boldsymbol{\nu}}\Bigg|_{\Gamma-0} = 0.$$

By continuity of the normal derivative of $V_{II}^0(\boldsymbol{x})$ on Γ, we get

$$\left.\frac{\partial V_{II}^0(\boldsymbol{x})}{\partial \boldsymbol{\nu}}\right|_{\Gamma+0} = 0.$$

We see the restriction of $V_{II}^0(\boldsymbol{x})$ on Ω_e is a solution to the following exterior Neumann problem,

$$\begin{cases} -\triangle V_{II}^0(\boldsymbol{x}) = 0, & \text{in } \Omega_e, \\ \left.\dfrac{\partial V_{II}^0(\boldsymbol{x})}{\partial \boldsymbol{\nu}}\right|_{\Gamma+0} = 0, & \\ V_{II}^0(\boldsymbol{x}) = \mathcal{O}(|\boldsymbol{x}|^{1-n}), & \text{as } |\boldsymbol{x}| \to \infty. \end{cases}$$

Therefore, $V_{II}^0(\boldsymbol{x}) \equiv 0$, $\forall \boldsymbol{x} \in \bar{\Omega}_e$ and

$$\psi_0(\boldsymbol{x}) = \left[V_{II}^0(\boldsymbol{x})\right]_\Gamma = 0.$$

This indicates that $\mathbb{N}(T_-) = 0$. By Fredholm's Alternative Theorem, the conclusions in (1) are proved completely.

2. Suppose that $\psi_0 \in \mathbb{N}(T_+)$. We have $\psi_0 \in L_2(\Gamma)$ and $\psi_0 \in C(\Gamma)$. Let

$$V_{II}^0(\boldsymbol{x}) = \int_\Gamma \frac{\partial}{\partial \boldsymbol{\nu}_y} E_n(\boldsymbol{x}, \boldsymbol{y}) \psi_0(\boldsymbol{y}) \, \mathrm{d}s_{\boldsymbol{y}}.$$

Then we have

$$\begin{cases} -\triangle V_{II}^0(\boldsymbol{x}) = 0, & \text{in } \Omega_e, \\ V_{II}^0\big|_{\Gamma+0} = 0, & \\ V_{II}^0(\boldsymbol{x}) = \mathcal{O}(|\boldsymbol{x}|^{1-n}), & \text{as } |\boldsymbol{x}| \to \infty. \end{cases}$$

By the uniqueness of the solution to the exterior Dirichlet problem (3.2.2), we know that

$$\begin{cases} V_{II}^0(\boldsymbol{x}) \equiv 0, & \forall \boldsymbol{x} \in \bar{\Omega}_e, \\ \left.\dfrac{\partial V_{II}^0(\boldsymbol{x})}{\partial \boldsymbol{\nu}_x}\right|_{\Gamma+0} = 0, & \\ \left.\dfrac{\partial V_{II}^0(\boldsymbol{x})}{\partial \boldsymbol{\nu}_x}\right|_{\Gamma-0} = 0. & \end{cases}$$

On the domain Ω, the solution $V_{II}^0(\boldsymbol{x})$ satisfies the interior Neumann problem

$$\begin{cases} -\triangle V_{II}^0(\boldsymbol{x}) = 0, & \text{in } \Omega, \\ \left.\dfrac{\partial V_{II}^0(\boldsymbol{x})}{\partial \boldsymbol{\nu}_x}\right|_{\Gamma-0} = 0. & \end{cases}$$

We know $V_{II}^0(\boldsymbol{x}) \equiv -C$ is the general solution to this interior Neumann problem. By jump condition of $V_{II}^0(\boldsymbol{x})$ on Γ, we get

$$\left[V_{II}^0(\boldsymbol{x})\right]_\Gamma = V_{II}^0(\boldsymbol{x})\bigg|_{\Gamma+0} - V_{II}^0(\boldsymbol{x})\bigg|_{\Gamma-0} = C = \psi_0(\boldsymbol{x}).$$

We arrive at

$$\dim \mathbb{N}(T_+) = 1 \quad \text{and} \quad \mathbb{N}(T_+) = \text{span}\{1\}.$$

By Fredholm's Alternative Theorem, we get

$$\dim \mathbb{N}(T_+^*) = 1.$$

We only need to find a base function φ_0 of $\mathbb{N}(T_+^*)$. Suppose that $\varphi_0 \in \mathbb{N}(T_+^*)$. We have $\varphi_0 \in C(\Gamma)$. Let

$$V_I^0(\boldsymbol{x}) = \int_\Gamma E_n(\boldsymbol{x}, \boldsymbol{y})\varphi_0(\boldsymbol{y})\, \mathrm{d}s_{\boldsymbol{y}},$$

and

$$\begin{aligned} \frac{\partial V_I^0}{\partial \boldsymbol{\nu}}\bigg|_{\Gamma-0} &= \frac{1}{2}\varphi_0(\boldsymbol{x}) + (\mathbb{K}_{II}^*\varphi_0)(\boldsymbol{x}) = 0, \\ \triangle V_I^0(\boldsymbol{x}) &= 0, \qquad \forall \boldsymbol{x} \in \mathbb{R}^n \setminus \Gamma, \end{aligned}$$

we get $V_I^0(\boldsymbol{x}) \equiv C$ in $\bar{\Omega}$ and $V_I^0(\boldsymbol{x})\big|_\Gamma = C$. Taking $C = 1$ yields

$$1 = \int_\Gamma E_n(\boldsymbol{x}, \boldsymbol{y})\varphi_0(\boldsymbol{y})\, \mathrm{d}s_{\boldsymbol{y}}. \tag{3.4.50}$$

For the case $n = 3$ and $n = 2$, $\gamma_\Gamma \neq 0$, the base function $\varphi_0(\boldsymbol{x})$ is given by the unique solution to the integral equation (3.4.50). Furthermore, we have

$$\int_\Gamma \varphi_0(\boldsymbol{y})\, \mathrm{d}s_{\boldsymbol{y}} \neq 0 \tag{3.4.51}$$

and

$$1 = \int_\Gamma E_n(0, \boldsymbol{y})\, \varphi_0(\boldsymbol{y})\, \mathrm{d}s_{\boldsymbol{y}}, \tag{3.4.52}$$

because the origin of coordinates is in Ω. For the rest case, $n = 2$ and $\gamma_\Gamma = 0$, we take $C = 0$. This yields

$$0 = \int_\Gamma E_n(\boldsymbol{x}, \boldsymbol{y})\varphi_0(\boldsymbol{y})\, \mathrm{d}s_{\boldsymbol{y}}. \tag{3.4.53}$$

The base function $\varphi_0(\boldsymbol{x})$ is given by the non-trivial solution with

$$\int_\Gamma \varphi_0(\boldsymbol{x})\,\mathrm{d}s_x = 1. \tag{3.4.54}$$

Therefore the null space $\mathbb{N}(T_+^*) = \mathrm{span}\{\varphi_0(\boldsymbol{x})\}$. The proof of the lemma is completed.

□

The Fredholm Alternative Theorem, Lemmas 3.4.3 and 3.3.8 yield.

Theorem 3.4.6 *1. Given $g \in C(\Gamma)$, the integral equation (3.4.38) has a unique solution $\psi \in C(\Gamma)$ and*

$$u(\boldsymbol{x}) = \int_\Gamma \frac{\partial E_n(\boldsymbol{x}, \boldsymbol{y})}{\partial \boldsymbol{\nu}_y}\psi(\boldsymbol{y})\,\mathrm{d}s_y, \qquad \forall \boldsymbol{x} \in \Omega$$

is the unique solution to the interior Dirichlet problem (3.2.1).

2. Given $g \in C(\Gamma)$, the integral equation (3.4.39) has a solution if and only if $\int_\Gamma g(\boldsymbol{x})\varphi_0(\boldsymbol{x})\,\mathrm{d}s_x = 0$ with the base function φ_0 of the null space $\mathbb{N}(T_+^*)$. Suppose that

$$\int_\Gamma g(\boldsymbol{x})\varphi_0(\boldsymbol{x})\,\mathrm{d}s_x = 0. \tag{3.4.55}$$

The general solution to the integral equation (3.4.39) is given by

$$\psi(\boldsymbol{x}) = \psi^*(\boldsymbol{x}) + C,$$

where C is an arbitrary constant and $\psi^*(\boldsymbol{x})$ is a special solution to (3.4.39). On the domain Ω_e,

$$u(\boldsymbol{x}) = \int_\Gamma \frac{\partial E_n(\boldsymbol{x}, \boldsymbol{y})}{\partial \boldsymbol{\nu}_y}\psi(\boldsymbol{y})\,\mathrm{d}s_y = \int_\Gamma \frac{\partial E_n(\boldsymbol{x}, \boldsymbol{y})}{\partial \boldsymbol{\nu}_y}\psi^*(\boldsymbol{y})\,\mathrm{d}s_y, \qquad \forall \boldsymbol{x} \in \Omega_e,$$

which is the unique solution to the exterior Dirichlet problem (3.2.2).

3. Given $h \in C(\Gamma)$, the integral equation (3.4.40) has a solution if and only if $\int_\Gamma h(\boldsymbol{x})\,\mathrm{d}s_x = 0$. Assume that

$$\int_\Gamma h(\boldsymbol{x})\,\mathrm{d}s_x = 0. \tag{3.4.56}$$

The general solution to the integral equation (3.4.40) is given by

$$\varphi(\boldsymbol{x}) = \varphi^*(\boldsymbol{x}) + C\varphi_0(\boldsymbol{x}).$$

Here, $\varphi^*(\boldsymbol{x})$ is a special solution to (3.4.40), $\varphi_0(\boldsymbol{x})$ is the base function of the null space $\mathbb{N}(T_+^*)$ and C is an arbitrary constant. On the domain Ω_e,

$$u(\boldsymbol{x}) = \int_\Gamma E_n(\boldsymbol{x}, \boldsymbol{y})\varphi(\boldsymbol{y})\, ds_y$$
$$= \int_\Gamma E_n(\boldsymbol{x}, \boldsymbol{y})\varphi^*(\boldsymbol{y})\, ds_y + \begin{Bmatrix} 0, & n = 2, \gamma_\Gamma = 0 \\ C, & \text{otherwise} \end{Bmatrix}, \qquad \forall \boldsymbol{x} \in \Omega, \tag{3.4.57}$$

which is a solution to the interior Neumann problem (3.2.4).

4. Given $h \in C(\Gamma)$, the integral equation (3.4.4) has a unique solution $\varphi \in C(\Gamma)$. On the domain Ω_e,

$$u(\boldsymbol{x}) = \int_\Gamma E_n(\boldsymbol{x}, \boldsymbol{y})\varphi(\boldsymbol{y})\, ds_y, \qquad \forall \boldsymbol{x} \in \Omega_e \tag{3.4.58}$$

is the unique solution to the exterior Neumann problem (3.2.6).

Remark 3.4.1 In the case $g \in C(\Gamma)$ and $\int_\Gamma g(\boldsymbol{x})\varphi_0(\boldsymbol{x})\, ds_x \neq 0$, the integral equation (3.4.39) does not have a solution. But this does not mean that the exterior Dirichlet problems (3.2.2) and (3.2.3) do not have solutions. It only means that in this case $\int_\Gamma g(\boldsymbol{x})\varphi_0(\boldsymbol{x})\, ds_x \neq 0$, the solution to the exterior Dirichlet problems can not be represented by a double layer potential $V_{II}(\boldsymbol{x})$ with density function $\psi(\boldsymbol{x})$.

1. The solution to the exterior Dirichlet problem (3.2.2) at the case $\int_\Gamma g(\boldsymbol{x})\varphi_0(\boldsymbol{x})\, ds_x \neq 0$. Let

$$g^*(\boldsymbol{x}) = g(\boldsymbol{x}) - C^* \tag{3.4.59}$$

with

$$C^* = \int_\Gamma g(\boldsymbol{x})\varphi_0(\boldsymbol{x})\, ds_x \Big/ \int_\Gamma \varphi_0(\boldsymbol{x})\, ds_x. \tag{3.4.60}$$

It is obvious that

$$\int_\Gamma g^*(\boldsymbol{x})\varphi_0(\boldsymbol{x})\, ds_x = 0. \tag{3.4.61}$$

Let

$$u(\boldsymbol{x}) = \int_\Gamma \frac{\partial E_n(\boldsymbol{x}, \boldsymbol{y})}{\partial \boldsymbol{\nu}_y}\psi(\boldsymbol{y})\, ds_y + C^*, \qquad \forall \boldsymbol{x} \in \Omega_e \tag{3.4.62}$$

be the solution to the exterior Dirichlet problem (3.2.2). By the Dirichlet boundary condition on Γ, we get

$$\frac{1}{2}\psi(\boldsymbol{x}) + (\mathbb{K}_{II}\psi)(\boldsymbol{x}) = g^*(\boldsymbol{x}), \qquad \forall \boldsymbol{x} \in \Gamma. \tag{3.4.63}$$

This integral equation has solution since $g^*(\boldsymbol{x})$ satisfies the condition (3.4.61). The general solution to the integral equation (3.4.63) is given by

$$\psi(\boldsymbol{x}) = \psi^*(\boldsymbol{x}) + C, \tag{3.4.64}$$

where $\psi^*(\boldsymbol{x})$ is a special solution to (3.4.63) and C is an arbitrary constant. Substituting (3.4.64) into (3.4.62) yields the unique solution to the exterior Dirichlet problem (3.2.2), namely,

$$u(\boldsymbol{x}) = \int_\Gamma \frac{\partial E_n(\boldsymbol{x}, \boldsymbol{y})}{\partial \boldsymbol{\nu}_y} \psi^*(\boldsymbol{y})\, \mathrm{d}s_y + C^*. \tag{3.4.65}$$

2. The solution to the exterior Dirichlet problem (3.2.3) in the case $\int_\Gamma g(\boldsymbol{x})\varphi_0(\boldsymbol{x})\mathrm{d}s_x \neq 0$. Let

$$g(\boldsymbol{x}) = g^*(\boldsymbol{x}) + C^* E_n(0, \boldsymbol{x}), \qquad \forall \boldsymbol{x} \in \Gamma. \tag{3.4.66}$$

Multiplying (3.4.66) with $\varphi_0(\boldsymbol{x})$ and integrating on Γ yields

$$\int_\Gamma g(\boldsymbol{x})\varphi_0(\boldsymbol{x})\, \mathrm{d}s_x = \int_\Gamma g^*(\boldsymbol{x})\varphi_0(\boldsymbol{x})\, \mathrm{d}s_x + C^* \int_\Gamma E_n(0, \boldsymbol{x})\varphi_0(\boldsymbol{x})\, \mathrm{d}s_x.$$

Note that the origin of coordinates is in Ω and

$$\int_\Gamma E_n(0, \boldsymbol{x})\varphi_0(\boldsymbol{x})\, \mathrm{d}s_x = \begin{cases} 0, & n = 2,\ \gamma_\Gamma = 0; \\ 1, & \text{otherwise}. \end{cases} \tag{3.4.67}$$

In the case $n = 3$ and $n = 2,\ \gamma_\Gamma \neq 0$, taking

$$C^* = \int_\Gamma g(\boldsymbol{x})\varphi_0(\boldsymbol{x})\, \mathrm{d}s_x,$$

we get

$$\int_\Gamma g^*(\boldsymbol{x})\varphi_0(\boldsymbol{x})\, \mathrm{d}s_x = 0.$$

Let

$$u(\boldsymbol{x}) = \int_\Gamma \frac{\partial E_n(\boldsymbol{x}, \boldsymbol{y})}{\partial \boldsymbol{\nu}_y} \psi(\boldsymbol{y})\, \mathrm{d}s_y + C^* E_n(0, \boldsymbol{x}), \qquad \forall \boldsymbol{x} \in \Omega_e \tag{3.4.68}$$

be the solution to the exterior Dirichlet problem (3.2.3). The density function ψ satisfies the integral equation below,

$$\frac{1}{2}\psi(\boldsymbol{x}) + (\mathbb{K}_{II}\psi)(\boldsymbol{x}) = g^*(\boldsymbol{x}), \qquad \forall \boldsymbol{x} \in \Gamma, \tag{3.4.69}$$

with

$$g^*(\boldsymbol{x}) = g(\boldsymbol{x}) - C^* E_n(0, \boldsymbol{x}).$$

The general solution to the integral equation (3.4.69) is given by

$$\psi(\boldsymbol{x}) = \psi^*(\boldsymbol{x}) + C, \tag{3.4.70}$$

where $\psi^*(\boldsymbol{x})$ is a special solution to the integral equation (3.4.69) and C is a constant. Substituting (3.4.70) into (3.4.68) leads to the unique solution to the exterior Dirichlet problem (3.2.3),

$$u(\boldsymbol{x}) = \int_\Gamma \frac{\partial E_n(\boldsymbol{x}, \boldsymbol{y})}{\partial \boldsymbol{\nu}_y} \psi^*(\boldsymbol{y})\, \mathrm{d}s_y + C^* E_n(0, \boldsymbol{x}), \tag{3.4.71}$$

in the case $n = 3$ and $n = 2$, $\gamma_\Gamma \neq 0$.

In the case $n = 2$, $\gamma_\Gamma = 0$ and $\int_\Gamma g(\boldsymbol{x})\varphi_0(\boldsymbol{x})\mathrm{d}s_x \neq 0$, the exterior Dirichlet problem (3.2.3) does not have solution.

The boundary reductions for the boundary value problems of partial differential equations to boundary integral (integro-differential) equations are the foundation of the boundary integral method for numerical solutions of partial differential equations. In Sect. 3.4, the boundary value problems of the Laplace equation are reduced to the integral (integro-differential) equations of the first kind or the integral equations of the second kind. Both of them can be applied to the boundary integral method for the numerical solutions of the Laplace equation. Each of them possesses its advantages and disadvantages. The integral (integro-differential) equations of the first kind maintain the symmetry of the original boundary value problems and only the weakly singular integral integrals involved in their equivalent variational forms.

References

Adams, R. A., & Fournier, J. J. (2003). *Sobolev spaces*. Academic.

Chen, J. T., Lin, S. R., & Chen K. (2005). Degenerate scale problem when solving laplace's equation by BEM and its treatment. *International Journal for Numerical Methods in Engineering, 62*(2), 233–261.

Colton, D., & Kress, R. (2013). *Integral equation methods in scattering theory*. SIAM.

Folland G. B. (1995). *Introduction to partial differential equations* (Vol. 102). Princeton University Press.

Gilbarg, D., & Trudinger, N. S. (2001). *Elliptic partial differential equations of second order* (Vol. 224). Springer Science & Business Media.

Grisvard P. (2011). *Elliptic problems in nonsmooth domains*. SIAM.

Günter, N. M. (1967). *Potential theory and its applications to basic problems of mathematical physics*. New York: Ungar.

Han, H. (1988). Boundary integro-differential equations of elliptic boundary value problems and their numerical solutions. *Scientia Sinica, Series A-Mathematical, Physical, Astronomical and Technical Sciences, 31*, 1153–1165.

Han, H. (1994). The boundary element method for solving variational inequalities. *Contemporary Mathematics, 163*, 55–70.

Han, H., Lee, Y., Yin, D., & Chen, J. (2015). The necessary and sufficient condition for the existence and uniqueness of a system of fredholm integral equations of the first kind. *Scientia Sinica Mathematica, 45*, 1231–1248.

Hsiao, G. C., & Wendland, W. L. (1977). A finite element method for some integral equations of the first kind. *Journal of Mathematical Analysis and Applications, 58*(3), 449–481.

Hsiao, G. C., & Wendland, W. L. (2008). *Boundary integral equations*. Springer.

Kondratév, V. A., & Oleinik, O. A. (1987). On the uniqueness of the solutions of boundary-value problems in unbounded domains, and on isolated singular points of solutions of the system of elasticity theory and second-order elliptic equations. *Russian Mathematical Surveys, 42*(4), 143.

Lax, P. D., & Milgram, A. N. (1954). Parabolic equations. Contributions to the theory of partial differential equations. *Annals of Mathematics Studies, 33*, 167–190.

Maue, A. W. (1949). Zur formulierung eines allgemeinen beugungs-problems durch eine integral-gleichung. *Zeitschrift für Physik, 126*(7–9), 601–618.

Petrovsky I. G. (2012). *Lectures on partial differential equations*. Dover Publications.

Yan, Y., & Sloan, I. H. (1988). On integral equations of the first kind with logarithmic kernels. *The Journal of Integral Equations and Applications*, 549–579.

Yosida K. (2012). *Functional analysis* (Vol. 123). Springer Science & Business Media.

Chapter 4
Boundary Value Problems of Modified Helmholtz Equation

In this chapter we discuss the boundary integro-differential equations for the boundary value problems of the modified Helmholtz equation given by

$$-\Delta u(\boldsymbol{x}) + k^2 u(\boldsymbol{x}) = 0, \tag{4.0.1}$$

with constant $k > 0$.

One of the sources of the modified Helmholtz equation (4.0.1) is from the convection diffusion equation:

$$-\epsilon^2 \Delta v(\boldsymbol{x}) + \boldsymbol{a} \cdot \nabla v(\boldsymbol{x}) = 0, \tag{4.0.2}$$

with constant $\epsilon > 0$ and real constant vector $\boldsymbol{a} = (a_1, a_2, \ldots, a_n)^{\mathrm{T}} \neq \boldsymbol{0}$.

Let

$$v(\boldsymbol{x}) = \mathrm{e}^{\boldsymbol{a}\cdot\boldsymbol{x}/2\epsilon^2} u(\boldsymbol{x}),$$

by a direct computing it arrives at

$$-\epsilon^2 \Delta u(\boldsymbol{x}) + \frac{\|\boldsymbol{a}\|^2}{4\epsilon^2} u(\boldsymbol{x}) = 0,$$

namely $u(\boldsymbol{x})$ satisfies the modified Helmholtz equation (4.0.1) with $k = \|\boldsymbol{a}\|/2\epsilon^2$. The convection dominated case ($0 < \epsilon^2 \ll \|\boldsymbol{a}\|$) of Eq. (4.0.2) is more interesting, which exports the singular perturbation problems (Hemker, 1996; Han et al., 2008).

H. Han and D. Yin, *Mathematical Foundation of the Boundary Integro-Differential Equation Method*, https://doi.org/10.1007/978-981-95-1088-7_4

4.1 The Dirichlet and Neumann Boundary Problems of Modified Helmholtz Equation

For given functions $g(\boldsymbol{x})$ and $h(\boldsymbol{x})$ on Γ consider the following four principal boundary value problems of modified Helmholtz equation.

The interior Dirichlet problem:

$$\begin{cases} -\Delta u(\boldsymbol{x}) + k^2 u(\boldsymbol{x}) = 0, & \forall \boldsymbol{x} \in \Omega, \quad (4.1.1a) \\ u(\boldsymbol{x})\big|_\Gamma = g(\boldsymbol{x}), & \forall \boldsymbol{x} \in \Gamma. \quad (4.1.1b) \end{cases}$$

The exterior Dirichlet problem:

$$\begin{cases} -\Delta u(\boldsymbol{x}) + k^2 u(\boldsymbol{x}) = 0, & \forall \boldsymbol{x} \in \Omega_e \quad (4.1.2a) \\ u(\boldsymbol{x})\big|_\Gamma = g(\boldsymbol{x}), & \forall \boldsymbol{x} \in \Gamma, \quad (4.1.2b) \\ u(\boldsymbol{x}) \to 0, \ \text{as} & |\boldsymbol{x}| \to +\infty. \quad (4.1.2c) \end{cases}$$

The interior Neumann problem:

$$\begin{cases} -\Delta u(\boldsymbol{x}) + k^2 u(\boldsymbol{x}) = 0, & \forall \boldsymbol{x} \in \Omega, \quad (4.1.3a) \\ \left.\dfrac{\partial u(\boldsymbol{x})}{\partial \boldsymbol{\nu}_x}\right|_\Gamma = h(\boldsymbol{x}), & \forall \boldsymbol{x} \in \Gamma. \quad (4.1.3b) \end{cases}$$

The exterior Neumann problem:

$$\begin{cases} -\Delta u(\boldsymbol{x}) + k^2 u(\boldsymbol{x}) = 0, & \forall \boldsymbol{x} \in \Omega_e, \quad (4.1.4a) \\ \left.\dfrac{\partial u(\boldsymbol{x})}{\partial \boldsymbol{\nu}_x}\right|_\Gamma = h(\boldsymbol{x}), \forall \boldsymbol{x} \in \Gamma, & \quad (4.1.4b) \\ u(\boldsymbol{x}) \to 0, \text{as} & |\boldsymbol{x}| \to +\infty. \quad (4.1.4c) \end{cases}$$

For given $g \in H^{\frac{1}{2}}(\Gamma)$ and $h \in H^{-\frac{1}{2}}(\Gamma)$, we discuss the equivalent variational problems corresponding to the above four principal boundary value problems of the modified Helmholtz equation. The equivalent variational problem of interior Dirichlet problem (4.1.1) is given by

$$\begin{cases} \text{Find} \quad u \in H_g^1(\Omega), \quad \text{such that} \\ \qquad a^+(u, v; k) = 0, \quad \forall v \in H_0^1(\Omega). \end{cases} \tag{4.1.5}$$

The bilinear form $a^+(u, v; k)$ on $H^1(\Omega) \times H^1(\Omega)$ is defined by

$$a^+(u, v; k) = \int_\Omega (\nabla u \cdot \nabla v + k^2 uv) \mathrm{d}\boldsymbol{x}, \quad \forall u, v \in H^1(\Omega). \tag{4.1.6}$$

The equivalent variational problem of interior Neumann problem (4.1.3) is in the following

$$\begin{cases} \text{Find } u \in H^1(\Omega), \text{ such that} \\ a^+(u, v; k) = \int_\Gamma hv \mathrm{d}s_x, \quad \forall v \in H^1(\Omega). \end{cases} \tag{4.1.7}$$

For discussing the equivalent variational problems of the exterior problems (4.1.2) and (4.1.4), introduce the following two spaces (Hsiao & Wendland, 2008):

$$H^1_{\text{loc}}(\Omega_e) = \left\{ v \middle| v \in H^1(G), \text{ for any bounded domain } G \subset \Omega_e \right\},$$

$$H^1_{\text{comp}}(\Omega_e) = \left\{ v \middle| v \in H^1_{\text{loc}}(\Omega_e) \text{ and the support of } v \text{ is compact} \right\}.$$

Obviously

$$H^1_{\text{comp}}(\Omega_e) \subset H^1(\Omega_e) \subset H^1_{\text{loc}}(\Omega_e).$$

Furthermore let

$$H^1_{\text{loc},g}(\Omega_e) = \left\{ v \middle| v \in H^1_{\text{loc}}(\Omega_e) \text{ and } v\Big|_\Gamma = g \right\},$$

$$H^1_{\text{loc},0}(\Omega_e) = \left\{ v \middle| v \in H^1_{\text{loc}}(\Omega_e) \text{ and } v\Big|_\Gamma = 0 \right\}.$$

The exterior Dirichlet problem (4.1.2) is equivalent to the following variational problem:

$$\begin{cases} \text{Find } u \in H^1_{\text{loc},g}(\Omega_e), \text{ such that} \\ a_e^+(u, v; k) = 0, \quad \forall v \in H^1_{\text{comp}}(\Omega_e), \\ u(\boldsymbol{x}) \to 0, \quad \text{as } |\boldsymbol{x}| \to +\infty, \end{cases} \tag{4.1.8}$$

with

$$a_e^+(u, v; k) = \int_{\Omega_e} \left(\nabla u \cdot \nabla v + k^2 uv \right) \mathrm{d}\boldsymbol{x}. \tag{4.1.9}$$

The exterior Neumann problem (4.1.4) is equivalent to the follow variational problem:

$$\begin{cases} \text{Find } u \in H^1_{\text{loc}}(\Omega_e), \text{ such that} \\ a_e^+(u, v; k) = -\int_\Gamma hv \mathrm{d}s_x, \quad \forall v \in H^1_{\text{comp}}(\Omega_e), \\ u(\boldsymbol{x}) \to 0, \quad \text{as } |\boldsymbol{x}| \to +\infty. \end{cases} \tag{4.1.10}$$

The exterior Dirichlet problem (4.1.2) has been studied by the artificial boundary method (Han & Wu, 2013).

4.2 Single and Double Layer Potentials of Modified Helmholtz Operator for the Continuous Densities

The fundamental solution of the modified Helmholtz operator $-\Delta + k^2$ is given by (2.2.25), namely

$$E_n^+(\boldsymbol{x}, \boldsymbol{y}; k) = \begin{cases} \dfrac{1}{2\pi} K_0\left(k|\boldsymbol{x} - \boldsymbol{y}|\right), & n = 2, \\ \dfrac{1}{4\pi} \dfrac{\mathrm{e}^{-k|\boldsymbol{x}-\boldsymbol{y}|}}{|\boldsymbol{x} - \boldsymbol{y}|}, & n = 3. \end{cases} \tag{4.2.1}$$

For given functions $\varphi(\boldsymbol{x})$, $\psi(\boldsymbol{x})$ on Γ, the single layer potential of modified Helmholtz operator is defined as

$$V_{I,k}^+(\boldsymbol{x}) = \int_\Gamma E_n^+(\boldsymbol{x}, \boldsymbol{y}; k)\varphi(\boldsymbol{y})\mathrm{d}s_{\boldsymbol{y}}, \quad \forall \boldsymbol{x} \in \mathbb{R}^n, \tag{4.2.2}$$

the double layer potential of modified Helmholtz operator is given by

$$V_{II,k}^+(\boldsymbol{x}) = \int_\Gamma \frac{\partial}{\partial \boldsymbol{\nu}_{\boldsymbol{y}}} E_n^+(\boldsymbol{x}, \boldsymbol{y}; k)\psi(\boldsymbol{y})\mathrm{d}s_{\boldsymbol{y}}, \quad \forall \boldsymbol{x} \in \mathbb{R}^n. \tag{4.2.3}$$

For discussing the properties of $V_{I,k}^+(\boldsymbol{x})$ and $V_{II,k}^+(\boldsymbol{x})$, let

$$E_n^+(\boldsymbol{x}, \boldsymbol{y}; k) = E_n(\boldsymbol{x}, \boldsymbol{y}) + B_n^+(\boldsymbol{x}, \boldsymbol{y}; k), \tag{4.2.4}$$

in which $E_n(\boldsymbol{x}, \boldsymbol{y})$, the fundamental solution of Laplace operator, is given by (3.3.3). Then we have

$$B_n^+(\boldsymbol{x}, \boldsymbol{y}; k) = \begin{cases} \dfrac{1}{2\pi} \left\{K_0\left(k|\boldsymbol{x} - \boldsymbol{y}|\right) + \ln|\boldsymbol{x} - \boldsymbol{y}|\right\}, & n = 2, \\ \dfrac{1}{4\pi} \dfrac{\mathrm{e}^{-k|\boldsymbol{x}-\boldsymbol{y}|} - 1}{|\boldsymbol{x} - \boldsymbol{y}|}, & n = 3. \end{cases} \tag{4.2.5}$$

For the function $B_n^+(\boldsymbol{x}, \boldsymbol{y}; k)$ we obtain

Lemma 4.2.1 *(i) $B_n^+(\boldsymbol{x}, \boldsymbol{y}; k)$ is continuous on $(\boldsymbol{x}, \boldsymbol{y}) \in \mathbb{R}^n \times \mathbb{R}^n$ and $B_n^+(\boldsymbol{x}, \boldsymbol{y}; k)$ is continuously differentiable for any times provided by $\boldsymbol{x} \neq \boldsymbol{y}$.*

(ii) Consider the first derivatives of $B_n^+(\boldsymbol{x}, \boldsymbol{y}; k)$, namely

$$\frac{\partial B_n^+(\boldsymbol{x}, \boldsymbol{y}; k)}{\partial x_i} = -\frac{\partial B_n^+(\boldsymbol{x}, \boldsymbol{y}; k)}{\partial y_i}, \quad 1 \le i \le n.$$

At the case $n = 2$, the first derivatives of $B_2^+(\boldsymbol{x}, \boldsymbol{y}; k)$ are continuous on $(\boldsymbol{x}, \boldsymbol{y}) \in \mathbb{R}^2 \times \mathbb{R}^2$.

At the case $n = 3$,

$$\frac{\partial B_n^+(\boldsymbol{x}, \boldsymbol{y}; k)}{\partial x_i} = \mathcal{O}(1), \quad as\ |\boldsymbol{x} - \boldsymbol{y}| \to 0, \ \ 1 \le i \le 3.$$

Proof The sum formulas of the zero order modified Bessel functions $I_0(z)$ and $K_0(z)$ are

$$I_0(z) = \sum_{m=0}^{\infty} \frac{\left(\frac{1}{4}z^2\right)^m}{(m!)^2},$$

$$K_0(z) = -I_0(z)\big[\ln(z/2) + \gamma)\big] + \sum_{m=0}^{\infty}\left(\sum_{l=1}^{m}\frac{1}{l}\right)\frac{\left(\frac{1}{4}z^2\right)^m}{(m!)^2},$$

in which the real number γ is the Euler constant (Andrews, 1998).

Therefore

$$\begin{aligned} B_2^+(\boldsymbol{x}, \boldsymbol{y}; k) = &\frac{1}{2\pi}\Bigg\{ - I_0\big(k|\boldsymbol{x} - \boldsymbol{y}|\big)\big(\ln|\boldsymbol{x} - \boldsymbol{y}| + \ln(k/2) + \gamma\big) + \\ &\sum_{m=0}^{\infty}\left(\sum_{l=1}^{m}\frac{1}{l}\right)\frac{\left(\frac{k^2}{4}|\boldsymbol{x} - \boldsymbol{y}|^2\right)^m}{(m!)^2} + \ln|\boldsymbol{x} - \boldsymbol{y}|\Bigg\} \\ = &\frac{1}{2\pi}\Bigg\{\big(1 - I_0(k|\boldsymbol{x} - \boldsymbol{y}|\big)\ln|\boldsymbol{x} - \boldsymbol{y}| - I_0(k|\boldsymbol{x} - \boldsymbol{y}|)(\ln(k/2) + \gamma) + \\ &\sum_{m=0}^{\infty}\left(\sum_{l=1}^{m}\frac{1}{l}\right)\frac{\left(\frac{k^2}{4}|\boldsymbol{x} - \boldsymbol{y}|^2\right)^m}{(m!)^2}\Bigg\}. \end{aligned} \tag{4.2.6}$$

Furthermore

$$B_3^+(\boldsymbol{x}, \boldsymbol{y}; k) = \frac{1}{4\pi}\sum_{m=1}^{\infty}\frac{(-k)^m|\boldsymbol{x} - \boldsymbol{y}|^{m-1}}{m!}. \tag{4.2.7}$$

By the formulas (4.2.6)–(4.2.7), the conclusions in Lemma 4.2.1 are proved directly. □

It is ready to discuss the properties of single layer potential $V^+_{I,k}(\boldsymbol{x})$ and double layer potential $V^+_{II,k}(\boldsymbol{x})$. For $V^+_{I,k}(\boldsymbol{x})$, we have

Theorem 4.2.1 *Assume that* $\Gamma \in C^2$.

For given density function $\varphi \in C(\Gamma)$, *the following results hold*

(i) $V^+_{I,k}(\boldsymbol{x})$ *is well defined on* $\mathbb{R}^n$, *which is continuous on* $\mathbb{R}^n$. $V^+_{I,k}(\boldsymbol{x})$ *is continuously differentiable for any times and satisfies the modified Helmholtz equation* (4.0.1) *on* $\mathbb{R}^n\backslash\Gamma$.

(ii)

$$\left.\begin{aligned} V^+_{I,k}(\boldsymbol{x}) &= \mathcal{O}\left(\frac{1}{|\boldsymbol{x}|^{(n-1)/2}}\mathrm{e}^{-k|\boldsymbol{x}|}\right), \\ \frac{\partial}{\partial r}V^+_{I,k}(\boldsymbol{x}) &= \mathcal{O}\left(\frac{1}{|\boldsymbol{x}|^{(n-1)/2}}\mathrm{e}^{-k|\boldsymbol{x}|}\right), \end{aligned}\right\} \quad as\ r = |\boldsymbol{x}| \to +\infty. \tag{4.2.8}$$

(iii) For $\varphi \in C^1(\Gamma)$, *the first derivatives of single layer potential* $V^+_{I,k}(\boldsymbol{x})$ *can be continuously extended from* $\Omega(\Omega_e)$ *to the boundary* Γ. *This is*

$$\left.\begin{aligned} \lim_{\substack{z\to x\in\Gamma\\ z\in\Omega}} \nabla_z V^+_{I,k}(z) &= \int_\Gamma \nabla_x E^+_n(\boldsymbol{x}, \boldsymbol{y}; k)\psi(\boldsymbol{y})\,\mathrm{d}s_y + \frac{1}{2}\boldsymbol{\nu}(\boldsymbol{x})\psi(\boldsymbol{x}), \\ \lim_{\substack{z\to x\in\Gamma\\ z\in\Omega_e}} \nabla_z V^+_{I,k}(z) &= \int_\Gamma \nabla_x E^+_n(\boldsymbol{x}, \boldsymbol{y}; k)\psi(\boldsymbol{y})\,\mathrm{d}s_y - \frac{1}{2}\boldsymbol{\nu}(\boldsymbol{x})\psi(\boldsymbol{x}). \end{aligned}\right\} \tag{4.2.9}$$

Moreover

$$\left.\begin{aligned} \left.\frac{\partial V^+_{I,k}(\boldsymbol{x})}{\partial \boldsymbol{\nu}_x}\right|_{\Gamma-0} &= \lim_{\substack{z\to x\in\Gamma\\ z\in\Omega}} \boldsymbol{\nu}(\boldsymbol{x})\cdot\nabla_z V^+_{I,k}(z) \\ &= \int_\Gamma \tfrac{\partial}{\partial \boldsymbol{\nu}_x} E^+_n(\boldsymbol{x}, \boldsymbol{y}; k)\varphi(\boldsymbol{y})\mathrm{d}s_y + \tfrac{1}{2}\varphi(\boldsymbol{x}), \quad \forall \boldsymbol{x} \in \Gamma, \\ \left.\frac{\partial V^+_{I,k}(\boldsymbol{x})}{\partial \boldsymbol{\nu}_x}\right|_{\Gamma+0} &= \lim_{\substack{z\to x\in\Gamma\\ z\in\Omega_e}} \boldsymbol{\nu}(\boldsymbol{x})\cdot\nabla_z V^+_{I,k}(z) \\ &= \int_\Gamma \tfrac{\partial}{\partial \boldsymbol{\nu}_x} E^+_n(\boldsymbol{x}, \boldsymbol{y}; k)\varphi(\boldsymbol{y})\mathrm{d}s_y - \tfrac{1}{2}\varphi(\boldsymbol{x}), \quad \forall \boldsymbol{x} \in \Gamma. \end{aligned}\right\} \tag{4.2.10}$$

Specially for $\varphi \in C(\Gamma)$, *the equations* (4.2.10) *are still correct in the sense given in (3.3.33).*

Proof (i) The fundamental solution of the modified Helmholtz operator $E^+_n(\boldsymbol{x}, \boldsymbol{y}; k)$ satisfies

$$|E^+_n(\boldsymbol{x}, \boldsymbol{y}; k)| \le C\begin{Bmatrix} \big|\ln|\boldsymbol{x}-\boldsymbol{y}|\big|, & n=2 \\ \dfrac{1}{|\boldsymbol{x}-\boldsymbol{y}|}, & n=3 \end{Bmatrix}, \quad \text{as } |\boldsymbol{x}-\boldsymbol{y}| \to 0.$$

Obviously it is a weakly singular kernel. The conclusion in (i) are obtained directly by the properties of $E_n^+(\boldsymbol{x}, \boldsymbol{y}; k)$.

(ii) At first consider the case $n = 2$, we recall the asymptotic expansion of the second modified Bessel function $K_\alpha(z)$ (Andrews, 1998), this is

$$K_\alpha(z) \sim \sqrt{\frac{\pi}{z}} \mathrm{e}^{-z} \left\{ 1 + \frac{4\alpha^2 - 1}{8z} + \frac{(4\alpha^2 - 1)(4\alpha^2 - 9)}{2!(8z)^2} + \cdots \right\},$$
$$|\arg z| < \frac{3\pi}{2}, \quad |z| \to +\infty. \tag{4.2.11}$$

Applying the asymptotic expansion (4.2.11) with $\alpha = 0, 1$, it directly arrives at the functions

$$|\boldsymbol{x}|^{\frac{1}{2}} \mathrm{e}^{k|\boldsymbol{x}|} E_2^+(\boldsymbol{x}, \boldsymbol{y}; k) \text{ and } |\boldsymbol{x}|^{\frac{1}{2}} \mathrm{e}^{k|\boldsymbol{x}|} \frac{\partial E_2^+(\boldsymbol{x}, \boldsymbol{y}; k)}{\partial |\boldsymbol{x}|}$$

are uniformly bounded for $\boldsymbol{y} \in \Gamma$ and $|\boldsymbol{x}| \to +\infty$.

For the case $n = 3$,

$$|\boldsymbol{x}| \mathrm{e}^{k|\boldsymbol{x}|} E_3^+(\boldsymbol{x}, \boldsymbol{y}; k) = \frac{1}{4\pi} \left(\frac{|\boldsymbol{x}|}{|\boldsymbol{x} - \boldsymbol{y}|} \mathrm{e}^{k|\boldsymbol{x}| - k|\boldsymbol{x} - \boldsymbol{y}|} \right),$$
$$|\boldsymbol{x}| \mathrm{e}^{k|\boldsymbol{x}|} \frac{\partial}{\partial |\boldsymbol{x}|} E_3^+(\boldsymbol{x}, \boldsymbol{y}; k) = \frac{1}{4\pi} |\boldsymbol{x}| \mathrm{e}^{k|\boldsymbol{x}|} \frac{\partial}{\partial |\boldsymbol{x}|} \left(\frac{1}{|\boldsymbol{x} - \boldsymbol{y}|} \mathrm{e}^{-k|\boldsymbol{x} - \boldsymbol{y}|} \right),$$

which are uniformly bounded for $\boldsymbol{y} \in \Gamma$ and $|\boldsymbol{x}| \to +\infty$. Therefore the results in (ii) are proved completely.

(iii) By the formula (4.2.4), $V_{I,k}^+(\boldsymbol{x})$ can be rewritten as

$$V_{I,k}^+(\boldsymbol{x}) = \int_\Gamma E_n(\boldsymbol{x}, \boldsymbol{y}) \varphi(\boldsymbol{y}) \mathrm{d}s_{\boldsymbol{y}} + \int_\Gamma B_n^+(\boldsymbol{x}, \boldsymbol{y}; k) \varphi(\boldsymbol{y}) \mathrm{d}s_{\boldsymbol{y}}$$
$$:= V_I(\boldsymbol{x}) + U_{I,k}^+(\boldsymbol{x})$$

The properties of $B_n^+(\boldsymbol{x}, \boldsymbol{y}; k)$ in Lemma 4.2.1 yield that $U_{I,k}^+(\boldsymbol{x})$ and its first derivatives are continuous on $\mathbb{R}^n$ with the given function $\varphi \in C(\Gamma)$. Furthermore combining the results in Theorem 3.3.5, the proof is completed. □

For the double layer potential $V_{II,k}^+(\boldsymbol{x})$, we obtain

Theorem 4.2.2 *Suppose that* $\Gamma \in C^2$.

For the density function $\psi \in C(\Gamma)$, *we have*

(i) The double layer potential $V_{II,k}^+(\boldsymbol{x})$ *is well defined on* $\mathbb{R}^n$. *On* $\mathbb{R}^n \backslash \Gamma$, $V_{II,k}^+(\boldsymbol{x})$ *is continuously differentiable for any times and satisfies the modified Helmholtz equation* (4.0.1).

(ii)

$$\left.\begin{aligned} V^+_{II,k}(\boldsymbol{x}) &= \mathcal{O}\left(\frac{1}{|\boldsymbol{x}|^{(n-1)/2}}\mathrm{e}^{-k|\boldsymbol{x}|}\right), \quad as \ |\boldsymbol{x}| \to \infty, \\ \frac{\partial V^+_{II,k}(\boldsymbol{x})}{\partial r} &= \mathcal{O}\left(\frac{1}{|\boldsymbol{x}|^{(n-1)/2}}\mathrm{e}^{-k|\boldsymbol{x}|}\right), \quad as \ r = |\boldsymbol{x}| \to \infty. \end{aligned}\right\} \tag{4.2.12}$$

(iii) The restriction of $V^+_{II,k}(\boldsymbol{x})$ in $\Omega(\Omega_e)$ has a continuous extension to $\bar{\Omega}(\bar{\Omega}_e)$. Moreover

$$\left.\begin{aligned} \lim_{\substack{z\to \boldsymbol{x}\in\Gamma \\ z\in\Omega}} V^+_{II,k}(z) &= -\frac{1}{2}\psi(\boldsymbol{x}) + \int_\Gamma \frac{\partial}{\partial \boldsymbol{\nu}_{\boldsymbol{y}}} E^+_n(\boldsymbol{x},\boldsymbol{y};k)\psi(\boldsymbol{y})\mathrm{d}s_{\boldsymbol{y}}, \quad \forall \boldsymbol{x}\in\Gamma, \\ \lim_{\substack{z\to \boldsymbol{x}\in\Gamma \\ z\in\Omega_e}} V^+_{II,k}(z) &= \frac{1}{2}\psi(\boldsymbol{x}) + \int_\Gamma \frac{\partial}{\partial \boldsymbol{\nu}_{\boldsymbol{y}}} E^+_n(\boldsymbol{x},\boldsymbol{y};k)\psi(\boldsymbol{y})\mathrm{d}s_{\boldsymbol{y}}, \quad \forall \boldsymbol{x}\in\Gamma. \end{aligned}\right\} \tag{4.2.13}$$

(iv) The first derivatives of double layer potential $V^+_{II,k}(\boldsymbol{x})$ can be continuously extended from $\Omega(\Omega_e)$ to the boundary Γ provided by the density function $\psi \in C^2(\Gamma)$. That are

$$\begin{aligned} \left.\frac{\partial}{\partial x_i}V^+_{II,k}(\boldsymbol{x})\right|_{\Gamma-0} &= \lim_{\substack{z\to \boldsymbol{x}\in\Gamma \\ z\in\Omega}} \frac{\partial}{\partial z_i}V^+_{II,k}(z) \\ &= -\sum_{j=1,j\neq i}^{n} \frac{\partial}{\partial x_j}\int_\Gamma E^+_n(\boldsymbol{x},\boldsymbol{y};k)\frac{\partial}{\partial \boldsymbol{\tau}^{j,i}_{\boldsymbol{y}}}\psi(\boldsymbol{y})\mathrm{d}s_{\boldsymbol{y}} - \\ & k^2\int_\Gamma E^+_n(\boldsymbol{x},\boldsymbol{y};k)\Big(\nu_i(\boldsymbol{y})\psi(\boldsymbol{y})\Big)\mathrm{d}s_{\boldsymbol{y}} - \\ & \frac{1}{2}\sum_{j=1,j\neq i}^{n} \nu_j(\boldsymbol{x})\frac{\partial}{\partial \boldsymbol{\tau}^{j,i}_{\boldsymbol{x}}}\psi(\boldsymbol{x}), \quad \forall \boldsymbol{x}\in\Gamma. \end{aligned} \tag{4.2.14}$$

$$\begin{aligned} \left.\frac{\partial}{\partial x_i}V^+_{II,k}(\boldsymbol{x})\right|_{\Gamma+0} &= \lim_{\substack{z\to \boldsymbol{x}\in\Gamma \\ z\in\Omega_e}} \frac{\partial}{\partial z_i}V^+_{II,k}(z) \\ &= -\sum_{j=1,j\neq i}^{n} \frac{\partial}{\partial x_j}\int_\Gamma E^+_n(\boldsymbol{x},\boldsymbol{y};k)\frac{\partial}{\partial \boldsymbol{\tau}^{j,i}_{\boldsymbol{y}}}\psi(\boldsymbol{y})\mathrm{d}s_{\boldsymbol{y}} - \\ & k^2\int_\Gamma E^+_n(\boldsymbol{x},\boldsymbol{y};k)\Big(\nu_i(\boldsymbol{y})\psi(\boldsymbol{y})\Big)\mathrm{d}s_{\boldsymbol{y}} + \\ & \frac{1}{2}\sum_{j=1,j\neq i}^{n} \nu_j(\boldsymbol{x})\frac{\partial}{\partial \boldsymbol{\tau}^{j,i}_{\boldsymbol{x}}}\psi(\boldsymbol{x}), \quad \forall \boldsymbol{x}\in\Gamma. \end{aligned} \tag{4.2.15}$$

(v) The first derivative of the double layer potential $V^+_{II,k}(\boldsymbol{x})$ on Γ with respect to $\boldsymbol{\nu}(\boldsymbol{x})$ is continuous and

$$
\begin{aligned}
\frac{\partial}{\partial \boldsymbol{\nu}_x} V^+_{II,k}(\boldsymbol{x})\bigg|_\Gamma &= \frac{\partial}{\partial \boldsymbol{\nu}_x}\int_\Gamma \frac{\partial}{\partial \boldsymbol{\nu}_y} E^+_n(\boldsymbol{x},\boldsymbol{y};k)\psi(\boldsymbol{y})\mathrm{d}s_y \\
&= \sum_{1\le i<j\le n} \frac{\partial}{\partial \boldsymbol{\tau}^{i,j}_x}\int_\Gamma E^+_n(\boldsymbol{x},\boldsymbol{y};k)\frac{\partial}{\partial \boldsymbol{\tau}^{i,j}_y}\psi(\boldsymbol{y})\mathrm{d}s_y - \\
& k^2\int_\Gamma E^+_n(\boldsymbol{x},\boldsymbol{y};k)\big(\boldsymbol{\nu}(\boldsymbol{x})\cdot\boldsymbol{\nu}(\boldsymbol{y})\big)\psi(\boldsymbol{y})\mathrm{d}s_y, \quad \forall \boldsymbol{x}\in\Gamma. \qquad (4.2.16)
\end{aligned}
$$

Proof (i) By the properties of fundamental solution $E^+_n(\boldsymbol{x},\boldsymbol{y};k)$ and the inequality (3.3.6), the conclusions are arrived at directly.

(ii) It is straightforward to prove that

$$
|\boldsymbol{x}|^{(n-1)/2}\mathrm{e}^{k|\boldsymbol{x}|}\frac{\partial}{\partial \boldsymbol{\nu}_y}E^+_n(\boldsymbol{x},\boldsymbol{y};k),
$$

$$
|\boldsymbol{x}|^{(n-1)/2}\mathrm{e}^{k|\boldsymbol{x}|}\frac{\partial}{\partial |\boldsymbol{x}|}\left(\frac{\partial}{\partial \boldsymbol{\nu}_y}E^+_n(\boldsymbol{x},\boldsymbol{y};k)\right),
$$

are uniformly bounded for $\boldsymbol{y}\in\Gamma$ and $r=|\boldsymbol{x}|\to+\infty$, in the proof for the case $n=2$, the following two equalities

$$
K_0'(z) = -K_1(z), \qquad K_0''(z) = K_0(z) - \frac{1}{z}K_1(z),
$$

and the asymptotic formula (4.2.11) are used.

(iii) Applying the formula (4.2.4), it arrives at

$$
\begin{aligned}
V^+_{II,k}(\boldsymbol{x}) &= \int_\Gamma \frac{\partial}{\partial \boldsymbol{\nu}_y}E_n(\boldsymbol{x},\boldsymbol{y})\psi(\boldsymbol{y})\mathrm{d}s_y + \int_\Gamma \frac{\partial}{\partial \boldsymbol{\nu}_y}E^+_n(\boldsymbol{x},\boldsymbol{y};k)\psi(\boldsymbol{y})\mathrm{d}s_y \\
&:= V_{II}(\boldsymbol{x}) + U^+_{II,k}(\boldsymbol{x}),
\end{aligned}
$$

where $V_{II}(\boldsymbol{x})$ is the double layer potential of Laplace operator defined by (3.3.11). By the results come form Lemma 4.2.1, we know that

$$
\frac{\partial}{\partial \boldsymbol{\nu}_y}B^+_n(\boldsymbol{x},\boldsymbol{y};k) = \sum_{i=1}^n \nu_i(\boldsymbol{y})\frac{\partial}{\partial y_i}B^+_n(\boldsymbol{x},\boldsymbol{y};k)
$$

is uniformly bounded for $\boldsymbol{y}\in\Gamma$ and $|\boldsymbol{x}-\boldsymbol{y}|\to 0$. It yields that $U^+_{II,k}(\boldsymbol{x})$ is well defined continuously function on $\mathbb{R}^n$.

Combining the properties of the double layer potential $V_{II}(\boldsymbol{x})$, which are given by Theorem 3.3.6, the conclusions are proved completely.

(iv) For $\boldsymbol{y}\in\Gamma$, $\boldsymbol{x}\in\mathbb{R}^n\backslash\Gamma$, we obtain

$$\left.\begin{aligned}&\frac{\partial}{\partial x_i}E_n^+(\boldsymbol{x},\boldsymbol{y};k)=-\frac{\partial}{\partial y_i}E_n^+(\boldsymbol{x},\boldsymbol{y};k),\quad 1\le i\le n,\\&\frac{\partial^2}{\partial x_i\partial y_j}E_n^+(\boldsymbol{x},\boldsymbol{y};k)=\frac{\partial^2}{\partial x_j\partial y_i}E_n^+(\boldsymbol{x},\boldsymbol{y};k)\\&\qquad\qquad=-\tfrac{\partial^2}{\partial x_i\partial x_j}E_n^+(\boldsymbol{x},\boldsymbol{y};k),\ 1\le i,j\le n,\end{aligned}\right\}\tag{4.2.17}$$

$$-\Delta_x E_n^+(\boldsymbol{x},\boldsymbol{y};k)+k^2E_n^+(\boldsymbol{x},\boldsymbol{y};k)=0.\tag{4.2.18}$$

Applying the equalities (4.2.17)–(4.2.18), it arrives at

$$\begin{aligned}\frac{\partial}{\partial x_i}\frac{\partial}{\partial \boldsymbol{\nu}_{\boldsymbol{y}}}E_n^+(\boldsymbol{x},\boldsymbol{y};k)&=\sum_{j=1}^n\nu_j(\boldsymbol{y})\frac{\partial^2}{\partial x_i\partial y_j}E_n^+(\boldsymbol{x},\boldsymbol{y};k)\\&=\sum_{j=1}^n\nu_j(\boldsymbol{y})\frac{\partial^2}{\partial x_j\partial y_i}E_n^+(\boldsymbol{x},\boldsymbol{y};k)\\&=\sum_{j=1}^n\frac{\partial}{\partial x_j}\left(\nu_j(\boldsymbol{y})\frac{\partial}{\partial y_i}E_n^+(\boldsymbol{x},\boldsymbol{y};k)\right)\\&=\sum_{j=1}^n\frac{\partial}{\partial x_j}\left(\nu_j(\boldsymbol{y})\frac{\partial}{\partial y_i}-\nu_i(\boldsymbol{y})\frac{\partial}{\partial y_j}\right)E_n^+(\boldsymbol{x},\boldsymbol{y};k)+\\&\nu_i(\boldsymbol{y})\sum_{j=1}^n\frac{\partial^2}{\partial x_j\partial y_j}E_n^+(\boldsymbol{x},\boldsymbol{y};k)\\&=\sum_{j=1,j\neq i}^n\frac{\partial}{\partial x_j}\frac{\partial}{\partial \boldsymbol{\tau}_{\boldsymbol{y}}^{j,i}}E_n^+(\boldsymbol{x},\boldsymbol{y};k)-k^2\nu_i(\boldsymbol{y})E_n^+(\boldsymbol{x},\boldsymbol{y};k).\end{aligned}\tag{4.2.19}$$

Therefore for $\boldsymbol{x}\in\mathbb{R}^n\backslash\Gamma$, the equality (4.2.19) yields

$$\begin{aligned}\frac{\partial}{\partial x_i}V_{\mathit{II},k}^+(\boldsymbol{x})&=\frac{\partial}{\partial x_i}\int_\Gamma\frac{\partial}{\partial \boldsymbol{\nu}_{\boldsymbol{y}}}E_n^+(\boldsymbol{x},\boldsymbol{y};k)\psi(\boldsymbol{y})\mathrm{d}s_{\boldsymbol{y}}\\&=\sum_{j=1,j\neq i}^n\frac{\partial}{\partial x_j}\int_\Gamma\frac{\partial}{\partial \boldsymbol{\tau}_{\boldsymbol{y}}^{j,i}}E_n^+(\boldsymbol{x},\boldsymbol{y};l)\psi(\boldsymbol{y})\mathrm{d}s_{\boldsymbol{y}}-\\&k^2\int_\Gamma E_n^+(\boldsymbol{x},\boldsymbol{y};k)\Big(\nu_i(\boldsymbol{y})\psi(\boldsymbol{y})\Big)\mathrm{d}s_{\boldsymbol{y}}\\&=-\sum_{j=1,j\neq i}^n\frac{\partial}{\partial x_j}\int_\Gamma E_n^+(\boldsymbol{x},\boldsymbol{y};k)\frac{\partial}{\partial \boldsymbol{\tau}_{\boldsymbol{y}}^{j,i}}\psi(\boldsymbol{y})\mathrm{d}s_{\boldsymbol{y}}-\\&k^2\int_\Gamma E_n^+(\boldsymbol{x},\boldsymbol{y};k)\Big(\nu_i(\boldsymbol{y})\psi(\boldsymbol{y})\Big)\mathrm{d}s_{\boldsymbol{y}}.\end{aligned}\tag{4.2.20}$$

The equality (4.2.20) shows that the first derivatives of the double layer potential $V^{+}_{II,k}(\boldsymbol{x})$ can be expressed as a summation of the first derivatives of the single layer potentials and one single layer potential.

By the properties of the single layer potential $V^{+}_{I,k}(\boldsymbol{x})$ and its first derivatives given by Theorem 4.2.1, the conclusion (iv) is proved directly.

(v) We know that

$$\begin{aligned}
\left.\frac{\partial}{\partial \boldsymbol{\nu}_{\boldsymbol{x}}} V^{+}_{II,k}(\boldsymbol{x})\right|_{\Gamma-0} &= \sum_{i=1}^{n} \nu_i(\boldsymbol{x}) \frac{\partial}{\partial x_i} \left. V^{+}_{II,k}(\boldsymbol{x})\right|_{\Gamma-0} \\
&= - \sum_{i,j=1,i\neq j}^{n} \nu_i(\boldsymbol{x}) \frac{\partial}{\partial x_j} \int_{\Gamma} E^{+}_{n}(\boldsymbol{x},\boldsymbol{y};k) \frac{\partial}{\partial \boldsymbol{\tau}^{j,i}_{\boldsymbol{y}}} \psi(\boldsymbol{y}) \mathrm{d}s_{\boldsymbol{y}} - \\
&\quad k^2 \int_{\Gamma} E^{+}_{n}(\boldsymbol{x},\boldsymbol{y};k) \big(\boldsymbol{\nu}(\boldsymbol{x}) \cdot \boldsymbol{\nu}(\boldsymbol{y})\big) \psi(\boldsymbol{y}) \mathrm{d}s_{\boldsymbol{y}} - \\
&\quad \frac{1}{2} \sum_{i,j=1,i\neq j}^{n} \nu_i(\boldsymbol{x}) \nu_j(\boldsymbol{x}) \frac{\partial}{\partial \boldsymbol{\tau}^{j,i}_{\boldsymbol{x}}} \psi(\boldsymbol{x}) \\
&= - \sum_{i,j=1,i\neq j}^{n} \nu_i(\boldsymbol{x}) \frac{\partial}{\partial x_j} \int_{\Gamma} E^{+}_{n}(\boldsymbol{x},\boldsymbol{y};k) \frac{\partial}{\partial \boldsymbol{\tau}^{j,i}_{\boldsymbol{y}}} \psi(\boldsymbol{y}) \mathrm{d}s_{\boldsymbol{y}} - \\
&\quad k^2 \int_{\Gamma} E^{+}_{n}(\boldsymbol{x},\boldsymbol{y};k) \big(\boldsymbol{\nu}(\boldsymbol{x}) \cdot \boldsymbol{\nu}(\boldsymbol{y})\big) \psi(\boldsymbol{y}) \mathrm{d}s_{\boldsymbol{y}}.
\end{aligned} \tag{4.2.21}$$

The last equality is from the following fact

$$\boldsymbol{\tau}^{j,i}_{\boldsymbol{x}} = -\boldsymbol{\tau}^{i,j}_{\boldsymbol{x}}, \quad 1 \leq i, j \leq n.$$

Meanwhile, the equality (4.2.21) can be rewritten as

$$\begin{aligned}
\left.\frac{\partial}{\partial \boldsymbol{\nu}_{\boldsymbol{x}}} V^{+}_{II,k}(\boldsymbol{x})\right|_{\Gamma-0} &= - \sum_{i,j=1,i\neq j}^{n} \nu_j(\boldsymbol{x}) \frac{\partial}{\partial x_j} \int_{\Gamma} E^{+}_{n}(\boldsymbol{x},\boldsymbol{y};k) \frac{\partial}{\partial \boldsymbol{\tau}^{i,j}_{\boldsymbol{y}}} \psi(\boldsymbol{y}) \mathrm{d}s_{\boldsymbol{y}} - \\
&\quad k^2 \int_{\Gamma} E^{+}_{n}(\boldsymbol{x},\boldsymbol{y};k) \big(\boldsymbol{\nu}(\boldsymbol{x}) \cdot \boldsymbol{\nu}(\boldsymbol{y})\big) \psi(\boldsymbol{y}) \mathrm{d}s_{\boldsymbol{y}} \\
&= - \sum_{i,j=1,i\neq j}^{n} \nu_j(\boldsymbol{x}) \frac{\partial}{\partial x_i} \int_{\Gamma} E^{+}_{n}(\boldsymbol{x},\boldsymbol{y};k) \frac{\partial}{\partial \boldsymbol{\tau}^{j,i}_{\boldsymbol{y}}} \psi(\boldsymbol{y}) \mathrm{d}s_{\boldsymbol{y}} - \\
&\quad k^2 \int_{\Gamma} E^{+}_{n}(\boldsymbol{x},\boldsymbol{y};k) \big(\boldsymbol{\nu}(\boldsymbol{x}) \cdot \boldsymbol{\nu}(\boldsymbol{y})\big) \psi(\boldsymbol{y}) \mathrm{d}s_{\boldsymbol{y}}.
\end{aligned} \tag{4.2.22}$$

Combining the equalities (4.2.21)–(4.2.22) we obtain

$$
\begin{aligned}
&\frac{\partial}{\partial \boldsymbol{\nu}_x} V_{II,k}^{+}(x)\bigg|_{\Gamma-0} \\
&= \frac{1}{2} \sum_{i,j=1,i\neq j}^{n} \left(\nu_j(x)\frac{\partial}{\partial x_i} - \nu_i(x)\frac{\partial}{\partial x_j} \right) \int_\Gamma E_n^{+}(x, y; k) \frac{\partial}{\partial \boldsymbol{\tau}_y^{j,i}} \psi(y) \mathrm{d}s_y - \\
&k^2 \int_\Gamma E_n^{+}(x, y; k)\big(\boldsymbol{\nu}(x)\cdot\boldsymbol{\nu}(y)\big)\psi(y)\mathrm{d}s_y \\
&= \frac{1}{2} \sum_{i,j=1,i\neq j}^{n} \frac{\partial}{\partial \boldsymbol{\tau}_x^{j,i}} \int_\Gamma E_n^{+}(x, y; k) \frac{\partial}{\partial \boldsymbol{\tau}_y^{j,i}} \psi(y) \mathrm{d}s_y - \\
&k^2 \int_\Gamma E_n^{+}(x, y; k)\big(\boldsymbol{\nu}(x)\cdot\boldsymbol{\nu}(y)\big)\psi(y)\mathrm{d}s_y \\
&= \sum_{1\le i<j\le n} \frac{\partial}{\partial \boldsymbol{\tau}_x^{i,j}} \int_\Gamma E_n^{+}(x, y; k) \frac{\partial}{\partial \boldsymbol{\tau}_y^{i,j}} \psi(y) \mathrm{d}s_y - \\
&k^2 \int_\Gamma E_n^{+}(x, y; k)\big(\boldsymbol{\nu}(x)\cdot\boldsymbol{\nu}(y)\big)\psi(y)\mathrm{d}s_y
\end{aligned}
\tag{4.2.23}
$$

Similarly we have

$$
\begin{aligned}
\frac{\partial}{\partial \boldsymbol{\nu}_x} V_{II,k}^{+}(x)\bigg|_{\Gamma+0} = &\sum_{1\le i<j\le n} \frac{\partial}{\partial \boldsymbol{\tau}_x^{i,j}} \int_\Gamma E_n^{+}(x, y; k) \frac{\partial}{\partial \boldsymbol{\tau}_y^{i,j}} \psi(y) \mathrm{d}s_y - \\
&k^2 \int_\Gamma E_n^{+}(x, y; k)\big(\boldsymbol{\nu}(x)\cdot\boldsymbol{\nu}(y)\big)\psi(y)\mathrm{d}s_y.
\end{aligned}
\tag{4.2.24}
$$

The conclusion (v) is obtained by the equalities (4.2.23)–(4.2.24) immediately.

In the case $n = 2$, the proof of conclusion (4.2.16) was given by Han (1988). □

4.3 Single Layer Potential $V_{I,k}^{+}(x)$ and Double Layer Potential $V_{II,k}^{+}(x)$ in Soblov Spaces

We now discuss the single layer potential $V_{I,k}^{+}(x)$ and the double layer potential $V_{II,k}^{+}(x)$ with given density $\varphi \in H^{-\frac{1}{2}}(\Gamma)$ and $\psi \in H^{\frac{1}{2}}(\Gamma)$ by the solvability and regularity of the corresponding transmission problem (Hsiao & Wendland, 2008).

For given density $\varphi \in C^1(\Gamma)$, Theorem 4.2.1 yields that $V_{I,k}^{+}(x)$ is the unique solution of the below transmission problem:

$$\begin{cases} -\Delta u(\boldsymbol{x}) + k^2 u(\boldsymbol{x}) = 0, \quad \forall \boldsymbol{x} \in \mathbb{R}^n \backslash \Gamma, & (4.3.1a) \\ \Big[u(\boldsymbol{x})\Big]_\Gamma = 0, \quad \forall \boldsymbol{x} \in \Gamma, & (4.3.1b) \\ \left[\dfrac{\partial u(\boldsymbol{x})}{\partial \boldsymbol{\nu}_x}\right]_\Gamma = -\varphi(\boldsymbol{x}), \quad \forall \boldsymbol{x} \in \Gamma, & (4.3.1c) \\ \left.\begin{aligned} u(\boldsymbol{x}) &= \mathcal{O}\left(\dfrac{1}{|\boldsymbol{x}|^{(n-1)/2}} \mathrm{e}^{-k|\boldsymbol{x}|}\right), \\ \dfrac{\partial u(\boldsymbol{x})}{\partial r} &= \mathcal{O}\left(\dfrac{1}{|\boldsymbol{x}|^{(n-1)/2}} \mathrm{e}^{-k|\boldsymbol{x}|}\right), \end{aligned}\right\} \quad \text{as } r = |\boldsymbol{x}| \to +\infty. & (4.3.1d) \end{cases}$$

In addition

$$\begin{aligned} \left.\frac{\partial V_{I,k}^{+}(\boldsymbol{x})}{\partial \boldsymbol{\nu}_x}\right|_\Gamma &= \frac{1}{2}\left[\left.\frac{\partial V_{I,k}^{+}(\boldsymbol{x})}{\partial \boldsymbol{\nu}_x}\right|_{\Gamma+0} + \left.\frac{\partial V_{I,k}^{+}(\boldsymbol{x})}{\partial \boldsymbol{\nu}_x}\right|_{\Gamma-0}\right] \\ &= \int_\Gamma \frac{\partial}{\partial \boldsymbol{\nu}_x} E_n^{+}(\boldsymbol{x}, \boldsymbol{y}; k) \varphi(\boldsymbol{y}) \mathrm{d}s_y, \quad \forall \boldsymbol{x} \in \Gamma. \end{aligned} \tag{4.3.2}$$

The equivalent variational problem of the transmission problem (4.3.1d) is

$$\begin{cases} \text{Find} \quad u \in H^1(\Omega), \quad \text{such that} \\ \qquad a_n^{+}(u, v; k) = -\int_\Gamma \varphi(\boldsymbol{x}) v(\boldsymbol{x}) \mathrm{d}s_x, \quad \forall v \in H^1(\mathbb{R}^n), \end{cases} \tag{4.3.3}$$

with

$$a_n^{+}(u, v; k) = \int_{\mathbb{R}^n} \nabla u \cdot \nabla v \mathrm{d}\boldsymbol{x} + k^2 \int_{\mathbb{R}^n} uv \mathrm{d}\boldsymbol{x}, \quad \forall u, v \in H^1(\mathbb{R}^n). \tag{4.3.4}$$

An application of Lax–Milgram theorem arrives at

Lemma 4.3.1 *For given $\varphi \in H^{-\frac{1}{2}}(\Gamma)$ the variational problem* (4.3.3) *exists a unique solution $u_{I,\varphi} \in H^1(\mathbb{R}^n)$. Moreover*

$$\left.\begin{aligned} &\|u_{I,\varphi}\|_{1,\mathbb{R}^n} \le C\|\varphi\|_{-\frac{1}{2},\Gamma}, \\ &\left\|u_{I,\varphi}\big|_\Gamma\right\|_{\frac{1}{2},\Gamma} \le C\|\varphi\|_{-\frac{1}{2},\Gamma}, \\ &\left\|\frac{\partial}{\partial \boldsymbol{\nu}_x} u_{I,\varphi}\Big|_{\Gamma+0}\right\|_{-\frac{1}{2},\Gamma} + \left\|\frac{\partial}{\partial \boldsymbol{\nu}_x} u_{I,\varphi}\Big|_{\Gamma-0}\right\|_{-\frac{1}{2},\Gamma} \le C\|\varphi\|_{-\frac{1}{2},\Gamma}. \end{aligned}\right\} \tag{4.3.5}$$

For given $\varphi \in H^{-\frac{1}{2}}(\Gamma)$, the corresponding single layer potential is defined by

$$V_{I,k}^{+}(\boldsymbol{x}) = \int_\Gamma E_n^{+}(\boldsymbol{x}, \boldsymbol{y}; k) \varphi(\boldsymbol{y}) \mathrm{d}s_y := u_{I,\varphi}(\boldsymbol{x}), \tag{4.3.6}$$

and

$$V_{I,k}^{+}(\boldsymbol{x})\Big|_{\Gamma} = u_{I,\varphi}(\boldsymbol{x})\Big|_{\Gamma} := \left(\mathbb{K}_{I,k}^{+}\varphi\right)(\boldsymbol{x}) = \int_{\Gamma} E_n^{+}(\boldsymbol{x}, \boldsymbol{y}; k)\varphi(\boldsymbol{y})\mathrm{d}s_{\boldsymbol{y}}, \quad \forall \boldsymbol{x} \in \Gamma, \tag{4.3.7}$$

$$\begin{aligned} \frac{\partial V_{I,k}^{+}(\boldsymbol{x})}{\partial \boldsymbol{\nu}_{\boldsymbol{x}}}\Bigg|_{\Gamma} &= \frac{1}{2}\left\{\frac{\partial V_{I,k}^{+}(\boldsymbol{x})}{\partial \boldsymbol{\nu}_{\boldsymbol{x}}}\Bigg|_{\Gamma+0} + \frac{\partial V_{I,k}^{+}(\boldsymbol{x})}{\partial \boldsymbol{\nu}_{\boldsymbol{x}}}\Bigg|_{\Gamma-0}\right\} \\ &:= \left(\mathbb{K}_{II,k}^{*}\varphi\right)(\boldsymbol{x}) = \int_{\Gamma} \frac{\partial}{\partial \boldsymbol{\nu}_{\boldsymbol{x}}} E_n^{+}(\boldsymbol{x}, \boldsymbol{y}; k)\varphi(\boldsymbol{y})\mathrm{d}s_{\boldsymbol{y}}, \quad \forall \boldsymbol{x} \in \Gamma, \end{aligned} \tag{4.3.8}$$

We already know that the last equalities in (4.3.6), (4.3.7) and (4.3.8) hold for $\varphi \in C^1(\Gamma)$. Due to the following two facts, space $C^1(\Gamma)$ is dense in $H^{-\frac{1}{2}}(\Gamma)$ and the estimate (4.3.5) holds for $\varphi \in H^{-\frac{1}{2}}(\Gamma)$, it arrives at the last equalities in (4.3.6), (4.3.7) and (4.3.8) hold for $\varphi \in H^{-\frac{1}{2}}(\Gamma)$.

Now we have

Theorem 4.3.1 *Suppose that $\Gamma \in C^2$, and the density $\varphi \in H^{-\frac{1}{2}}(\Gamma)$.*

(i) The single layer potential $V_{I,k}^{+}(\boldsymbol{x})$ is well defined on $\mathbb{R}^n$ and $V_{I,k}^{+} \in H^1(\mathbb{R}^n)$. On $\mathbb{R}^n \setminus \Gamma$, $V_{I,k}^{+}(\boldsymbol{x})$ satisfies the modified Helmholtz equation (4.0.1).

(ii)

$$V_{I,k}^{+}(\boldsymbol{x})\Big|_{\Gamma-0} = V_{I,k}^{+}(\boldsymbol{x})\Big|_{\Gamma+0} = \int_{\Gamma} E_n^{+}(\boldsymbol{x}, \boldsymbol{y}; k)\varphi(\boldsymbol{y})\mathrm{d}s_{\boldsymbol{y}}, \quad \forall \boldsymbol{x} \in \Gamma. \tag{4.3.9}$$

(iii)

$$\left.\begin{aligned} \frac{\partial V_{I,k}^{+}(\boldsymbol{x})}{\partial \boldsymbol{\nu}_{\boldsymbol{x}}}\Bigg|_{\Gamma-0} &= \frac{1}{2}\varphi(\boldsymbol{x}) + \int_{\Gamma} E_n^{+}(\boldsymbol{x}, \boldsymbol{y}; k)\varphi(\boldsymbol{y})\mathrm{d}s_{\boldsymbol{y}}, \\ \frac{\partial V_{I,k}^{+}(\boldsymbol{x})}{\partial \boldsymbol{\nu}_{\boldsymbol{x}}}\Bigg|_{\Gamma+0} &= -\frac{1}{2}\varphi(\boldsymbol{x}) + \int_{\Gamma} E_n^{+}(\boldsymbol{x}, \boldsymbol{y}; k)\varphi(\boldsymbol{y})\mathrm{d}s_{\boldsymbol{y}}. \end{aligned}\right\} \quad \forall \boldsymbol{x} \in \Gamma. \tag{4.3.10}$$

(iv)

$$\left.\begin{aligned} V_{I,k}^{+}(\boldsymbol{x}) &= \mathcal{O}\left(\frac{1}{|\boldsymbol{x}|^{(n-1)/2}}\mathrm{e}^{-k|\boldsymbol{x}|}\right), \\ \frac{\partial V_{I,k}^{+}(\boldsymbol{x})}{\partial r} &= \mathcal{O}\left(\frac{1}{|\boldsymbol{x}|^{(n-1)/2}}\mathrm{e}^{-k|\boldsymbol{x}|}\right), \end{aligned}\right\} \quad \text{as} \quad r = |\boldsymbol{x}| \to +\infty. \tag{4.3.11}$$

(v) The operators $\mathbb{K}_{I,k}^{+}$ *and* $\mathbb{K}_{II,k}^{+,*}$ *are continuous and bounded*

$$\mathbb{K}_{I,k}^{+}: \qquad H^{-\frac{1}{2}+\gamma}(\Gamma) \longrightarrow H^{\frac{1}{2}+\gamma}(\Gamma), \tag{4.3.12}$$

$$\mathbb{K}_{II,k}^{+,*}: \qquad H^{-\frac{1}{2}+\gamma}(\Gamma) \longrightarrow H^{-\frac{1}{2}+\gamma}(\Gamma), \tag{4.3.13}$$

provided by $0 \leq \gamma \leq \dfrac{1}{2}$.

(vi) Let

$$\mathbb{A}^{+}(\varphi, \mu; k) = \int_{\Gamma}\int_{\Gamma} E_n^{+}(\boldsymbol{x}, \boldsymbol{y}; k)\varphi(\boldsymbol{y})\mu(\boldsymbol{x})\mathrm{d}s_y\mathrm{d}s_x. \tag{4.3.14}$$

Then $\mathbb{A}^{+}(\varphi, \mu; k)$ *is bounded and coercive bilinear form on* $H^{-\frac{1}{2}}(\Gamma) \times H^{-\frac{1}{2}}(\Gamma)$, *namely there exist two positive constants* M *and* α *such that*

$$|\mathbb{A}^{+}(\varphi, \mu; k)| \leq M\|\varphi\|_{-\frac{1}{2},\Gamma}\|\mu\|_{-\frac{1}{2},\Gamma}, \quad \forall \varphi, \mu \in H^{-\frac{1}{2}}(\Gamma), \tag{4.3.15}$$

$$\mathbb{A}^{+}(\varphi, \varphi; k) \geq \alpha\|\varphi\|^2_{-\frac{1}{2},\Gamma}, \qquad \forall \varphi \in H^{-\frac{1}{2}}(\Gamma). \tag{4.3.16}$$

Proof The conclusion (i)–(v) can be obtained by the definition of single layer potential $V_{I,k}^{+}(\boldsymbol{x})$ with density $\varphi \in H^{-\frac{1}{2}}(\Gamma)$ and the regularity of the transmission problem (4.3.1d).

In the following we only give the proof of conclusion (vi). For $\varphi \in H^{-\frac{1}{2}}(\Gamma)$, we have

$$\begin{aligned}
\left|\mathbb{A}^{+}(\varphi, \mu; k)\right| &= \left|\int_{\Gamma} u_{I,\varphi}(\boldsymbol{x})\mu(\boldsymbol{x})\mathrm{d}s_x\right| \\
&= \left|\int_{\Gamma} (\mathbb{K}_{I,k}^{+}\varphi)(\boldsymbol{x})\mu(\boldsymbol{x})\mathrm{d}s_x\right| \\
&\leq \left\|(\mathbb{K}_{I,k}^{+}\varphi)\right\|_{\frac{1}{2},\Gamma}\left\|\mu\right\|_{-\frac{1}{2},\Gamma} \\
&\leq M\left\|\varphi\right\|_{-\frac{1}{2},\Gamma}\left\|\mu\right\|_{-\frac{1}{2},\Gamma}
\end{aligned}$$

The last inequality is from that the operator $\mathbb{K}_{I,k}^{+}: H^{-\frac{1}{2}}(\Gamma) \longrightarrow H^{\frac{1}{2}}(\Gamma)$ is bounded.

Furthermore

$$\mathbb{A}^{+}(\varphi, \mu; k) = \int_{\Gamma} u_{I,\varphi}(\boldsymbol{x})\mu(\boldsymbol{x})\mathrm{d}s_x.$$

Since $u_{I,\varphi}(\boldsymbol{x})$ is the unique solution of the transmission problem (4.3.2), it arrives at

$$\mathbb{A}^{+}(\varphi, \varphi; k) = a_n^{+}(u_{I,\varphi}, u_{I,\varphi}; k). \tag{4.3.17}$$

On the other hand we have

$$
\begin{aligned}
\|\varphi\|_{\frac{1}{2},\Gamma} &\le \left\| \frac{\partial u_{I,\varphi}}{\partial \boldsymbol{\nu}_{\boldsymbol{x}}}\bigg|_{\Gamma+0} \right\|_{-\frac{1}{2},\Gamma} + \left\| \frac{\partial u_{I,\varphi}}{\partial \boldsymbol{\nu}_{\boldsymbol{x}}}\bigg|_{\Gamma-0} \right\|_{-\frac{1}{2},\Gamma} \\
&\le C \left\{ a_n^{+}(u_{I,\varphi}, u_{I,\varphi}; k) \right\}^{\frac{1}{2}}, \qquad (4.3.18)
\end{aligned}
$$

with a constant $C > 0$. The last inequality is form the trace Theorem.

The equality (4.3.17) and inequality (4.3.18) yield the inequality (4.3.16) directly. □

Similarly for given density $\psi \in C^2(\Gamma)$, the double layer potential $V_{II,k}^{+}(\boldsymbol{x})$ is the unique solution of the following transmission problem:

$$
\begin{cases}
-\Delta u(\boldsymbol{x}) + k^2 u(\boldsymbol{x}) = 0, \quad \forall \boldsymbol{x} \in \mathbb{R}^n \backslash \Gamma, & (4.3.19\text{a}) \\
\Big[u(\boldsymbol{x}) \Big]_{\Gamma} = \psi(\boldsymbol{x}), \quad \forall \boldsymbol{x} \in \Gamma, & (4.3.19\text{b}) \\
\left[\dfrac{\partial u(\boldsymbol{x})}{\partial \boldsymbol{\nu}_{\boldsymbol{x}}} \right]_{\Gamma} = 0, \quad \forall \boldsymbol{x} \in \Gamma, & (4.3.19\text{c}) \\
\left. \begin{aligned} u(\boldsymbol{x}) &= \mathcal{O}\left(\frac{1}{|\boldsymbol{x}|^{(n-1)/2}} \mathrm{e}^{-k|\boldsymbol{x}|} \right), \\ \frac{\partial u(\boldsymbol{x})}{\partial r} &= \mathcal{O}\left(\frac{1}{|\boldsymbol{x}|^{(n-1)/2}} \mathrm{e}^{-k|\boldsymbol{x}|} \right), \end{aligned} \right\} \quad \text{as} \quad r = |\boldsymbol{x}| \to +\infty. & (4.3.19\text{d})
\end{cases}
$$

Besides

$$
\begin{aligned}
V_{II,k}^{+}(\boldsymbol{x})\Big|_{\Gamma} &= \frac{1}{2} \left\{ V_{II,k}^{+}(\boldsymbol{x})\Big|_{\Gamma+0} + V_{II,k}^{+}(\boldsymbol{x})\Big|_{\Gamma-0} \right\} \\
&= \int_{\Gamma} \frac{\partial}{\partial \boldsymbol{\nu}_{\boldsymbol{y}}} E_n^{+}(\boldsymbol{x}, \boldsymbol{y}; k) \psi(\boldsymbol{y}) \mathrm{d}s_{\boldsymbol{y}}, \quad \forall \boldsymbol{x} \in \Gamma. \qquad (4.3.20)
\end{aligned}
$$

For discussing the solvability of the transmission problem (4.3.19d) with given density $\psi \in H^{\frac{1}{2}}(\Gamma)$, introduce the space

$$
H^1(\mathbb{R}^n \backslash \Gamma) = \left\{ v\Big|_{\Omega} \in H^1(\Omega) \text{ and } v\Big|_{\Omega_e} \in H^1(\Omega_e) \right\}. \qquad (4.3.21)
$$

The equivalent variational problem of the transmission problem (4.3.19d) is given by

$$\begin{cases} \text{Find } u \in H^1(\mathbb{R}^n\backslash\Gamma), \text{ such that} \\ a^+(u, v; k) + a_e^+(u, v; k) \\ \quad = \displaystyle\int_\Gamma \left(\frac{\partial u}{\partial \boldsymbol{\nu}_x} v\right)\bigg|_{\Gamma-0} \mathrm{d}s_x - \int_\Gamma \left(\frac{\partial u}{\partial \boldsymbol{\nu}_x} v\right)\bigg|_{\Gamma+0} \mathrm{d}s_x, \ \forall v \in H^1(\mathbb{R}^n\backslash\Gamma), & (4.3.22a) \\ \Big[u(\boldsymbol{x})\Big]_\Gamma = \psi(\boldsymbol{x}), \qquad \forall \boldsymbol{x} \in \Gamma, & (4.3.22b) \\ \left[\dfrac{\partial u(\boldsymbol{x})}{\partial \boldsymbol{\nu}_x}\right]_\Gamma = 0, & (4.3.22c) \end{cases}$$

in which the bilinear form $a^+(u, v; k)$ and $a_e^+(u, v; k)$ are defined by (4.1.6) and (4.1.9).

Lemma 4.3.2 *For given $\psi \in H^{\frac{1}{2}}(\Gamma)$, the variational problem (4.3.22c) exists a unique solution $u_{II,\psi} \in H^1(\mathbb{R}^n\backslash\Gamma)$, and the following estimate hold:*

$$\left.\begin{aligned} &\|u_{II,\psi}\|_{1,\Omega} \le C\|\psi\|_{\frac{1}{2},\Gamma}, \\ &\|u_{II,\psi}\|_{1,\Omega_e} \le C\|\psi\|_{\frac{1}{2},\Gamma}, \\ &\left\|u_{II,\psi}\Big|_{\Gamma-0}\right\|_{\frac{1}{2},\Gamma} \le C\left\|\psi\right\|_{\frac{1}{2},\Gamma}, \\ &\left\|u_{II,\psi}\Big|_{\Gamma+0}\right\|_{\frac{1}{2},\Gamma} \le C\left\|\psi\right\|_{\frac{1}{2},\Gamma}, \\ &\left\|\frac{\partial u_{II,\psi}}{\partial \boldsymbol{\nu}_x}\Big|_\Gamma\right\|_{-\frac{1}{2},\Gamma} \le C\left\|\psi\right\|_{\frac{1}{2},\Gamma}. \end{aligned}\right\} \tag{4.3.23}$$

Proof Suppose $u \in H^1(\mathbb{R}^n\backslash\Gamma)$ is a solution of problem (4.3.22c) with $\psi(\boldsymbol{x}) = 0$. Taking the test function $v(\boldsymbol{x}) = u(\boldsymbol{x})$ in problem (4.3.22c) and combining the continuities of function $u(\boldsymbol{x})$ and its derivative with respect to $\boldsymbol{\nu}_x$ on Γ, it arrives at

$$a^+(u, u; k) + a_e^+(u, u; k) = 0,$$

namely

$$u(\boldsymbol{x}) \equiv 0, \qquad \forall \boldsymbol{x} \in \mathbb{R}^n.$$

The uniqueness of problem (4.3.22c) is proved.

Secondly for proving the existence of (4.3.22c), introduce several auxiliary functions. For given $\psi \in H^{\frac{1}{2}}(\Gamma)$ consider the following interior Dirichlet problem

$$\begin{cases} -\Delta u(\boldsymbol{x}) + k^2 u(\boldsymbol{x}) = 0, & \forall \boldsymbol{x} \in \Omega, \\ u(\boldsymbol{x})\Big|_\Gamma = -\psi(\boldsymbol{x}), & \forall \boldsymbol{x} \in \Gamma. \end{cases} \tag{4.3.24}$$

We know that (Chen & Wu, 2004) the problem (4.3.24) exists a unique weak solution $u_0 \in H^1(\Omega)$, which satisfies

$$
\begin{cases} a^+(u_0, v; k) = \displaystyle\int_\Gamma \frac{\partial u_0(\boldsymbol{x})}{\partial \boldsymbol{\nu}_x}\Big|_{\Gamma-0} v(\boldsymbol{x})\Big|_{\Gamma-0} \mathrm{d}s_x, & \forall v \in H^1(\Omega), \\ u_0(\boldsymbol{x})\Big|_\Gamma = -\psi(\boldsymbol{x}), & \forall \boldsymbol{x} \in \Gamma. \end{cases} \tag{4.3.25}
$$

In addition

$$
\|u_0\|_{1,\Omega} \le C\|\psi\|_{\frac{1}{2},\Gamma}, \tag{4.3.26}
$$

$$
\left\| \frac{\partial u_0}{\partial \boldsymbol{\nu}_x}\Big|_{\Gamma-0} \right\|_{-\frac{1}{2},\Gamma} \le C\left\|\psi\right\|_{\frac{1}{2},\Gamma}. \tag{4.3.27}
$$

Let

$$
u_1(\boldsymbol{x}) = \begin{cases} u_0(\boldsymbol{x}), & \forall \boldsymbol{x} \in \Omega, \\ \frac{1}{2}u_0(\boldsymbol{x}), & \forall \boldsymbol{x} \in \Gamma, \\ 0, & \forall \boldsymbol{x} \in \Omega_e. \end{cases} \tag{4.3.28}
$$

Obviously $u_1 \in H^1(\mathbb{R}^n \backslash \Gamma)$ satisfied

$$
\begin{cases} a^+(u_1, v; k) + a_e^+(u_1, v; k) = \displaystyle\int_\Gamma \frac{\partial u_0(\boldsymbol{x})}{\partial \boldsymbol{\nu}_x}\Big|_{\Gamma-0} v(\boldsymbol{x})\Big|_{\Gamma-0} \mathrm{d}s_x, \\ \qquad\qquad\qquad\qquad \forall v \in H^1(\mathbb{R}^n \backslash \Gamma), & (4.3.29\text{a}) \\ [u_1(\boldsymbol{x})]_\Gamma = \psi(\boldsymbol{x}), & (4.3.29\text{b}) \\ \left[\dfrac{\partial u_1(\boldsymbol{x})}{\partial \boldsymbol{\nu}_x}\right]_\Gamma = -\dfrac{\partial u_0(\boldsymbol{x})}{\partial \boldsymbol{\nu}_x}\Big|_{\Gamma-0}. & (4.3.29\text{c}) \end{cases}
$$

Next consider the transmission problem (4.3.1d) with the density $\varphi = -\dfrac{\partial u_0}{\partial \boldsymbol{\nu}_x}\Big|_{\Gamma-0} \in H^{-\frac{1}{2}}(\Gamma)$, its equivalent variational problem (4.3.3) exists a unique solution $u_2 \in H^1(\mathbb{R}^n)$ satisfies

$$
\begin{cases} a^+(u_2, v; k) + a_e^+(u_2, v; k) = \displaystyle\int_\Gamma \left(\frac{\partial u_2(\boldsymbol{x})}{\partial \boldsymbol{\nu}_x} v(\boldsymbol{x})\right)\Big|_{\Gamma-0} \mathrm{d}s_x - \\ \qquad \displaystyle\int_\Gamma \left(\frac{\partial u_2(\boldsymbol{x})}{\partial \boldsymbol{\nu}_x} v(\boldsymbol{x})\right)\Big|_{\Gamma+0} \mathrm{d}s_x, \ \forall v \in H^1(\mathbb{R}^n \backslash \Gamma), & (4.3.30\text{a}) \\ \left[u_2(\boldsymbol{x})\right]_\Gamma = 0, & (4.3.30\text{b}) \\ \left[\dfrac{\partial u_2(\boldsymbol{x})}{\partial \boldsymbol{\nu}_x}\right]_\Gamma = \dfrac{\partial u_0(\boldsymbol{x})}{\partial \boldsymbol{\nu}_x}\Big|_{\Gamma-0}. & (4.3.30\text{c}) \end{cases}
$$

Moreover we obtain

$$\left.\begin{aligned}
&\left\|u_2\right\|_{1,\mathbb{R}^n} \le C\left\|\frac{\partial u_0}{\partial \boldsymbol{\nu}_x}\Big|_{\Gamma-0}\right\|_{-\frac{1}{2},\Gamma} \le C\left\|\psi\right\|_{\frac{1}{2},\Gamma},\\
&\left\|u_2\big|_\Gamma\right\|_{\frac{1}{2},\Gamma} \le C\left\|\psi\right\|_{\frac{1}{2},\Gamma},\\
&\left\|\frac{\partial u_2}{\partial \boldsymbol{\nu}_x}\Big|_{\Gamma-0}\right\|_{-\frac{1}{2},\Gamma} \le C\left\|\psi\right\|_{\frac{1}{2},\Gamma},\\
&\left\|\frac{\partial u_2}{\partial \boldsymbol{\nu}_x}\Big|_{\Gamma+0}\right\|_{-\frac{1}{2},\Gamma} \le C\left\|\psi\right\|_{\frac{1}{2},\Gamma}.
\end{aligned}\right\} \tag{4.3.31}$$

Define

$$u_{II,\psi}(\boldsymbol{x}) = u_1(\boldsymbol{x}) + u_2(\boldsymbol{x}), \quad \forall \boldsymbol{x} \in \mathbb{R}^n, \tag{4.3.32}$$

then $u_{II,\psi} \in H^1(\mathbb{R}^n \backslash \Gamma)$. It is straightforward to check that $u_{II,\psi}(\boldsymbol{x})$ is the unique solution of (4.3.22c) with given $\psi \in H^{\frac{1}{2}}(\Gamma)$ and the estimates (4.3.23) hold. The proof is completed. □

The double layer potential of modified Helmholtz operator with given density $\psi \in H^{\frac{1}{2}}(\Gamma)$ is defined by

$$V^+_{II,k}(\boldsymbol{x}) = u_{II,\psi}(\boldsymbol{x}) = \int_\Gamma \frac{\partial}{\partial \boldsymbol{\nu}_x} E^+_n(\boldsymbol{x}, \boldsymbol{y}; k)\psi(\boldsymbol{y})\mathrm{d}s_y, \quad \forall \boldsymbol{x} \in \mathbb{R}^n \backslash \Gamma, \tag{4.3.33}$$

$$\begin{aligned}
V^+_{II,k}(\boldsymbol{x})\Big|_\Gamma &= \frac{1}{2}\left\{u_{II,\psi}(\boldsymbol{x})\Big|_{\Gamma+0} + u_{II,\psi}(\boldsymbol{x})\Big|_{\Gamma-0}\right\} := \left(\mathbb{K}^+_{II,k}\psi\right)(\boldsymbol{x})\\
&= \int_\Gamma \frac{\partial}{\partial \boldsymbol{\nu}_y} E^+_n(\boldsymbol{x}, \boldsymbol{y}; k)\psi(\boldsymbol{y})\mathrm{d}s_y, \quad \forall \boldsymbol{x} \in \Gamma,
\end{aligned} \tag{4.3.34}$$

$$\begin{aligned}
\frac{\partial}{\partial \boldsymbol{\nu}_x} V^+_{II,k}(\boldsymbol{x})\Big|_\Gamma &= \frac{\partial}{\partial \boldsymbol{\nu}_x} u_{II,\psi}(\boldsymbol{x})\Big|_\Gamma := \left(\mathbb{K}^+_{III,k}\psi\right)(\boldsymbol{x})\\
&= \frac{\partial}{\partial \boldsymbol{\nu}_x}\int_\Gamma \frac{\partial}{\partial \boldsymbol{\nu}_y} E^+_n(\boldsymbol{x}, \boldsymbol{y}; k)\psi(\boldsymbol{y})\mathrm{d}s_y\\
&= \sum_{1\le i<j\le n} \frac{\partial}{\partial \boldsymbol{\tau}^{i,j}_x}\int_\Gamma E^+_n(\boldsymbol{x}, \boldsymbol{y}; k)\frac{\partial}{\partial \boldsymbol{\tau}^{i,j}_y}\psi(\boldsymbol{y})\mathrm{d}s_y -\\
&\quad k^2\int_\Gamma E^+_n(\boldsymbol{x}, \boldsymbol{y}; k)\big(\boldsymbol{\nu}(\boldsymbol{x})\cdot\boldsymbol{\nu}(\boldsymbol{y})\big)\psi(\boldsymbol{y})\mathrm{d}s_y, \quad \forall \boldsymbol{x} \in \Gamma.
\end{aligned} \tag{4.3.35}$$

We have known that for the given density $\psi \in C^2(\Gamma)$, the last equalities in (4.3.33), (4.3.34) and (4.3.35) hold, for $\psi \in H^{\frac{1}{2}}(\Gamma)$ the last equalities in (4.3.33), (4.3.34) and (4.3.35) still hold, provided by the space $C^2(\Gamma)$ is dense in $H^{\frac{1}{2}}(\Gamma)$ and the estimates (4.3.23) hold for $\psi \in H^{\frac{1}{2}}(\Gamma)$.

Furthermore we have

Theorem 4.3.2 *Assume that* $\Gamma \in C^2$, *and the density* $\psi \in H^{\frac{1}{2}}(\Gamma)$.

(i) The double layer potential $V^+_{II,k}(\boldsymbol{x})$ *is well defined on* $\mathbb{R}^n$ *and* $V^+_{II,k} \in H^1(\mathbb{R}^n \backslash \Gamma)$. *On the domain* $\mathbb{R}^n \backslash \Gamma$, $V^+_{II,k}(\boldsymbol{x})$ *satisfies the modified Helmholtz equation* (4.0.1).

(ii)

$$\left.\begin{aligned} V^+_{II,k}(\boldsymbol{x})\Big|_{\Gamma+0} &= \frac{1}{2}\psi(\boldsymbol{x}) + \int_\Gamma \frac{\partial}{\partial \boldsymbol{v}_y} E^+_n(\boldsymbol{x}, \boldsymbol{y}; k)\psi(\boldsymbol{y})\mathrm{d}s_y, \quad \forall \boldsymbol{x} \in \Gamma, \\ V^+_{II,k}(\boldsymbol{x})\Big|_{\Gamma-0} &= -\frac{1}{2}\psi(\boldsymbol{x}) + \int_\Gamma \frac{\partial}{\partial \boldsymbol{v}_y} E^+_n(\boldsymbol{x}, \boldsymbol{y}; k)\psi(\boldsymbol{y})\mathrm{d}s_y, \quad \forall \boldsymbol{x} \in \Gamma, \end{aligned}\right\} \tag{4.3.36}$$

$$V^+_{II,k}(\boldsymbol{x})\Big|_\Gamma = \int_\Gamma \frac{\partial}{\partial \boldsymbol{v}_y} E^+_n(\boldsymbol{x}, \boldsymbol{y}; k)\psi(\boldsymbol{y})\mathrm{d}s_y := \left(\mathbb{K}^+_{II,k}\psi\right)(\boldsymbol{x}), \quad \forall \boldsymbol{x} \in \Gamma. \tag{4.3.37}$$

(iii)

$$\begin{aligned} \frac{\partial}{\partial \boldsymbol{v}_x}(\boldsymbol{x}) V^+_{II,k}(\boldsymbol{x})\Big|_{\Gamma+0} &= \frac{\partial}{\partial \boldsymbol{v}_x} V^+_{II,k}(\boldsymbol{x})\Big|_{\Gamma-0} = \frac{\partial}{\partial \boldsymbol{v}_x} V^+_{II,k}(\boldsymbol{x})\Big|_\Gamma \\ &= \frac{\partial}{\partial \boldsymbol{v}_x} \int_\Gamma \frac{\partial}{\partial \boldsymbol{v}_y} E^+_n(\boldsymbol{x}, \boldsymbol{y}; k)\psi(\boldsymbol{y})\mathrm{d}s_y \\ &= \sum_{1\le i<j\le n} \frac{\partial}{\partial \boldsymbol{\tau}^{i,j}_x} \int_\Gamma E^+_n(\boldsymbol{x}, \boldsymbol{y}; k) \frac{\partial}{\partial \boldsymbol{\tau}^{i,j}_y} \psi(\boldsymbol{y})\mathrm{d}s_y - \\ &\quad k^2 \int_\Gamma E^+_n(\boldsymbol{x}, \boldsymbol{y}; k)\big(\boldsymbol{v}(\boldsymbol{x}) \cdot \boldsymbol{v}(\boldsymbol{y})\big)\psi(\boldsymbol{y})\mathrm{d}s_y \\ &:= \left(\mathbb{K}^+_{III,k}\psi\right)(\boldsymbol{x}), \quad \forall \boldsymbol{x} \in \Gamma. \end{aligned} \tag{4.3.38}$$

(iv)

$$\left.\begin{aligned} V^+_{II,k}(\boldsymbol{x}) &= \mathcal{O}\left(\frac{1}{|\boldsymbol{x}|^{(n-1)/2}} \mathrm{e}^{-k|\boldsymbol{x}|}\right), \\ \frac{\partial V^+_{II,k}(\boldsymbol{x})}{\partial r} &= \mathcal{O}\left(\frac{1}{|\boldsymbol{x}|^{(n-1)/2}} \mathrm{e}^{-k|\boldsymbol{x}|}\right), \end{aligned}\right\} \quad \text{as} \quad r = |\boldsymbol{x}| \to +\infty. \tag{4.3.39}$$

(v) The operators $\mathbb{K}^+_{II,k}$ *and* $\mathbb{K}^+_{III,k}$ *are continuous and bounded*

$$\begin{aligned} \mathbb{K}^+_{II,k} &: H^{\frac{1}{2}+\gamma}(\Gamma) \longrightarrow H^{\frac{1}{2}+\gamma}(\Gamma), \\ \mathbb{K}^+_{III,k} &: H^{\frac{1}{2}+\gamma}(\Gamma) \longrightarrow H^{-\frac{1}{2}+\gamma}(\Gamma), \end{aligned}$$

provided by $0 \le \gamma \le \dfrac{1}{2}$.

(vi) Let

$$\mathbb{B}^{+}(\psi, \rho; k) = \sum_{1 \le i < j \le n} \int_{\Gamma} \int_{\Gamma} E_n^{+}(\boldsymbol{x}, \boldsymbol{y}; k) \frac{\partial}{\partial \boldsymbol{\tau}_{\boldsymbol{y}}^{i,j}} \psi(\boldsymbol{y}) \frac{\partial}{\partial \boldsymbol{\tau}_{\boldsymbol{x}}^{i,j}} \rho(\boldsymbol{x}) \mathrm{d}s_y \mathrm{d}s_x +$$

$$k^2 \int_{\Gamma} \int_{\Gamma} E_n^{+}(\boldsymbol{x}, \boldsymbol{y}; k) \big(\boldsymbol{\nu}(\boldsymbol{x}) \cdot \boldsymbol{\nu}(\boldsymbol{y})\big) \psi(\boldsymbol{y}) \rho(\boldsymbol{x}) \mathrm{d}s_y \mathrm{d}s_x,$$

$$\forall \psi, \rho \in H^{\frac{1}{2}}(\Gamma). \qquad (4.3.40)$$

Then $\mathbb{B}^{+}(\psi, \rho; k)$ *is bounded and coercive bilinear form on* $H^{\frac{1}{2}}(\Gamma) \times H^{\frac{1}{2}}(\Gamma)$, *namely there exist two positive constants* M *and* α *such that*

$$|\mathbb{B}^{+}(\psi, \rho; k)| \le M \|\psi\|_{\frac{1}{2},\Gamma} \|\rho\|_{\frac{1}{2},\Gamma}, \quad \forall \psi, \rho \in H^{\frac{1}{2}}(\Gamma), \qquad (4.3.41)$$

$$\mathbb{B}^{+}(\psi, \psi; k) \ge \alpha \|\psi\|_{\frac{1}{2},\Gamma}^2, \qquad \forall \psi \in H^{\frac{1}{2}}(\Gamma). \qquad (4.3.42)$$

4.4 Boundary Reduction for the Boundary Value Problems of Modified Helmholtz Equation

4.4.1 Boundary Integral Equation and Integro-Differential Equation of the First Kind

At first we consider the interior and exterior Dirichlet boundary value problems (4.1.1) and (4.1.2). Let the single layer potential with an unknown density $\varphi(\boldsymbol{x})$ be the solution of problem (4.1.1) on $\overline{\Omega}$ and the solution of problem (4.1.2) on $\overline{\Omega_e}$, namely

$$u_{I,\varphi}(\boldsymbol{x}) = \int_{\Gamma} E_n^{+}(\boldsymbol{x}, \boldsymbol{y}; k) \varphi(\boldsymbol{x}) \mathrm{d}s_y, \quad \forall \boldsymbol{x} \in \mathbb{R}^n. \qquad (4.4.1)$$

The unknown density $\varphi(\boldsymbol{x})$ is to be determined by Dirichlet boundary condition in the problems, that is

$$\left(\mathbb{K}_{I,k}^{+} \varphi\right)(\boldsymbol{x}) = \int_{\Gamma} E_n^{+}(\boldsymbol{x}, \boldsymbol{y}; k) \varphi(\boldsymbol{y}) \mathrm{d}s_y = g(\boldsymbol{x}), \quad \forall \boldsymbol{x} \in \Gamma, \qquad (4.4.2)$$

in which $g \in H^{\frac{1}{2}}(\Gamma)$ is given.

For the boundary integral equation of the first kind (4.4.2), we obtain

Theorem 4.4.1 *(i) For given* $g \in H^{\frac{1}{2}}(\Gamma)$, *the problem* (4.4.2) *exists a unique solution* $\varphi \in H^{-\frac{1}{2}}(\Gamma)$, *and*

$$\|\varphi\|_{-\frac{1}{2},\Gamma} \le C \|g\|_{\frac{1}{2},\Gamma}. \qquad (4.4.3)$$

(ii) The operator $\mathbb{K}_{I,k}^{+}$ *is invertible and its inverse operator* $\left(\mathbb{K}_{I,k}^{+}\right)^{-1}$ *is continuous and bounded*

$$\left(\mathbb{K}_{I,k}^{+}\right)^{-1} : \; H^{\frac{1}{2}+\gamma}(\Gamma) \longrightarrow H^{-\frac{1}{2}+\gamma}(\Gamma), \tag{4.4.4}$$

provided by $0 \le \gamma \le \dfrac{1}{2}$.

Proof The equivalent variational problem of problem (4.4.2) is given by

$$\begin{cases} \text{Find } \varphi \in H^{-\frac{1}{2}}(\Gamma), \quad \text{such that} \\ \mathbb{A}^{+}(\varphi, \mu; k) = \displaystyle\int_{\Gamma} g(\boldsymbol{x})\mu(\boldsymbol{x})\mathrm{d}s_{\boldsymbol{x}}, \quad \forall \mu \in H^{-\frac{1}{2}}(\Gamma). \end{cases} \tag{4.4.5}$$

From the Theorem 4.3.1, we know that $\mathbb{A}^{+}(\varphi, \mu; k)$ is a bounded coercive bilinear form on $H^{-\frac{1}{2}}(\Gamma) \times H^{-\frac{1}{2}}(\Gamma)$, furthermore the fact, given function $g \in H^{\frac{1}{2}}(\Gamma)$, yields that the right side of the variational equality (4.4.5) is a given linear functional on $H^{-\frac{1}{2}}(\Gamma)$. An application of Lax–Milgram Theorem, it arrives at the problem (4.4.5) (problem (4.4.2)) exists a unique solution $\varphi \in H^{-\frac{1}{2}}(\Gamma)$ and the estimate (4.4.3) holds. It means that the operator $\left(\mathbb{K}_{I,k}^{+}\right)^{-1}$ is continuous and bounded

$$\left(\mathbb{K}_{I,k}^{+}\right)^{-1} : \quad H^{\frac{1}{2}+\gamma}(\Gamma) \longrightarrow H^{-\frac{1}{2}+\gamma}(\Gamma),$$

provided by $0 \le \gamma \le 1/2$, and the regularity of the solution of Dirichlet boundary value problem (4.1.1). □

Secondly we discuss the interior and exterior Neumann boundary value problem (4.1.3) and (4.1.4). Let the double layer potential with an unknown density $\psi \in H^{\frac{1}{2}}(\Gamma)$ be the solution of problem (4.1.3) in Ω and the solution of problem (4.1.4) in Ω_e, namely

$$u_{II,\psi}(\boldsymbol{x}) = \int_{\Gamma} \frac{\partial}{\partial \boldsymbol{\nu}_{\boldsymbol{y}}} E_n^{+}(\boldsymbol{x}, \boldsymbol{y}; k)\psi(\boldsymbol{y})\mathrm{d}s_{\boldsymbol{y}}, \quad \forall \boldsymbol{x} \in \mathbb{R}^n. \tag{4.4.6}$$

The unknown density $\psi(\boldsymbol{x})$ is to be determined by the Neumann boundary condition in the problems (4.1.3) and (4.1.4), that is the hypersingular boundary integral equation

$$-\frac{\partial}{\partial \boldsymbol{\nu}_{\boldsymbol{x}}} \int_{\Gamma} \frac{\partial}{\partial \boldsymbol{\nu}_{\boldsymbol{x}}} E_n^{+}(\boldsymbol{x}, \boldsymbol{y}; k)\psi(\boldsymbol{x})\mathrm{d}s_{\boldsymbol{y}} = -h(\boldsymbol{x}), \quad \forall \boldsymbol{x} \in \Gamma, \tag{4.4.7}$$

in which $h \in H^{-\frac{1}{2}}(\Gamma)$ is given.

The equalities (4.3.35) yield that the hypersingularity boundary integral equation (4.4.7) is equivalent to the following boundary integro-differential equation of the first kind

$$
\begin{aligned}
-\left(\mathbb{K}^{+}_{\mathit{III},k}\psi\right)(\boldsymbol{x}) = &- \sum_{1\le i<j\le n} \frac{\partial}{\partial \boldsymbol{\tau}^{i,j}_{\boldsymbol{x}}} \int_{\Gamma} E^{+}_{n}(\boldsymbol{x},\boldsymbol{y};k) \frac{\partial}{\partial \boldsymbol{\tau}^{i,j}_{\boldsymbol{y}}} \psi(\boldsymbol{y})\mathrm{d}s_{\boldsymbol{y}} + \\
& k^2 \int_{\Gamma} E^{+}_{n}(\boldsymbol{x},\boldsymbol{y};k)\big(\boldsymbol{\nu}(\boldsymbol{x})\cdot\boldsymbol{\nu}(\boldsymbol{y})\big)\psi(\boldsymbol{y})\mathrm{d}s_{\boldsymbol{y}} \\
&= -h(\boldsymbol{x}), \quad \forall \boldsymbol{x}\in\Gamma.
\end{aligned} \tag{4.4.8}
$$

Furthermore we have

Theorem 4.4.2 *(i) For given $h \in H^{-\frac{1}{2}}(\Gamma)$, the problem* (4.4.8) *exists a unique solution $\psi \in H^{\frac{1}{2}}(\Gamma)$, and*

$$\|\psi\|_{\frac{1}{2},\Gamma} \le C\|h\|_{-\frac{1}{2},\Gamma}. \tag{4.4.9}$$

(ii) The operator $\mathbb{K}^{+}_{\mathit{III},k}$ is invertible and its inverse operator $\left(\mathbb{K}^{+}_{\mathit{III},k}\right)^{-1}$ is continuous and bounded

$$\left(\mathbb{K}^{+}_{\mathit{III},k}\right)^{-1} : \quad H^{-\frac{1}{2}+\gamma}(\Gamma) \longrightarrow H^{\frac{1}{2}+\gamma}(\Gamma), \tag{4.4.10}$$

provided by $0 \le \gamma \le \dfrac{1}{2}$.

Proof For given $h \in H^{-\frac{1}{2}}(\Gamma)$, the boundary integro-differential equation (4.4.8) is equivalent to the below variational problem:

$$
\begin{cases}
\text{Find } \psi \in H^{\frac{1}{2}}(\Gamma), \quad \text{such that} \\[2ex]
\qquad \mathbb{B}^{+}(\psi,\rho;k) = -\displaystyle\int_{\Gamma} h(\boldsymbol{x})\rho(\boldsymbol{x})\mathrm{d}s_{\boldsymbol{x}}, \quad \forall \rho \in H^{\frac{1}{2}}(\Gamma).
\end{cases} \tag{4.4.11}
$$

in which the bilinear form $\mathbb{B}^{+}(\psi,\rho;k)$ is defined by (4.3.40). From Theorem 4.3.2 we know that $\mathbb{B}(\psi,\rho;k)$ is a bounded coercive bilinear form on $H^{\frac{1}{2}}(\Gamma)\times H^{\frac{1}{2}}(\Gamma)$. For the given function $h \in H^{-\frac{1}{2}}(\Gamma)$, the right side of the variational equality (4.4.11) is a bounded linear functional on $H^{\frac{1}{2}}(\Gamma)$. The conclusion (i) is proved directly by Lax–Milgram Theorem. It means that operator $\mathbb{K}^{+}_{\mathit{III},k}$ is invertible and its inverse operator $\left(\mathbb{K}^{+}_{\mathit{III},k}\right)^{-1}$ is continuous and bounded:

$$\left(\mathbb{K}^{+}_{\mathit{III},k}\right)^{-1} : \quad H^{-\frac{1}{2}}(\Gamma) \longrightarrow H^{\frac{1}{2}}(\Gamma).$$

Moreover by the regularity of the solution of Neumann boundary value problem (4.1.3) the condition (ii) is proved completely. □

4.4.2 Boundary Integral Equations of the Second Kind

Let the double layer potential with unknown density $\psi \in H^{\frac{1}{2}}(\Gamma)$ be the solution of interior Dirichlet problem (4.1.1) in Ω and exterior Dirichlet problem (4.1.2) in Ω_e, namely

$$u_{\mathit{II},\psi}(\boldsymbol{x}) = \int_\Gamma \frac{\partial}{\partial \boldsymbol{v}_y} E_n^+(\boldsymbol{x}, \boldsymbol{y}; k)\psi(\boldsymbol{y})\mathrm{d}s_y, \quad \forall \boldsymbol{x} \in \mathbb{R}^n. \tag{4.4.12}$$

For the interior Dirichlet problem (4.1.1), the density $\psi(\boldsymbol{x})$ is to be determined by the Dirichlet boundary condition (4.1.1b), namely

$$-\frac{1}{2}\psi(\boldsymbol{x}) + \int_\Gamma \frac{\partial}{\partial \boldsymbol{v}_y} E_n^+(\boldsymbol{x}, \boldsymbol{y}; k)\psi(\boldsymbol{y})\mathrm{d}s_y = g(\boldsymbol{x}), \quad \forall \boldsymbol{x} \in \Gamma. \tag{4.4.13}$$

Using the operator $K_{\mathit{II},k}^+$ the boundary integral equation of second kind (4.4.13) can be rewritten as

$$-\frac{1}{2}\psi(\boldsymbol{x}) + \left(\mathbb{K}_{\mathit{II},k}^+\psi\right)(\boldsymbol{x}) = g(\boldsymbol{x}), \quad \forall \boldsymbol{x} \in \Gamma. \tag{4.4.14}$$

For the exterior Dirichlet problem (4.1.2), the density $\psi(\boldsymbol{x})$ is to be determined by the Dirichlet condition (4.1.2c), we obtain

$$\frac{1}{2}\psi(\boldsymbol{x}) + \int_\Gamma \frac{\partial}{\partial \boldsymbol{v}_y} E_n^+(\boldsymbol{x}, \boldsymbol{y}; k)\psi(\boldsymbol{y})\mathrm{d}s_y = g(\boldsymbol{x}), \quad \forall \boldsymbol{x} \in \Gamma, \tag{4.4.15}$$

namely

$$\frac{1}{2}\psi(\boldsymbol{x}) + \left(\mathbb{K}_{\mathit{II},k}^+\psi\right)(\boldsymbol{x}) = g(\boldsymbol{x}), \quad \forall \boldsymbol{x} \in \Gamma. \tag{4.4.16}$$

For the boundary integral equation (4.4.13) and (4.4.15), we have

Theorem 4.4.3 *Suppose that the given function $g \in H^{\frac{1}{2}}(\Gamma)$.*

(i) Boundary integral equation (4.4.13) *exists a unique solution $\psi \in H^{\frac{1}{2}}(\Gamma)$, and*

$$\|\psi\|_{\frac{1}{2},\Gamma} \le C\|g\|_{\frac{1}{2},\Gamma}. \tag{4.4.17}$$

(ii) Boundary integral equation (4.4.15) *exists a unique solution $\psi \in H^{\frac{1}{2}}(\Gamma)$, and*

$$\|\psi\|_{\frac{1}{2},\Gamma} \le C\|g\|_{\frac{1}{2},\Gamma}. \tag{4.4.18}$$

(iii) The operator $\left(-\frac{1}{2}I + \mathbb{K}_{\mathit{II},k}^+\right)$ and $\left(\frac{1}{2}I + \mathbb{K}_{\mathit{II},k}^+\right)$ are invertible and the inverse operator $\left(-\frac{1}{2}I + \mathbb{K}_{\mathit{II},k}^+\right)^{-1}$ and $\left(\frac{1}{2}I + \mathbb{K}_{\mathit{II},k}^+\right)^{-1}$ are continuous and bounded

$$\left.\begin{aligned}\left(-\frac{1}{2}I+\mathbb{K}^{+}_{II,k}\right)^{-1} &: \quad H^{\frac{1}{2}+\gamma}(\Gamma)\longrightarrow H^{\frac{1}{2}+\gamma}(\Gamma),\\ \left(\frac{1}{2}I+\mathbb{K}^{+}_{II,k}\right)^{-1} &: \quad H^{\frac{1}{2}+\gamma}(\Gamma)\longrightarrow H^{\frac{1}{2}+\gamma}(\Gamma),\end{aligned}\right\} \tag{4.4.19}$$

provided by $0\le\gamma\le\dfrac{1}{2}$.

Proof An equivalent variational problem of the boundary integral equation (4.4.13) is given by

$$\begin{cases}\text{Find } \psi\in H^{\frac{1}{2}}(\Gamma), \quad \text{such that}\\ \displaystyle\int_\Gamma\left\{-\frac{1}{2}\psi(\boldsymbol{x})+\left(\mathbb{K}^{+}_{II,k}\psi\right)(\boldsymbol{x})\right\}\mu(\boldsymbol{x})\mathrm{d}s_{\boldsymbol{x}}=\int_\Gamma g(\boldsymbol{x})\mu(\boldsymbol{x})\mathrm{d}s_{\boldsymbol{x}},\ \forall\mu\in H^{-\frac{1}{2}}(\Gamma).\end{cases} \tag{4.4.20}$$

By the Theorem 4.4.2 we know that for any $\mu\in H^{-\frac{1}{2}}(\Gamma)$ exists a unique $\rho\in H^{\frac{1}{2}}(\Gamma)$, such that

$$\mu(\boldsymbol{x})=\left(\mathbb{K}^{+}_{III,k}\rho\right)(\boldsymbol{x}). \tag{4.4.21}$$

Taking (4.4.21) in the variational problem (4.4.20), it yields that

$$\begin{cases}\text{Find } \psi\in H^{\frac{1}{2}}(\Gamma), \quad \text{such that}\\ \displaystyle\mathbb{C}^{+}(\psi,\rho;k):=\int_\Gamma\left\{-\frac{1}{2}\psi(\boldsymbol{x})+\left(\mathbb{K}^{+}_{II,k}\psi\right)(\boldsymbol{x})\right\}\left(\mathbb{K}^{+}_{III,k}\rho\right)(\boldsymbol{x})\mathrm{d}s_{\boldsymbol{x}}\\ \displaystyle\qquad=\int_\Gamma g(\boldsymbol{x})\left(\mathbb{K}^{+}_{III,k}\rho\right)(\boldsymbol{x})\mathrm{d}s_{\boldsymbol{x}},\quad \forall\rho\in H^{\frac{1}{2}}(\Gamma).\end{cases} \tag{4.4.22}$$

Similarly we obtain the following two equivalent variational problem of the boundary integral equation (4.4.15),

$$\begin{cases}\text{Find } \psi\in H^{\frac{1}{2}}(\Gamma), \quad \text{such that}\\ \displaystyle -\int_\Gamma\left\{\frac{1}{2}\psi(\boldsymbol{x})+\left(\mathbb{K}^{+}_{II,k}\psi\right)(\boldsymbol{x})\right\}\mu(\boldsymbol{x})\mathrm{d}s_{\boldsymbol{x}}=-\int_\Gamma g(\boldsymbol{x})\mu(\boldsymbol{x})\mathrm{d}s_{\boldsymbol{x}},\ \forall\mu\in H^{-\frac{1}{2}}(\Gamma),\end{cases} \tag{4.4.23}$$

and

$$
\begin{cases}
\text{Find } \psi \in H^{\frac{1}{2}}(\Gamma), \quad \text{such that} \\
\mathbb{C}_e^+(\psi, \rho; k) := -\int_\Gamma \left\{ \frac{1}{2}\psi(\boldsymbol{x}) + \left(\mathbb{K}_{II,k}^+ \psi\right)(\boldsymbol{x}) \right\} \left(\mathbb{K}_{III,k}^+ \rho\right)(\boldsymbol{x}) \mathrm{d}s_x \\
\qquad = -\int_\Gamma g(\boldsymbol{x}) \left(\mathbb{K}_{III,k}^+ \rho\right)(\boldsymbol{x}) \mathrm{d}s_x, \quad \forall \rho \in H^{\frac{1}{2}}(\Gamma).
\end{cases} \tag{4.4.24}
$$

Obviously $\mathbb{C}^+(\psi, \rho; k)$ and $\mathbb{C}_e^+(\psi, \rho; k)$ are two bounded bilinear forms on $H^{\frac{1}{2}}(\Gamma) \times H^{\frac{1}{2}}(\Gamma)$. For any $\rho \in H^{\frac{1}{2}}(\Gamma)$, let

$$
u_{II,\rho}(\boldsymbol{x}) = \int_\Gamma \frac{\partial}{\partial \boldsymbol{\nu}_x} E_n^+(\boldsymbol{x}, \boldsymbol{y}; k)\rho(\boldsymbol{y}) \mathrm{d}s_y, \quad \forall \boldsymbol{x} \in \mathbb{R}^n. \tag{4.4.25}
$$

Then $u_{II,\rho} \in H^1(\mathbb{R}^n \backslash \Gamma)$. In the domain Ω, $u_{II,\rho}(\boldsymbol{x})$ is the unique solution of the following interior Neumann problem:

$$
\begin{cases}
-\Delta u_{II,\rho}(\boldsymbol{x}) + k^2 u_{II,\rho}(\boldsymbol{x}) = 0, \quad \forall \boldsymbol{x} \in \Omega, & (4.4.26\text{a}) \\
\dfrac{\partial}{\partial \boldsymbol{\nu}_x} u_{II,\rho}(\boldsymbol{x})\Big|_\Gamma = \left(K_{III,k}^+ \rho\right)(\boldsymbol{x}), \quad \forall \boldsymbol{x} \in \Gamma. & (4.4.26\text{b})
\end{cases}
$$

In the domain Ω_e, $u_{II,\rho}(\boldsymbol{x})$ is the unique solution of the below exterior Neumann problem:

$$
\begin{cases}
-\Delta u_{II,\rho}(\boldsymbol{x}) + k^2 u_{II,\rho}(\boldsymbol{x}) = 0, \quad \forall \boldsymbol{x} \in \Omega_e, & (4.4.27\text{a}) \\
\dfrac{\partial}{\partial \boldsymbol{\nu}_x} u_{II,\rho}(\boldsymbol{x}) = \left(\mathbb{K}_{III,k}^+ \rho\right)(\boldsymbol{x}), \quad \forall \boldsymbol{x} \in \Gamma, & (4.4.27\text{b}) \\
\left. \begin{aligned} & u_{II,\rho}(\boldsymbol{x}) = \mathcal{O}\left(\frac{1}{|\boldsymbol{x}|^{(n-1)/2}} \mathrm{e}^{-k|\boldsymbol{x}|}\right), \\ & \frac{\partial}{\partial r} u_{II,\rho}(\boldsymbol{x}) = \mathcal{O}\left(\frac{1}{|\boldsymbol{x}|^{(n-1)/2}} \mathrm{e}^{-k|\boldsymbol{x}|}\right), \end{aligned} \right\} \quad \text{as } r = |\boldsymbol{x}| \to +\infty. & (4.4.27\text{c})
\end{cases}
$$

The problem (4.4.26) yields that

$$
\begin{aligned}
a^+(u_{II,\psi}, u_{II,\rho}; k) &= \int_\Omega \nabla u_{II,\psi}(\boldsymbol{x}) \cdot \nabla u_{II,\rho}(\boldsymbol{x}) \mathrm{d}\boldsymbol{x} + k^2 \int_\Omega u_{II,\psi}(\boldsymbol{x}) u_{II,\rho}(\boldsymbol{x}) \mathrm{d}\boldsymbol{x} \\
&= \int_\Gamma u_{II,\psi}(\boldsymbol{x})\Big|_{\Gamma-0} \frac{\partial}{\partial \boldsymbol{\nu}_x} u_{II,\rho}(\boldsymbol{x})\Big|_\Gamma \mathrm{d}s_x \\
&= \mathbb{C}^+(\psi, \rho; k).
\end{aligned} \tag{4.4.28}
$$

Similarly the problem (4.4.27) yield that

$$
\begin{aligned}
a_e^+(u_{II,\psi}, u_{II,\rho}; k) &= \int_{\Omega_e} \nabla u_{II,\psi}(\boldsymbol{x}) \cdot \nabla u_{II,\rho}(\boldsymbol{x})\mathrm{d}\boldsymbol{x} + k^2 \int_{\Omega_e} u_{II,\psi}(\boldsymbol{x}) u_{II,\rho}(\boldsymbol{x})\mathrm{d}\boldsymbol{x} \\
&= -\int_\Gamma u_{II,\psi}(\boldsymbol{x})\Big|_{\Gamma+0} \frac{\partial}{\partial \boldsymbol{\nu}_x} u_{II,\rho}(\boldsymbol{x})\Big|_\Gamma \mathrm{d}s_x \\
&= \mathbb{C}_e^+(\psi, \rho; k).
\end{aligned} \tag{4.4.29}
$$

On the other hand we have

$$
\begin{aligned}
\left\|\psi\right\|^2_{\frac{1}{2},\Gamma} &= \left\|\left(\mathbb{K}^+_{III,k}\right)^{-1}\left(\mathbb{K}_{III,k}\psi\right)\right\|^2_{\frac{1}{2},\Gamma} \\
&\le C\left\|\left(\mathbb{K}_{III,k}\psi\right)\right\|^2_{-\frac{1}{2},\Gamma} \\
&\le Ca^+(u_{II,\psi}, u_{II,\psi}; k) \\
&= C\mathbb{C}^+(\psi, \psi; k),
\end{aligned}
$$

in which the boundness of operator $\left(\mathbb{K}^+_{III,k}\right)^{-1}$ produces the first inequality and the trace theorem of $u_{II,\psi} \in H^1(\Omega)$ arrives at the second inequality. Namely there is a constant $\alpha > 0$, such that

$$
\mathbb{C}^+(\psi, \psi; k) \ge \alpha\|\psi\|^2_{\frac{1}{2},\Gamma}, \quad \forall \psi \in H^{\frac{1}{2}}(\Gamma). \tag{4.4.30}
$$

Similarly we get

$$
\mathbb{C}_e^+(\psi, \psi; k) \ge \alpha\|\psi\|^2_{\frac{1}{2},\Gamma}, \quad \forall \psi \in H^{\frac{1}{2}}(\Gamma), \tag{4.4.31}
$$

with a constant $\alpha > 0$.

Combining the results on $\mathbb{C}^+(\psi, \rho; k)$ and $\mathbb{C}_e^+(\psi, \rho; k)$ obtain above, we know that $\mathbb{C}^+(\psi, \rho; k)$ and $\mathbb{C}_e^+(\psi, \rho; k)$ are two bounded and symmetrical bilinear forms on $H^{\frac{1}{2}}(\Gamma) \times H^{\frac{1}{2}}(\Gamma)$ and they are coercive. Since $g \in H^{\frac{1}{2}}(\Gamma)$, the right sides of the variational equalities (4.4.22) and (4.4.24) are two bounded functionals on $H^{\frac{1}{2}}(\Gamma)$. The conclusion (i) and (ii) are proved by the Lax–Milgram Theorem directly. Furthermore the conclusion (iii) is proved by the conclusion (i)–(ii) and the regularity of the solutions of the boundary value problems (4.1.1) and (4.1.2). □

We now return to the interior and exterior Neumann problems of Modified Helmholtz equation, let single layer potential with unknown density $\varphi \in H^{-\frac{1}{2}}(\Gamma)$ be the solution of which. Namely

$$
u_{I,\varphi}(\boldsymbol{x}) = \int_\Gamma E_n^+(\boldsymbol{x}, \boldsymbol{y}; k)\varphi(\boldsymbol{y})\mathrm{d}s_y, \quad \forall \boldsymbol{x} \in \mathbb{R}^n. \tag{4.4.32}
$$

For the interior Neumann problem (4.1.3), the unknown density $\varphi \in H^{-\frac{1}{2}}(\Gamma)$ is to be determined by the Neumann condition (4.1.3b), that is

$$\frac{1}{2}\varphi(\boldsymbol{x}) + \frac{\partial}{\partial \boldsymbol{v}_x}\int_\Gamma E_n^+(\boldsymbol{x}, \boldsymbol{y}; k)\varphi(\boldsymbol{y})\mathrm{d}s_y = h(\boldsymbol{x}), \quad \forall \boldsymbol{x} \in \Gamma. \tag{4.4.33}$$

It can be rewritten as

$$\frac{1}{2}\varphi(\boldsymbol{x})+\left(\mathbb{K}_{II,k}^{+,*}\varphi\right)(\boldsymbol{x}) = h(\boldsymbol{x}), \ \ \forall \boldsymbol{x} \in \Gamma. \tag{4.4.34}$$

For the exterior Neumann problem (4.1.4), the unknown density $\varphi \in H^{-\frac{1}{2}}(\Gamma)$ is to be determined by the Neumann boundary condition (4.1.4c), that is

$$-\frac{1}{2}\varphi(\boldsymbol{x}) + \frac{\partial}{\partial \boldsymbol{v}_x}\int_\Gamma E_n^+(\boldsymbol{x}, \boldsymbol{y}; k)\varphi(\boldsymbol{y})\mathrm{d}s_y = h(\boldsymbol{x}), \quad \forall \boldsymbol{x} \in \Gamma. \tag{4.4.35}$$

Namely

$$-\frac{1}{2}\varphi(\boldsymbol{x})+\left(\mathbb{K}_{II,k}^{+,*}\varphi\right)(\boldsymbol{x}) = h(\boldsymbol{x}), \ \ \forall \boldsymbol{x} \in \Gamma. \tag{4.4.36}$$

The equivalent variational problems of boundary integral equations (4.4.33) and (4.4.35) are given by

$$\begin{cases} \text{Find } \varphi \in H^{-\frac{1}{2}}(\Gamma), \quad \text{such that} \\ \displaystyle\int_\Gamma \left\{\frac{1}{2}\varphi(\boldsymbol{x})+\left(\mathbb{K}_{II,k}^{+,*}\varphi\right)(\boldsymbol{x})\right\}\rho(\boldsymbol{x})\mathrm{d}s_x = \int_\Gamma h(\boldsymbol{x})\rho(\boldsymbol{x})\mathrm{d}s_x, \ \forall \rho \in H^{\frac{1}{2}}(\Gamma), \end{cases} \tag{4.4.37}$$

and

$$\begin{cases} \text{Find } \varphi \in H^{-\frac{1}{2}}(\Gamma), \quad \text{such that} \\ \displaystyle\int_\Gamma \left\{-\frac{1}{2}\varphi(\boldsymbol{x})+\left(\mathbb{K}_{II,k}^{+,*}\varphi\right)(\boldsymbol{x})\right\}\rho(\boldsymbol{x})\mathrm{d}s_x = \int_\Gamma h(\boldsymbol{x})\rho(\boldsymbol{x})\mathrm{d}s_x, \ \forall \rho \in H^{\frac{1}{2}}(\Gamma). \end{cases} \tag{4.4.38}$$

On the other hand, Theorem 4.4.1 shows that for any $\rho \in H^{\frac{1}{2}}(\Gamma)$ there exists a unique $\mu \in H^{-\frac{1}{2}}(\Gamma)$, such that

$$\rho(\boldsymbol{x}) = \left(\mathbb{K}_{I,k}^{+}\mu\right)(\boldsymbol{x}). \tag{4.4.39}$$

Taking (4.4.39) in the variational problem (4.4.37), we obtain the second equivalent variational problem of boundary integral equation (4.4.33), that is

$$\begin{cases} \text{Find } \varphi \in H^{-\frac{1}{2}}(\Gamma), \quad \text{such that} \\ \qquad \mathbb{D}^+(\varphi, \mu; k) := \int_\Gamma \left\{ \frac{1}{2}\varphi(\boldsymbol{x}) + \left(\mathbb{K}^{+,*}_{II,k}\varphi\right)(\boldsymbol{x}) \right\} \left(\mathbb{K}^+_{I,k}\mu\right)(\boldsymbol{x}) \mathrm{d}s_x \\ \qquad\qquad = \int_\Gamma h(\boldsymbol{x}) \left(\mathbb{K}^+_{I,k}\mu\right)(\boldsymbol{x}) \mathrm{d}s_x, \ \forall \mu \in H^{-\frac{1}{2}}(\Gamma). \end{cases} \tag{4.4.40}$$

Similarly we get the second equivalent variational problem of boundary integral equation (4.4.35) in the following:

$$\begin{cases} \text{Find } \varphi \in H^{-\frac{1}{2}}(\Gamma), \quad \text{such that} \\ \qquad \mathbb{D}^+_e(\varphi, \mu; k) := -\int_\Gamma \left\{ -\frac{1}{2}\varphi(\boldsymbol{x}) + \left(\mathbb{K}^{+,*}_{II,k}\varphi\right)(\boldsymbol{x}) \right\} \left(\mathbb{K}^+_{I,k}\mu\right)(\boldsymbol{x}) \mathrm{d}s_x \\ \qquad\qquad = -\int_\Gamma h(\boldsymbol{x}) \left(\mathbb{K}^+_{I,k}\mu\right)(\boldsymbol{x}) \mathrm{d}s_x, \ \forall \mu \in H^{-\frac{1}{2}}(\Gamma). \end{cases} \tag{4.4.41}$$

Theorem 4.4.4 *Assume that the given function* $h \in H^{-\frac{1}{2}}(\Gamma)$.

(i) Boundary integral equation (4.4.33) *there is a unique solution* $\varphi \in H^{-\frac{1}{2}}(\Gamma)$, *and*

$$\|\varphi\|_{-\frac{1}{2},\Gamma} \le C\|h\|_{-\frac{1}{2},\Gamma}. \tag{4.4.42}$$

(ii) Boundary integral equation (4.4.35) *there is a unique solution* $\varphi \in H^{-\frac{1}{2}}(\Gamma)$, *and*

$$\|\varphi\|_{-\frac{1}{2},\Gamma} \le C\|h\|_{-\frac{1}{2},\Gamma}. \tag{4.4.43}$$

(iii) The operators $\left(\frac{1}{2}I + \mathbb{K}^{+,*}_{II,k}\right)$ *and* $\left(-\frac{1}{2}I + \mathbb{K}^{+,*}_{II,k}\right)$ *are invertible and the inverse* $\left(\frac{1}{2}I + \mathbb{K}^{+,*}_{II,k}\right)^{-1}$ *and* $\left(-\frac{1}{2}I + \mathbb{K}^{+,*}_{II,k}\right)^{-1}$ *are continuous and bounded:*

$$\left.\begin{aligned} \left(\frac{1}{2}I + \mathbb{K}^{+,*}_{II,k}\right)^{-1} &: \quad H^{-\frac{1}{2}+\gamma}(\Gamma) \longrightarrow H^{-\frac{1}{2}+\gamma}(\Gamma), \\ \left(-\frac{1}{2}I + \mathbb{K}^{+,*}_{II,k}\right)^{-1} &: \quad H^{-\frac{1}{2}+\gamma}(\Gamma) \longrightarrow H^{-\frac{1}{2}+\gamma}(\Gamma), \end{aligned}\right\} \tag{4.4.44}$$

with $0 \le \gamma \le \frac{1}{2}$.

Proof Due to the equivalence of problem (4.4.33) between (4.4.40) and problem (4.4.35) between (4.4.41), we only need to discuss problem (4.4.40) and problem (4.4.41).

$$\begin{cases} \text{For } \mu \in H^{-\frac{1}{2}}(\Gamma), \quad \text{let} \\ \qquad u_{I,\mu}(\boldsymbol{x}) = \displaystyle\int_\Gamma E_n^+(\boldsymbol{x}, \boldsymbol{y}; k)\mu(\boldsymbol{y})\mathrm{d}s_{\boldsymbol{y}}, \quad \forall \boldsymbol{x} \in \mathbb{R}^n. \end{cases} \tag{4.4.45}$$

$u_{I,\varphi}(\boldsymbol{x})$ and $u_{I,\mu}(\boldsymbol{x})$ are two single layer potentials of the modified Helmholtz operator with density $\varphi \in H^{-\frac{1}{2}}(\Gamma)$ and $\mu \in H^{-\frac{1}{2}}(\Gamma)$. By the Theorem 4.3.1 we know that $u_{I,\varphi} \in H^1(\mathbb{R}^n)$ and $u_{I,\mu} \in H^1(\mathbb{R}^n)$. Furthermore in the domain Ω, $u_{I,\varphi}(\boldsymbol{x})$ is the unique solution of the following interior Neumann problem:

$$\begin{cases} -\Delta u_{I,\varphi}(\boldsymbol{x}) + k^2 u_{I,\varphi}(\boldsymbol{x}) = 0, \quad \forall \boldsymbol{x} \in \Omega, & (4.4.46a) \\ \dfrac{\partial}{\partial \boldsymbol{v}_{\boldsymbol{x}}} u_{I,\varphi}(\boldsymbol{x})\Big|_\Gamma = \dfrac{1}{2}\varphi(\boldsymbol{x}) + \left(\mathbb{K}_{II,k}^{+,*}\varphi\right)(\boldsymbol{x}), \quad \forall \boldsymbol{x} \in \Gamma, & (4.4.46b) \end{cases}$$

in the domain Ω_e, $u_{I,\varphi}(\boldsymbol{x})$ is the unique solution of the below exterior Neumann problem:

$$\begin{cases} -\Delta u_{I,\varphi}(\boldsymbol{x}) + k^2 u_{I,\varphi}(\boldsymbol{x}) = 0, \quad \forall \boldsymbol{x} \in \Omega_e, & (4.4.47a) \\ \dfrac{\partial}{\partial \boldsymbol{v}_{\boldsymbol{x}}} u_{I,\varphi}(\boldsymbol{x})\Big|_\Gamma = -\dfrac{1}{2}\varphi(\boldsymbol{x}) + \left(\mathbb{K}_{II,k}^{+,*}\varphi\right)(\boldsymbol{x}), \quad \forall \boldsymbol{x} \in \Gamma, & (4.4.47b) \\ \left.\begin{array}{l} u_{I,\varphi}(\boldsymbol{x}) = \mathcal{O}\left(\dfrac{1}{|\boldsymbol{x}|^{(n-1)/2}} \mathrm{e}^{-k|\boldsymbol{x}|}\right), \\ \dfrac{\partial}{\partial r} u_{I,\varphi}(\boldsymbol{x}) = \mathcal{O}\left(\dfrac{1}{|\boldsymbol{x}|^{(n-1)/2}} \mathrm{e}^{-k|\boldsymbol{x}|}\right), \end{array}\right\} \quad \text{as } r = |\boldsymbol{x}| \to +\infty. & (4.4.47c) \end{cases}$$

The problem (4.4.46) yields that

$$\begin{aligned} a^+(u_{I,\varphi}, u_{I,\mu}; k) &= \int_\Omega \nabla u_{I,\varphi}(\boldsymbol{x}) \cdot \nabla u_{I,\mu}(\boldsymbol{x})\mathrm{d}\boldsymbol{x} + k^2 \int_\Omega u_{I,\varphi}(\boldsymbol{x}) u_{I,\mu}(\boldsymbol{x})\mathrm{d}\boldsymbol{x} \\ &= \int_\Gamma \frac{\partial}{\partial \boldsymbol{v}_{\boldsymbol{x}}} u_{I,\varphi}(\boldsymbol{x})\Big|_{\Gamma-0} u_{I,\mu}(\boldsymbol{x})\Big|_\Gamma \mathrm{d}s_{\boldsymbol{x}} \\ &:= \mathbb{D}^+(\varphi, \mu; k). \end{aligned} \tag{4.4.48}$$

On the other hand we have

$$\begin{aligned} \left\|\varphi\right\|^2_{-\frac{1}{2},\Gamma} &= \left\|\left(\mathbb{K}_{I,k}^+\right)^{-1}\left(\mathbb{K}_{I,k}^+\varphi\right)\right\|^2_{-\frac{1}{2},\Gamma} \\ &\le C\left\|\mathbb{K}_{I,k}^+\varphi\right\|^2_{\frac{1}{2},\Gamma} \\ &= C\left\|u_{I,\varphi}\right\|^2_{\frac{1}{2},\Gamma} \\ &\le Ca^+(\varphi, \varphi; k), \quad \forall \varphi \in H^{-\frac{1}{2}}(\Gamma). \end{aligned}$$

Namely the exists a constant $\alpha > 0$, such that

$$\mathbb{D}^{+}(\varphi, \varphi; k) \geq \alpha \|\varphi\|^2_{-\frac{1}{2},\Gamma}, \quad \forall \varphi \in H^{-\frac{1}{2}}(\Gamma). \tag{4.4.49}$$

Similarly the exterior Neumann problem (4.4.47) yields that

$$\begin{aligned} a_e^{+}(u_{I,\varphi}, u_{I,\mu}; k) &= \int_{\Omega_e} \nabla u_{I,\varphi}(\boldsymbol{x}) \cdot \nabla u_{I,\mu}(\boldsymbol{x}) \mathrm{d}\boldsymbol{x} + k^2 \int_{\Omega} u_{I,\varphi}(\boldsymbol{x}) u_{I,\mu}(\boldsymbol{x}) \mathrm{d}\boldsymbol{x} \\ &= -\int_{\Gamma} \frac{\partial}{\partial \boldsymbol{\nu}_{\boldsymbol{x}}} u_{I,\varphi}(\boldsymbol{x})\Big|_{\Gamma+0} u_{I,\mu}(\boldsymbol{x})\Big|_{\Gamma} \mathrm{d}s_{\boldsymbol{x}} \\ &:= \mathbb{D}_e^{+}(\varphi, \mu; k), \quad \forall \varphi, \mu \in H^{-\frac{1}{2}}(\Gamma), \end{aligned} \tag{4.4.50}$$

and

$$\mathbb{D}_e^{+}(\varphi, \varphi; k) \geq \alpha \|\varphi\|^2_{-\frac{1}{2},\Gamma}, \quad \forall \varphi, \mu \in H^{-\frac{1}{2}}(\Gamma). \tag{4.4.51}$$

Combining the results on $\mathbb{D}^{+}(\varphi, \mu; k)$ and $\mathbb{D}_e^{+}(\varphi, \mu; k)$ obtained above, we have that $\mathbb{D}^{+}(\varphi, \mu; k)$ and $\mathbb{D}_e^{+}(\varphi, \mu; k)$ are two bounded and symmetrical bilinear forms on $H^{-\frac{1}{2}}(\Gamma) \times H^{-\frac{1}{2}}(\Gamma)$ and they are coercive. Due to the given function $h \in H^{-\frac{1}{2}}(\Gamma)$, the right sides of the variational equalities (4.4.40) and (4.4.41) are two bounded functionals on $H^{-\frac{1}{2}}(\Gamma)$. The conclusion (i)–(ii) are obtained by the Lax–Milgram Theorem directly and the conclusion (iii) holds by the conclusions (i)–(ii) and the regularity of the solutions of the boundary value problems (4.1.10) and (4.1.4). The proof of Theorem 4.4.4 is completed. □

References

Andrews L. C. (1998). *Special functions of mathematics for engineers* (Vol. 49). SPIE.

Chen Y. Z. & Wu L. C. (2004). *Second order elliptic equations and elliptic systems* (Vol. 174). American Mathematical Soc.

Han, H. (1988). Boundary integro-differential equations of elliptic boundary value problems and their numerical solutions. *Scientia Sinica, Series A-Mathematical, Physical, Astronomical and Technical Sciences, 31*, 1153–1165.

Han, H., Huang, Z., & Kellogg, R. B. (2008). A tailored Finite point method for a singular perturbation problem on an unbounded domain. *Journal of Scientific Computing, 36*(2), 243–261.

Han H. & Wu X. (2013). *Artificial boundary method*. Tsinghua University Press & Springer.

Hemker, P. W. (1996). A singularly perturbed model problem for numerical computa-tion. *Journal of computational and applied mathematics, 76*(1–2), 277–285.

Hsiao, G. C., & Wendland, W. L. (2008). *Boundary integral equations*. Springer.

Chapter 5
Boundary Value Problems of Helmholtz Equation

In this chapter, we discuss the boundary value problems of Helmholtz equation, which is from the time-harmonic acoustic scattering (Colton & Kress, 2013). Consider the acoustic wave propagation in a homogeneous isotropic medium with density ρ_0, speed of sound c_0 and damping coefficient γ_0 in $\mathbb{R}^n$ ($n = 2,\ 3$), where the constant $\rho_0 > 0$, $c_0 > 0$ and $\gamma_0 \geqslant 0$. The wave motion can be expressed by a velocity potential $w = w(\boldsymbol{x}, t)$, namely the velocity field is given by

$$\boldsymbol{v} = \frac{1}{\rho_0}\nabla w,$$

and the pressure p is obtained by

$$p - p_0 = -\frac{\partial w}{\partial t} - \gamma_0 w,$$

where p_0 denotes the pressure of the undisturbed medium. In the linearized theory the velocity potential w satisfies the following wave equation

$$\frac{\partial^2 w}{\partial t^2} + \gamma_0 \frac{\partial w}{\partial t} - c_0^2 \Delta w = 0.$$

Furthermore consider the time-harmonic acoustic waves of the above wave equation in the form $w(\boldsymbol{x}, t) = u(\boldsymbol{x})\mathrm{e}^{-\mathrm{i}\omega t}$ with frequency $\omega > 0$, then the function $u(\boldsymbol{x})$ satisfies the reduced wave equation, namely Helmholtz equation

$$-\Delta u - k^2 u = 0, \tag{5.0.1}$$

where the wave number $k \neq 0$ is obtain by $k^2 = \omega(\omega + \mathrm{i}\gamma_0)/c_0^2$.

We choose the sign of k such that

H. Han and D. Yin, *Mathematical Foundation of the Boundary Integro-Differential Equation Method*, https://doi.org/10.1007/978-981-95-1088-7_5

$$\operatorname{Re} k > 0, \quad \operatorname{Im} k \geqslant 0, \tag{5.0.2}$$

That is

$$\left.\begin{aligned} \operatorname{Re} k &= \left[\frac{1}{2c_0^2}\left(\sqrt{\omega^4 + \gamma_0^2\omega^2} + \omega^2\right)\right]^{1/2} \\ \operatorname{Im} k &= \left[\frac{1}{2c_0^2}\left(\sqrt{\omega^4 + \gamma_0^2\omega^2} - \omega^2\right)\right]^{1/2} \end{aligned}\right\}, \tag{5.0.3}$$

when the damping coefficient $\gamma_0 = 0$, the wave number k is a positive real number.

In this chapter wave number k always satisfies the condition (5.0.2).

5.1 Interior and Exterior Boundary Value Problems of Helmholtz Equation

We consider time-harmonic acoustic wave propagation in a bounded domain Ω and the corresponding exterior domain $\Omega_e = \mathbb{R}^n \backslash \overline{\Omega}$ in the homogeneous isotropic medium separately. The velocity potential $u(\boldsymbol{x})$ satisfies the Helmholtz equation (5.0.1) with the wave number k given by (5.0.2) in Ω and Ω_e. The boundary Γ of Ω (and Ω_e) is an obstacle, which leads to boundary conditions of $u(\boldsymbol{x})$ on the boundary Γ for interior problems and exterior problems (Colton & Kress, 2013):

(i) Γ is a sound-soft obstacle, which leads to the Dirichlet boundary condition on Γ of the velocity potential $u(\boldsymbol{x})$.

(ii) Γ is a sound-hard obstacle, it leads to the Neumann boundary condition on Γ of the velocity potential $u(\boldsymbol{x})$.

(iii) When Γ is a impedance obstacle, that is, the excess pressure is proportional to the normal velocity on the boundary Γ. It yields impedance boundary condition on Γ, which is also called Robin boundary condition.

In the exterior problems, the scattered wave must satisfies the additional Sommefeld radiation condition at infinity

$$\frac{\partial u(\boldsymbol{x})}{\partial r} - iku(\boldsymbol{x}) = \boldsymbol{o}\left(|\boldsymbol{x}|^{-(n-1)/2}\right), \quad \text{as } r = |\boldsymbol{x}| \to +\infty.$$

This condition selects the outgoing wave, which is needed for the uniqueness of the exterior boundary value problems.

Now we consider the following principal boundary value problems of Helmholtz equation, in which the complex functions $g(\boldsymbol{x}), h(\boldsymbol{x})$ are given on the boundary Γ and $\boldsymbol{\nu}_x$ is the unit outward normal vector at $\boldsymbol{x} \in \Gamma$ of Ω defined by (3.1.9). In general, the unknown function $u(\boldsymbol{x})$ is complex.

Interior Dirichlet problem:

$$\begin{cases} -\Delta u(\boldsymbol{x}) - k^2 u(\boldsymbol{x}) = 0, \quad \forall \boldsymbol{x} \in \Omega, & (5.1.1a) \\ u(\boldsymbol{x})\Big|_{\Gamma} = g(\boldsymbol{x}), \quad \forall \boldsymbol{x} \in \Gamma. & (5.1.1b) \end{cases}$$

Exterior Dirichlet problem:

$$\begin{cases} -\Delta u(\boldsymbol{x}) - k^2 u(\boldsymbol{x}) = 0, \quad \forall \boldsymbol{x} \in \Omega_e, & (5.1.2a) \\ u(\boldsymbol{x})\Big|_{\Gamma} = g(\boldsymbol{x}), \quad \forall \boldsymbol{x} \in \Gamma, & (5.1.2b) \\ \dfrac{\partial u(\boldsymbol{x})}{\partial r} - \mathrm{i}ku(\boldsymbol{x}) = o\Big(|\boldsymbol{x}|^{-(n-1)/2}\Big), \quad \text{as } r = |\boldsymbol{x}| \to +\infty. & (5.1.2c) \end{cases}$$

Interior Neumann problem:

$$\begin{cases} -\Delta u(\boldsymbol{x}) - k^2 u(\boldsymbol{x}) = 0, \quad \forall \boldsymbol{x} \in \Omega, & (5.1.3a) \\ \dfrac{\partial u(\boldsymbol{x})}{\partial \nu_{\boldsymbol{x}}}\Big|_{\Gamma} = h(\boldsymbol{x}), \quad \forall \boldsymbol{x} \in \Gamma. & (5.1.3b) \end{cases}$$

Exterior Neumann problem:

$$\begin{cases} -\Delta u(\boldsymbol{x}) - k^2 u(\boldsymbol{x}) = 0, \quad \forall \boldsymbol{x} \in \Omega_e, & (5.1.4a) \\ \dfrac{\partial u(\boldsymbol{x})}{\partial \nu_{\boldsymbol{x}}}\Big|_{\Gamma} = h(\boldsymbol{x}), \quad \forall \boldsymbol{x} \in \Gamma, & (5.1.4b) \\ \dfrac{\partial u(\boldsymbol{x})}{\partial r} - iku(\boldsymbol{x}) = o\Big(|\boldsymbol{x}|^{-(n-1)/2}\Big), \quad \text{as } r = |\boldsymbol{x}| \to +\infty. & (5.1.4c) \end{cases}$$

Exterior impedance boundary value problem:

$$\begin{cases} -\Delta u(\boldsymbol{x}) - k^2 u(\boldsymbol{x}) = 0, \quad \forall \boldsymbol{x} \in \Omega_e, & (5.1.5a) \\ \dfrac{\partial u(\boldsymbol{x})}{\partial \nu_{\boldsymbol{x}}}\Big|_{\Gamma} + \lambda u(\boldsymbol{x}) = g(\boldsymbol{x}), \quad \forall \boldsymbol{x} \in \Gamma, & (5.1.5b) \\ \dfrac{\partial u(\boldsymbol{x})}{\partial r} - \mathrm{i}ku(\boldsymbol{x}) = o\Big(|\boldsymbol{x}|^{-(n-1)/2}\Big), \quad \text{as } r = |\boldsymbol{x}| \to +\infty, & (5.1.5c) \end{cases}$$

in which the constant λ is given by Colton and Kress (2013),

$$\lambda = \mathrm{i}\chi\rho_0(\omega_0 + \mathrm{i}\gamma_0), \tag{5.1.6}$$

with $\chi > 0$, the acoustic impedance of the obstacle Γ.

The classical solutions of the five principal boundary value problems of Helmholtz equation are defined by

Definition 5.1.1 (i) A complex function $u \in C^0(\overline{\Omega}) \cap C^2(\Omega)$, satisfies Helmholtz equation in Ω and the boundary condition (5.1.1b) on Γ with given function $g \in C^0(\Gamma)$, is called a classical solution of interior Dirichlet problem (5.1.1).

(ii) A complex function $u \in C^0(\overline{\Omega}_e) \cap C^2(\Omega_e)$, satisfies Helmholtz equation in Ω_e, the boundary condition (5.1.2b) on Γ and Sommerfeld radiation condition (5.1.2c), is called a classical solution of exterior Dirichlet problem (5.1.2).

(iii) A complex function $u \in C_\nu(\overline{\Omega}) \cap C^2(\Omega)$, satisfies Helmholtz equation in Ω and the Neumann boundary condition (5.1.3b) on Γ with given function $h \in C^0(\Gamma)$, is called a classical solution of interior Neumann problem (5.1.3).

(iv) A complex function $u \in C_\nu(\overline{\Omega}_e) \cap C^2(\Omega_e)$, satisfies Helmholtz equation in Ω_e, Neumann boundary condition (5.1.4b) and Sommerfeld radiation condition (5.1.4c) at infinity, is called a classical solution of exterior Neumann problem (5.1.4).

(v) A complex function $u \in C_\nu(\overline{\Omega}_e) \cap C^2(\Omega_e)$, satisfies Helmholtz equation in Ω_e, impedance boundary condition (5.1.5b) and Sommerfeld radiation condition (5.1.5c) at infinity, is called a classical solution of exterior impedance problem (5.1.5).

Furthermore we discuss the weak solution of the five principal boundary value problem of Helmholtz equation. For the interior problems, introduce the bilinear form $a(u, v; k)$ on $H^1(\Omega) \times H^1(\Omega)$,

$$a(u, v; k) = \int_\Omega \nabla u(\boldsymbol{x}) \cdot \nabla \overline{v(\boldsymbol{x})} \mathrm{d}\boldsymbol{x} - k^2 \int_\Omega u(\boldsymbol{x}) \overline{v(\boldsymbol{x})} \mathrm{d}\boldsymbol{x}, \quad \forall u, v \in H^1(\Omega), \tag{5.1.7}$$

and

$$H_g^1(\Omega) = \left\{ v \middle| v \in H^1(\Omega) \text{ and } v\Big|_\Gamma = g \right\}, \tag{5.1.8}$$

with given function $g \in H^{1/2}(\Gamma)$.

For the exterior problems, we introduce the following two spaces (Hsiao & Wendland, 2008):

$$H_{\mathrm{loc}}^1(\Omega_e) = \left\{ v \middle| v \in H^1(G), \text{ for every bounded domain } G \subset \Omega_e \right\}, \tag{5.1.9}$$

$$H_{\mathrm{comp}}^1(\Omega_e) = \left\{ v \middle| v \in H_{\mathrm{loc}}^1(\Omega_e) \text{ and the support of } v \text{is compact} \right\}. \tag{5.1.10}$$

Moreover let

$$H_{\mathrm{loc},g}^1(\Omega_e) = \left\{ v \middle| v \in H_{\mathrm{loc}}^1(\Omega_e), \text{ and } v|_\Gamma = g \right\}, \tag{5.1.11}$$

$$H_{\mathrm{comp},0}^1(\Omega_e) = \left\{ v \middle| v \in H_{\mathrm{comp}}^1(\Omega_e) \text{ and } v|_\Gamma = 0 \right\}, \tag{5.1.12}$$

$$a_e(u, v; k) = \int_{\Omega_e} \nabla u(\boldsymbol{x}) \cdot \nabla \overline{v(\boldsymbol{x})} \mathrm{d}\boldsymbol{x} - k^2 \int_{\Omega_e} u(\boldsymbol{x}) \overline{v(\boldsymbol{x})} \mathrm{d}\boldsymbol{x}, \quad \forall u \in H_{\mathrm{loc}}^1(\Omega_e), v \in H_{\mathrm{comp}}^1(\Omega_e). \tag{5.1.13}$$

Definition 5.1.2 (i) For given $g \in H^{1/2}(\Gamma)$, find $u \in H_g^1(\Omega)$, such that

$$a(u, v; k) = 0, \quad \forall v \in H_0^1(\Omega). \tag{5.1.14}$$

Then $u(\boldsymbol{x})$, the solution of variational problem (5.1.14), is called a weak solution of interior Dirichlet problem (5.1.1).

(ii) For given $g \in H^{1/2}(\Gamma)$, find $u \in H^1_{\mathrm{loc},g}(\Omega_e)$, such that

$$\begin{cases} a_e(u, v; k) = 0, \quad \forall v \in H^1_{\mathrm{comp},0}(\Omega_e), & (5.1.15\mathrm{a}) \\ \dfrac{\partial u(\boldsymbol{x})}{\partial r} - iku(\boldsymbol{x}) = o\left(|\boldsymbol{x}|^{-(n-1)/2}\right), \quad \text{as } r = |\boldsymbol{x}| \to +\infty. & (5.1.15\mathrm{b}) \end{cases}$$

Then $u(\boldsymbol{x})$, the solution of variational problem (5.1.15), is called a weak solution of exterior Dirichlet problem (5.1.2).

(iii) For given $h \in H^{-1/2}(\Gamma)$, find $u \in H^1(\Omega)$, such that

$$a(u, v; k) = \int_\Gamma h(\boldsymbol{x})\overline{v(\boldsymbol{x})}\mathrm{d}s_{\boldsymbol{x}}, \quad \forall v \in H^1(\Omega). \tag{5.1.16}$$

Then $u(\boldsymbol{x})$, the solution of variational problem (5.1.16), is called a weak solution of exterior Neumann problem (5.1.3).

(iv) For given $h \in H^{-1/2}(\Gamma)$, find $u \in H^1_{\mathrm{loc}}(\Omega_e)$, such that

$$\begin{cases} a_e(u, v; k) = -\displaystyle\int_\Gamma h(\boldsymbol{x})\overline{v(\boldsymbol{x})}\mathrm{d}s_{\boldsymbol{x}}, \quad \forall v \in H^1_{\mathrm{comp}}(\Omega_e), & (5.1.17\mathrm{a}) \\ \dfrac{\partial u(\boldsymbol{x})}{\partial r} - iku(\boldsymbol{x}) = o\left(|\boldsymbol{x}|^{-(n-1)/2}\right), \quad \text{as } r = |\boldsymbol{x}| \to +\infty. & (5.1.17\mathrm{b}) \end{cases}$$

Then $u(\boldsymbol{x})$, the solution of variational problem (5.1.17), is called a weak solution of exterior Neumann problem (5.1.4).

(v) For given $h \in H^{-1/2}(\Gamma)$, find $u \in H^1_{\mathrm{loc}}(\Omega_e)$, such that

$$\begin{cases} a_e(u, v; k) - \lambda\displaystyle\int_\Gamma u(\boldsymbol{x})\overline{v(\boldsymbol{x})}\mathrm{d}s_{\boldsymbol{x}} = -\displaystyle\int_\Gamma h(\boldsymbol{x})\overline{v(\boldsymbol{x})}\mathrm{d}s_{\boldsymbol{x}}, \quad \forall v \in H^1_{\mathrm{comp}}(\Omega_e), & (5.1.18\mathrm{a}) \\ \dfrac{\partial u(\boldsymbol{x})}{\partial r} - iku(\boldsymbol{x}) = o\left(|\boldsymbol{x}|^{-(n-1)/2}\right), \quad \text{as } r = |\boldsymbol{x}| \to +\infty. & (5.1.18\mathrm{b}) \end{cases}$$

Then $u(\boldsymbol{x})$, the solution of variational problem (5.1.18), is called a weak solution of exterior impedance boundary value problem (5.1.5).

The above five principal boundary value problems of Helmholtz equation are our main concern in this chapter. The solvability of the interior Dirichlet problem (5.1.1) and the interior Neumann problems (5.1.3) are related to the following two eigenvalue problems of Laplace operator on the bounded domain Ω closely. Namely, they are:

Interior Dirichlet eigenvalue problem of Laplace operator:

$$\begin{cases} \text{Find constant } \lambda \text{ and nontrivial function } u(\boldsymbol{x}) \text{ such that} \\ -\Delta u(\boldsymbol{x}) = \lambda u(\boldsymbol{x}), \quad \forall \boldsymbol{x} \in \Omega, & (5.1.19a) \\ u(\boldsymbol{x})\Big|_{\Gamma} = 0, \quad \forall \boldsymbol{x} \in \Gamma. & (5.1.19b) \end{cases}$$

Interior Neumann eigenvalue problem of Laplace operator:

$$\begin{cases} \text{Find constant } \lambda \text{ and nontrivial function } u(\boldsymbol{x}) \text{ such that} \\ -\Delta u(\boldsymbol{x}) = \lambda u(\boldsymbol{x}), \quad \forall \boldsymbol{x} \in \Omega, & (5.1.20a) \\ \dfrac{\partial u(\boldsymbol{x})}{\partial \boldsymbol{\nu}_x}\Big|_{\Gamma} = 0, \quad \forall \boldsymbol{x} \in \Gamma. & (5.1.20b) \end{cases}$$

Recall the main results of the above two eigenvalue problems (Courant & Hilbert, 2008; Stakgold & Holst, 2011), it yields:

Lemma 5.1.1 (*i*) *The eigenvalues of the interior Dirichlet eigenvalue problem (5.1.19) are real, positive and have no limit point, thus they can be arranged in increasing order*

$$0 < \lambda_1^D < \lambda_2^D < \cdots < \lambda_j^D < \cdots, \quad j \to +\infty. \tag{5.1.21}$$

The eigenfunctions corresponding to the eigenvalue λ_j^D *constitute a finite dimensional function space* V_j^D. *Let real eigenfunction* $\left\{u_{j,l}^D,\ l = 1, 2, \cdots, J^D\right\}$ *be a set of basis functions of* V_j^D, *that is*

$$V_j^D = \left\{ u_j^D(\boldsymbol{x}) \middle| u_j^D(\boldsymbol{x}) = \sum_{l=1}^{J^D} C_l u_{j,l}^D(\boldsymbol{x}),\ \forall C_l \in \mathbb{C},\ l = 1, 2, \cdots, J^D \right\}. \tag{5.1.22}$$

(*ii*) *The eigenvalues of the interior Neumann eigenvalue problem (5.1.20) are real, nonnegative and have no limit point, therefore they can be arranged in increasing order*

$$0 = \lambda_1^N < \lambda_2^N < \cdots < \lambda_j^N < \cdots, \quad j \to +\infty. \tag{5.1.23}$$

The eigenfunctions corresponding to the eigenvalue λ_j^N *constitute a finite dimensional function space* V_j^N. *Let real eigenfunctions* $\{u_{j,l}^N(\boldsymbol{x}),\ l = 1, 2, \cdots, J^N\}$ *be a set of basis of* V_j^N, *namely*

$$V_j^N = \left\{ u_j^N(\boldsymbol{x}) \middle| u_j^N(\boldsymbol{x}) = \sum_{l=1}^{J^N} C_l u_{j,l}^N(\boldsymbol{x}),\ \forall C_l \in \mathbb{C},\ l = 1, 2, \cdots, J^N \right\}. \tag{5.1.24}$$

For the uniqueness of the five principal boundary value problems of Helmholtz equation, the following results hold

Theorem 5.1.1 *(i) The interior Dirichlet boundary value problem (5.1.1) has at most one solution provided that k^2 is not an eigenvalue of the interior Dirichlet eigenvalue problem (5.1.19).*

(ii) The exterior Dirichlet boundary value problem (5.1.2) has at most one solution.

(iii) The interior Neumann boundary value problem (5.1.3) has at most one solution provided that k^2 is not a positive eigenvalue of the interior Neumann eigenvalue problem (5.1.20).

(iv) The exterior Neumann boundary value problem (5.1.4) has at most one solution.

(v) The exterior impedance boundary value problem (5.1.5) has at most one solution provided that

$$\operatorname{Im}(\overline{k\lambda}) \geqslant 0. \tag{5.1.25}$$

The conditions (i) and (iii) follow directly form the Lemma 5.1.1. The proofs of conclusions (ii),(iv) and (v) can be found in the book by Colton and Kress (2013).

The existences of the five principal boundary value problem will be discussed in the coming Sect. 5.3.

5.2 Single and Double Layers Potentials of Helmholtz Equation

The fundamental solution of Helmholtz operator is given by (2.2.22), namely

$$E_n(\boldsymbol{x}, \boldsymbol{y}; k) = \begin{cases} \dfrac{\mathrm{i}}{4} H_0^1(k|\boldsymbol{x}-\boldsymbol{y}|), & n = 2 \\ \dfrac{1}{4\pi} \dfrac{\mathrm{e}^{\mathrm{i}k|\boldsymbol{x}-\boldsymbol{y}|}}{|\boldsymbol{x}-\boldsymbol{y}|}, & n = 3. \end{cases} \tag{5.2.1}$$

For given functions $\varphi(\boldsymbol{x})$ and $\psi(\boldsymbol{x})$ on Γ, the single layer potential and the double layer potential of Helmholtz operator are defined by

$$V_{I,k}(\boldsymbol{x}) = \int_\Gamma E_n(\boldsymbol{x}, \boldsymbol{y}; k)\varphi(\boldsymbol{y})\mathrm{d}s_{\boldsymbol{y}}, \quad \forall \boldsymbol{x} \in \mathbb{R}^n, \tag{5.2.2}$$

$$V_{II,k}(\boldsymbol{x}) = \int_\Gamma \frac{\partial}{\partial \boldsymbol{\nu}_{\boldsymbol{y}}} E_n(\boldsymbol{x}, \boldsymbol{y}; k)\psi(\boldsymbol{y})\mathrm{d}s_{\boldsymbol{y}}, \quad \forall \boldsymbol{x} \in \mathbb{R}^n. \tag{5.2.3}$$

To study the properties of $V_{I,k}(\boldsymbol{x})$ and $V_{II,k}(\boldsymbol{x})$, the fundamental solution $E_n(\boldsymbol{x}, \boldsymbol{y}; k)$ is decomposed into the following form

$$E_n(\boldsymbol{x}, \boldsymbol{y}; k) = E_n^+(\boldsymbol{x}, \boldsymbol{y}; 1) + B_n(\boldsymbol{x}, \boldsymbol{y}; k), \tag{5.2.4}$$

in which $E_n^+(\boldsymbol{x}, \boldsymbol{y}; 1)$ is the fundamental solution of modified Helmholtz operator (4.2.1), then

$$B_n(\boldsymbol{x}, \boldsymbol{y}; k) = E_n(\boldsymbol{x}, \boldsymbol{y}; k) - E_n^+(\boldsymbol{x}, \boldsymbol{y}; 1) = \begin{cases} \dfrac{\mathrm{i}}{4} H_0^1(k|\boldsymbol{x}-\boldsymbol{y}|) - \dfrac{1}{2\pi} K_0(|\boldsymbol{x}-\boldsymbol{y}|), & n=2 \\ \dfrac{1}{4\pi} \dfrac{\mathrm{e}^{\mathrm{i}k|\boldsymbol{x}-\boldsymbol{y}|} - \mathrm{e}^{-|\boldsymbol{x}-\boldsymbol{y}|}}{|\boldsymbol{x}-\boldsymbol{y}|}, & n = 3. \end{cases} \tag{5.2.5}$$

For the case $n = 2$,

$$B_2(\boldsymbol{x}, \boldsymbol{y}; k) = \frac{\mathrm{i}}{4} J_0(k|\boldsymbol{x} - \boldsymbol{y}|) - \frac{1}{4} Y_0(k|\boldsymbol{x} - \boldsymbol{y}|) - \frac{1}{2\pi} K_0(|\boldsymbol{x} - \boldsymbol{y}|),$$

where $J_0(z)$ and $Y_0(z)$ are the zero order Bessel functions of the first and second kind:

$$J_0(z) = 1 + \sum_{m=1}^{\infty} \frac{(-1)^m}{(m!)^2} \left(\frac{z^2}{4}\right)^m,$$

$$Y_0(z) = \frac{2}{\pi} J_0(z)\Big(\ln z + \gamma - \ln 2\Big) - \frac{2}{\pi} \sum_{m=1}^{\infty} \left(\sum_{l=1}^{m} \frac{1}{l}\right) \frac{(-1)^m}{(m!)^2} \left(\frac{z^2}{4}\right)^2.$$

Furthermore combining the sum formula of $K_0(z)$ given in Chap. 4, it arrives at

$$\begin{aligned} B_2(\boldsymbol{x}, \boldsymbol{y}; k) = {} & \frac{\mathrm{i}}{4} J_0\big(k|\boldsymbol{x} - \boldsymbol{y}|\big) - \frac{1}{2\pi} J_0(k|\boldsymbol{x} - \boldsymbol{y}|) \ln k - \\ & \frac{1}{2\pi} \Big(J_0(k|\boldsymbol{x} - \boldsymbol{y}|) - 1\Big)\Big(\ln|\boldsymbol{x} - \boldsymbol{y}| + \gamma - \ln 2\Big) - \\ & \frac{1}{2\pi} \sum_{m=1}^{\infty} \left(\sum_{l=1}^{m} \frac{1}{l}\right) \frac{(-1)^m}{(m!)^2} \left(\frac{k^2|\boldsymbol{x} - \boldsymbol{y}|^2}{4}\right)^m + \\ & \frac{1}{2\pi} \Big(I_0\big(k|\boldsymbol{x} - \boldsymbol{y}|\big) - 1\Big)\Big(\ln|\boldsymbol{x} - \boldsymbol{y}| + \gamma - \ln 2\Big) + \\ & \frac{1}{2\pi} \sum_{m=1}^{\infty} \left(\sum_{l=1}^{m} \frac{1}{l}\right) \frac{1}{(m!)^2} \left(\frac{|\boldsymbol{x} - \boldsymbol{y}|^2}{4}\right)^m. \end{aligned} \tag{5.2.6}$$

For the case $n = 3$,

$$B_3(\boldsymbol{x}, \boldsymbol{y}; k) = \frac{1}{4\pi} \frac{1}{|\boldsymbol{x} - \boldsymbol{y}|} \left(\mathrm{e}^{\mathrm{i}k|\boldsymbol{x}-\boldsymbol{y}|} - \mathrm{e}^{-|\boldsymbol{x}-\boldsymbol{y}|}\right)$$

$$= \frac{1}{4\pi}\sum_{m=1}^{\infty}\frac{(ik)^m - (-1)^m}{m!}|\boldsymbol{x}-\boldsymbol{y}|^{m-1}. \tag{5.2.7}$$

For the function $B_n(\boldsymbol{x}, \boldsymbol{y}; k)$ we have

Lemma 5.2.1 *(i) $B_n(\boldsymbol{x}, \boldsymbol{y}; k)$ is continuous on $(\boldsymbol{x}, \boldsymbol{y}) \in \mathbb{R}^n \times \mathbb{R}^n$ and infinitely continuously differentiable provided by $|\boldsymbol{x}-\boldsymbol{y}| \neq 0$.*

(ii) The first derivatives of $B_n(\boldsymbol{x}, \boldsymbol{y}; k)$, namely consider

$$\frac{\partial}{\partial x_i}B_n(\boldsymbol{x}, \boldsymbol{y}; k) = -\frac{\partial}{\partial y_i}B_n(\boldsymbol{x}, \boldsymbol{y}; k), \quad 1 \leqslant i \leqslant n.$$

For the case $n = 2$, the first derivatives of $B_2(\boldsymbol{x}, \boldsymbol{y}; k)$ are continuous on $(\boldsymbol{x}, \boldsymbol{y}) \in \mathbb{R}^n \times \mathbb{R}^n$.

For the case $n = 3$,

$$\frac{\partial}{\partial x_i}B_3(\boldsymbol{x}, \boldsymbol{y}; k) = \mathcal{O}(1), \quad as \quad |\boldsymbol{x}-\boldsymbol{y}| \to 0, \ 1 \leqslant i \leqslant 3.$$

(iii) Consider the second derivatives of $B_n(\boldsymbol{x}, \boldsymbol{y}; k)$, namely

$$\frac{\partial^2}{\partial x_i \partial x_j}B_n(\boldsymbol{x}, \boldsymbol{y}; k) = \frac{\partial^2}{\partial y_i \partial y_j}B_n(\boldsymbol{x}, \boldsymbol{y}; k) = -\frac{\partial^2}{\partial x_i \partial y_j}B_n(\boldsymbol{x}, \boldsymbol{y}; k), \ 1 \leqslant i, j \leqslant n,$$

then

$$\frac{\partial^2}{\partial x_i \partial x_j}B_n(\boldsymbol{x}, \boldsymbol{y}; k) = \begin{cases} \mathcal{O}\left(\ln\dfrac{1}{|\boldsymbol{x}-\boldsymbol{y}|}\right), & as\ |\boldsymbol{x}-\boldsymbol{y}| \to 0, n = 2, \\ \mathcal{O}\left(\dfrac{1}{|\boldsymbol{x}-\boldsymbol{y}|}\right), & as\ |\boldsymbol{x}-\boldsymbol{y}| \to 0, n = 3. \end{cases}$$

(iv)

$$-\Delta_y B_n(\boldsymbol{x}, \boldsymbol{y}; k) - k^2 B_n(\boldsymbol{x}, \boldsymbol{y}; k) = (1+k^2)E_n^+(\boldsymbol{x}, \boldsymbol{y}; 1), \tag{5.2.8}$$

$$\Delta_y^2\left(\frac{1}{1+k^2}B_n(\boldsymbol{x}, \boldsymbol{y}; k)\right) + (k-1)^2\Delta_y\left(\frac{1}{1+k^2}B_n(\boldsymbol{x}, \boldsymbol{y}; k)\right) - k^2\left(\frac{1}{1+k^2}B_n(\boldsymbol{x}, \boldsymbol{y}; k)\right) = \delta(\boldsymbol{x}-\boldsymbol{y}). \tag{5.2.9}$$

Proof The conclusion (i)–(iii) are obtained directly form the equalities (5.2.6)–(5.2.7).

By the definition of $B_n(\boldsymbol{x}, \boldsymbol{y}; k)$, we have

$$\begin{aligned} & -\Delta_y B_n(\boldsymbol{x}, \boldsymbol{y}; k) - k^2 B_n(\boldsymbol{x}, \boldsymbol{y}; k) \\ & = -\Delta_y E_n(\boldsymbol{x}, \boldsymbol{y}; k) - k^2 E_n(\boldsymbol{x}, \boldsymbol{y}; k) + \Delta_y E_n^+(\boldsymbol{x}, \boldsymbol{y}; 1) + k^2 E_n^+(\boldsymbol{x}, \boldsymbol{y}; 1) \end{aligned}$$

$$= \delta(\boldsymbol{x}-\boldsymbol{y}) - \Big(- \Delta_y E_n^+(\boldsymbol{x}, \boldsymbol{y}; 1) + E_n^+(\boldsymbol{x}, \boldsymbol{y}; 1)\Big) + (1+k^2) E_n^+(\boldsymbol{x}, \boldsymbol{y}; 1)$$
$$= (1+k^2) E_n^+(\boldsymbol{x}, \boldsymbol{y}; 1).$$

We have known

$$- \Delta_y E_n^+(\boldsymbol{x}, \boldsymbol{y}; 1) + E_n^+(\boldsymbol{x}, \boldsymbol{y}; 1) = \delta(\boldsymbol{x}-\boldsymbol{y}).$$

Combining the above two equalities, the equality (5.2.9) is obtained immediately. The equality (5.2.9) means that $\dfrac{1}{1+k^2} B_n(\boldsymbol{x}, \boldsymbol{y}; k)$ is a fundamental solution of the fourth order partial differential operator in (5.2.9). The proof is completed. □

It is ready to study the single and double layer potential $V_{I,k}(\boldsymbol{x})$ and $V_{II,k}(\boldsymbol{x})$.

5.2.1 Single Layer Potential $V_{I,k}(\boldsymbol{x})$

By the equality (5.2.4), the single layer potential $V_{I,k}(x)$ can be rewritten as

$$\begin{aligned} V_{I,k}(\boldsymbol{x}) &= \int_\Gamma E_n^+(\boldsymbol{x}, \boldsymbol{y}; 1)\varphi(\boldsymbol{y}) \mathrm{d}s_y + \int_\Gamma B_n(\boldsymbol{x}, \boldsymbol{y}; k)\varphi(\boldsymbol{y}) \mathrm{d}s_y \\ &= V_{I,1}^+(\boldsymbol{x}) + U_{I,k}(\boldsymbol{x}), \quad \forall \boldsymbol{x} \in \mathbb{R}^n, \end{aligned} \tag{5.2.10}$$

in which $V_{I,1}^+(\boldsymbol{x})$ is the single layer potential of modified Helmholtz operator defined by (4.2.2) and $U_{I,k}(\boldsymbol{x})$ is defined by

$$U_{I,k}(\boldsymbol{x}) = \int_\Gamma B_n(\boldsymbol{x}, \boldsymbol{y}; k)\varphi(\boldsymbol{y}) \mathrm{d}s_y, \quad \forall \boldsymbol{x} \in \mathbb{R}^n. \tag{5.2.11}$$

The properties of the single layer potential $V_{I,1}^+(\boldsymbol{x})$ with given density $\varphi(\boldsymbol{x})$ are given in Theorems 4.2.1 and 4.3.2. In the following we focus our attention on the layer potential $U_{I,k}(\boldsymbol{x})$, we have

Lemma 5.2.2 *Suppose that $\Gamma \in C^2$ and $\varphi \in C(\Gamma)$.*
(i) $U_{I,k}(\boldsymbol{x})$ is well defined on $\mathbb{R}^n$ and $U_{I,k} \in C^2(\mathbb{R}^n)$ satisfies

$$- \Delta U_{I,k}(\boldsymbol{x}) - k^2 U_{I,k}(\boldsymbol{x}) = (1+k^2) V_{I,1}^+(\boldsymbol{x}), \quad \forall \boldsymbol{x} \in \mathbb{R}^n. \tag{5.2.12}$$

(ii)

$$\frac{\partial}{\partial |\boldsymbol{x}|} U_{I,k}(\boldsymbol{x}) - \mathrm{i}k U_{I,k}(\boldsymbol{x}) = \boldsymbol{o}\Big(\|\varphi\|_{-1/2,\Gamma} |\boldsymbol{x}|^{-(n-1)/2}\Big), \quad \textit{as } |\boldsymbol{x}| \to \infty. \tag{5.2.13}$$

(iii) $U_{I,k} \in H^2_{\mathrm{loc}}(\mathbb{R}^n)$ and there exists a positive constant C, such that

$$\left\|U_{I,k}\right\|_{3/2,\Gamma} \leqslant C\left\|\varphi\right\|_{-1/2,\Gamma}, \tag{5.2.14}$$

$$\left\|\frac{\partial}{\partial \boldsymbol{\nu}_x}U_{I,k}\right\|_{1/2,\Gamma} \leqslant C\left\|\varphi\right\|_{-1/2,\Gamma}. \tag{5.2.15}$$

Furthermore for any bounded domain G there is a positive constant C, such that

$$\left\|U_{I,k}\right\|_{2,G} \leqslant C\left\|\varphi\right\|_{1/2,\Gamma}. \tag{5.2.16}$$

Proof (i) The conclusion (i) follows directly from Lemma 5.2.1.

(ii) Let

$$\begin{aligned} W_I(\boldsymbol{x}, \boldsymbol{y}; k) &= \frac{\partial}{\partial|\boldsymbol{x}|}B_n(\boldsymbol{x}, \boldsymbol{y}; k) - \mathrm{i}kB_n(\boldsymbol{x}, \boldsymbol{y}; k) \\ &= \frac{\partial}{\partial|\boldsymbol{x}|}E_n(\boldsymbol{x}, \boldsymbol{y}; k) - \mathrm{i}kE_n(\boldsymbol{x}, \boldsymbol{y}; k) - \left(\frac{\partial}{\partial|\boldsymbol{x}|}E_n^+(\boldsymbol{x}, \boldsymbol{y}; 1) - \mathrm{i}kE_n^+(\boldsymbol{x}, \boldsymbol{y}; 1)\right), \end{aligned}$$

then we have

$$\begin{aligned} \left|\frac{\partial}{\partial|\boldsymbol{x}|}U_{I,k}(\boldsymbol{x}) - \mathrm{i}kU_{I,k}(\boldsymbol{x})\right| &= \left|\int_\Gamma W_I(\boldsymbol{x}, \boldsymbol{y}; k)\varphi(\boldsymbol{y})\mathrm{d}s_y\right| \\ &\leqslant \|W_I(\boldsymbol{x}, \cdot; k)\|_{1/2,\Gamma}\,\|\varphi\|_{-1/2,\Gamma} \\ &\leqslant C\,\|W_I(\boldsymbol{x}, \cdot; k)\|_{1,\Omega}\,\|\varphi\|_{-1/2,\Gamma}, \quad \forall \boldsymbol{x} \in \Omega_e. \end{aligned}$$

The last inequality is from the trace theorem of Sobolev space $H^1(\Omega)$ and

$$\|W_I(\boldsymbol{x}, \cdot; k)\|_{1,\Omega}^2 = \int_\Omega \left(\left|\nabla_y W_I(\boldsymbol{x}, \boldsymbol{y}; k)\right|^2 + \left|W_I(\boldsymbol{x}, \boldsymbol{y}; k)\right|^2\right)\mathrm{d}\boldsymbol{y}, \quad \forall \boldsymbol{x} \in \Omega_e.$$

It is straightforward to check that

$$|\boldsymbol{x}|^{n-1}\left(\left|\nabla_y W_I(\boldsymbol{x}, \boldsymbol{y}; k)\right|^2 + \left|W_I(\boldsymbol{x}, \boldsymbol{y}; k)\right|^2\right) = \boldsymbol{o}(1), \quad \text{as } |\boldsymbol{x}| \to \infty,$$

uniformly for all $\boldsymbol{y} \in \Omega$ and all direction $\dfrac{\boldsymbol{x}}{|\boldsymbol{x}|}$. Combining the above results conclusion (ii) is proved.

(iii) Since $U_{I,k} \in C^2(\mathbb{R}^n)$, obviously, $U_{I,k} \in H^2_{\mathrm{loc}}(\mathbb{R}^n)$. There exists a positive constant R, such that

$$\overline{\Omega} \in \Omega_R = \left\{\boldsymbol{x}\,\middle|\,|\boldsymbol{x}| < R\right\}.$$

Let

$$\Gamma_R = \left\{\boldsymbol{x}\,\middle|\,|\boldsymbol{x}| = R\right\},$$

which is the boundary of Ω_R. Obviously $\Gamma \cap \Gamma_R = \varnothing$.

For layer potential $U_{I,k}(\boldsymbol{x})$ on the domain Ω_R, we have

$$\begin{aligned}\left|U_{I,k}(\boldsymbol{x})\right| &= \left|\int_\Gamma B_n(\boldsymbol{x}, \boldsymbol{y}; k)\varphi(\boldsymbol{y})\mathrm{d}s_{\boldsymbol{y}}\right| \\ &\leqslant \left\|B_n(\boldsymbol{x}, \cdot; k)\right\|_{1/2,\Gamma}\left\|\varphi\right\|_{-1/2,\Gamma} \\ &\leqslant C\left\|B_n(\boldsymbol{x}, \cdot; k)\right\|_{1,\Omega}\left\|\varphi\right\|_{-1/2,\Gamma}, \quad \forall \boldsymbol{x} \in \Omega_R, \end{aligned} \tag{5.2.17}$$

and

$$\left\|B_n(\boldsymbol{x}, \cdot; k)\right\|_{1,\Omega}^2 = \int_\Omega \left(\left|\nabla_y B_n(\boldsymbol{x}, \boldsymbol{y}; k)\right|^2 + \left|B_n(\boldsymbol{x}, \boldsymbol{y}; k)\right|^2\right)\mathrm{d}\boldsymbol{y} \leqslant C, \quad \forall \boldsymbol{x} \in \Omega_R. \tag{5.2.18}$$

The estimates (5.2.17)–(5.2.18) yield

$$\|U_{I,k}\|_{0,\Omega_R} \leqslant C\|\varphi\|_{-1/2,\Gamma}. \tag{5.2.19}$$

Recall the estimate (4.3.5) in Lemma 4.3.1, we obtain

$$\|V_{I,1}^+\|_{0,\Omega_R} \leqslant C\|\varphi\|_{-1/2,\Gamma}. \tag{5.2.20}$$

On the boundary Γ_R,

$$U_{I,k}(\boldsymbol{x})\Big|_{\Gamma_R} = \int_\Gamma B_n(\boldsymbol{x}, \boldsymbol{y}; k)\varphi(\boldsymbol{y})\mathrm{d}s_{\boldsymbol{y}}, \quad \forall \boldsymbol{x} \in \Gamma_R,$$

is differentiable on $(\boldsymbol{x}, \boldsymbol{y}) \in \Gamma_R \times \Gamma$. It is straightforward to check that the following estimate holds

$$\|U_{I,k}\|_{3/2,\Gamma_R} \leqslant C\|\varphi\|_{-1/2,\Gamma}. \tag{5.2.21}$$

On the domain Ω_R, $U_{I,k}(\boldsymbol{x})$ satisfies

$$\left.\begin{aligned}&-\Delta U_{I,k}(\boldsymbol{x}) = k^2 U_{I,k}(\boldsymbol{x}) + (1+k^2)V_{I,1}^+(\boldsymbol{x}) := F_{I,k}(\boldsymbol{x}), \quad \forall \boldsymbol{x} \in \Omega_R, \\ &U_{I,k}(\boldsymbol{x})\Big|_\Gamma = U_{I,k}(\boldsymbol{x})\Big|_\Gamma,\end{aligned}\right\}$$

which is a Dirichlet boundary value problem of Poisson equation. The estimates (5.2.19)–(5.2.20) yield that

$$\|F_{I,k}\|_{0,\Omega_R} \leqslant C\|\varphi\|_{-1/2,\Gamma}. \tag{5.2.22}$$

By the regular estimate of $U_{I,k}(\boldsymbol{x})$ on Ω_R, we obtain

$$\begin{aligned}\left\|U_{I,k}\right\|_{2,\Omega_R} &\leqslant C\left\{\left\|U_{I,k}\right\|_{3/2,\Gamma_R}+\left\|F_{I,k}\right\|_{0,\Omega_R}\right\}\\ &\leqslant C\left\|\varphi\right\|_{-1/2,\Gamma}.\end{aligned}$$

The last inequality is from inequalities (5.2.21)–(5.2.22). By the trace theorem of Sobolev space $H^2(\Omega_R)$, we have

$$\begin{aligned}\left\|U_{I,k}\right\|_{3/2,\Gamma} &\leqslant \left\|U_{I,k}\right\|_{2,\Omega_R} \leqslant C\left\|\varphi\right\|_{-1/2,\Gamma},\\ \left\|\frac{\partial}{\partial \boldsymbol{\nu}_{\boldsymbol{x}}}U_{I,k}\right\|_{1/2,\Gamma} &\leqslant \left\|U_{I,k}\right\|_{2,\Omega_R} \leqslant C\left\|\varphi\right\|_{-1/2,\Gamma},\end{aligned}$$

namely the estimates (5.2.14) and (5.2.15) are proved.

For any bounded domain G, there is a constant $R > 0$, such that

$$\overline{\Omega}\cup\overline{G}\subset\Omega_R=\left\{\boldsymbol{x}\,\middle|\,|\boldsymbol{x}|<R\right\},$$

then

$$\left\|U_{I,k}\right\|_{2,G} \leqslant \left\|U_{I,k}\right\|_{2,\Omega_R} \leqslant C\left\|\varphi\right\|_{-1/2,\Gamma}.$$

This Lemma is proved completely. □

For given $\varphi \in H^{-1/2}(\Gamma)$, there exists function series $\{\varphi_m \in C(\Gamma), m = 1, 2\cdots\}$, such that

$$\lim_{m\to\infty}\|\varphi-\varphi_m\|_{-1/2,\Gamma}=0,$$

provided by $C(\Gamma)$ is dense in $H^{-1/2}(\Gamma)$. Then we have

$$U_{I,k}(\boldsymbol{x})=\int_\Gamma B_n(\boldsymbol{x},\boldsymbol{y};k)\varphi(\boldsymbol{y})\mathrm{d}s_{\boldsymbol{y}}=\lim_{m\to\infty}\int_\Gamma B_n(\boldsymbol{x},\boldsymbol{y};k)\varphi_m(\boldsymbol{y})\mathrm{d}s_{\boldsymbol{y}},\quad \forall\boldsymbol{x}\in\mathbb{R}^n.$$

The Lemmas 5.2.1 and 5.2.2 yield the following results.

Lemma 5.2.3 *Assume that $\Gamma \in C^2$ and $\varphi \in H^{-1/2}(\Gamma)$.*

(i) $U_{I,k}$ is well defined on $\mathbb{R}^n$ and $U_{I,k} \in C^2(\mathbb{R}^n\backslash\Gamma)$ satisfies

$$-\Delta U_{I,k}(\boldsymbol{x})-k^2U_{I,k}(\boldsymbol{x})=(1+k^2)V_{I,1}^{+}(\boldsymbol{x}),\quad \forall\boldsymbol{x}\in\mathbb{R}^n\backslash\Gamma. \tag{5.2.23}$$

(ii)

$$\frac{\partial}{\partial|\boldsymbol{x}|}U_{I,k}(\boldsymbol{x})-\mathrm{i}kU_{I,k}(\boldsymbol{x})=\boldsymbol{o}\left(\left\|\varphi\right\|_{-1/2,\Gamma}|\boldsymbol{x}|^{-(n-1)/2}\right),\quad as\ |\boldsymbol{x}|\to\infty. \tag{5.2.24}$$

(iii) $U_{I,k} \in H^2_{\text{loc}}(\mathbb{R}^n)$ *and*

$$\left\| U_{I,k} \right\|_{3/2,\Gamma} \leqslant C \left\| \varphi \right\|_{-1/2,\Gamma}, \tag{5.2.25}$$

$$\left\| \frac{\partial}{\partial \boldsymbol{\nu}_x} U_{I,k} \right\|_{1/2,\Gamma} \leqslant C \left\| \varphi \right\|_{-1/2,\Gamma}. \tag{5.2.26}$$

The Lemma 5.2.2 and Theorem 4.2.1 yield the following results

Theorem 5.2.1 *Suppose that* $\Gamma \in C^2$.

(i) For given density $\varphi \in C(\Gamma)$*, the single layer potential* $V_{I,k}(\boldsymbol{x})$ *is well defined on* $\mathbb{R}^n$*, which is continuous on* $\mathbb{R}^n$. $V_{I,k}(\boldsymbol{x})$ *is infinitely continuously differentiable and satisfies the Helmholtz equation (5.0.1) on* $\mathbb{R}^n \backslash \Gamma$.

(ii) For $\varphi \in C(\Gamma)$,

$$\frac{\partial}{\partial |\boldsymbol{x}|} V_{I,k}(\boldsymbol{x}) - \mathrm{i}k V_{I,k}(\boldsymbol{x}) = \boldsymbol{o}\left(|\boldsymbol{x}|^{-(n-1)/2} \right), \quad as \ |\boldsymbol{x}| \to \infty. \tag{5.2.27}$$

(iii) For $\varphi \in C^1(\Gamma)$*, the first derivatives of* $V_{I,k}(\boldsymbol{x})$ *can be continuously extended from* $\Omega\,(\Omega_e)$ *to the boundary* Γ*. Namely*

$$\left.\begin{aligned} \lim_{\substack{z \to x \in \Gamma \\ x \in \Omega}} \nabla_z V_{I,k}(z) &= \int_\Gamma \nabla_x E_n(\boldsymbol{x}, \boldsymbol{y}; k)\varphi(\boldsymbol{y})\mathrm{d}s_y + \frac{1}{2}\boldsymbol{\nu}(\boldsymbol{x})\varphi(\boldsymbol{x}), \ \forall \boldsymbol{x} \in \Gamma, \\ \lim_{\substack{z \to x \in \Gamma \\ x \in \Omega_e}} \nabla_z V_{I,k}(z) &= \int_\Gamma \nabla_x E_n(\boldsymbol{x}, \boldsymbol{y}; k)\varphi(\boldsymbol{y})\mathrm{d}s_y - \frac{1}{2}\boldsymbol{\nu}(\boldsymbol{x})\varphi(\boldsymbol{x}), \ \forall \boldsymbol{x} \in \Gamma. \end{aligned}\right\} \tag{5.2.28}$$

Furthermore

$$\left.\begin{aligned} \frac{\partial}{\partial \boldsymbol{\nu}_x} V_{I,k}(\boldsymbol{x})\Big|_{\Gamma-0} &= \frac{1}{2}\varphi(\boldsymbol{x}) + \int_\Gamma \frac{\partial}{\partial \boldsymbol{\nu}_x} E_n(\boldsymbol{x}, \boldsymbol{y}; k)\varphi(\boldsymbol{y})\mathrm{d}s_y, \quad \forall \boldsymbol{x} \in \Gamma, \\ \frac{\partial}{\partial \boldsymbol{\nu}_x} V_{I,k}(\boldsymbol{x})\Big|_{\Gamma+0} &= -\frac{1}{2}\varphi(\boldsymbol{x}) + \int_\Gamma \frac{\partial}{\partial \boldsymbol{\nu}_x} E_n(\boldsymbol{x}, \boldsymbol{y}; k)\varphi(\boldsymbol{y})\mathrm{d}s_y, \quad \forall \boldsymbol{x} \in \Gamma. \end{aligned}\right\} \tag{5.2.29}$$

Specially for $\varphi \in C(\Gamma)$*, the equalities (5.2.29) are still correct in the sense given in (3.3.33).*

Introduce two operators $\mathbb{K}_{I,k}$ and $\mathbb{K}^*_{II,k}$, there are

$$\left.\begin{aligned} \left(\mathbb{K}_{I,k}\varphi\right)(\boldsymbol{x}) &= \int_\Gamma E_n(\boldsymbol{x}, \boldsymbol{y}; k)\varphi(\boldsymbol{y})\mathrm{d}s_y, \quad \forall \boldsymbol{x} \in \Gamma, \\ \left(\mathbb{K}^*_{II,k}\varphi\right)(\boldsymbol{x}) &= \int_\Gamma \frac{\partial}{\partial \boldsymbol{\nu}_x} E_n(\boldsymbol{x}, \boldsymbol{y}; k)\varphi(\boldsymbol{y})\mathrm{d}s_y, \quad \forall \boldsymbol{x} \in \Gamma, \end{aligned}\right\} \tag{5.2.30}$$

Then we have

$$\left.\begin{aligned}(\mathbb{K}_{I,k}\varphi)(\boldsymbol{x}) &= \int_\Gamma E_n^+(\boldsymbol{x},\boldsymbol{y};1)\varphi(\boldsymbol{y})\mathrm{d}s_y + \int_\Gamma B_n(\boldsymbol{x},\boldsymbol{y};k)\varphi(\boldsymbol{y})\mathrm{d}s_y\\ &= (\mathbb{K}_{I,1}^+\varphi)(\boldsymbol{x}) + (\Pi_{I,k}\varphi)(\boldsymbol{x}),\\ (\mathbb{K}_{II,k}^*\varphi)(\boldsymbol{x}) &= \int_\Gamma \frac{\partial}{\partial\boldsymbol{\nu}_x}E_n^+(\boldsymbol{x},\boldsymbol{y};1)\varphi(\boldsymbol{y})\mathrm{d}s_y + \int_\Gamma \frac{\partial}{\partial\boldsymbol{\nu}_x}B_n(\boldsymbol{x},\boldsymbol{y};k)\varphi(\boldsymbol{y})\mathrm{d}s_y\\ &= (\mathbb{K}_{II,1}^{+,*}\varphi)(\boldsymbol{x}) + (\Pi_{II,k}^*\varphi)(\boldsymbol{x}),\end{aligned}\right\} \tag{5.2.31}$$

with

$$\left.\begin{aligned}(\Pi_{I,k}\varphi) &= \int_\Gamma B_n(\boldsymbol{x},\boldsymbol{y};k)\varphi(\boldsymbol{y})\mathrm{d}s_y, \quad \forall\boldsymbol{x}\in\Gamma,\\ (\Pi_{II,k}^*\varphi)(\boldsymbol{x}) &= \int_\Gamma \frac{\partial}{\partial\boldsymbol{\nu}_x}B_n(\boldsymbol{x},\boldsymbol{y};k)\varphi(\boldsymbol{x})\mathrm{d}s_y, \quad \forall\boldsymbol{x}\in\Gamma.\end{aligned}\right\} \tag{5.2.32}$$

Similarly the Lemma 5.2.3 and Theorem 4.3.1 yield the following results

Theorem 5.2.2 *Presume that $\Gamma\in C^2$ and the density $\varphi\in H^{-1/2}(\Gamma)$.*

(i) The single layer potential $V_{I,k}(\boldsymbol{x})$ is well defined on $\mathbb{R}^n$ and $V_{I,k}\in H^1_{\mathrm{loc}}(\mathbb{R}^n)$. Furthermore $V_{I,k}\in C^2(\mathbb{R}^n\backslash\Gamma)$ satisfies the Helmholtz equation (5.0.1) in $\mathbb{R}^n\backslash\Gamma$.

(ii)

$$\frac{\partial}{\partial|\boldsymbol{x}|}V_{I,k}(\boldsymbol{x}) - \mathrm{i}kV_{I,k}(\boldsymbol{x}) = o\left(|\boldsymbol{x}|^{-(n-1)/2}\right), \quad \text{as } |\boldsymbol{x}|\to\infty.$$

(iii)

$$\begin{aligned}\left[V_{I,k}(\boldsymbol{x})\right]_\Gamma &= 0,\\ V_{I,k}(\boldsymbol{x})\Big|_\Gamma &= \int_\Gamma E_n(\boldsymbol{x},\boldsymbol{y};k)\varphi(\boldsymbol{y})\mathrm{d}s_y, \quad \forall\boldsymbol{x}\in\Gamma.\end{aligned}$$

(iv)

$$\left.\begin{aligned}\frac{\partial}{\partial\boldsymbol{\nu}_x}V_{I,k}(\boldsymbol{x})\Big|_{\Gamma-0} &= \frac{1}{2}\varphi(\boldsymbol{x}) + \int_\Gamma \frac{\partial}{\partial\boldsymbol{\nu}_x}E_n(\boldsymbol{x},\boldsymbol{y};k)\varphi(\boldsymbol{y})\mathrm{d}s_y, \quad \forall\boldsymbol{x}\in\Gamma,\\ \frac{\partial}{\partial\boldsymbol{\nu}_x}V_{I,k}(\boldsymbol{x})\Big|_{\Gamma+0} &= -\frac{1}{2}\varphi(\boldsymbol{x}) + \int_\Gamma \frac{\partial}{\partial\boldsymbol{\nu}_x}E_n(\boldsymbol{x},\boldsymbol{y};k)\varphi(\boldsymbol{y})\mathrm{d}s_y, \quad \forall\boldsymbol{x}\in\Gamma.\end{aligned}\right\}$$

(v) The operators $\Pi_{I,k}$ and $\Pi_{II,k}^*$ are continuous and bounded

$$\Pi_{I,k}:\ H^{-1/2}(\Gamma)\to H^{3/2}(\Gamma),$$

$$\Pi_{II,k}^*:\ H^{-1/2}(\Gamma)\to H^{1/2}(\Gamma).$$

Moreover $\mathbb{K}_{I,k} = \mathbb{K}^{+}_{I,1} + \Pi_{I,k}$ *and* $\mathbb{K}^{*}_{II,k} = \mathbb{K}^{+,*}_{II,1} + \Pi^{*}_{II,k}$ *are continuous and bounded*

$$\begin{aligned}\mathbb{K}_{I,k}:\ & H^{-1/2+\gamma}(\Gamma) \to H^{1/2+\gamma}(\Gamma),\\ \mathbb{K}^{*}_{II,k}:\ & H^{-1/2+\gamma}(\Gamma) \to H^{-1/2+\gamma}(\Gamma),\end{aligned}$$

provided by $0 \leqslant \gamma \leqslant 1/2$.

5.2.2 The Double Layer Potential $V_{II,k}(x)$

Return to the double layer potential $V_{II,k}(\boldsymbol{x})$ of Helmholtz operator, it can be rewritten as

$$\begin{aligned}V_{II,k}(\boldsymbol{x}) &= \int_{\Gamma} \frac{\partial}{\partial \boldsymbol{\nu}_{\boldsymbol{y}}} E^{+}_{n}(\boldsymbol{x},\boldsymbol{y};1)\varphi(\boldsymbol{y})\mathrm{d}s_{\boldsymbol{y}} + \int_{\Gamma} \frac{\partial}{\partial \boldsymbol{\nu}_{\boldsymbol{y}}} B_{n}(\boldsymbol{x},\boldsymbol{y};k)\varphi(\boldsymbol{y})\mathrm{d}s_{\boldsymbol{y}}\\ &= V^{+}_{II,1}(\boldsymbol{x}) + U_{II,k}(\boldsymbol{x}), \quad \forall \boldsymbol{x} \in \mathbb{R}^{n},\end{aligned} \tag{5.2.33}$$

in which $V^{+}_{II,1}(\boldsymbol{x})$ is the double layer potential of the modified Helmholtz operator defined by (4.3.20) and layer potential $U_{II,k}(\boldsymbol{x})$ is defined by

$$U_{II,k}(\boldsymbol{x}) = \int_{\Gamma} \frac{\partial}{\partial \boldsymbol{\nu}_{\boldsymbol{y}}} B_{n}(\boldsymbol{x},\boldsymbol{y};k)\psi(\boldsymbol{y})\mathrm{d}s_{\boldsymbol{y}}, \quad \forall \boldsymbol{x} \in \mathbb{R}^{n}. \tag{5.2.34}$$

The properties of $V^{+}_{II,1}(\boldsymbol{x})$ have been given in Chap. 4, for the layer potential $U_{II,k}(\boldsymbol{x})$ we have

Lemma 5.2.4 *Suppose that* $\Gamma \in C^{2}$ *and* $\psi(\boldsymbol{x}) \in C^{1}(\Gamma)$.

(*i*) $U_{II,k}(\boldsymbol{x})$ *is well defined on* $\mathbb{R}^{n}$ *and* $U_{II,k} \in C^{1}(\mathbb{R}^{n}) \cap C^{2}(\mathbb{R}^{n}\backslash\Gamma)$ *satisfies*

$$-\Delta U_{II,k}(\boldsymbol{x}) - k^{2}U_{II,k}(\boldsymbol{x}) = (1+k^{2})V^{+}_{II,1}(\boldsymbol{x}), \quad \forall\, \boldsymbol{x} \in \mathbb{R}^{n}\backslash\Gamma.$$

(*ii*)

$$\frac{\partial}{\partial|\boldsymbol{x}|}U_{II,k}(\boldsymbol{x}) - \mathrm{i}kU_{II,k}(\boldsymbol{x}) = \boldsymbol{o}\left(\|\psi\|_{1/2,\Gamma}|\boldsymbol{x}|^{-(n-1)/2}\right), \quad \text{as } |\boldsymbol{x}| \to \infty.$$

(*iii*) $U_{II,k} \in H^{2}_{\mathrm{loc}}(\mathbb{R}^{n})$ *and there exists a positive constant* C, *such that*

$$\begin{aligned}\|U_{II,k}\|_{3/2,\Gamma} &\leqslant C\|\psi\|_{1/2,\Gamma},\\ \left\|\frac{\partial}{\partial \boldsymbol{\nu}_{\boldsymbol{x}}}U_{II,k}\right\|_{1/2,\Gamma} &\leqslant C\|\psi\|_{1/2,\Gamma}.\end{aligned}$$

Furthermore for any bounded domain G, there exists a positive constant C such that

$$\|U_{II,k}\|_{2,G} \leqslant C\|\psi\|_{1/2,\Gamma}.$$

(*iv*) *On the boundary* Γ,

$$\frac{\partial U_{II,k}}{\partial \boldsymbol{\nu}_x}(\boldsymbol{x}) = \sum_{1\leqslant i<j\leqslant n} \frac{\partial}{\partial \boldsymbol{\tau}_x^{i,j}} \int_\Gamma B_n(\boldsymbol{x},\boldsymbol{y};k)\frac{\partial}{\partial \boldsymbol{\tau}_y^{i,j}}\psi(\boldsymbol{y})\mathrm{d}s_y + \int_\Gamma \boldsymbol{\nu}(\boldsymbol{x})\cdot$$
$$\boldsymbol{\nu}(\boldsymbol{y})\Big\{k^2 B_n(\boldsymbol{x},\boldsymbol{y};k)+(1+k^2)E_n^+(\boldsymbol{x},\boldsymbol{y};1)\Big\}\psi(\boldsymbol{y})\mathrm{d}s_y,\ \ \forall\, \boldsymbol{x}\in\Gamma.$$

Proof (i) Note

$$\begin{aligned} U_{II,k}(\boldsymbol{x}) &= \int_\Gamma \frac{\partial}{\partial \boldsymbol{\nu}_y} B_n(\boldsymbol{x},\boldsymbol{y};k)\psi(\boldsymbol{y})\mathrm{d}s_y \\ &= \sum_{j=1}^n \int_\Gamma \nu_j(\boldsymbol{y})\frac{\partial}{\partial y_j} B_n(\boldsymbol{x},\boldsymbol{y};k)\psi(\boldsymbol{y})\mathrm{d}s_y,\ \forall\,\boldsymbol{x}\in\mathbb{R}^n. \end{aligned}$$

By the properties of $B_n(\boldsymbol{x},\boldsymbol{y};k)$ given in Lemma 5.2.1 and $\{\nu_j(\boldsymbol{x})\in C^1(\Gamma),\ 1\leqslant j\leqslant n\}$, the conclusion (i) is proved straightly.

(ii) Let

$$\begin{aligned} W_{II}(\boldsymbol{x},\boldsymbol{y};k) &= \frac{\partial}{\partial|\boldsymbol{x}|}\Big(\frac{\partial}{\partial \boldsymbol{\nu}_y}B_n(\boldsymbol{x},\boldsymbol{y};k)\Big) - \mathrm{i}k\frac{\partial}{\partial \boldsymbol{\nu}_y}B_n(\boldsymbol{x},\boldsymbol{y};k) \\ &= \sum_{j=1}^n \nu_j(\boldsymbol{y})\frac{\partial}{\partial y_j}\Big(\frac{\partial}{\partial|\boldsymbol{x}|}B_n(\boldsymbol{x},\boldsymbol{y};k) - \mathrm{i}kB_n(\boldsymbol{x},\boldsymbol{y};k)\Big),\ \ \forall\boldsymbol{y}\in\Gamma,\boldsymbol{x}\in\mathbb{R}^n. \end{aligned}$$

Then

$$\begin{aligned} \Big|\frac{\partial U_{II,k}}{\partial|\boldsymbol{x}|}(\boldsymbol{x}) - \mathrm{i}kU_{II,k}(\boldsymbol{x})\Big| &= \Big|\int_\Gamma W_{II}(\boldsymbol{x},\boldsymbol{y};k)\psi(\boldsymbol{y})\mathrm{d}s_y\Big| \\ &\leqslant \big\|W_{II}(\boldsymbol{x},\cdot;k)\big\|_{0,\Gamma}\ \|\psi\|_{0,\Gamma} \\ &\leqslant C\big\|W_{II}(\boldsymbol{x},\cdot;k)\big\|_{0,\Gamma}\ \|\psi\|_{1/2,\Gamma}, \end{aligned}$$

where

$$\big\|W_{II}(\boldsymbol{x},\cdot;k)\big\|_{0,\Gamma}^2 = \int_\Gamma \Big|W_{II}(\boldsymbol{x},\boldsymbol{y};k)\Big|^2\mathrm{d}s_y.$$

It is straightforward to check that

$$|\boldsymbol{x}|^{(n-1)/2}\Big|W_{II}(\boldsymbol{x},\boldsymbol{y};k)\Big| = o(1),\ \text{ as } |\boldsymbol{x}|\to\infty,$$

uniformly for $\boldsymbol{y} \in \Gamma$ and all directions $\dfrac{\boldsymbol{x}}{|\boldsymbol{x}|}$. Then the conclusion (ii) is proved.

(iii) Find $R > 0$, such that $\overline{\Omega} \subset \Omega_R = \{\boldsymbol{x} \,\big|\, |\boldsymbol{x}| < R\}$. Let $\Gamma_R = \partial\Omega_R$, then $\Gamma \cap \Gamma_R = \varnothing$. Consider

$$\begin{aligned} U_{II,k}(\boldsymbol{x}) &= \int_\Gamma \frac{\partial}{\partial \nu_y} B_n(\boldsymbol{x}, \boldsymbol{y}; k)\psi(\boldsymbol{y})\mathrm{d}s_y, \quad \forall\, \boldsymbol{x} \in \Gamma_R, \\ &:= \int_\Gamma D_{II}(\boldsymbol{x}, \boldsymbol{y}; k)\psi(\boldsymbol{y})\mathrm{d}s_y. \end{aligned}$$

The function $D_{II}(\boldsymbol{x}, \boldsymbol{y}; k)$ and its derivatives

$$\frac{\partial}{\partial x_i} D_{II}(\boldsymbol{x}, \boldsymbol{y}; k)\ (1 \leqslant i \leqslant n), \quad \frac{\partial^2}{\partial x_i \partial x_j} D_{II}(\boldsymbol{x}, \boldsymbol{y}; k)\ \ (1 \leqslant i, j \leqslant n),$$

are bounded on $(\boldsymbol{x}, \boldsymbol{y}) \in \Gamma_R \times \Gamma$, namely

$$\left.\begin{aligned} &\left|D_{II}(\boldsymbol{x}, \boldsymbol{y}; k)\right| < C, \\ &\left|\frac{\partial}{\partial x_j} D_{II}(\boldsymbol{x}, \boldsymbol{y}; k)\right| < C, \quad 1 \leqslant i \leqslant n, \\ &\left|\frac{\partial^2}{\partial x_i \partial x_j} D_{II}(\boldsymbol{x}, \boldsymbol{y}; k)\right| < C, \quad 1 \leqslant i \leqslant n, \end{aligned}\right\} \quad \forall (\boldsymbol{x}, \boldsymbol{y}) \in \Gamma_R \times \Gamma.$$

It directly arrives at

$$\left\|U_{II,k}\right\|_{3/2,\Gamma_R} \leqslant C \left\|\psi\right\|_{1/2,\Gamma}. \tag{5.2.35}$$

On the domain Ω_R,

$$\begin{aligned} \left|U_{II,k}\right| &= \left|\sum_{j=1}^n \int_\Gamma \nu_j(\boldsymbol{y}) \frac{\partial}{\partial y_j} B_n(\boldsymbol{x}, \boldsymbol{y}; k)\psi(\boldsymbol{y})\mathrm{d}s_y\right| \\ &\leqslant C \sum_{j=1}^n \left\|\frac{\partial}{\partial y_j} B_n(\boldsymbol{x}, \boldsymbol{y}; k)\right\|_{0,\Gamma} \left\|\psi\right\|_{0,\Gamma}, \end{aligned}$$

in which

$$\left\|\frac{\partial}{\partial y_j} B_n(\boldsymbol{x}, \boldsymbol{y}; k)\right\|_{0,\Gamma}^2 = \int_\Gamma \left|\frac{\partial}{\partial y_j} B_n(\boldsymbol{x}, \boldsymbol{y}; k)\right|^2 \mathrm{d}s_y, \quad 1 \leqslant j \leqslant n$$

are bounded uniformly for all $\boldsymbol{y} \in \Gamma$ and $\boldsymbol{x} \in \Omega_R$. It yields

$$\left.\begin{aligned} \left|U_{II,k}\right| &\leqslant C\|\psi\|_{1/2,\Gamma}, \quad \forall\, \boldsymbol{x} \in \Omega_R, \\ \left\|U_{II,k}\right\|_{0,\Omega_R} &\leqslant C\|\psi\|_{1/2,\Gamma}. \end{aligned}\right\} \tag{5.2.36}$$

Furthermore on the domain Ω_R, $U_{II,k}(\boldsymbol{x})$ satisfies

$$\left.\begin{aligned} &-\Delta U_{II,k}(\boldsymbol{x}) = k^2 U_{II,k}(\boldsymbol{x}) + (1+k^2)V_{II,1}^{+}(\boldsymbol{x}) := F_{II,k}(\boldsymbol{x}), \ \forall\, \boldsymbol{x} \in \Omega_R\backslash\Gamma, \\ &\left[U_{II,k}(\boldsymbol{x})\right]_{\Gamma} = 0, \\ &\left[\frac{\partial U_{II,k}}{\partial \boldsymbol{\nu}_{\boldsymbol{x}}}(\boldsymbol{x})\right]_{\Gamma} = 0, \\ &U_{II,k}(\boldsymbol{x})\Big|_{\Gamma_R} = U_{II,k}(\boldsymbol{x}), \quad \forall\, \boldsymbol{x} \in \Gamma_R. \end{aligned}\right\} \tag{5.2.37}$$

By variational equality (4.4.30) we obtain

$$\|V_{II,1}^{+}\|_{0,\mathbb{R}^n} \leqslant C\|\psi\|_{1/2,\Gamma}. \tag{5.2.38}$$

The estimates (5.2.36) and (5.2.38) yield that

$$\|F_{II,k}\|_{0,\Omega_R} \leqslant C\|\psi\|_{1/2,\Gamma}, \tag{5.2.39}$$

then the discontinuous boundary value problem (5.2.37) is equivalent to

$$\left.\begin{aligned} -\Delta U_{II,k}(\boldsymbol{x}) &= F_{II,k}(\boldsymbol{x}), \quad \forall\, \boldsymbol{x} \in \Omega_R, \\ U_{II,k}(\boldsymbol{x})\Big|_{\Gamma_R} &= U_{II,k}(\boldsymbol{x}), \quad \forall\, \boldsymbol{x} \in \Gamma. \end{aligned}\right\} \tag{5.2.40}$$

By the regular estimates of problem (5.2.40), we obtain:

$$\left\|U_{II,k}\right\|_{2,\Omega_R} \leqslant C\left\{\left\|U_{II,k}\right\|_{3/2,\Gamma_R} + \left\|F_{II,k}\right\|_{0,\Omega_R}\right\} \leqslant C\|\psi\|_{1/2,\Gamma}.$$

For any given bounded domain G, there is a constant $R > 0$, such that

$$\left(\overline{\Omega} \cup \overline{G}\right) \in \Omega_R = \left\{\boldsymbol{x}\,\Big|\,|\boldsymbol{x}| < R\right\},$$

then we have

$$\|U_{II,k}\|_{2,G} \leqslant \|U_{II,k}\|_{2,\Omega_R} \leqslant C\|\psi\|_{1/2,\Gamma}.$$

It means that $U_{II,k} \in H^2_{\mathrm{loc}}(\mathbb{R}^n)$.

(iv) Consider the first derivatives of $U_{II,k}(\boldsymbol{x})$.

$$
\begin{aligned}
\frac{\partial U_{II,k}}{\partial x_i}(\boldsymbol{x}) &= \frac{\partial}{\partial x_i}\int_\Gamma \frac{\partial}{\partial \boldsymbol{\nu}_{\boldsymbol{y}}} B_n(\boldsymbol{x},\boldsymbol{y};k)\psi(\boldsymbol{y})\mathrm{d}s_{\boldsymbol{y}} \\
&= \sum_{j=1}^n \int_\Gamma \nu_j(\boldsymbol{y}) \frac{\partial^2}{\partial x_i \partial y_j} B_n(\boldsymbol{x},\boldsymbol{y};k)\psi(\boldsymbol{y})\mathrm{d}s_{\boldsymbol{y}} \\
&= \sum_{j=1}^n \int_\Gamma \nu_j(\boldsymbol{y}) \frac{\partial^2}{\partial x_j \partial y_i} B_n(\boldsymbol{x},\boldsymbol{y};k)\psi(\boldsymbol{y})\mathrm{d}s_{\boldsymbol{y}} \\
&= \sum_{j=1}^n \frac{\partial}{\partial x_j}\int_\Gamma \Big(-\nu_i(\boldsymbol{y})\frac{\partial}{\partial y_j} + \nu_j(\boldsymbol{y})\frac{\partial}{\partial y_i}\Big) B_n(\boldsymbol{x},\boldsymbol{y};k)\psi(\boldsymbol{y})\mathrm{d}s_{\boldsymbol{y}} + \\
&\quad \sum_{j=1}^n \int_\Gamma \nu_i(\boldsymbol{y}) \frac{\partial^2}{\partial x_j \partial y_j} B_n(\boldsymbol{x},\boldsymbol{y};k)\psi(\boldsymbol{y})\mathrm{d}s_{\boldsymbol{y}} \\
&= -\sum_{j=1}^n \frac{\partial}{\partial x_j}\int_\Gamma B_n(\boldsymbol{x},\boldsymbol{y};k)\frac{\partial}{\partial \boldsymbol{\tau}_{\boldsymbol{y}}^{i,j}}\psi(\boldsymbol{y})\mathrm{d}s_{\boldsymbol{y}} - \\
&\quad \int_\Gamma \nu_i(\boldsymbol{y})\Delta_{\boldsymbol{y}} B_n(\boldsymbol{x},\boldsymbol{y};k)\psi(\boldsymbol{y})\mathrm{d}s_{\boldsymbol{y}} \\
&= \int_\Gamma \nu_i(\boldsymbol{y})\Big\{k^2 B_n(\boldsymbol{x},\boldsymbol{y};k) + (1+k^2)E_n^+(\boldsymbol{x},\boldsymbol{y};1)\Big\}\psi(\boldsymbol{y})\mathrm{d}s_{\boldsymbol{y}} - \\
&\quad \sum_{j=1}^n \frac{\partial}{\partial x_j}\int_\Gamma B_n(\boldsymbol{x},\boldsymbol{y};k)\frac{\partial}{\partial \boldsymbol{\tau}_{\boldsymbol{y}}^{i,j}}\psi(\boldsymbol{y})\mathrm{d}s_{\boldsymbol{y}}, \quad 1 \leqslant i \leqslant 3,\ \forall\, \boldsymbol{x} \in \mathbb{R}^n.
\end{aligned}
$$

The last equality is from (5.2.8).

On the boundary Γ, we have

$$
\begin{aligned}
\frac{\partial U_{II,k}}{\partial \boldsymbol{\nu}_{\boldsymbol{x}}}(\boldsymbol{x}) &= \sum_{i=1}^n \nu_i(\boldsymbol{x})\frac{\partial}{\partial x_i} U_{II,k}(\boldsymbol{x}) \\
&= -\sum_{j=1}^n \nu_i(\boldsymbol{x})\frac{\partial}{\partial x_j}\int_\Gamma B_n(\boldsymbol{x},\boldsymbol{y};k)\frac{\partial}{\partial \boldsymbol{\tau}_{\boldsymbol{y}}^{i,j}}\psi(\boldsymbol{y})\mathrm{d}s_{\boldsymbol{y}} + \\
&\quad \int_\Gamma \boldsymbol{\nu}(\boldsymbol{x})\cdot\boldsymbol{\nu}(\boldsymbol{y})\Big\{k^2 B_n(\boldsymbol{x},\boldsymbol{y};k) + (1+k^2)E_n^+(\boldsymbol{x},\boldsymbol{y};1)\Big\}\psi(\boldsymbol{y})\mathrm{d}s_{\boldsymbol{y}} \\
&= -\frac{1}{2}\sum_{i,j=1}^n \Big(-\nu_i(\boldsymbol{x})\frac{\partial}{\partial x_j} + \nu_j(\boldsymbol{x})\frac{\partial}{\partial x_i}\int_\Gamma\Big) B_n(\boldsymbol{x},\boldsymbol{y};k)\frac{\partial}{\partial \boldsymbol{\tau}_{\boldsymbol{y}}^{i,j}}\psi(\boldsymbol{y})\mathrm{d}s_{\boldsymbol{y}} + \\
&\quad \int_\Gamma \boldsymbol{\nu}(\boldsymbol{x})\cdot\boldsymbol{\nu}(\boldsymbol{y})\Big\{k^2 B_n(\boldsymbol{x},\boldsymbol{y};k) + (1+k^2)E_n^+(\boldsymbol{x},\boldsymbol{y};1)\Big\}\psi(\boldsymbol{y})\mathrm{d}s_{\boldsymbol{y}}
\end{aligned}
$$

$$= \sum_{1 \leqslant i,j \leqslant n} \frac{\partial}{\partial \boldsymbol{\tau}_x^{i,j}} \int_\Gamma B_n(\boldsymbol{x}, \boldsymbol{y}; k) \frac{\partial}{\partial \boldsymbol{\tau}_y^{i,j}} \psi(\boldsymbol{y}) \mathrm{d}s_y + \int_\Gamma \boldsymbol{\nu}(\boldsymbol{x}) \cdot$$

$$\boldsymbol{\nu}(\boldsymbol{y}) \Big\{ k^2 B_n(\boldsymbol{x}, \boldsymbol{y}; k) + (1+k^2) E_n^+(\boldsymbol{x}, \boldsymbol{y}; 1) \Big\} \psi(\boldsymbol{y}) \mathrm{d}s_y, \ \forall \boldsymbol{x} \in \Gamma.$$

The proof of the Lemma 5.2.4 is completed. □

For given $\psi \in H^{1/2}(\Gamma)$, there exsits a function series $\{\psi_m \in C^1(\Gamma),\ m = 1, 2, \cdots\}$, such that

$$\lim_{m \to \infty} \left\| \psi - \psi_m \right\|_{1/2,\Gamma} = 0,$$

provided by $C^1(\Gamma)$ is dense in $H^{1/2}(\Gamma)$. Then we have

$$U_{II,k}(\boldsymbol{x}) = \int_\Gamma \frac{\partial}{\partial \boldsymbol{\nu}_y} B_n(\boldsymbol{x}, \boldsymbol{y}; k) \psi(\boldsymbol{y}) \mathrm{d}s_y = \lim_{m \to \infty} \int_\Gamma \frac{\partial}{\partial \boldsymbol{\nu}_y} B_n(\boldsymbol{x}, \boldsymbol{y}; k) \psi_m(\boldsymbol{y}) \mathrm{d}s_y.$$

The Lemmas 5.2.1 and 5.2.4 yield the following results.

Lemma 5.2.5 *Suppose that* $\Gamma \in C^2$ *and* $\psi \in H^{1/2}(\Gamma)$.

(*i*) $U_{II,k}(\boldsymbol{x})$ *is well defined on* $\mathbb{R}^n$ *and* $U_{II,k} \in C^2(\mathbb{R}^n \backslash \Gamma)$ *satisfies*

$$-\Delta U_{II,k}(\boldsymbol{x}) - k^2 U_{II,k}(\boldsymbol{x}) = (1+k^2) V_{II,1}^+(\boldsymbol{x}), \quad \forall \boldsymbol{x} \in \mathbb{R}^n \backslash \Gamma.$$

(*ii*)

$$\frac{\partial}{\partial |\boldsymbol{x}|} U_{II,k}(\boldsymbol{x}) - \mathrm{i} k U_{II,k}(\boldsymbol{x}) = \boldsymbol{o}\left(|\boldsymbol{x}|^{-(n-1)/2}\right), \quad as \ |\boldsymbol{x}| \to +\infty.$$

(*iii*) $U_{II,k} \in H^2_{\mathrm{loc}}(\mathbb{R}^n)$ *and there exists a positive constant* C, *such that*

$$\left\| U_{II,k} \right\|_{3/2,\Gamma} \leqslant C \left\| \psi \right\|_{1/2,\Gamma},$$

$$\left\| \frac{\partial}{\partial \boldsymbol{\nu}_x} U_{II,k} \right\|_{1/2,\Gamma} \leqslant C \left\| \psi \right\|_{1/2,\Gamma}.$$

(*iv*) *On the boundary* Γ,

$$\left. \frac{\partial U_{II,k}}{\partial \boldsymbol{\nu}_x}(\boldsymbol{x}) \right|_\Gamma = \sum_{1 \leqslant i < j \leqslant n} \frac{\partial}{\partial \boldsymbol{\tau}_x^{i,j}} \int_\Gamma B_n(\boldsymbol{x}, \boldsymbol{y}; k) \frac{\partial}{\partial \boldsymbol{\tau}_y^{i,j}} \psi(\boldsymbol{y}) \mathrm{d}s_y +$$

$$\int_\Gamma \boldsymbol{\nu}(\boldsymbol{x}) \cdot \boldsymbol{\nu}(\boldsymbol{y}) \left\{ k^2 B_n(\boldsymbol{x}, \boldsymbol{y}; k) + (1+k^2) E_n^+(\boldsymbol{x}, \boldsymbol{y}; 1) \right\} \psi(\boldsymbol{y}) \mathrm{d}s_y.$$

Introduce the following operators.

$$\left.\begin{aligned}
(\mathbb{K}_{\mathit{II},k}\psi)(\boldsymbol{x}) &= \int_\Gamma \frac{\partial}{\partial \boldsymbol{\nu}_{\boldsymbol{y}}} E_n(\boldsymbol{x}, \boldsymbol{y}; k)\psi(\boldsymbol{y})\mathrm{d}s_{\boldsymbol{y}} \\
&= \int_\Gamma \frac{\partial}{\partial \boldsymbol{\nu}_{\boldsymbol{y}}} E_n^+(\boldsymbol{x}, \boldsymbol{y}; 1)\psi(\boldsymbol{y})\mathrm{d}s_{\boldsymbol{y}} + \int_\Gamma \frac{\partial}{\partial \boldsymbol{\nu}_{\boldsymbol{y}}} B_n(\boldsymbol{x}, \boldsymbol{y}; k)\psi(\boldsymbol{y})\mathrm{d}s_{\boldsymbol{y}} \\
&= (\mathbb{K}^+_{\mathit{II},1}\psi)(\boldsymbol{x}) + (\Pi_{\mathit{II},k}\psi)(\boldsymbol{x}), \quad \forall \boldsymbol{x} \in \Gamma, \\
(\mathbb{K}_{\mathit{III},k}\psi)(\boldsymbol{x}) &= \frac{\partial}{\partial \boldsymbol{\nu}_{\boldsymbol{x}}} \int_\Gamma \frac{\partial}{\partial \boldsymbol{\nu}_{\boldsymbol{y}}} E_n(\boldsymbol{x}, \boldsymbol{y}; k)\psi(\boldsymbol{y})\mathrm{d}s_{\boldsymbol{y}} \\
&= \frac{\partial}{\partial \boldsymbol{\nu}_{\boldsymbol{x}}} \int_\Gamma \frac{\partial}{\partial \boldsymbol{\nu}_{\boldsymbol{y}}} E_n^+(\boldsymbol{x}, \boldsymbol{y}; 1)\psi(\boldsymbol{y})\mathrm{d}s_{\boldsymbol{y}} + \\
&\quad \frac{\partial}{\partial \boldsymbol{\nu}_{\boldsymbol{x}}} \int_\Gamma \frac{\partial}{\partial \boldsymbol{\nu}_{\boldsymbol{y}}} B_n(\boldsymbol{x}, \boldsymbol{y}; k)\psi(\boldsymbol{y})\mathrm{d}s_{\boldsymbol{y}} \\
&= (\mathbb{K}^+_{\mathit{III},1}\psi)(\boldsymbol{x}) + (\Pi_{\mathit{III},k}\psi)(\boldsymbol{x}), \quad \forall\, \boldsymbol{x} \in \Gamma,
\end{aligned}\right\} \tag{5.2.41}$$

in which

$$\left.\begin{aligned}
(\Pi_{\mathit{II},k}\psi)(\boldsymbol{x}) &= \int_\Gamma \frac{\partial}{\partial \boldsymbol{\nu}_{\boldsymbol{y}}} B_n(\boldsymbol{x}, \boldsymbol{y}; k)\psi(\boldsymbol{y})\mathrm{d}s_{\boldsymbol{y}}, \quad \forall\, \boldsymbol{x} \in \Gamma, \\
(\Pi_{\mathit{III},k}\psi)(\boldsymbol{x}) &= \frac{\partial}{\partial \boldsymbol{\nu}_{\boldsymbol{x}}} \int_\Gamma \frac{\partial}{\partial \boldsymbol{\nu}_{\boldsymbol{y}}} B_n(\boldsymbol{x}, \boldsymbol{y}; k)\psi(\boldsymbol{y})\mathrm{d}s_{\boldsymbol{y}}, \quad \forall\, \boldsymbol{x} \in \Gamma.
\end{aligned}\right\} \tag{5.2.42}$$

Obviously

$$\left.\begin{aligned}
\mathbb{K}_{\mathit{II},k} &= \mathbb{K}^+_{\mathit{II},1} + \Pi_{\mathit{II},k}, \\
\mathbb{K}_{\mathit{III},k} &= \mathbb{K}^+_{\mathit{III},1} + \Pi_{\mathit{III},k}.
\end{aligned}\right\} \tag{5.2.43}$$

The Lemma 5.2.5 and Theorem 4.3.2 arrive at the following theorem.

Theorem 5.2.3 *Presume that $\Gamma \in C^2$ and the density $\psi \in H^{1/2}(\Gamma)$.*

(*i*) *The double layer potential $V_{\mathit{II},k}(\boldsymbol{x})$ is well defined on $\mathbb{R}^n$ and $V_{\mathit{II},k} \in H^1_{\text{loc}}(\mathbb{R}^n \backslash \Gamma) \cap C^2(\mathbb{R}^n \backslash \Gamma)$ satisfies the Helmholtz equation (5.0.1) on $\mathbb{R}^n \backslash \Gamma$.*

(*ii*)

$$\frac{\partial}{\partial |\boldsymbol{x}|} V_{\mathit{II},k}(\boldsymbol{x}) - \mathrm{i}k V_{\mathit{II},k}(\boldsymbol{x}) = o\left(|\boldsymbol{x}|^{-(n-1)/2}\right), \quad \textit{as} \ |\boldsymbol{x}| \to +\infty.$$

(*iii*)

$$\left.\begin{aligned}
V_{\mathit{II},k}(\boldsymbol{x})\Big|_{\Gamma+0} &= \frac{1}{2}\psi(\boldsymbol{x}) + \int_\Gamma \frac{\partial}{\partial \boldsymbol{\nu}_{\boldsymbol{y}}} E_n(\boldsymbol{x}, \boldsymbol{y}; k)\psi(\boldsymbol{y})\mathrm{d}s_{\boldsymbol{y}} \\
&= \frac{1}{2}\psi(\boldsymbol{x}) + (\mathbb{K}_{\mathit{II},k}\psi)(\boldsymbol{x}), \quad \forall \boldsymbol{x} \in \Gamma, \\
V_{\mathit{II},k}(\boldsymbol{x})\Big|_{\Gamma-0} &= -\frac{1}{2}\psi(\boldsymbol{x}) + \int_\Gamma \frac{\partial}{\partial \boldsymbol{\nu}_{\boldsymbol{y}}} E_n(\boldsymbol{x}, \boldsymbol{y}; k)\psi(\boldsymbol{y})\mathrm{d}s_{\boldsymbol{y}} \\
&= -\frac{1}{2}\psi(\boldsymbol{x}) + (\mathbb{K}_{\mathit{II},k}\psi)(\boldsymbol{x}), \quad \forall \boldsymbol{x} \in \Gamma.
\end{aligned}\right\}$$

(*iv*)

$$
\begin{aligned}
&\frac{\partial}{\partial \boldsymbol{\nu}_x} V_{\mathit{II},k}(\boldsymbol{x})\Big|_{\Gamma+0} \\
&= \frac{\partial}{\partial \boldsymbol{\nu}_x} V_{\mathit{II},k}(\boldsymbol{x})\Big|_{\Gamma-0} = \frac{\partial}{\partial \boldsymbol{\nu}_x} V_{\mathit{II},k}(\boldsymbol{x}) \\
&= \frac{\partial}{\partial \boldsymbol{\nu}_x} \int_\Gamma \frac{\partial}{\partial \boldsymbol{\nu}_y} E_n(\boldsymbol{x}, \boldsymbol{y}; k)\psi(\boldsymbol{y}) \mathrm{d}s_y \\
&= \sum_{1 \leqslant i < j \leqslant n} \frac{\partial}{\partial \boldsymbol{\tau}_x^{i,j}} \int_\Gamma E_n(\boldsymbol{x}, \boldsymbol{y}; k) \frac{\partial}{\partial \boldsymbol{\tau}_y^{i,j}} \psi(\boldsymbol{y}) \mathrm{d}s_y + \\
&k^2 \int_\Gamma E_n(\boldsymbol{x}, \boldsymbol{y}; k)\big(\boldsymbol{\nu}(\boldsymbol{x}) \cdot \boldsymbol{\nu}(\boldsymbol{y})\big)\psi(\boldsymbol{y}) \mathrm{d}s_y = \big(\mathbb{K}_{\mathit{III},k}\psi\big)(\boldsymbol{x}), \quad \forall\, \boldsymbol{x} \in \Gamma.
\end{aligned}
$$

(v) *The operators*

$$
\begin{aligned}
\Pi_{\mathit{II},k} &: H^{1/2}(\Gamma) \to H^{3/2}(\Gamma), \\
\Pi_{\mathit{III},k} &: H^{1/2}(\Gamma) \to H^{1/2}(\Gamma),
\end{aligned}
$$

are continuous and bounded. Furthermore the operators

$$
\begin{aligned}
\mathbb{K}_{\mathit{II},k} &: H^{1/2+\gamma}(\Gamma) \to H^{1/2+\gamma}(\Gamma), \\
\mathbb{K}_{\mathit{III},k} &: H^{1/2+\gamma}(\Gamma) \to H^{-1/2+\gamma}(\Gamma),
\end{aligned}
$$

are continuous and bounded provided by $0 \leqslant \gamma \leqslant 1/2$.

5.3 Boundary Reduction for the Principal Boundary Value Problems of Helmholtz Equation

The boundary reduction of the five principal boundary value problems of Helmh- oltz equation are considered in this section. "Traditionally, the use of integral equations of the first kind for studying boundary-value problems in acoustic scattering theory has been neglected due to the lack of a Riesz–Fredholm theory for the equation of the first kind and the fact that integral equation of the first kind is improperly posed" mentioned by Colton and Kress (2013). But in this section boundary integral equation of first kind and boundary integro-differentiable equation of the first kind is also used for the boundary value problem in acoustic scattering theory due to the corresponding Reisz–Fredholm theorems for them being established on the complex function spaces $H^{-1/2}(\Gamma)$ and $H^{1/2}(\Gamma)$.

Let

$$
\langle \varphi, \rho \rangle_{-1/2,\Gamma} = \int_\Gamma \int_\Gamma E_n^+(\boldsymbol{x}, \boldsymbol{y}; 1)\varphi(\boldsymbol{x})\overline{\rho(\boldsymbol{x})} \mathrm{d}s_y \mathrm{d}s_x, \quad \forall\, \varphi, \rho \in H^{-1/2}(\Gamma). \tag{5.3.1}
$$

It is straightforward to check that $\langle\varphi,\rho\rangle_{-1/2,\Gamma}$ is an inner product of complex function space $H^{-1/2}(\Gamma)$ by Theorem 4.3.1, it means that $H^{-1/2}(\Gamma)$ is a Hilbert space. Furthermore

$$\begin{aligned}\langle\varphi,\rho\rangle_{-1/2,\Gamma} &= \int_\Gamma \Big(\mathbb{K}^+_{I,1}\varphi\Big)(\boldsymbol{x})\overline{\rho(\boldsymbol{x})}\mathrm{d}s_{\boldsymbol{x}} \\ &= \int_\Gamma \varphi(\boldsymbol{x})\Big(\mathbb{K}^+_{I,1}\overline{\rho}\Big)(\boldsymbol{y})\mathrm{d}s_{\boldsymbol{y}}, \quad \forall\,\varphi,\rho\in H^{-1/2}(\Gamma).\end{aligned} \tag{5.3.2}$$

For the complex function space $H^{1/2}(\Gamma)$, let

$$\langle\psi,\mu\rangle_{1/2,\Gamma} = -\int_\Gamma\int_\Gamma \frac{\partial^2}{\partial\boldsymbol{\nu}_{\boldsymbol{x}}\partial\boldsymbol{\nu}_{\boldsymbol{y}}}E_n^+(\boldsymbol{x},\boldsymbol{y};1)\psi(\boldsymbol{y})\overline{\mu(\boldsymbol{x})}\mathrm{d}s_{\boldsymbol{y}}\mathrm{d}s_{\boldsymbol{x}}, \quad \forall\,\psi,\mu\in H^{1/2}(\Gamma). \tag{5.3.3}$$

The Theorem 4.3.2 yields that $\langle\psi,\mu\rangle_{1/2,\Gamma}$ is an inner product of space $H^{1/2}(\Gamma)$, and

$$\begin{aligned}\langle\psi,\mu\rangle_{1/2,\Gamma} &= -\int_\Gamma \big(\mathbb{K}^+_{III,1}\psi\big)(\boldsymbol{x})\overline{\mu(\boldsymbol{x})}\mathrm{d}s_{\boldsymbol{x}} \\ &= -\int_\Gamma \psi(\boldsymbol{y})\big(\mathbb{K}^+_{III,1}\overline{\mu}\big)(\boldsymbol{y})\mathrm{d}s_{\boldsymbol{y}} \\ &= \sum_{1\leqslant i<j\leqslant n}\int_\Gamma\int_\Gamma E_n^+(\boldsymbol{x},\boldsymbol{y};1)\left(\frac{\partial}{\partial\boldsymbol{\tau}_{\boldsymbol{y}}^{i,j}}\psi(\boldsymbol{y})\right)\left(\frac{\partial}{\partial\boldsymbol{\tau}_{\boldsymbol{x}}^{i,j}}\overline{\mu(\boldsymbol{x})}\right)\mathrm{d}s_{\boldsymbol{y}}\mathrm{d}s_{\boldsymbol{x}}+ \\ &\quad \int_\Gamma\int_\Gamma E_n^+(\boldsymbol{x},\boldsymbol{y};1)\boldsymbol{\nu}(\boldsymbol{x})\cdot\boldsymbol{\nu}(\boldsymbol{y})\psi(\boldsymbol{y})\overline{\mu(\boldsymbol{x})}\mathrm{d}s_{\boldsymbol{y}}\mathrm{d}s_{\boldsymbol{x}}, \quad \forall\psi,\mu\in H^{1/2}(\Gamma).\end{aligned} \tag{5.3.4}$$

Obviously $H^{1/2}(\Gamma)$ is a Hilbert space.

We recall the Green's representation for the solutions of Helmholtz equation in bounded domain Ω and in the exterior domain Ω_e with Sommefeld radiation condition at infinity given by Colton and Kress (2013).

Theorem 5.3.1 *Suppose that $u\in H^1(\Omega)\cap C^2(\Omega)$ satisfies Helmholtz equation in the bounded domain Ω:*

$$-\Delta u(\boldsymbol{x}) - k^2u(\boldsymbol{x}) = 0, \quad \forall\,\boldsymbol{x}\in\Omega.$$

Then

$$\int_\Gamma E_n(\boldsymbol{x},\boldsymbol{y};k)\frac{\partial}{\partial\boldsymbol{\nu}_{\boldsymbol{y}}}u(\boldsymbol{y})\mathrm{d}s_{\boldsymbol{y}} - \int_\Gamma\frac{\partial}{\partial\boldsymbol{\nu}_{\boldsymbol{y}}}E_n(\boldsymbol{x},\boldsymbol{y};k)u(\boldsymbol{y})\mathrm{d}s_{\boldsymbol{y}} = \begin{cases}u(\boldsymbol{x}), & \forall\,\boldsymbol{x}\in\Omega,\\ 0, & \forall\,\boldsymbol{x}\in\Omega_e.\end{cases} \tag{5.3.5}$$

Theorem 5.3.2 *Assume that $u\in H^1_{\mathrm{loc}}(\Omega_e)\cap C^2(\Omega_e)$ satisfies Helmholtz equation in Ω_e and Sommefeld radiation condition at infinity:*

$$\left.\begin{aligned} & -\Delta u(\boldsymbol{x})-k^2 u(\boldsymbol{x})=0, \quad \forall\, \boldsymbol{x} \in \Omega_e, \\ & \frac{\partial}{\partial|\boldsymbol{x}|} u(\boldsymbol{x})-\mathrm{i} k u(\boldsymbol{x})=\boldsymbol{o}\Big(|\boldsymbol{x}|^{-(n-1)/2}\Big), \quad \text{as } |\boldsymbol{x}| \rightarrow+\infty. \end{aligned}\right\}$$

Then

$$\int_\Gamma \frac{\partial}{\partial \boldsymbol{\nu}_y} E_n(\boldsymbol{x}, \boldsymbol{y}; k) u(\boldsymbol{y}) \mathrm{d} s_y-\int_\Gamma E_n(\boldsymbol{x}, \boldsymbol{y}; k) \frac{\partial}{\partial \boldsymbol{\nu}_y} u(\boldsymbol{y}) \mathrm{d} s_y=\begin{cases}0, & \forall\, \boldsymbol{x} \in \Omega, \\ u(\boldsymbol{x}), & \forall\, \boldsymbol{x} \in \Omega_e .\end{cases} \tag{5.3.6}$$

The Green's representations (5.3.5) and (5.3.6) are the foundation of the direct boundary value integral equation method for the boundary value problems of Helmholtz equation.

5.3.1 Boundary Integral Equation of the First Kind

Consider the interior and exterior Dirichlet problem (5.1.1) and (5.1.2) for given function $g \in H^{1/2}(\Gamma)$. The single layer potential

$$u(\boldsymbol{x})=\int_\Gamma E_n(\boldsymbol{x}, \boldsymbol{y}; k) \varphi(\boldsymbol{y}) \mathrm{d} s_y, \quad \forall \boldsymbol{x} \in \mathbb{R}^n, \tag{5.3.7}$$

solves the interior Dirichlet problem (5.1.1) in Ω and the exterior Dirichlet problem (5.1.2) in Ω_e provided that $\varphi \in H^{-1/2}(\Gamma)$ is a solution of the following boundary integral equation of first kind.

$$\begin{cases}\text{Find} \quad \varphi \in H^{-\frac{1}{2}}(\Gamma), \quad \text{such that} \\ \Big(\mathbb{K}_{I,k} \varphi\Big)(\boldsymbol{x})=\int_\Gamma E_n(\boldsymbol{x}, \boldsymbol{y}; k) \varphi(\boldsymbol{y}) \mathrm{d} s_y=g(\boldsymbol{x}), \quad \forall \boldsymbol{x} \in \Gamma .\end{cases} \tag{5.3.8}$$

Introduce operator $\overline{\mathbb{K}}_{I,k}$ defined by

$$\Big(\overline{\mathbb{K}}_{I,k} \varphi\Big)(\boldsymbol{x})=\int_\Gamma \overline{E_n(\boldsymbol{x}, \boldsymbol{y}; k)} \varphi(\boldsymbol{y}) \mathrm{d} s_y, \quad \forall \varphi \in H^{-1/2}(\Gamma). \tag{5.3.9}$$

Apparently the operator $\overline{\mathbb{K}}_{I,k}: \quad H^{-1/2+\gamma}(\Gamma) \rightarrow H^{1/2+\gamma}(\Gamma)$ is continuous and bounded provided by $0 \leqslant \gamma \leqslant 1/2$.

Consider $\mathbb{N}\big(\mathbb{K}_{I,k}\big)$ and $\mathbb{N}\big(\overline{\mathbb{K}}_{I,k}\big)$, the null space of operators $\mathbb{K}_{I,k}$ and $\overline{\mathbb{K}}_{I,k}$. Suppose that $\rho \in \mathbb{N}\big(\mathbb{K}_{I,k}\big)$, then $\rho \in H^{-1/2}(\Gamma)$ and

$$\big(\mathbb{K}_{I,k} \rho\big)(\boldsymbol{x})=0, \quad \forall\, \boldsymbol{x} \in \Gamma.$$

Let

$$u_\rho(\boldsymbol{x}) = \int_\Gamma E_n(\boldsymbol{x}, \boldsymbol{y}; k)\rho(\boldsymbol{x})\mathrm{d}s_{\boldsymbol{x}}, \quad \forall\, \boldsymbol{x} \in \mathbb{R}^n.$$

On the domain Ω_e, $u_\rho(\boldsymbol{x})$ is a solution of exterior Dirichlet problem (5.1.2) with given $g(\boldsymbol{x}) \equiv 0$. The uniqueness of the problem (5.1.2) arrives at

$$\begin{aligned} &u_\rho(\boldsymbol{x}) \equiv 0, \quad \forall\, \boldsymbol{x} \in \Omega_e, \\ &0 = \frac{\partial u_\rho}{\partial \boldsymbol{\nu}_{\boldsymbol{x}}}(\boldsymbol{x})\bigg|_{\Gamma+0} = -\frac{1}{2}\rho(\boldsymbol{x}) + \int_\Gamma \frac{\partial}{\partial \boldsymbol{\nu}_{\boldsymbol{x}}} E_n(\boldsymbol{x}, \boldsymbol{y}; k)\rho(\boldsymbol{y})\mathrm{d}s_{\boldsymbol{y}}, \quad \forall\, \boldsymbol{x} \in \Gamma. \end{aligned}$$

Furthermore we obtain

$$\frac{1}{2}\rho(\boldsymbol{x}) = \int_\Gamma \frac{\partial}{\partial \boldsymbol{\nu}_{\boldsymbol{x}}} E_n(\boldsymbol{x}, \boldsymbol{y}; k)\rho(\boldsymbol{y})\mathrm{d}s_{\boldsymbol{y}}, \quad \forall\, \boldsymbol{x} \in \Gamma. \tag{5.3.10}$$

On the domain Ω, $u_\rho(\boldsymbol{x})$ satisfies

$$\left.\begin{aligned} -\Delta u_\rho(\boldsymbol{x}) = k^2 u_\rho(\boldsymbol{x}), \;\; \forall\, \boldsymbol{x} \in \Omega, \\ u_\rho(\boldsymbol{x})\Big|_\Gamma = 0, \;\; \forall \boldsymbol{x} \in \Gamma, \end{aligned}\right\} \tag{5.3.11}$$

which is an interior Dirichlet eigenvalue problem.

(i) If k^2 is not an interior Dirichlet eigenvalue, namely $k^2 \neq \lambda_j^D \;\; (j = 1, 2, \cdots)$, then

$$\begin{aligned} &u_\rho(\boldsymbol{x}) \equiv 0, \quad \forall\, \boldsymbol{x} \in \Omega, \\ &0 = \frac{\partial u_\rho}{\partial \boldsymbol{\nu}_{\boldsymbol{x}}}(\boldsymbol{x})\bigg|_{\Gamma-0} = \frac{1}{2}\rho(\boldsymbol{x}) + \int_\Gamma \frac{\partial}{\partial \boldsymbol{\nu}_{\boldsymbol{x}}} E_n(\boldsymbol{x}, \boldsymbol{y}; k)\rho(\boldsymbol{y})\mathrm{d}s_{\boldsymbol{y}} = \rho(\boldsymbol{x}), \quad \forall \boldsymbol{x} \in \Gamma. \end{aligned}$$

It means that

$$\mathbb{N}\Big(\mathbb{K}_{I,k}\Big) = \Big\{0\Big\}, \quad \mathbb{N}\Big(\overline{\mathbb{K}}_{I,k}\Big) = \Big\{0\Big\}.$$

(ii) If k^2 is an eigenvalue of the interior Dirichlet eigenvalue problem (5.3.11), namely $k^2 = \lambda_j^D \;\; (1 \leqslant j < +\infty)$, then problem (5.3.11) exists nontrivial solution $u_\rho(\boldsymbol{x})$, which is an eigenfunction corresponding to the eigenvalue λ_j^D. For any $u_\rho \in V_j^D$, $u_\rho(\boldsymbol{x})$ is a nontrivial solution of problem (5.3.11) with $k^2 = \lambda_j^D$. On the boundary Γ,

$$\frac{\partial u_\rho}{\partial \boldsymbol{\nu}_{\boldsymbol{x}}}(\boldsymbol{x})\bigg|_{\Gamma-0} = \rho(\boldsymbol{x}), \quad \forall\, \boldsymbol{x} \in \Gamma,$$

and $\rho(\boldsymbol{x})$ is nontrivial.

Let

$$\rho_{j,l}^D(\boldsymbol{x}) = \frac{\partial}{\partial \boldsymbol{\nu}_{\boldsymbol{x}}} u_{j,l}^D(\boldsymbol{x})\bigg|_{\Gamma-0}, \quad \forall\, \boldsymbol{x} \in \Gamma, \; 1 \leqslant l \leqslant J^D,$$

where $\left\{u_{j,l}^D(\boldsymbol{x}),\ l = 1, 2, \cdots, J^D\right\}$ is the set of basis functions of V_j^D.

We obtain

$$\mathbb{N}\left(\mathbb{K}_{I,k}\right) = \left\{\rho \,\middle|\, \rho(\boldsymbol{x}) = \sum_{l=1}^{J^D} C_l \rho_{j,l}^D(\boldsymbol{x}),\ \forall\, C_l \in \mathbb{C},\ l = 1, 2, \cdots, J^D\right\}$$

and

$$\mathbb{N}\left(\overline{\mathbb{K}}_{I,k}\right) = \left\{\overline{\rho} \,\middle|\, \rho(\boldsymbol{x}) \in \mathbb{N}\left(\mathbb{K}_{I,k}\right)\right\}.$$

It is ready to study the solvability of boundary integral equation of the first kind (5.3.8). We have

Theorem 5.3.3 *Suppose that the given function $g \in H^{1/2}(\Gamma)$, either*

(i) k^2 is not an eigenvalue of the interior Dirichlet eigenvalue problem (5.1.19), the boundary integral equation of the first kind (5.3.8) exists a unique solution $\varphi \in H^{-1/2}(\Gamma)$, or

(ii) $k^2 = \lambda_j^D$ $(1 \leqslant j < +\infty)$ is an eigenvalue of the interior Dirichlet eigenvalue problem (5.1.19), then the boundary integral equation of the first kind (5.3.8) exists solution if and only if

$$\int_\Gamma g(\boldsymbol{x})\rho(\boldsymbol{x})\mathrm{d}s_x = 0, \quad \forall\, \rho \in \mathbb{N}\left(\mathbb{K}_{I,k}\right). \tag{5.3.12}$$

Proof Rewrite the boundary integral equation of the first kind (5.3.8) into the following equivalent form:

$$\int_\Gamma E_n(\boldsymbol{x}, \boldsymbol{y}; k)\varphi(\boldsymbol{y})\mathrm{d}s_y + \int_\Gamma B_n(\boldsymbol{x}, \boldsymbol{y}; k)\varphi(\boldsymbol{y})\mathrm{d}s_y = g(\boldsymbol{x}), \quad \forall \boldsymbol{x} \in \Gamma.$$

That is

$$\mathbb{K}_{I,1}^+\varphi + \Pi_{I,k}\varphi = g.$$

By the Theorem 4.4.1, we know that the operator $\mathbb{K}_{I,1}^+$ is invertible, therefore the boundary integral equation of the first kind (5.3.8) is equivalent to the following second kind operator equation on Hilbert space $H^{-1/2}(\Gamma)$:

$$\varphi + \left(\mathbb{K}_{I,1}^+\right)^{-1}\Pi_{I,k}\varphi = \left(\mathbb{K}_{I,1}^+\right)^{-1} g,$$

namely

$$\left(\mathbb{I} + \left(\mathbb{K}_{I,1}^+\right)^{-1}\Pi_{I,k}\right)\varphi = \left(\mathbb{K}_{I,1}^+\right)^{-1} g, \tag{5.3.13}$$

in which $\mathbb{I}$ is the identity operator.

The properties of operator $\Pi_{I,k}$ and $\left(\mathbb{K}_{I,1}^{+}\right)^{-1}$ given in Theorems 5.2.2 and 4.4.1 yield that the operator

$$\left(\mathbb{K}_{I,1}^{+}\right)^{-1}\Pi_{I,k}: \quad H^{-1/2}(\Gamma) \to H^{-1/2}(\Gamma)$$

is compact.

Let

$$\left(\overline{\Pi}_{I,k}\varphi\right)(\boldsymbol{x}) = \int_{\Gamma} \overline{B_n(\boldsymbol{x}, \boldsymbol{y}; k)}\varphi(\boldsymbol{y})\mathrm{d}s_{\boldsymbol{y}}, \quad \forall\, \boldsymbol{x} \in \Gamma.$$

Then

$$\begin{aligned}\left\langle\left(\mathbb{K}_{I,1}^{+}\right)^{-1}\Pi_{I,k}\varphi, \rho\right\rangle_{-1/2,\Gamma} &= \int_{\Gamma} \left(\Pi_{I,k}\varphi\right)\overline{\rho(\boldsymbol{x})}\mathrm{d}s_{\boldsymbol{x}} \\ &= \int_{\Gamma}\int_{\Gamma} B_n(\boldsymbol{x}, \boldsymbol{y}; k)\varphi(\boldsymbol{y})\overline{\rho(\boldsymbol{x})}\mathrm{d}s_{\boldsymbol{y}}\mathrm{d}s_{\boldsymbol{x}} \\ &= \int_{\Gamma} \varphi(\boldsymbol{y})\overline{\left(\overline{\Pi}_{I,k}\rho\right)(\boldsymbol{y})}\mathrm{d}s_{\boldsymbol{y}} \\ &= \left\langle\varphi, \left(\mathbb{K}_{I,1}^{+}\right)^{-1}\overline{\Pi}_{I,k}\rho\right\rangle_{-1/2,\Gamma}.\end{aligned}$$

It means that $\left(\mathbb{K}_{I,1}^{+}\right)^{-1}\overline{\Pi}_{I,k}$ is the adjoint operator of $\left(\mathbb{K}_{I,1}^{+}\right)^{-1}\Pi_{I,k}$. Furthermore we have

$$\left.\begin{aligned}\mathbb{N}\left(\mathbb{I} + \left(\mathbb{K}_{I,1}^{+}\right)^{-1}\Pi_{I,k}\right) &= \mathbb{N}\left(\mathbb{K}_{I,k}\right), \\ \mathbb{N}\left(\mathbb{I} + \left(\mathbb{K}_{I,1}^{+}\right)^{-1}\overline{\Pi}_{I,k}\right) &= \mathbb{N}\left(\overline{\mathbb{K}}_{I,k}\right).\end{aligned}\right\} \tag{5.3.14}$$

A direct application of Fredholm's Alternative Theorem (Theorem 3.4.5) arrives at the second operator equation (5.3.13) satisfies the following alternative:

Either

(i)

$$\mathbb{N}\left(\mathbb{I} + \left(\mathbb{K}_{I,1}^{+}\right)^{-1}\Pi_{I,k}\right) = \left\{0\right\},$$

and Eq. (5.3.13) exists a unique solution for any given $\left(\mathbb{K}_{I,1}^{+}\right)^{-1}g \in H^{-1/2}(\Gamma)$, or

(ii) $\dim \mathbb{N}\left(\mathbb{I} + \left(\mathbb{K}_{I,1}^{+}\right)^{-1}\Pi_{I,k}\right) > 0$. In this case the Eq. (5.3.13) exists solution if and only if

$$\left\langle\left(\mathbb{K}_{I,1}^{+}\right)^{-1}g, \rho\right\rangle_{-1/2,\Gamma} = 0, \quad \forall\, \rho \in \mathbb{N}\left(\mathbb{I} + \left(\mathbb{K}_{I,1}^{+}\right)^{-1}\overline{\Pi}_{I,k}\right). \tag{5.3.15}$$

By equality (5.3.14) and $\mathbb{N}\left(\overline{\mathbb{K}}_{I,k}\right) = \left\{\overline{\rho}\,\middle|\, \rho \in \mathbb{N}\left(\mathbb{K}_{I,k}\right)\right\}$, the condition (5.3.15) is equivalent to

$$0 = \left\langle \left(\mathbb{K}_{I,1}^{+}\right)^{-1} g, \overline{\rho} \right\rangle_{-1/2,\Gamma} = \int_{\Gamma} g(\boldsymbol{x})\rho(\boldsymbol{x})\mathrm{d}s_{\boldsymbol{x}}, \quad \forall\, \rho(\boldsymbol{x}) \in \mathbb{N}\big(\mathbb{K}_{I,k}\big).$$

Combining the above results, the proof of this theorem is completed. □

Remark 5.3.1 The Theorem 5.3.3 shows that when $k^2 = \mu_j^D$ $(1 \leqslant j < +\infty)$ is an interior Dirichlet eigenvalue of problem (5.1.19) and the given $g \in H^{1/2}(\Gamma)$ does not satisfy the condition (5.3.12), then the solution of the boundary integral equation of the first kind (5.3.12) does not exist. It means that in this case the solution of exterior Dirichlet boundary value problem (5.1.2) can not be obtained from the boundary integral equation of the first kind (5.3.8). One way to solve this puzzle, introduce a new representation for the solution of problem (5.1.2). Let

$$u(\boldsymbol{x}) = \int_{\Gamma} E_n(\boldsymbol{x}, \boldsymbol{y}; k)\varphi(\boldsymbol{y})\mathrm{d}s_{\boldsymbol{y}} + \int_{\Gamma} \frac{\partial}{\partial \boldsymbol{\nu}_{\boldsymbol{y}}} E_n(\boldsymbol{x}, \boldsymbol{y}; k) g(\boldsymbol{y})\mathrm{d}s_{\boldsymbol{y}}, \quad \forall\, \boldsymbol{x} \in \Omega_e, \tag{5.3.16}$$

be the solution of exterior Dirichlet boundary value problem (5.1.2). The Dirichlet boundary condition (5.1.2b) yields the following boundary integral equation of the first kind

$$\begin{cases} \text{Find} \quad \varphi \in H^{-\frac{1}{2}}(\Gamma), \quad \text{such that} \\ \int_{\Gamma} E_n(\boldsymbol{x}, \boldsymbol{y}; k)\varphi(\boldsymbol{y})\mathrm{d}s_{\boldsymbol{y}} \\ \quad = \frac{1}{2} g(\boldsymbol{x}) - \int_{\Gamma} \frac{\partial}{\partial \boldsymbol{\nu}_{\boldsymbol{y}}} E_n(\boldsymbol{x}, \boldsymbol{y}; k) g(\boldsymbol{y})\mathrm{d}s_{\boldsymbol{y}} := g^*(\boldsymbol{x}), \quad \forall \boldsymbol{x} \in \Gamma. \end{cases} \tag{5.3.17}$$

with $g^* \in H^{1/2}(\Gamma)$ provided by $g \in H^{1/2}(\Gamma)$.

In case $k^2 = \mu_j^D$ $(1 \leqslant j < +\infty)$ is an interior Dirichlet eigenvalue, for any $\rho \in \mathbb{N}\big(\mathbb{K}_{I,k}\big)$, we have

$$\begin{aligned} \int_{\Gamma} g^*(\boldsymbol{x})\rho(\boldsymbol{x})\mathrm{d}s_{\boldsymbol{x}} &= \frac{1}{2}\int_{\Gamma} g(\boldsymbol{x})\rho(\boldsymbol{x})\mathrm{d}s_{\boldsymbol{x}} - \int_{\Gamma}\int_{\Gamma} \frac{\partial}{\partial \boldsymbol{\nu}_{\boldsymbol{y}}} E_n(\boldsymbol{x}, \boldsymbol{y}; k) g(\boldsymbol{y})\rho(\boldsymbol{x})\mathrm{d}s_{\boldsymbol{y}}\mathrm{d}s_{\boldsymbol{x}} \\ &= \frac{1}{2}\int_{\Gamma} g(\boldsymbol{x})\rho(\boldsymbol{x})\mathrm{d}s_{\boldsymbol{x}} - \frac{1}{2}\int_{\Gamma} g(\boldsymbol{y})\rho(\boldsymbol{y})\mathrm{d}s_{\boldsymbol{y}} = 0, \end{aligned}$$

the penultimate equality is from equality (5.3.10). Therefore the boundary integral equation of the first kind (5.1.9) exists solutions:

$$\varphi(\boldsymbol{x}) = \varphi^*(\boldsymbol{x}) + \rho(\boldsymbol{x}), \quad \forall\, \rho \in \mathbb{N}\big(\mathbb{K}_{I,k}\big),$$

where $\varphi^*(\boldsymbol{x})$ is a solution of problem (5.3.17), then

$$u(\boldsymbol{x}) = \int_{\Gamma} E_n(\boldsymbol{x}, \boldsymbol{y}; k)\varphi^*(\boldsymbol{y})\mathrm{d}s_{\boldsymbol{y}} + \int_{\Gamma} \frac{\partial}{\partial \boldsymbol{\nu}_{\boldsymbol{y}}} E_n(\boldsymbol{x}, \boldsymbol{y}; k) g(\boldsymbol{y})\mathrm{d}s_{\boldsymbol{y}}, \quad \forall\, \boldsymbol{x} \in \Omega_e,$$

is a solution of exterior Dirichlet boundary value problem (5.1.2).

The Theorems 5.1.1, 5.3.3 and Remark 5.3.1 yield the following results.

Theorem 5.3.4 *Suppose that wave number k satisfies condition (5.0.2) and the given function $g \in H^{1/2}(\Gamma)$.*

(i) The interior Dirichlet boundary value problem (5.1.1) exists a unique solution provided by k^2 is not an eigenvalue of the interior Dirichlet eigenvalue problem (5.1.19). In case $k^2 = \lambda_j^D$ $(1 \leqslant j < +\infty)$ is an eigenvalue of problem (5.1.19), the problem (5.1.1) exists solutions if and only if

$$\int_\Gamma g(\boldsymbol{x})\rho(\boldsymbol{x})\mathrm{d}s_x = 0, \quad \forall\, \rho \in \mathbb{N}\big(\mathbb{K}_{I,k}\big).$$

(ii) The exterior Dirichelt boundary value problem (5.1.2) exist a unique solution.

5.3.2 Boundary Integro-Differential Equations of the First Kind

For given function $h \in H^{-1/2}(\Gamma)$, consider the interior and exterior Neumann problem (5.1.3) and (5.1.4). Let the double layer potential

$$u(\boldsymbol{x}) = \int_\Gamma \frac{\partial}{\partial \boldsymbol{\nu}_y} E_n(\boldsymbol{x}, \boldsymbol{y}; k)\psi(\boldsymbol{y})\mathrm{d}s_y, \quad \forall\, \boldsymbol{x} \in \mathbb{R}^n,$$

solves the interior Neumann problem (5.1.3) in Ω and the exterior Neumann problem (5.1.4) in Ω_e provided that $\psi \in H^{1/2}(\Gamma)$ is a solution of the following hypersingular integral equation of the first kind

$$\frac{\partial}{\partial \boldsymbol{\nu}_x}\int_\Gamma \frac{\partial}{\partial \boldsymbol{\nu}_y} E_n(\boldsymbol{x}, \boldsymbol{y}; k)\psi(\boldsymbol{y})\mathrm{d}s_y = h(\boldsymbol{x}), \quad \forall\, \boldsymbol{x} \in \Gamma, \tag{5.3.18}$$

which is equivalent to the boundary integro-differential equation of the first kind

$$\begin{aligned}\left(\mathbb{K}_{III,k}\psi\right)(\boldsymbol{x}) = & \sum_{1\leqslant i<j\leqslant n} \frac{\partial}{\partial \tau_x^{i,j}} \int_\Gamma E_n(\boldsymbol{x}, \boldsymbol{y}; k)\frac{\partial}{\partial \tau_y^{i,j}}\psi(\boldsymbol{y})\mathrm{d}s_y + \\ & k^2 \int_\Gamma E_n(\boldsymbol{x}, \boldsymbol{y}; k)\big(\boldsymbol{\nu}(\boldsymbol{x})\cdot\boldsymbol{\nu}(\boldsymbol{y})\big)\psi(\boldsymbol{y})\mathrm{d}s_y = h(\boldsymbol{x}), \quad \forall\, \boldsymbol{x} \in \Gamma.\end{aligned} \tag{5.3.19}$$

Let

$$\begin{aligned}\left(\overline{\mathbb{K}}_{III,k}\psi\right)(\boldsymbol{x}) = & \sum_{1\leqslant i<j\leqslant n} \frac{\partial}{\partial \tau_x^{i,j}} \int_\Gamma \overline{E_n(\boldsymbol{x}, \boldsymbol{y}; k)}\frac{\partial}{\partial \tau_y^{i,j}}\psi(\boldsymbol{y})\mathrm{d}s_y + \\ & k^2 \int_\Gamma \overline{E_n(\boldsymbol{x}, \boldsymbol{y}; k)}\big(\boldsymbol{\nu}(\boldsymbol{x})\cdot\boldsymbol{\nu}(\boldsymbol{y})\big)\psi(\boldsymbol{y})\mathrm{d}s_y,\end{aligned}$$

then we have

$$\left.\begin{aligned}\mathbb{K}_{III,k} &= \mathbb{K}^+_{III,1} + \Pi_{III,k},\\ \overline{\mathbb{K}}_{III,k} &= \mathbb{K}^+_{III,1} + \overline{\Pi}_{III,k},\end{aligned}\right\} \tag{5.3.20}$$

with

$$\left(\overline{\Pi}_{III,k}\psi\right)(\boldsymbol{x}) = \sum_{1\leqslant i<j\leqslant n} \frac{\partial}{\partial \boldsymbol{\tau}_{\boldsymbol{x}}^{i,j}} \int_\Gamma \overline{B_n(\boldsymbol{x}, \boldsymbol{y}; k)} \frac{\partial}{\partial \boldsymbol{\tau}_{\boldsymbol{y}}^{i,j}} \psi(\boldsymbol{y}) \mathrm{d}s_{\boldsymbol{y}} + \\ \int_\Gamma \boldsymbol{\nu}(\boldsymbol{x}) \cdot \boldsymbol{\nu}(\boldsymbol{y}) \left\{k^2 \overline{B_n(\boldsymbol{x}, \boldsymbol{y}; k)} + (1+k^2) E_n^+(\boldsymbol{x}, \boldsymbol{y}; 1)\right\} \psi(\boldsymbol{y}) \mathrm{d}s_{\boldsymbol{y}}.$$

Consider the null space of the operators $\mathbb{K}_{III,k}$ and $\overline{\mathbb{K}}_{III,k}$.

If $\mu \in \mathbb{N}\big(\mathbb{K}_{III,k}\big)$, then $\mu \in H^{1/2}(\Gamma)$ and

$$\left(\mathbb{K}_{III,k}\mu\right)(\boldsymbol{x}) = 0, \quad \forall\, \boldsymbol{x} \in \Gamma.$$

Let

$$u_\mu(\boldsymbol{x}) = \int_\Gamma \frac{\partial}{\partial \boldsymbol{\nu}_{\boldsymbol{y}}} E_n(\boldsymbol{x}, \boldsymbol{y}; k) \mu(\boldsymbol{y}) \mathrm{d}s_{\boldsymbol{y}}, \quad \forall\, \boldsymbol{x} \in \mathbb{R}^n.$$

In the exterior domain Ω_e, $u_\mu(\boldsymbol{x})$ is the solution of exterior Neumann problem with Neumann boundary condition

$$\left.\frac{\partial u_\mu(\boldsymbol{x})}{\partial \boldsymbol{\nu}_{\boldsymbol{x}}}\right|_{\Gamma+0} = 0, \quad \forall\, \boldsymbol{x} \in \Gamma.$$

By the uniqueness of exterior Neumann problem we obtain

$$u_\mu(\boldsymbol{x}) \equiv 0, \quad \forall\, \boldsymbol{x} \in \Omega_e$$

and

$$\left.u_\mu(\boldsymbol{x})\right|_{\Gamma+0} = 0.$$

On the domain Ω, $u_\mu(\boldsymbol{x})$ satisfies

$$\left.\begin{aligned} -\Delta u_\mu(\boldsymbol{x}) - k^2 u_\mu(\boldsymbol{x}) = 0, &\quad \forall\, \boldsymbol{x} \in \Omega,\\ \left.\frac{\partial}{\partial \boldsymbol{\nu}_{\boldsymbol{x}}} u_\mu(\boldsymbol{x})\right|_{\Gamma-0} = 0, &\quad \forall\, \boldsymbol{x} \in \Gamma.\end{aligned}\right\} \tag{5.3.21}$$

(i) k^2 is not a positive eigenvalue of the interior Neumann eigenvalue problem, then $u_\mu(\boldsymbol{x}) \equiv 0,\ \forall\, \boldsymbol{x} \in \Omega$ and $\left.u_\mu(\boldsymbol{x})\right|_{\Gamma-0} = 0$. Furthermore

$$\mu(\boldsymbol{x}) = u_\mu(\boldsymbol{x})\Big|_{\Gamma+0} - u_\mu(\boldsymbol{x})\Big|_{\Gamma-0} \equiv 0, \quad \forall\, \boldsymbol{x} \in \Gamma,$$

it arrives at

$$\mathbb{N}\Big(\mathbb{K}_{III,k}\Big) = \Big\{0\Big\} \text{ and } \mathbb{N}\Big(\overline{\mathbb{K}}_{III,k}\Big) = \Big\{0\Big\}.$$

(ii) $k^2 = \lambda_j^N$ $(2 \leqslant j < +\infty)$ is an eigenvalue of the interior Neumann eigenvalue problem, the problem (5.3.21) exists nontrivial solutions $u_\mu(\boldsymbol{x})$, which are the eigenfunctions of the interior Neuamnn eigenvalue problem corresponding to eigenvalue $k^2 = \lambda_j^N$, and

$$\mu(\boldsymbol{x}) = u_\mu(\boldsymbol{x})\Big|_{\Gamma+0} - u_\mu(\boldsymbol{x})\Big|_{\Gamma-0} = -u_\mu(\boldsymbol{x})\Big|_{\Gamma-0} \neq 0, \quad \forall\, \boldsymbol{x} \in \Gamma.$$

Let

$$\mu_{j,l}^N(\boldsymbol{x}) = u_{j,l}^N(\boldsymbol{x})\Big|_{\Gamma-0}, \quad \forall\, l = 1, 2, \cdots, J^N,$$

where $\Big\{u_{j,l}^N(\boldsymbol{x}),\ l = 1, 2, \cdots, J^N\Big\}$ is the set of basis of space V_j^N. In this case

$$\mathbb{N}\Big(\mathbb{K}_{III,k}\Big) = \left\{\mu \,\middle|\, \mu(\boldsymbol{x}) = \sum_{l=1}^{J^N} C_l \mu_{j,l}^N(\boldsymbol{x}),\ C_l \in \mathbb{C}, l = 1, 2, \cdots, J^N\right\}$$

and

$$\mathbb{N}\Big(\overline{\mathbb{K}}_{III,k}\Big) = \Big\{\overline{\mu} \,\Big|\, \mu \in \mathbb{N}\Big(\mathbb{K}_{III,k}\Big)\Big\}.$$

Obviously

$$\dim \mathbb{N}\Big(\mathbb{K}_{III,k}\Big) = \dim \mathbb{N}\Big(\overline{\mathbb{K}}_{III,k}\Big) > 0.$$

For the boundary integro-differential equation of the first kind (5.3.19) the following alternative theorem holds.

Theorem 5.3.5 *Assume that the wave number k satisfies condition (5.0.2) and the given function $h \in H^{-1/2}(\Gamma)$. Either*

(i) k^2 is not an eigenvalue of the interior Neumann eigenvalue problem (5.1.20), then the boundary integro-differential equation of the first kind (5.3.19) exists a unique solution $\psi \in H^{1/2}(\Gamma)$, or

(ii) $k^2 = \lambda_j^N$ $(2 \leqslant j < +\infty)$ is a positive interior Neumann eigenvalue of problem (5.1.20), then the boundary integro-differential of the first kind (5.3.19) exists solutions if and only if

$$\int_\Gamma h(\boldsymbol{x})\mu(\boldsymbol{x})\mathrm{d}s_{\boldsymbol{x}} = 0, \quad \forall\, \mu \in \mathbb{N}\Big(\mathbb{K}_{III,k}\Big). \tag{5.3.22}$$

Proof Rewrite the boundary integro-differential equation of the first kind (5.3.19) into the following equivalent form:

$$\left(\mathbb{K}^{+}_{\mathit{III},1}\psi\right)(\boldsymbol{x})+\left(\Pi_{\mathit{III},k}\psi\right)(\boldsymbol{x})=h(\boldsymbol{x}),\quad \forall\,\boldsymbol{x}\in\Gamma. \tag{5.3.23}$$

The operator $\mathbb{K}^{+}_{\mathit{III},1}$: $H^{1/2}(\Gamma)\to H^{-1/2}(\Gamma)$ is invertible and $\left(\mathbb{K}^{+}_{\mathit{III},1}\right)^{-1}$: $H^{-1/2+\gamma}(\Gamma)\to H^{1/2+\gamma}(\Gamma)$, $(0\leqslant\gamma\leqslant 1/2)$ is bounded provided by Theorem 4.4.2. Therefore the integro-differential equation (5.3.23) is equivalent to the second kind operator equation on the Hilbert space $H^{1/2}(\Gamma)$:

$$\psi+\left(\mathbb{K}^{+}_{\mathit{III},1}\right)^{-1}\Pi_{\mathit{III},k}\psi=\left(\mathbb{K}^{+}_{\mathit{III},1}\right)^{-1}h. \tag{5.3.24}$$

For $\psi,\mu\in H^{1/2}(\Gamma)$, we have

$$\begin{aligned}\left\langle\left(\mathbb{K}^{+}_{\mathit{III},1}\right)^{-1}\Pi_{\mathit{III},k}\psi,\mu\right\rangle_{1/2,\Gamma}&=\int_{\Gamma}\left(\Pi_{\mathit{III},k}\psi\right)(\boldsymbol{x})\overline{\mu(\boldsymbol{x})}\mathrm{d}s_x\\&=\int_{\Gamma}\int_{\Gamma}\frac{\partial^2}{\partial\boldsymbol{\nu}_x\partial\boldsymbol{\nu}_y}B_n(\boldsymbol{x},\boldsymbol{y};k)\psi(\boldsymbol{y})\overline{\mu(\boldsymbol{x})}\mathrm{d}s_x\mathrm{d}s_y\\&=\int_{\Gamma}\psi(\boldsymbol{y})\left\{\overline{\int_{\Gamma}\frac{\partial^2}{\partial\boldsymbol{\nu}_x\partial\boldsymbol{\nu}_y}\overline{B_n(\boldsymbol{x},\boldsymbol{y};k)}\mu(\boldsymbol{x})\mathrm{d}s_x}\right\}\mathrm{d}s_y\\&=\int_{\Gamma}\psi(\boldsymbol{y})\overline{\left(\overline{\Pi}_{\mathit{III},k}\mu\right)(\boldsymbol{y})}\mathrm{d}s_y\\&=\left\langle\psi,\left(\mathbb{K}^{+}_{\mathit{III},1}\right)^{-1}\overline{\Pi}_{\mathit{III},k}\mu\right\rangle_{1/2,\Gamma}.\end{aligned}$$

It means that $\left(\mathbb{K}^{+}_{\mathit{III},1}\right)^{-1}\overline{\Pi}_{\mathit{III},k}\mu$ is the adjoint operator of $\left(\mathbb{K}^{+}_{\mathit{III},1}\right)^{-1}\Pi_{\mathit{III},k}$ and

$$\begin{aligned}&\mathbb{N}\left(\mathbb{I}+\left(\mathbb{K}^{+}_{\mathit{III},1}\right)^{-1}\Pi_{\mathit{III},k}\right)=\mathbb{N}\left(\mathbb{K}_{\mathit{III},k}\right),\\&\mathbb{N}\left(\mathbb{I}+\left(\mathbb{K}^{+}_{\mathit{III},1}\right)^{-1}\overline{\Pi}_{\mathit{III},k}\right)=\mathbb{N}\left(\overline{\mathbb{K}}_{\mathit{III},k}\right)=\left\{\overline{\mu}\,\middle|\,\mu\in\mathbb{N}\left(\mathbb{K}_{\mathit{III},k}\right)\right\}.\end{aligned}$$

An application of Fredholm's Alternative Theorem (Theorem 3.4.5) yields that the second operator equation (5.3.24) satisfies the following alternative:

Either

(i)

$$\mathbb{N}\left(\mathbb{I}+\left(\mathbb{K}^{+}_{\mathit{III},1}\right)^{-1}\Pi_{\mathit{III},k}\right)=\{0\},$$

and Eq. (5.3.24) exists a unique solution for $\left(\mathbb{K}^+_{III,1}\right)^{-1}h \in H^{1/2}(\Gamma)$, or

(ii)

$$\dim \mathbb{N}\left(\left(\mathbb{K}^+_{III,1}\right)^{-1}\Pi_{III,k}\right) > 0.$$

In this case the Eq. (5.3.24) exists solution if and only if

$$\left\langle\left(\mathbb{K}^+_{III,1}\right)^{-1}h,\mu\right\rangle_{1/2,\Gamma} = 0, \quad \forall\,\mu \in \mathbb{N}\left(\mathbb{I} + \left(\mathbb{K}^+_{III,1}\right)^{-1}\overline{\Pi}_{III,k}\right). \tag{5.3.25}$$

Since

$$0 = \left\langle\left(\mathbb{K}^+_{III,1}\right)^{-1}h,\overline{\mu}\right\rangle_{1/2,\Gamma} = \int_\Gamma h(\boldsymbol{x})\mu(\boldsymbol{x})\mathrm{d}s_y, \quad \forall\,\mu(\boldsymbol{x}) \in \mathbb{N}\left(\mathbb{K}_{III,k}\right).$$

Combining the results obtained above, the proof of Theorem 5.3.5 is completed. □

Remark 5.3.2 In the case $k^2 = \lambda_j^N$ $(2 \leqslant j < +\infty)$ is a positive interior Neumann eigenvalue of problem (5.1.20) and condition (5.3.22) does not satisfy. Therefore the boundary integro-differential equation (5.3.19) does not exist any solution. It means that the solution of exterior Neumann problem (5.1.4) can not be obtained by solving the boundary integro-differential equation (5.3.19). Similarly introduce a new representation for the solution of problem (5.1.4). Let

$$u(\boldsymbol{x}) = \int_\Gamma \frac{\partial}{\partial \boldsymbol{\nu}_y} E_n(\boldsymbol{x},\boldsymbol{y};k)\psi(\boldsymbol{y})\mathrm{d}s_y - \int_\Gamma E_n(\boldsymbol{x},\boldsymbol{y};k)h(\boldsymbol{y})\mathrm{d}s_y, \quad \forall\,\boldsymbol{x} \in \Omega_e, \tag{5.3.26}$$

by the solution of exterior Neumann problem (5.1.4). The Neumann boundary conditon (5.1.4b) yields the following boundary integro-differential equation of the first kind

$$\left(\mathbb{K}_{III,k}\psi\right)(\boldsymbol{x}) = -\frac{1}{2}h(\boldsymbol{x}) + \int_\Gamma \frac{\partial}{\partial \boldsymbol{\nu}_x} E_n(\boldsymbol{x},\boldsymbol{y};k)h(\boldsymbol{y})\mathrm{d}s_y := h^*(\boldsymbol{x}), \quad \forall\,\boldsymbol{x} \in \Gamma. \tag{5.3.27}$$

For $\mu \in \mathbb{N}\left(\mathbb{K}_{III,k}\right)$, let

$$u_\mu(\boldsymbol{x}) = \int_\Gamma \frac{\partial}{\partial \boldsymbol{\nu}_y} E_n(\boldsymbol{x},\boldsymbol{y};k)\mu(\boldsymbol{y})\mathrm{d}s_y, \quad \forall\,\boldsymbol{x} \in \mathbb{R}^n.$$

Then

$$u_\mu(\boldsymbol{x}) \equiv 0, \quad \forall\,\boldsymbol{x} \in \Omega_e$$

and

$$u_\mu(\boldsymbol{x})\Big|_{\Gamma+0} = -\frac{1}{2}\mu(\boldsymbol{x}) + \int_\Gamma \frac{\partial}{\partial \boldsymbol{\nu}_y} E_n(\boldsymbol{x},\boldsymbol{y};k)\mu(\boldsymbol{y})\mathrm{d}s_y, \quad \forall\,\boldsymbol{x} \in \Gamma.$$

That is

$$\frac{1}{2}\mu(\boldsymbol{x}) = \int_\Gamma \frac{\partial}{\partial \boldsymbol{\nu}_y} E_n(\boldsymbol{x}, \boldsymbol{y}; k)\mu(\boldsymbol{y})\mathrm{d}s_y, \quad \forall\, \boldsymbol{x} \in \Gamma.$$

Therefore

$$\begin{aligned}\int_\Gamma h^*(\boldsymbol{x})\mu(\boldsymbol{x})\mathrm{d}s_x &= -\frac{1}{2}\int_\Gamma h(\boldsymbol{x})\mu(\boldsymbol{x})\mathrm{d}s_x + \int_\Gamma \int_\Gamma \frac{\partial}{\partial \boldsymbol{\nu}_x} E_n(\boldsymbol{x}, \boldsymbol{y}; k) h(\boldsymbol{y})\mu(\boldsymbol{x})\mathrm{d}s_y \mathrm{d}s_x \\ &= -\frac{1}{2}\int_\Gamma h(\boldsymbol{x})\mu(\boldsymbol{x})\mathrm{d}s_x + \int_\Gamma h(\boldsymbol{y})\left\{\int_\Gamma \frac{\partial}{\partial \boldsymbol{\nu}_x} E_n(\boldsymbol{x}, \boldsymbol{y}; k)\mu(\boldsymbol{x})\mathrm{d}s_x\right\}\mathrm{d}s_y \\ &= 0, \quad \forall\, \mu \in \mathbb{N}\big(\mathbb{K}_{\mathit{III},k}\big).\end{aligned}$$

It means that Eq. (5.3.27) exists solution in the case $k^2 = \lambda_j^N$ $(2 \leqslant j < +\infty)$, a positive eigenvalue of problem (5.1.20) and for any given $h \in H^{-1/2}(\Gamma)$.

The results in Theorems 5.1.1, 5.3.5 and Remark 5.3.2 arrive at

Theorem 5.3.6 *Assume that wave number k satisfies condition (5.0.2) and the given function* $h \in H^{-1/2}(\Gamma)$.

(i) The interior Neumann boundary value problem (5.1.3) exists a unique solution provided by k^2 *is not a positive eigenvalue of the interior Neumann eigenvalue problem (5.1.20). In case* $k^2 = \lambda_j^N$ $(2 \leqslant j < +\infty)$ *is a positive eigenvalue of problem (5.1.20), the problem (5.1.3) exists solutions if and only if*

$$\int_\Gamma h(\boldsymbol{x})\mu(\boldsymbol{x})\mathrm{d}s_x = 0, \quad \forall\, \mu \in \mathbb{N}\Big(\mathbb{K}_{\mathit{III},k}\Big).$$

(ii) The exterior Neumann boundary value problem (5.1.4) exist a unique solution.

5.3.3 Boundary Integral Equations of the Second Kind

Let

$$u(\boldsymbol{x}) = \int_\Gamma \frac{\partial}{\partial \boldsymbol{\nu}_y} E_n(\boldsymbol{x}, \boldsymbol{y}; k)\psi(\boldsymbol{y})\mathrm{d}s_y, \quad \forall\, \boldsymbol{x} \in \mathbb{R}^n \backslash \Gamma. \tag{5.3.28}$$

be the solution of interior Dirichlet problem (5.1.1) in Ω and exterior Dirichlet problem (5.1.2) in Ω_e with density $\psi \in H^{1/2}(\Gamma)$ to be determined. For the interior Dirichlet problem, the Dirichlet boundary condition (5.1.1b) yields

$$-\frac{1}{2}\psi(\boldsymbol{x}) + \int_\Gamma \frac{\partial}{\partial \boldsymbol{\nu}_y} E_n(\boldsymbol{x}, \boldsymbol{y}; k)\psi(\boldsymbol{y})\mathrm{d}s_y = g(\boldsymbol{x}), \quad \forall\, \boldsymbol{x} \in \Gamma. \tag{5.3.29}$$

For the exterior Dirichlet problem, the Dirichlet boundary condition (5.1.2b) yields

$$\frac{1}{2}\psi(\boldsymbol{x}) + \int_{\Gamma} \frac{\partial}{\partial \boldsymbol{\nu}_{\boldsymbol{y}}} E_n(\boldsymbol{x}, \boldsymbol{y}; k)\psi(\boldsymbol{y})\mathrm{d}s_{\boldsymbol{y}} = g(\boldsymbol{x}), \quad \forall\, \boldsymbol{x} \in \Gamma. \tag{5.3.30}$$

Furthermore let

$$u(\boldsymbol{x}) = \int_{\Gamma} E_n(\boldsymbol{x}, \boldsymbol{y}; k)\varphi(\boldsymbol{y})\mathrm{d}s_{\boldsymbol{y}}, \quad \forall\, \boldsymbol{x} \in \mathbb{R}^n, \tag{5.3.31}$$

be the solution of interior Neumann problem (5.1.3) in Ω and exterior Neumann problem (5.1.4) in Ω_e with density $\varphi \in H^{-1/2}(\Gamma)$ to be determined. For the interior Neumann problem, the Neumann boundary condition (5.1.3b) yields

$$\frac{1}{2}\varphi(\boldsymbol{x}) + \int_{\Gamma} \frac{\partial}{\partial \boldsymbol{\nu}_{\boldsymbol{x}}} E_n(\boldsymbol{x}, \boldsymbol{y}; k)\varphi(\boldsymbol{y})\mathrm{d}s_{\boldsymbol{y}} = h(\boldsymbol{x}), \quad \forall\, \boldsymbol{x} \in \Gamma, \tag{5.3.32}$$

and for the exterior Neumann problem, the Neumann boundary condition (5.1.4b) yields

$$-\frac{1}{2}\varphi(\boldsymbol{x}) + \int_{\Gamma} \frac{\partial}{\partial \boldsymbol{\nu}_{\boldsymbol{x}}} E_n(\boldsymbol{x}, \boldsymbol{y}; k)\varphi(\boldsymbol{y})\mathrm{d}s_{\boldsymbol{y}} = h(\boldsymbol{x}), \quad \forall\, \boldsymbol{x} \in \Gamma. \tag{5.3.33}$$

Recalling the operators $\mathbb{K}_{II,k} : H^{1/2}(\Gamma) \to H^{1/2}(\Gamma)$ and $\mathbb{K}^*_{II,k} : H^{-1/2}(\Gamma) \to H^{-1/2}(\Gamma)$ defined by (5.2.30), the four boundary integral equations (5.3.29)–(5.3.33) can be rewritten in the following forms:

$$\left(-\frac{1}{2}\mathbb{I} + \mathbb{K}_{II,k}\right)\psi = g, \tag{5.3.34}$$

$$\left(\frac{1}{2}\mathbb{I} + \mathbb{K}_{II,k}\right)\psi = g, \tag{5.3.35}$$

$$\left(\frac{1}{2}\mathbb{I} + \mathbb{K}^*_{II,k}\right)\varphi = h, \tag{5.3.36}$$

$$\left(-\frac{1}{2}\mathbb{I} + \mathbb{K}^*_{II,k}\right)\varphi = h. \tag{5.3.37}$$

For the null space $\mathbb{N}\left(-\frac{1}{2}\mathbb{I} + \mathbb{K}_{II,k}\right)$, $\mathbb{N}\left(\frac{1}{2}\mathbb{I} + \mathbb{K}_{II,k}\right)$, $\mathbb{N}\left(\frac{1}{2}\mathbb{I} + \mathbb{K}^*_{II,k}\right)$ and $\mathbb{N}\left(-\frac{1}{2}\mathbb{I} + \mathbb{K}^*_{II,k}\right)$ the following results hold

Theorem 5.3.7 *Suppose that the wave number k satisfies condition (5.0.2), the following results hold.*

(a) In the case k^2 is not an eigenvalue of the interior Dirichlet eigenvalue problem (5.1.19), then

$$\mathbb{N}\left(-\frac{1}{2}\mathbb{I}+\mathbb{K}_{II,k}\right)=\left\{0\right\}.$$

In the case $k^2=\lambda_j^D$ $(1\leqslant j<+\infty)$ is an eigenvalue of the interior Dirichlet eigenvalue problem, then

$$\dim\mathbb{N}\left(-\frac{1}{2}\mathbb{I}+\mathbb{K}_{II,k}\right)>0$$

and

$$\mathbb{N}\left(-\frac{1}{2}\mathbb{I}+\mathbb{K}_{II,k}\right)=\left\{\mu\;\middle|\;\mu(\boldsymbol{x})=\sum_{l=1}^{J^D}C_l\mu_{j,l}^D(\boldsymbol{x}),\,C_l\in\mathbb{C},l=1,2,\cdots,J^D\right\},$$

in which the basis functions

$$\mu_{j,l}^D(\boldsymbol{x})=\left(\mathbb{K}_{III,k}\right)^{-1}\left(\frac{\partial}{\partial\boldsymbol{\nu}_{\boldsymbol{x}}}u_{j,l}^D(\boldsymbol{x})\bigg|_{\Gamma-0}\right),\quad 1\leqslant l\leqslant J^D,$$

with $\left\{u_{j,l}^D,\ l=1,2,\cdots,J^D\right\}$ are the basis functions of space V_j^D, the eigenfunction space of problem (5.1.19) corresponding to the eigenvalue λ_j^D.

(b) In the case k^2 is not a positive eigenvalue of interior Neumann eigenvalue problem (5.1.20), then

$$\mathbb{N}\left(\frac{1}{2}\mathbb{I}+\mathbb{K}_{II,k}\right)=\left\{0\right\}.$$

In the case $k^2=\lambda_j^N$ $(2\leqslant j<+\infty)$ is a positive eigenvalue of the interior Neumann eigenvalue problem, then

$$\dim\mathbb{N}\left(\frac{1}{2}\mathbb{I}+\mathbb{K}_{II,k}\right)>0$$

and

$$\mathbb{N}\left(\frac{1}{2}\mathbb{I}+\mathbb{K}_{II,k}\right)=\left\{\mu(\boldsymbol{x})\middle|\;\mu(\boldsymbol{x})=\sum_{l=1}^{J^N}C_l\mu_{j,l}^N(\boldsymbol{x}),\ C_l\in\mathbb{C},l=1,2,\cdots,J^N\right\},$$

in which

$$\mu_{j,l}^N(\boldsymbol{x})=-u_{j,l}^N(\boldsymbol{x})\Big|_{\Gamma},\quad\forall\,\boldsymbol{x}\in\Gamma,\ 1\leqslant l\leqslant J^N,$$

and $\left\{u_{j,l}^{N},\ l=1,2,\cdots,J^{N}\right\}$ *are the basis functions of space* V_j^N, *the eigenvalue space of problem (5.1.20) corresponding to the eigenvalue* λ_j^N.

(c) *In the case* k^2 *is not a positive eigenvalue of interior Neumann eigenvalue problem (5.1.20), then*

$$\mathbb{N}\left(\frac{1}{2}\mathbb{I}+\mathbb{K}_{II,k}^{*}\right)=\left\{0\right\}.$$

In the case $k^2=\lambda_j^N$ $(2\leqslant j<+\infty)$ *is a positive eigenvalue of the interior Neumann eigenvalue problem, then*

$$\dim\mathbb{N}\left(\frac{1}{2}\mathbb{I}+\mathbb{K}_{II,k}^{*}\right)>0$$

and

$$\mathbb{N}\left(\frac{1}{2}\mathbb{I}+\mathbb{K}_{II,k}^{*}\right)=\left\{\rho\ \middle|\ \rho(\boldsymbol{x})=\sum_{l=1}^{J^N}C_l\left(\mathbb{K}_{I,k}\right)^{-1}\rho_{j,l}^{N}(\boldsymbol{x}),\ C_l\in\mathbb{C},l=1,2,\cdots,J^{N}\right\},$$

in which

$$\rho_{j,l}^{N}(\boldsymbol{x})=u_{j,l}^{N}(\boldsymbol{x})\Big|_{\Gamma-0},\quad 1\leqslant l\leqslant J^{N}.$$

(d) *In the case* k^2 *is not an eigenvalue of the interior Dirichlet eigenvalue problem, then*

$$\mathbb{N}\left(-\frac{1}{2}\mathbb{I}+\mathbb{K}_{II,k}^{*}\right)=\left\{0\right\}.$$

In the case $k^2=\lambda_j^D$ $(1\leqslant j<+\infty)$ *is an eigenvalue of the interior Dirichlet eigenvalue problem, then*

$$\dim\mathbb{N}\left(-\frac{1}{2}\mathbb{I}+\mathbb{K}_{II,k}^{*}\right)>0$$

and

$$\mathbb{N}\left(-\frac{1}{2}\mathbb{I}+\mathbb{K}_{II,k}^{*}\right)=\left\{\rho\ \middle|\ \rho(\boldsymbol{x})=\sum_{l=1}^{J^D}C_l\rho_{j,l}^{D}(\boldsymbol{x}),\ C_l\in\mathbb{C},l=1,2,\cdots,J^{D}\right\},$$

with

$$\rho_{j,l}^{D}(\boldsymbol{x})=\frac{\partial}{\partial\boldsymbol{\nu}_{\boldsymbol{x}}}u_{j,l}^{D}(\boldsymbol{x})\Big|_{\Gamma-0},\quad 1\leqslant l\leqslant J^{D}.$$

Proof (*a*) For any $\mu \in \mathbb{N}\left(-\frac{1}{2}\mathbb{I} + \mathbb{K}_{II,k}\right)$, namely $\mu \in H^{1/2}(\Gamma)$ and

$$\left(-\frac{1}{2}\mathbb{I} + \mathbb{K}_{II,k}\right)\mu = 0.$$

Let

$$u_\mu(\boldsymbol{x}) = \int_\Gamma \frac{\partial}{\partial \boldsymbol{\nu}_{\boldsymbol{y}}} E_n(\boldsymbol{x}, \boldsymbol{y}; k)\mu(\boldsymbol{y})\mathrm{d}s_{\boldsymbol{y}}, \quad \forall\, \boldsymbol{x} \in \mathbb{R}^n,$$

then

$$u_\mu(\boldsymbol{x})\Big|_{\Gamma-0} = 0, \quad \forall\, \boldsymbol{x} \in \Gamma.$$

On the domain Ω, $u_\mu(\boldsymbol{x})$ satisfies

$$\left.\begin{aligned} -\Delta u_\mu(\boldsymbol{x}) - k^2 u_\mu(\boldsymbol{x}) &= 0, \quad \forall\, \boldsymbol{x} \in \Omega, \\ u_\mu(\boldsymbol{x})\Big|_{\Gamma-0} &= 0, \quad \forall\, \boldsymbol{x} \in \Gamma. \end{aligned}\right\} \tag{5.3.38}$$

(i) In the case k^2 is not an eigenvalue of the interior Dirichlet eigenvalue problem, then

$$u_\mu(\boldsymbol{x}) \equiv 0, \quad \forall\, \boldsymbol{x} \in \Omega,$$

$$u_\mu(\boldsymbol{x})\Big|_{\Gamma-0} = 0, \quad \frac{\partial}{\partial \boldsymbol{\nu}_{\boldsymbol{x}}} u_\mu(\boldsymbol{x})\Big|_\Gamma = 0.$$

On the domain Ω_e, $u_\mu(\boldsymbol{x})$ is the solution of exterior Neumann problem (5.1.4) with given function $h(\boldsymbol{x}) \equiv 0$. Therefore

$$u_\mu(\boldsymbol{x}) \equiv 0, \quad \forall\, \boldsymbol{x} \in \Omega_e$$

and

$$u_\mu(\boldsymbol{x})\Big|_{\Gamma+0} = 0,$$

then

$$\mu(\boldsymbol{x}) = u_\mu(\boldsymbol{x})\Big|_{\Gamma+0} - u_\mu(\boldsymbol{x})\Big|_{\Gamma-0} \equiv 0, \quad \forall\, \boldsymbol{x} \in \Gamma.$$

It means that

$$\mathbb{N}\left(-\frac{1}{2}\mathbb{I} + \mathbb{K}_{II,k}\right) = \{0\}.$$

(ii) In the case $k^2 = \lambda_j^D$ $(1 \leqslant j < +\infty)$ is an eigenvalue of the interior Dirichlet eigenvalue problem, then $u_\mu(\boldsymbol{x})$ is an eigenfunction of the interior Dirichlet eigenvalue problem corresponding to the eigenvalue λ_j^D. That is

$$u_\mu \in V_j^D = \left\{ u_j^D(\boldsymbol{x}) = \sum_{l=1}^{J^D} C_l u_{j,l}^D(\boldsymbol{x}),\ \forall\, C_l \in \mathbb{C}, l = 1, 2, \cdots, J^D \right\}$$

defined by (5.1.22), namely

$$u_\mu(\boldsymbol{x}) = u_j^D(\boldsymbol{x}),\ \ \forall\, \boldsymbol{x} \in \Omega,$$

then

$$\left.\frac{\partial}{\partial \boldsymbol{\nu}_{\boldsymbol{x}}} u_j^D(\boldsymbol{x})\right|_{\Gamma-0} = \left.\frac{\partial}{\partial \boldsymbol{\nu}_{\boldsymbol{x}}} u_\mu(\boldsymbol{x})\right|_{\Gamma} = \left(\mathbb{K}_{III,k}\mu\right)(\boldsymbol{x}),\ \ \forall\, \boldsymbol{x} \in \Gamma. \tag{5.3.39}$$

The operator $\mathbb{K}_{III,k}\colon H^{1/2}(\Gamma) \to H^{-1/2}(\Gamma)$ is invertible and $\left(\mathbb{K}_{III,k}\right)^{-1}\colon H^{-1/2}(\Gamma) \to H^{1/2}(\Gamma)$ is bounded provided by $k^2 = \lambda_j^D$. Equation (5.3.39) yields

$$\mu(\boldsymbol{x}) = \left(\mathbb{K}_{III,k}\right)^{-1} \left(\left.\frac{\partial}{\partial \boldsymbol{\nu}_{\boldsymbol{x}}} u_j^D(\boldsymbol{x})\right|_{\Gamma-0} \right)$$

and

$$\mathbb{N}\left(-\frac{1}{2}\mathbb{I} + \mathbb{K}_{II,k}\right) = \left\{ \mu \;\middle|\; \mu(\boldsymbol{x}) = \sum_{l=1}^{J^D} C_l \left(\mathbb{K}_{III,k}\right)^{-1} \left(\left.\frac{\partial}{\partial \boldsymbol{\nu}_{\boldsymbol{x}}} u_{j,l}^D(\boldsymbol{x})\right|_{\Gamma-0} \right),\ \forall\, C_l \in \mathbb{C} \right\}.$$

Obviously

$$\dim \mathbb{N}\left(-\frac{1}{2}\mathbb{I} + \mathbb{K}_{II,k}\right) > 0.$$

(b) For any $\mu \in \mathbb{N}\left(\frac{1}{2}\mathbb{I} + \mathbb{K}_{II,k}\right)$, that is $\mu \in H^{1/2}(\Gamma)$ and

$$\left(\frac{1}{2}\mathbb{I} + \mathbb{K}_{II,k}\right)\mu = 0.$$

Let

$$u_\mu(\boldsymbol{x}) = \int_\Gamma \frac{\partial}{\partial \boldsymbol{\nu}_{\boldsymbol{y}}} E_n(\boldsymbol{x}, \boldsymbol{y}; k)\mu(\boldsymbol{y}) \mathrm{d}s_{\boldsymbol{y}},\ \ \ \forall\, \boldsymbol{x} \in \mathbb{R}^n \backslash \Gamma.$$

On the domain Ω_e, $u_\mu(\boldsymbol{x})$ is the solution of exterior Dirichlet problem (5.1.2) with given $g(\boldsymbol{x}) \equiv 0$, then we obtain

$$u_\mu(\boldsymbol{x}) \equiv 0, \ \ \forall\, \boldsymbol{x} \in \Omega_e$$

and

$$u_\mu(\boldsymbol{x})\Big|_{\Gamma+0} = 0, \quad \frac{\partial}{\partial \boldsymbol{\nu}_x} u_\mu(\boldsymbol{x})\Big|_{\Gamma} = 0.$$

On the domain Ω, $u_\mu(\boldsymbol{x})$ satisfies

$$\left.\begin{aligned} -\Delta u_\mu(\boldsymbol{x}) - k^2 u_\mu(\boldsymbol{x}) = 0, &\quad \forall\, \boldsymbol{x} \in \Omega, \\ \frac{\partial}{\partial \boldsymbol{\nu}_x} u_\mu(\boldsymbol{x})\Big|_{\Gamma} = 0, &\quad \forall\, \boldsymbol{x} \in \Gamma. \end{aligned}\right\} \tag{5.3.40}$$

(i) In the case k^2 is not a positive eigenvalue of the interior Neumann eigenvalue problem, then

$$u_\mu(\boldsymbol{x}) \equiv 0, \ \ \forall\, \boldsymbol{x} \in \Omega$$

and

$$\begin{aligned} &u_\mu(\boldsymbol{x})\Big|_{\Gamma-0} = 0, \\ &\mu(\boldsymbol{x}) = u_\mu(\boldsymbol{x})\Big|_{\Gamma+0} - u_\mu(\boldsymbol{x})\Big|_{\Gamma-0} = 0, \ \ \forall\, \boldsymbol{x} \in \Gamma. \end{aligned}$$

Therefore

$$\mathbb{N}\left(\frac{1}{2}\mathbb{I} + \mathbb{K}_{I\!I,k}\right) = \left\{0\right\}.$$

(ii) In the case $k^2 = \lambda_j^N \ \ (2 \leqslant j < \infty)$ is a positive eigenvalue of the interior Neumann problem, then

$$u_\mu \in V_j^N = \left\{ u_j^N \,\middle|\, u_j^N(\boldsymbol{x}) = \sum_{l=1}^{J^N} C_l u_{j,l}^N(\boldsymbol{x}), \ C_l \in \mathbb{C}, l = 1, 2, \cdots, J^N \right\}$$

defined by (5.1.24).

That is

$$u_\mu(\boldsymbol{x}) = \sum_{l=1}^{J^N} C_l u_{j,l}^N(\boldsymbol{x}), \ \ \forall\, \boldsymbol{x} \in \Omega.$$

Let

$$\mu_{j,l}^N(\boldsymbol{x}) = -u_{j,l}^N(\boldsymbol{x})\Big|_{\Gamma}, \ \ \forall\, \boldsymbol{x} \in \Gamma,$$

then

$$\mu(\boldsymbol{x}) = u_\mu(\boldsymbol{x})\Big|_{\Gamma+0} - u_\mu(\boldsymbol{x})\Big|_{\Gamma-0} = \sum_{l=1}^{J^N} C_l \mu_{j,l}^N(\boldsymbol{x}),$$

namely

$$\mathbb{N}\left(\frac{1}{2}\mathbb{I} + \mathbb{K}_{\mathit{II},k}\right) = \left\{\mu \,\middle|\, \mu(\boldsymbol{x}) = \sum_{l=1}^{J^N} C_l \mu_{j,l}^N(\boldsymbol{x}),\ C_l \in \mathbb{C}, l = 1, 2, \cdots, J^N\right\}.$$

(c) Suppose that $\rho \in \mathbb{N}\left(\frac{1}{2}\mathbb{I} + \mathbb{K}_{\mathit{II},k}^*\right)$, then $\rho \in H^{-1/2}(\Gamma)$ and

$$\left(\frac{1}{2}\mathbb{I} + \mathbb{K}_{\mathit{II},k}^*\right)\rho = 0.$$

Let

$$u_\rho(\boldsymbol{x}) = \int_\Gamma E_n(\boldsymbol{x}, \boldsymbol{y}; k)\rho(\boldsymbol{y})\mathrm{d}s_{\boldsymbol{y}}, \quad \forall\, \boldsymbol{x} \in \mathbb{R}^n.$$

On the domain Ω, $u_\rho(\boldsymbol{x})$ satisfies

$$\left.\begin{aligned} -\Delta u_\rho(\boldsymbol{x}) - k^2 u_\rho(\boldsymbol{x}) &= 0, \quad \forall\, \boldsymbol{x} \in \Omega, \\ \frac{\partial}{\partial \boldsymbol{\nu}_x} u_\rho(\boldsymbol{x})\Big|_\Gamma &= 0, \quad \forall\, \boldsymbol{x} \in \Gamma. \end{aligned}\right\} \tag{5.3.41}$$

(i) In the case k^2 is not a positive eigenvalue of interior Neumann eigenvalue problem, then

$$u_\rho(\boldsymbol{x}) \equiv 0, \quad \forall\, \boldsymbol{x} \in \Omega$$

and

$$\frac{\partial}{\partial \boldsymbol{\nu}_x} u_\rho(\boldsymbol{x})\Big|_{\Gamma-0} = 0, \ \ \forall\, \boldsymbol{x} \in \Gamma.$$

On the domain Ω_e, $u_\rho(\boldsymbol{x})$ is the solution of exterior Dirichlet problem (5.1.2) with given $g(\boldsymbol{x}) \equiv 0$, then

$$u_\rho(\boldsymbol{x}) \equiv 0, \ \ \forall\, \boldsymbol{x} \in \Omega_e,$$

and

$$\frac{\partial}{\partial \boldsymbol{\nu}_x} u_\rho(\boldsymbol{x})\Big|_{\Gamma+0} = 0, \ \ \forall\, \boldsymbol{x} \in \Gamma.$$

Therefore we obtain

$$\mu(\boldsymbol{x}) = -\frac{\partial}{\partial \boldsymbol{\nu}_x} u_\rho(\boldsymbol{x})\bigg|_{\Gamma-0} + \frac{\partial}{\partial \boldsymbol{\nu}_x} u_\rho(\boldsymbol{x})\bigg|_{\Gamma+0} \equiv 0, \quad \forall\, \boldsymbol{x} \in \Gamma,$$

namely

$$\mathbb{N}\left(\frac{1}{2}\mathbb{I} + \mathbb{K}^*_{II,k}\right) = \left\{0\right\}.$$

(ii) $k^2 = \lambda_j^N$ $(2 \leqslant j < +\infty)$ is a positive eigenvalue of the interior Neumann eigenvalue problem. On the domain Ω we have

$$u_\rho(\boldsymbol{x}) \in V_j^N,$$

that is

$$u_\rho(\boldsymbol{x}) = \sum_{l=1}^{J^N} C_l u_{j,l}^N(\boldsymbol{x}), \quad \forall\, \boldsymbol{x} \in \overline{\Omega}.$$

Let

$$\rho_{j,l}^N(\boldsymbol{x}) = u_{j,l}^N(\boldsymbol{x})\bigg|_{\Gamma}, \quad \forall\, \boldsymbol{x} \in \Gamma,$$

then

$$\left(\mathbb{K}_{I,k}\rho\right)(\boldsymbol{x}) = \int_\Gamma E_n(\boldsymbol{x}, \boldsymbol{y}; k)\varphi(\boldsymbol{y})\mathrm{d}s_y = \sum_{l=1}^{J^N} C_l \rho_{j,l}^N(\boldsymbol{x}), \quad \forall\, \boldsymbol{x} \in \Gamma.$$

In this case $k^2 = \lambda_j^N$, the operator $\mathbb{K}_{I,k} : \; H^{-1/2}(\Gamma) \to H^{1/2}(\Gamma)$ is invertible, it arrives at

$$\rho = \sum_{l=1}^{J^N} C_l \left(\mathbb{K}_{I,k}\right)^{-1} \rho_{j,l}^N(\boldsymbol{x}),$$

namely

$$\mathbb{N}\left(\frac{1}{2}\mathbb{I} + \mathbb{K}^*_{II,k}\right) = \left\{\rho \;\middle|\; \rho(\boldsymbol{x}) = \sum_{l=1}^{J^N} C_l \left(\mathbb{K}_{I,k}\right)^{-1} \rho_{j,l}^N(\boldsymbol{x}),\, C_l \in \mathbb{C}, l = 1, 2, \cdots, J^N\right\}.$$

(d) Suppose that $\rho \in \mathbb{N}\left(-\frac{1}{2}\mathbb{I} + \mathbb{K}^*_{II,k}\right)$, then $\rho \in H^{-1/2}(\Gamma)$ and

$$\left(-\frac{1}{2}\mathbb{I} + \mathbb{K}^*_{II,k}\right)\rho = 0.$$

Let

$$u_\rho(\boldsymbol{x}) = \int_\Gamma E_n(\boldsymbol{x}, \boldsymbol{y}; k)\rho(\boldsymbol{y})\mathrm{d}s_{\boldsymbol{y}}, \quad \forall\, \boldsymbol{x} \in \mathbb{R}^n,$$

then

$$\frac{\partial}{\partial \boldsymbol{\nu}_{\boldsymbol{x}}} u_\rho(\boldsymbol{x})\bigg|_{\Gamma+0} = 0, \quad \forall\, \boldsymbol{x} \in \Gamma.$$

On the domain Ω_e we obtain

$$u_\rho(\boldsymbol{x}) \equiv 0, \quad \forall\, \boldsymbol{x} \in \Omega_e$$

and

$$u_\rho(\boldsymbol{x})\bigg|_{\Gamma} = 0, \quad \forall\, \boldsymbol{x} \in \Gamma,$$

provided by the uniqueness of exterior Neumann problem (5.1.4), On the domain Ω, $u_\rho(\boldsymbol{x})$ satisfies

$$\left.\begin{aligned} -\Delta u_\rho(\boldsymbol{x}) - k^2 u_\rho(\boldsymbol{x}) = 0, \quad \forall\, \boldsymbol{x} \in \Omega, \\ u_\rho(\boldsymbol{x})\bigg|_{\Gamma} = 0, \quad \forall\, \boldsymbol{x} \in \Gamma. \end{aligned}\right\} \tag{5.3.42}$$

(i) In the case k^2 is not an eigenvalue of the interior Dirichlet eigenvalue problem, then

$$u_\rho(\boldsymbol{x}) \equiv 0, \quad \forall\, \boldsymbol{x} \in \Omega$$

and

$$\frac{\partial}{\partial \boldsymbol{\nu}_{\boldsymbol{x}}} u_\rho(\boldsymbol{x})\bigg|_{\Gamma-0} = 0, \quad \forall\, \boldsymbol{x} \in \Gamma.$$

Then we have

$$\rho(\boldsymbol{x}) = -\frac{\partial}{\partial \boldsymbol{\nu}_{\boldsymbol{x}}} u_\rho(\boldsymbol{x})\bigg|_{\Gamma+0} + \frac{\partial}{\partial \boldsymbol{\nu}_{\boldsymbol{x}}} u_\rho(\boldsymbol{x})\bigg|_{\Gamma-0} \equiv 0, \quad \forall\, \boldsymbol{x} \in \Gamma,$$

That is

$$\mathbb{N}\left(-\frac{1}{2}\mathbb{I} + \mathbb{K}^*_{\mathit{II},k}\right) = \left\{0\right\}.$$

(ii) $k^2 = \lambda_j^D \ (1 \leqslant j < +\infty)$, then on the domain Ω,

$$u_\rho \in V_j^D,$$

namely

$$u_\rho(\boldsymbol{x}) = \sum_{l=1}^{J^D} C_l u_{j,l}^D(\boldsymbol{x}), \quad \forall\, \boldsymbol{x} \in \overline{\Omega}$$

and

$$\left.\frac{\partial}{\partial \boldsymbol{\nu}_{\boldsymbol{x}}} u_\rho(\boldsymbol{x})\right|_{\Gamma-0} = \sum_{l=1}^{J^D} C_l \left(\left.\frac{\partial}{\partial \boldsymbol{\nu}_{\boldsymbol{x}}} u_{j,l}^D(\boldsymbol{x})\right|_{\Gamma-0} \right).$$

Let

$$\rho_{j,l}^D(\boldsymbol{x}) = \left(\left.\frac{\partial}{\partial \boldsymbol{\nu}_{\boldsymbol{x}}} u_{j,l}^D(\boldsymbol{x})\right|_{\Gamma-0} \right), \quad 1 \leqslant l \leqslant J^D, \ \forall\, \boldsymbol{x} \in \Gamma.$$

It arrives at

$$\rho(\boldsymbol{x}) = \sum_{l=1}^{J^D} C_l \rho_{j,l}^D(\boldsymbol{x}), \quad \forall\, \boldsymbol{x} \in \Gamma$$

and

$$\mathbb{N}\left(-\frac{1}{2}\mathbb{I} + \mathbb{K}_{\mathit{II},k}^* \right) = \left\{ \rho \,\middle|\, \rho(\boldsymbol{x}) = \sum_{l=1}^{J^D} C_l \rho_{j,l}^D(\boldsymbol{x}), \ C_l \in \mathbb{C}, l = 1, 2, \cdots, J^D \right\}.$$

The proof of Theorem 5.3.6 is completed. □

Now return to the four second kind integral equations (5.3.34)–(5.3.37) on spaces $H^{1/2}(\Gamma)$ or $H^{-1/2}(\Gamma)$ correspondingly. It is known that $\left\langle H^{1/2}(\Gamma), H^{-1/2}(\Gamma) \right\rangle$ is dual system corresponding to the following bilinear form

$$\langle \psi, \mu \rangle = \int_\Gamma \psi(\boldsymbol{x}) \mu(\boldsymbol{x}) \mathrm{d}s_{\boldsymbol{x}}, \quad \forall\, \psi \in H^{1/2}(\Gamma), \ \mu \in H^{-1/2}(\Gamma). \tag{5.3.43}$$

It is straightforward to check that the two operators $\mathbb{K}_{\mathit{II},k}: \ H^{1/2}(\Gamma) \to H^{1/2}(\Gamma)$ and $\mathbb{K}_{\mathit{II},k}^*: H^{-1/2}(\Gamma) \to H^{-1/2}(\Gamma)$ are adjoint. Are the two operators $\mathbb{K}_{\mathit{II},k}$ and $\mathbb{K}_{\mathit{II},k}^*$ compact? We have no answers yet, therefore the Fredholm Theorem (Colton & Kress, 2013, pp. 16–22) can not be applied to the second kind integral equation (5.3.34)–(5.3.37) directly. But for the Eqs. (5.3.34)–(5.3.37), the following alternative theorem still holds.

Theorem 5.3.8 *Assume that the wave number k satisfies condition (5.0.2) and the given functions $g \in H^{1/2}(\Gamma)$ and $h \in H^{-1/2}(\Gamma)$, then the following results hold.*

(a) Either k is not an eigenvalue of the interior Dirichlet eigenvalue problem, the second kind integral equation (5.3.34) exists a unique solution $\psi \in H^{1/2}(\Gamma)$ and the second kind integral equation (5.3.37) exists a unique solution $\varphi \in H^{-1/2}(\Gamma)$;

or $k^2 = \lambda_j^D \quad (1 \leqslant j < +\infty)$ *is an eigenvalue of the interior Dirichlet eigenvalue problem, the Eq. (5.3.34) exists solutions if and only if*

$$\langle g, \rho \rangle = 0, \quad \forall\, \rho \in \mathbb{N}\left(-\frac{1}{2}\mathbb{I} + \mathbb{K}_{\mathit{II},k}^{*}\right)$$

and the Eq. (5.3.37) exists solutions if and only if

$$\langle h, \mu \rangle = 0, \quad \forall\, \mu \in \mathbb{N}\left(-\frac{1}{2}\mathbb{I} + \mathbb{K}_{\mathit{II},k}\right).$$

(*b*) *Either* k^2 *is not a positive eigenvalue of the interior Neumann eigenvalue problem, the second kind integral equation (5.3.35) exists a unique solution* $\psi \in H^{1/2}(\Gamma)$ *and the second kind integral equation (5.3.36) exists a unique solution* $\varphi \in H^{-1/2}(\Gamma)$; *or* $k^2 = \lambda_j^N \quad (2 \leqslant j < +\infty)$ *is a positive eigenvalue of the interior Neumann eigenvalue problem, the Eq. (5.3.35) exists solution if and only if*

$$\langle g, \rho \rangle = 0, \quad \forall\, \rho \in \mathbb{N}\left(\frac{1}{2}\mathbb{I} + \mathbb{K}_{\mathit{II},k}^{*}\right)$$

and the Eq. (5.3.36) exists solutions if and only if

$$\langle h, \mu \rangle = 0, \quad \forall\, \mu \in \mathbb{N}\left(\frac{1}{2}\mathbb{I} + \mathbb{K}_{\mathit{II},k}\right).$$

Proof (*a*) Equations (5.3.34) and (5.3.37) can be rewritten as the following equivalent forms:

$$\left(-\frac{1}{2}\mathbb{I} + \mathbb{K}_{\mathit{II},1}^{+}\right)\psi + \Pi_{\mathit{II},k}\psi = g, \tag{5.3.44}$$

$$\left(-\frac{1}{2}\mathbb{I} + \mathbb{K}_{\mathit{II},1}^{+,*}\right)\varphi + \Pi_{\mathit{II},k}^{*}\varphi = h, \tag{5.3.45}$$

provided by equalities (5.2.31). Let

$$L = -\frac{1}{2}\mathbb{I} + \mathbb{K}_{\mathit{II},1}^{+},$$
$$L^{*} = -\frac{1}{2}\mathbb{I} + \mathbb{K}_{\mathit{II},1}^{+,*},$$

by the Theorems 4.3.2 and 4.4.4 we know that

$$L: \; H^{1/2}(\Gamma) \longrightarrow H^{1/2}(\Gamma),$$
$$L^{*}: \; H^{-1/2}(\Gamma) \longrightarrow H^{-1/2}(\Gamma)$$

are invertible and $(L)^{-1}$, $(L^*)^{-1}$ are bounded. Therefore the Eq. (5.3.34) is equivalent to

$$\psi + (L)^{-1}\Pi_{II,k}\psi = (L)^{-1}g, \tag{5.3.46}$$

and the Eq. (5.3.37) is equivalent to

$$\varphi + (L^*)^{-1}\Pi^*_{II,k}\varphi = (L^*)^{-1}h. \tag{5.3.47}$$

By Theorems 5.2.2, 5.2.4, we know that the operators

$$\begin{aligned} (L)^{-1}\Pi_{II,k}: \; & H^{1/2}(\Gamma) \longrightarrow H^{1/2}(\Gamma), \\ (L^*)^{-1}\Pi^*_{II,k}: \; & H^{-1/2}(\Gamma) \longrightarrow H^{-1/2}(\Gamma) \end{aligned}$$

are compact, but $(L)^{-1}\Pi_{II,k}$ and $(L^*)^{-1}\Pi^*_{II,k}$ are not adjoint. It is straightforward to check that the operators $(L)^{-1}\Pi_{II,k}$, $\Pi^*_{II,k}(L^*)^{-1}$ are adjoint and the operators $\Pi_{II,k}(L)^{-1}$, $(L^*)^{-1}\Pi^*_{II,k}$ are adjoint.

Furthermore,

$$\begin{aligned} \mathbb{N}\left(\mathbb{I} + \Pi^*_{II,k}(L^*)^{-1}\right) &= \mathbb{N}\left(\left(L^* + \Pi^*_{II,k}\right)\left(L^*\right)^{-1}\right) \\ &= \mathbb{N}\left(\left(-\frac{1}{2}\mathbb{I} + \mathbb{K}^*_{II,k}\right)\left(L^*\right)^{-1}\right) \\ &= \left\{L^*\rho \,\middle|\, \rho \in \mathbb{N}\left(-\frac{1}{2}\mathbb{I} + \mathbb{K}^*_{II,k}\right)\right\} \end{aligned} \tag{5.3.48}$$

and

$$\begin{aligned} \mathbb{N}\left(\mathbb{I} + \Pi_{II,k}(L)^{-1}\right) &= \mathbb{N}\left(\left(L + \Pi_{II,k}\right)\left(L\right)^{-1}\right) \\ &= \mathbb{N}\left(\left(-\frac{1}{2}\mathbb{I} + \mathbb{K}_{II,k}\right)\left(L\right)^{-1}\right) \\ &= \left\{L\mu \,\middle|\, \mu \in \mathbb{N}\left(-\frac{1}{2}\mathbb{I} + \mathbb{K}_{II,k}\right)\right\}. \end{aligned} \tag{5.3.49}$$

The invertibility of the operators L^* and L arrives at

$$\mathbb{N}\left(\mathbb{I} + \Pi^*_{II,k}\left(L^*\right)^{-1}\right) = \left\{0\right\} \text{if and only if} \mathbb{N}\left(\frac{1}{2}\mathbb{I} + \mathbb{K}^*_{II,k}\right) = \left\{0\right\}, \tag{5.3.50}$$

$$\mathbb{N}\left(\mathbb{I} + \Pi_{II,k}\left(L\right)^{-1}\right) = \left\{0\right\} \text{if and only if} \mathbb{N}\left(\frac{1}{2}\mathbb{I} + \mathbb{K}_{II,k}\right) = \left\{0\right\}. \tag{5.3.51}$$

For given $g \in H^{1/2}(\Gamma)$ and the second kind equation (5.3.46), a direct application of the Fredholm Alternative (Colton & Kress, 2013, p. 22) yields that either

$\mathbb{N}\left(\frac{1}{2}\mathbb{I}+\mathbb{K}^*_{II,k}\right)=\left\{0\right\}$, the Eq. (5.3.46) exists a unique solution $\psi\in H^{1/2}(\Gamma)$, or $\dim\mathbb{N}\left(\frac{1}{2}\mathbb{I}+\mathbb{K}^*_{II,k}\right)>0$, the Eq. (5.3.46) exists solutions if and only if

$$\langle g,\rho\rangle=0,\quad \forall\,\rho\in\mathbb{N}\left(-\frac{1}{2}\mathbb{I}+\mathbb{K}^*_{II,k}\right).$$

For given $h\in H^{-1/2}(\Gamma)$ and the second kind equation (5.3.47), similarly we obtain that either $\mathbb{N}\left(\frac{1}{2}\mathbb{I}+\mathbb{K}_{II,k}\right)=\left\{0\right\}$, the Eq. (5.3.47) exists a unique solution $\varphi\in H^{-1/2}(\Gamma)$, or $\dim\mathbb{N}\left(\frac{1}{2}\mathbb{I}+\mathbb{K}_{II,k}\right)>0$, the Eq. (5.3.47) exists solutions if and only if

$$\langle h,\mu\rangle=0,\quad \forall\,\mu\in\mathbb{N}\left(-\frac{1}{2}\mathbb{I}+\mathbb{K}_{II,k}\right).$$

Due to the equivalence of the Eqs. (5.3.34) and (5.3.46) (the Eqs. (5.3.37) and (5.3.47)) and the properties of $\mathbb{N}\left(\frac{1}{2}\mathbb{I}+\mathbb{K}^*_{II,k}\right)\left(\mathbb{N}\left(\frac{1}{2}\mathbb{I}+\mathbb{K}_{II,k}\right)\right)$ given in Theorem 5.3.6, the proof of conclusion (a) is completed.

The proof of conclusion (b) is similar to the proof of conclusion (a), which is omitted here. □

Let

$$u(\boldsymbol{x})=\int_\Gamma E_n(\boldsymbol{x},\boldsymbol{y};k)\varphi(\boldsymbol{y})\mathrm{d}s_y,\quad \forall\,\boldsymbol{x}\in\mathbb{R}^n,$$

solve the exterior impedance boundary value problem (5.1.5) with $\mathrm{Im}(\bar{k}\lambda)>0$, then the impedance boundary condition (5.1.5b) yields the following boundary integral equation of the second kind

$$-\frac{1}{2}\varphi+\mathbb{K}^*_{II,k}\varphi+\lambda\mathbb{K}_{I,k}\varphi=h. \tag{5.3.52}$$

The operator $-\frac{1}{2}\mathbb{I}+\mathbb{K}_{II,k}+\lambda\mathbb{K}_{I,k}$ is the adjoint operator of $-\frac{1}{2}\mathbb{I}+\mathbb{K}^*_{II,k}+\lambda\mathbb{K}_{I,k}$. Analogous to Theorem 5.3.7, we obtain

Theorem 5.3.9 *Suppose that the wave number k satisfies (5.0.2) and* Im $(\bar{k}\lambda)\geqslant 0$, *then*

(*i*) *In the case k^2 is not an eigenvalue of interior Dirichlet eigenvalue problem, the*

$$\mathbb{N}\left(-\frac{1}{2}\mathbb{I}+\mathbb{K}^*_{II,k}+\lambda\mathbb{K}_{I,k}\right)=\left\{0\right\},$$

$$\mathbb{N}\left(-\frac{1}{2}\mathbb{I}+\mathbb{K}_{II,k}+\lambda\mathbb{K}_{I,k}\right)=\left\{0\right\}.$$

(ii) In the case $k^2 = \lambda_j^D$ $(1 \leqslant j < +\infty)$ is an eigenvalue of the interior Dirichlet eigenvalue problem, then

$$\mathbb{N}\left(-\frac{1}{2}\mathbb{I} + \mathbb{K}^*_{II,k} + \lambda\mathbb{K}_{I,k}\right) = \left\{\rho \middle| \rho(\boldsymbol{x}) = \sum_{l=1}^{J^D} C_l \rho^D_{j,l}(\boldsymbol{x}), C_l \in \mathbb{C}, l = 1, 2, \cdots, J^D\right\}$$

and

$$\dim \mathbb{N}\left(-\frac{1}{2}\mathbb{I} + \mathbb{K}_{II,k} + \lambda\mathbb{K}_{I,k}\right) = \dim \mathbb{N}\left(-\frac{1}{2}\mathbb{I} + \mathbb{K}^*_{II,k} + \lambda\mathbb{K}_{I,k}\right) > 0.$$

Furthermore analogous to Theorem 5.3.8, we have

Theorem 5.3.10 *Assume that wave number k satisfies (5.0.2), $\text{Im}(\bar{k}\lambda) \geqslant 0$ and the given function $h \in H^{-1/2}(\Gamma)$, then either k^2 is not an eigenvalue of interior Dirichlet eigenvalue problem, the integral equation (5.3.52) exists a unique solution $\varphi \in H^{-1/2}(\Gamma)$, or $k^2 = \lambda_j^D$ $(1 \leqslant j < +\infty)$ is an eigenvalue of interior Dirichlet eigenvalue problem, the integral equation (5.3.52) exists solutions if and only if*

$$\langle h, \mu \rangle = 0, \quad \forall\, \mu \in \mathbb{N}\left(-\frac{1}{2}\mathbb{I} + \mathbb{K}_{II,k} + \lambda\mathbb{K}_{I,k}\right).$$

Remark 5.3.3 (i) In case $k^2 = \lambda_j^D$ $(1 \leqslant j < +\infty)$ is an eigenvalue of the interior Dirichlet eigenvalue problem and exists a function $\mu \in \mathbb{N}\left(-\frac{1}{2}\mathbb{I} + \mathbb{K}_{II,k}\right)$, such that

$$\langle h, \mu \rangle \neq 0,$$

then the second kind integral equation (5.3.37) exists no solution provided by Theorem 5.3.8. It means that the second kind integral equation (5.3.37) can not solve the exterior Neumann boundary problem (5.1.4) with $k^2 = \lambda_j^D$. But in this case the corresponding boundary integro-differential equation of the first kind (5.3.18) exists a unique solution, which solves the exterior Neumann boundary problem (5.1.4).

(ii) In case $k^2 = \lambda_j^N$ $(2 \leqslant j < +\infty)$ and exists a function $\rho \in \mathbb{N}\left(\frac{1}{2}\mathbb{I} + \mathbb{K}^*_{II,k}\right)$ such that

$$\langle g, \rho \rangle \neq 0,$$

then the second kind integral equation (5.3.35) exists no solution. Therefore the integral equation (5.3.35) can not solve the exterior Dirichlet boundary problem (5.1.2). But in this case the corresponding boundary integral equation of the first kind exists a unique solution, which solves the exterior Dirichlet boundary value problem.

5.3.4 Modified Integral and Integro-Differential Equations

We have known that the exterior boundary value problems (5.1.2) and (5.1.4) with all wave numbers k satisfied (5.0.2) is uniqueness. But nonuniqueness of the corresponding boundary integral equations (5.3.35) and (5.3.37) arise for some wave numbers, which means that in this case, the exterior boundary value problem (5.1.2) and (5.1.4) is not equivalent to the corresponding boundary integral equation (5.3.35) and (5.3.37). Is it possible to find the equivalent boundary integral equation of exterior boundary value problem (5.1.2) and (5.1.4) for all wave number k satisfied (5.0.2)? Leis (1965), Brakhage and Werner (1965), and Panich (1965) independently proposed to find the solution of exterior Dirichlet problem (5.1.2) in the form of a combined double and single layer potential

$$u(\boldsymbol{x}) = \int_\Gamma \left\{ \frac{\partial}{\partial \boldsymbol{\nu}_y} E_n(\boldsymbol{x}, \boldsymbol{y}; k) - \mathrm{i}\eta E_n(\boldsymbol{x}, \boldsymbol{y}; k) \right\} \psi(\boldsymbol{y}) \mathrm{d}s_y, \quad \forall \boldsymbol{x} \in \mathbb{R}^n \backslash \Gamma. \quad (5.3.53)$$

where $\eta > 0$ is a positive real number and the unknown density $\psi \in C(\Gamma)$. In the following the combined double and single layer potential (5.3.53) with unknown density $\psi \in H^{1/2}(\Gamma)$ solves the exterior Dirichlet problem (5.1.2), the exterior Neumann problem (5.1.4) and the exterior impedance problem (5.1.5) respectively.

For the problem (5.1.2), the Dirichlet boundary condition (5.1.2b) yields

$$\frac{1}{2}\psi + \mathbb{K}_{II,k}\psi - \mathrm{i}\eta \mathbb{K}_{I,k}\psi = g. \quad (5.3.54)$$

For the problem (5.1.4), the Neumann boundary condition (5.1.4b) yields

$$\mathbb{K}_{III,k}\psi + \frac{\mathrm{i}}{2}\eta\psi - \mathrm{i}\eta \mathbb{K}^*_{II,k}\psi = h. \quad (5.3.55)$$

For the problem (5.1.5), the impedance boundary condition (5.1.5b) yields

$$\mathbb{K}_{III,k}\psi + \frac{\mathrm{i}}{2}\eta\psi - \mathrm{i}\eta \mathbb{K}^*_{II,k}\psi + \lambda \left\{ \frac{1}{2}\psi + \mathbb{K}_{II,k}\psi - \mathrm{i}\eta \mathbb{K}_{II,k}\psi \right\} = h. \quad (5.3.56)$$

For the boundary integral equation (5.3.54) and the boundary integro-differential equations (5.3.55) and (5.3.56), we have

Theorem 5.3.11 *(i) For all wave numbers k satisfied (5.0.2) and $\eta > 0$, then*

$$\mathbb{N}\left(\frac{1}{2}\mathbb{I} + \mathbb{K}_{II,k} - i\eta \mathbb{K}_{I,k}\right) = \{0\},$$

$$\mathbb{N}\left(\mathbb{K}_{III,k} + \frac{i}{2}\eta\mathbb{I} - i\eta \mathbb{K}^*_{II,k}\right) = \{0\}.$$

(ii) For all wave number k satisfied (5.0.2), $\eta > 0$ *and* $\mathrm{Im}(\bar{k}\lambda) \geqslant 0$*, then*

$$\mathbb{N}\left(\mathbb{K}_{\mathit{III},k} + \frac{i}{2}\eta\mathbb{I} - i\eta\mathbb{K}^*_{\mathit{II},k} + \lambda\left(\frac{1}{2}\mathbb{I} + \mathbb{K}_{\mathit{II},k} - i\eta\mathbb{K}_{I,k}\right)\right) = \{0\}.$$

Proof Suppose that $\psi \in \mathbb{N}\left(\frac{1}{2}\mathbb{I} + \mathbb{K}_{\mathit{II},k} - \mathrm{i}\eta\mathbb{K}_{I,k}\right)$, let

$$u_\psi(\boldsymbol{x}) = \int_\Gamma \left(\frac{\partial}{\partial \boldsymbol{\nu}_{\boldsymbol{y}}} E_n(\boldsymbol{x}, \boldsymbol{y}; k) - \mathrm{i}\eta E_n(\boldsymbol{x}, \boldsymbol{y}; k)\right) \psi(\boldsymbol{y})\mathrm{d}s_{\boldsymbol{y}}, \quad \forall\, \boldsymbol{x} \in \mathbb{R}^n \backslash \Gamma.$$

On the domain Ω_e,

$$u_\psi(\boldsymbol{x})\Big|_{\Gamma+0} = 0, \quad \forall\, \boldsymbol{x} \in \Gamma,$$

and $u_\psi(\boldsymbol{x})$ is a solution of exterior Dirichlet problem (5.1.2), the uniqueness of which arrives at $u_\psi(\boldsymbol{x}) \equiv 0$, $\forall \boldsymbol{x} \in \Omega_e$, and

$$\frac{\partial}{\partial \boldsymbol{\nu}_{\boldsymbol{x}}} u_\psi(\boldsymbol{x})\Big|_{\Gamma+0} = \mathbb{K}_{\mathit{III},k}\psi + \frac{\mathrm{i}}{2}\eta\psi - \mathrm{i}\eta\mathbb{K}^*_{\mathit{II},k}\psi = 0.$$

On the domain Ω, $u_\psi(\boldsymbol{x})$ satisfies

$$\left.\begin{aligned} &-\Delta u_\psi(\boldsymbol{x}) - k^2 u_\psi(\boldsymbol{x}) = 0, && \text{in } \Omega, \\ &u_\psi(\boldsymbol{x})\Big|_{\Gamma-0} = -\psi(\boldsymbol{x}), && \forall\, \boldsymbol{x} \in \Gamma, \\ &\frac{\partial}{\partial \boldsymbol{\nu}_{\boldsymbol{x}}} u_\psi(\boldsymbol{x})\Big|_{\Gamma-0} = -\mathrm{i}\eta\psi(\boldsymbol{x}), && \forall\, \boldsymbol{x} \in \Gamma. \end{aligned}\right\}$$

The first Green's theorem yields that (Colton & Kress, 2013)

$$\begin{aligned} \mathrm{i}\eta \int_\Gamma \left|\psi(\boldsymbol{x})\right|^2 \mathrm{d}s_{\boldsymbol{x}} &= \int_\Gamma \overline{u_\psi(\boldsymbol{x})}\Big|_{\Gamma-0} \frac{\partial}{\partial \boldsymbol{\nu}_{\boldsymbol{x}}} u_\psi(\boldsymbol{x})\Big|_{\Gamma-0} \mathrm{d}s_{\boldsymbol{x}} \\ &= \int_\Omega \left(\left|\nabla u_\psi(\boldsymbol{x})\right|^2 - k^2\left|u_\psi(\boldsymbol{x})\right|^2\right) \mathrm{d}\boldsymbol{x}. \end{aligned}$$

Taking imaginary part of this equation, we have

$$\eta \int_\Gamma \left|\psi(\boldsymbol{x})\right|^2 \mathrm{d}s_{\boldsymbol{x}} + 2\mathrm{Re}(k)\mathrm{Im}(k) \int_\Omega \left|u_\psi(\boldsymbol{x})\right|^2 \mathrm{d}\boldsymbol{x} = 0.$$

It arrives at

$$\psi(\boldsymbol{x}) \equiv 0, \quad \forall\, \boldsymbol{x} \in \Gamma.$$

Therefore

$$\mathbb{N}\left(\frac{1}{2}\mathbb{I}+\mathbb{K}_{II,k}-\mathrm{i}\eta\mathbb{K}_{I,k}\right)=\left\{0\right\}.$$

Similarly, we have

$$\mathbb{N}\left(\mathbb{K}_{III,k}+\frac{\mathrm{i}}{2}\eta\mathbb{I}-\mathrm{i}\eta\mathbb{K}^*_{II,k}\right)=\left\{0\right\},$$

provided by all wave numbers satisfied (5.0.2) and $\eta > 0$. The proof of conclusion (ii) is similar, which is omitted here. □

Analogous to Theorems 5.3.8 and 5.3.5, we obtain

Theorem 5.3.12 *Suppose that the wave numbers k satisfied (5.0.2).*

(i) For given $g \in H^{1/2}(\Gamma)$, the boundary integral equation (5.3.54) exists a unique solution $\psi \in H^{1/2}(\Gamma)$.

(ii) For given $h \in H^{-1/2}(\Gamma)$, the boundary integro-differential equation (5.3.55) exists a unique solution $\psi \in H^{1/2}(\Gamma)$.

(iii) For given $h \in H^{-1/2}(\Gamma)$ and $\mathrm{Im}(\overline{K}\lambda) \geqslant 0$, the boundary integro-differential equation (5.3.56) exists a unique solution $\psi \in H^{1/2}(\Gamma)$.

There are many scholars who worked on this subject, we recommend the following literature for interested readers, Kupradze (1967), Leis (1958), Vekua (1967), Jones (1979), Colton and Kress (2013), Hsiao and Wendland (2008), Nédélec (2001).

5.4 The Boundary Integro-Differential Equation Method for Interior Dirichlet and Neumann Eigenvalue Problems of Laplace Operator

Steinbach and Unger (2009) presented and analyzed a boundary element method for the solution of the interior Dirichlet eigenvalue problem of Laplace operator in $\mathbb{R}^3$. In this section, we discuss the boundary integro-differential equation method for the interior Dirichlet and Neumann eigenvalue problems (5.1.19) and (5.1.20) in $\mathbb{R}^n (n = 2, 3)$.

5.4.1 Interior Dirichlet Eigenvalue Problems of Laplace Operator

Suppose that λ_j^D is an eigenvalue of interior Dirichlet eigenvalue problem (5.1.19) corresponding to real eigenfunction $u_j^D \in V_j^D$ and satisfies

$$\left.\begin{aligned} -\Delta u_j^D(\boldsymbol{x}) - \lambda_j^D u_j^D(\boldsymbol{x}) &= 0, \quad \forall\, \boldsymbol{x} \in \Omega, \\ u_j^D(\boldsymbol{x})\Big|_{\Gamma-0} &= 0, \quad \forall\, \boldsymbol{x} \in \Gamma. \end{aligned}\right\} \tag{5.4.1}$$

Namely on the domain Ω, $u_j^D(\boldsymbol{x})$ is a solution of Helmholtz equation with wave number $k = \sqrt{\lambda_j^D}$, let

$$\left.\frac{\partial u_j^D(\boldsymbol{x})}{\partial \boldsymbol{\nu}_{\boldsymbol{x}}}\right|_{\Gamma-0} = \rho_j^D(\boldsymbol{x}), \quad \forall\, \boldsymbol{x} \in \Gamma, \tag{5.4.2}$$

$\rho_j^D \in H^{-1/2}(\Gamma)$ and ρ_j^D is a nontrivial real function.

Then the representation Theorem 5.3.5 yields that the wave number $k = \sqrt{\lambda_j^D}$ and the nontrivial real function $\rho_j^D \in H^{-1/2}(\Gamma)$ satisfies

$$\int_\Gamma E_n(\boldsymbol{x}, \boldsymbol{y}; k)\rho_j^D(\boldsymbol{y})\mathrm{d}s_{\boldsymbol{y}} = 0, \quad \forall\, \boldsymbol{x} \in \Gamma.$$

It means that we obtain the following nonlinear boundary eigenvalue problem

$$\begin{cases} \text{Find positive wave number } k \text{ and nontrivial real function } \rho \in H^{-1/2}(\Gamma) \\ \qquad \text{such that} \\ \displaystyle\int_\Gamma E_n(\boldsymbol{x}, \boldsymbol{y}; k)\rho(\boldsymbol{y})\mathrm{d}s_{\boldsymbol{y}} = 0, \quad \forall\, \boldsymbol{x} \in \Gamma. \end{cases} \tag{5.4.3}$$

Theorem 5.4.1 *The interior Dirichlet eigenvalue problem (5.1.19) is equivalent to the nonlinear boundary eigenvalue problem (5.4.3).*

Proof We have shown that if λ_j^D $(1 \leqslant j < +\infty)$ is an eigenvalue of problem (5.1.19) corresponding to the eigenfunction $u_j^D \in H_0^1(\Omega)$, the $k = \sqrt{\lambda_j^D}$ is an eigenvalue of problem (5.4.3) corresponding to the eigenfunction $\rho = \left.\dfrac{\partial u_j^D}{\partial \boldsymbol{\nu}_{\boldsymbol{x}}}\right|_\Gamma \in H^{-1/2}(\Gamma)$. Otherwise suppose that $k > 0$ is an eigenvalue of problem (5.4.3) corresponding to the eigenfunction $\rho \in H^{-1/2}(\Gamma)$. Let

$$u_\rho(\boldsymbol{x}) = \int_\Gamma E_n(\boldsymbol{x}, \boldsymbol{y}; k)\rho(\boldsymbol{y})\mathrm{d}s_{\boldsymbol{y}}, \quad \forall\, \boldsymbol{x} \in \mathbb{R}^n. \tag{5.4.4}$$

On the domain Ω_e, $u_\rho(\boldsymbol{x})$ is the solution of exterior Dirichlet boundary value problem (5.1.2) with given $g \equiv 0$, $\forall \boldsymbol{x} \in \Gamma$. The uniqueness of (5.1.2) arrives at

$$u_\rho(\boldsymbol{x}) \equiv 0, \quad \forall\, \boldsymbol{x} \in \overline{\Omega}_e \tag{5.4.5}$$

and

$$\left.\frac{\partial u_\rho(\boldsymbol{x})}{\partial \boldsymbol{\nu}_x}\right|_{\Gamma+0} = -\frac{1}{2}\rho(\boldsymbol{x}) + \int_\Gamma \frac{\partial}{\partial \boldsymbol{\nu}_x} E_n(\boldsymbol{x}, \boldsymbol{y}; k)\rho(\boldsymbol{y})\mathrm{d}s_y = 0, \quad \forall\, \boldsymbol{x} \in \Gamma. \tag{5.4.6}$$

On the domain Ω, $u_\rho(\boldsymbol{x})$ satisfies

$$\left.\begin{aligned} -\Delta u_\rho(\boldsymbol{x}) - k^2 u_\rho(\boldsymbol{x}) &= 0, \quad \text{in } \Omega, \\ \left.u_\rho(\boldsymbol{x})\right|_\Gamma &= 0, \quad \forall\, \boldsymbol{x} \in \Gamma, \\ \left.\frac{\partial u_\rho(\boldsymbol{x})}{\partial \boldsymbol{\nu}_x}\right|_{\Gamma-0} &= \rho(\boldsymbol{x}), \quad \forall\, \boldsymbol{x} \in \Gamma. \end{aligned}\right\} \tag{5.4.7}$$

Obviously $k^2 > 0$ is an eigenvalue of interior Dirichlet eigenvalue problem (5.1.19) corresponding eigenfunction $u_\rho \in H_0^1(\Omega)$. The proof is completed. □

Taking the real part and the imaginary part of Eq. (5.4.3), we obtain its equivalent form involved only the real functions:

$$\left.\begin{aligned} &\text{Find positive wave number } k \text{ and nontrivial real function } \rho \in H^{-1/2}(\Gamma) \\ &\text{such that} \\ &\qquad \int_\Gamma \mathrm{Re}\Big\{E_n(\boldsymbol{x}, \boldsymbol{y}; k)\Big\}\rho(\boldsymbol{y})\mathrm{d}s_y = 0, \quad \forall\, \boldsymbol{x} \in \Gamma \\ &\text{and} \\ &\qquad \int_\Gamma \mathrm{Im}\Big\{E_n(\boldsymbol{x}, \boldsymbol{y}; k)\Big\}\rho(\boldsymbol{y})\mathrm{d}s_y = 0, \quad \forall\, \boldsymbol{x} \in \Gamma. \end{aligned}\right\} \tag{5.4.8}$$

We have known that for positive wave number k,

$$\mathrm{Re}\Big\{E_n(\boldsymbol{x}, \boldsymbol{y}; k)\Big\} = \begin{cases} -\dfrac{1}{4} Y_0(k|\boldsymbol{x} - \boldsymbol{y}|), & n = 2, \\ \dfrac{\cos(k|\boldsymbol{x} - \boldsymbol{y}|)}{4\pi|\boldsymbol{x} - \boldsymbol{y}|}, & n = 3, \end{cases}$$

$$\mathrm{Im}\Big\{E_n(\boldsymbol{x}, \boldsymbol{y}; k)\Big\} = \begin{cases} -\dfrac{1}{4} J_0(k|\boldsymbol{x} - \boldsymbol{y}|), & n = 2, \\ \dfrac{\sin(k|\boldsymbol{x} - \boldsymbol{y}|)}{4\pi|\boldsymbol{x} - \boldsymbol{y}|}, & n = 3. \end{cases}$$

Therefore the nonlinear boundary eigenvalue problem (5.4.8) can be written as following:

For the case $n = 2$:

$$\left.\begin{aligned}
&\text{Find positive wave number } k \text{ and nontrivial real function } \rho \in H^{-1/2}(\Gamma)\\
&\text{such that}\\
&\qquad -\int_\Gamma \frac{1}{4} Y_0(k|\boldsymbol{x}-\boldsymbol{y}|)\rho(\boldsymbol{y})\mathrm{d}s_y = 0, \quad \forall\, \boldsymbol{x} \in \Gamma\\
&\text{and}\\
&\qquad \int_\Gamma \frac{1}{4} J_0(k|\boldsymbol{x}-\boldsymbol{y}|)\rho(\boldsymbol{y})\mathrm{d}s_y = 0, \quad \forall\, \boldsymbol{x} \in \Gamma.
\end{aligned}\right\} \tag{5.4.9}$$

For the case $n = 3$ (Steinbach & Unger, 2009),

$$\left.\begin{aligned}
&\text{Find positive wave number } k \text{ and nontrivial real function } \rho \in H^{-1/2}(\Gamma)\\
&\text{such that}\\
&\qquad \int_\Gamma \frac{\cos(k|\boldsymbol{x}-\boldsymbol{y}|)}{4\pi|\boldsymbol{x}-\boldsymbol{y}|}\rho(\boldsymbol{y})\mathrm{d}s_y = 0, \quad \forall\, \boldsymbol{x} \in \Gamma\\
&\text{and}\\
&\qquad \int_\Gamma \frac{\sin(k|\boldsymbol{x}-\boldsymbol{y}|)}{4\pi|\boldsymbol{x}-\boldsymbol{y}|}\rho(\boldsymbol{y})\mathrm{d}s_y = 0, \quad \forall\, \boldsymbol{x} \in \Gamma.
\end{aligned}\right\} \tag{5.4.10}$$

Furthermore we obtain the following nonlinear boundary eigenvalue problems involved only the real part of the fundamental solution $E_n(\boldsymbol{x}, \boldsymbol{y}; k)$.

For the case $n = 2$,

$$\left.\begin{aligned}
&\text{Find positive wave number } k \text{ and nontrivial real function } \rho \in H^{-1/2}(\Gamma)\\
&\text{such that}\\
&\qquad -\int_\Gamma \frac{1}{4} Y_0(k|\boldsymbol{x}-\boldsymbol{y}|)\rho(\boldsymbol{y})\mathrm{d}s_y = 0, \quad \forall\, \boldsymbol{x} \in \Gamma\\
&\text{and}\\
&\qquad -\frac{1}{2}\rho(\boldsymbol{x}) + \int_\Gamma \frac{1}{4}\frac{\partial}{\partial \boldsymbol{\nu}_x} Y_0(k|\boldsymbol{x}-\boldsymbol{y}|)\rho(\boldsymbol{y})\mathrm{d}s_y = 0, \quad \forall\, \boldsymbol{x} \in \Gamma.
\end{aligned}\right\} \tag{5.4.11}$$

For the case $n = 3$,

$$\left.\begin{array}{l}\text{Find positive wave number } k \text{ and nontrivial real function } \rho \in H^{-1/2}(\Gamma) \\ \text{such that} \\ \qquad \displaystyle\int_\Gamma \frac{\cos(k|\boldsymbol{x}-\boldsymbol{y}|)}{4\pi|\boldsymbol{x}-\boldsymbol{y}|}\rho(\boldsymbol{y})\mathrm{d}s_y = 0, \ \ \forall\, \boldsymbol{x} \in \Gamma \\ \text{and} \\ \qquad \displaystyle -\frac{1}{2}\rho(\boldsymbol{x}) + \int_\Gamma \frac{\partial}{\boldsymbol{\nu}_x}\frac{\cos(k|\boldsymbol{x}-\boldsymbol{y}|)}{4\pi|\boldsymbol{x}-\boldsymbol{y}|}\rho(\boldsymbol{y})\mathrm{d}s_y = 0, \ \ \forall\, \boldsymbol{x} \in \Gamma. \end{array}\right\} \tag{5.4.12}$$

Theorem 5.4.2 *The problem (5.4.11) (n=2) and problem (5.4.12) (n=3) are equivalent to the interior Dirichlet eigenvalue problem (5.1.19).*

This proof is similar to the proof of the Theorem 5.4.1, which is omitted here.

5.4.2 *Interior Neuamann Eigenvalue Problem of Laplace Operator*

Assume that $\lambda_j^N > 0\,(2 \leqslant j < +\infty)$ is an eigenvalue of interior Neumann eigenvalue problem (5.1.20) corresponding to real eigenfunction $u_j^N \in V_j^N$ and $u_j^N \in H^1(\Omega)$. On the domain Ω, the nontrivial real function $u_j^N(\boldsymbol{x})$ satisfies

$$\begin{cases} -\Delta u_j^N(\boldsymbol{x}) - \lambda_j^N u_j^N(\boldsymbol{x}) = 0, & \forall\, \boldsymbol{x} \in \Omega, \\ \left.\dfrac{\partial u_j^N(\boldsymbol{x})}{\partial \boldsymbol{\nu}_x}\right|_{\Gamma-0} = 0, & \forall\, \boldsymbol{x} \in \Gamma. \end{cases} \tag{5.4.13}$$

It means that $u_j^N \in H^1(\Omega)$ is a solution of Helmholtz equation with wave number $k = \sqrt{\lambda_j^N}$ on the domain Ω. Let

$$\left. u_j^N(\boldsymbol{x})\right|_{\Gamma-0} = -\mu_j^N(\boldsymbol{x}), \quad \forall\, \boldsymbol{x} \in \Gamma, \tag{5.4.14}$$

$\mu_j^N \in H^{1/2}(\Gamma)$ is a nontrivial real function.

Again the representation theorem (5.3.5) yields that $k = \sqrt{\lambda_j^N} > 0$ and the nontrivial real function $\mu_j^N \in H^{1/2}(\Gamma)$ satisfies

$$\frac{\partial}{\partial \boldsymbol{\nu}_x}\int_\Gamma \frac{\partial}{\partial \boldsymbol{\nu}_y} E_n(\boldsymbol{x}, \boldsymbol{y}; k)\mu_j^N(\boldsymbol{y})\mathrm{d}s_y = 0, \ \ \forall\, \boldsymbol{x} \in \Gamma.$$

Namely we obtain the following nonlinear boundary eigenvalue problem:

$$\left.\begin{array}{l}\text{Find positive wave number } k \text{ and nontrivial real function } \mu \in H^{1/2}(\Gamma) \\ \text{such that} \\ \qquad \dfrac{\partial}{\partial \boldsymbol{\nu}_x}\displaystyle\int_\Gamma \frac{\partial}{\partial \boldsymbol{\nu}_y} E_n(\boldsymbol{x}, \boldsymbol{y}; k)\mu_j^N(\boldsymbol{y})\mathrm{d}s_y = 0, \quad \forall\, \boldsymbol{x} \in \Gamma.\end{array}\right\} \tag{5.4.15}$$

Theorem 5.4.3 *Suppose that $k > 0$ is an eigenvalue of problem (5.4.15) corresponding to the eigenfunction $\mu \in H^{1/2}(\Gamma)$, then $k^2 > 0$ is an eigenvalue of interior Neumann eigenvalue problem (5.1.20) corresponding to eigenfunction*

$$u_\mu(\boldsymbol{x}) = \int_\Gamma \frac{\partial}{\partial \boldsymbol{\nu}_y} E_n(\boldsymbol{x}, \boldsymbol{y}; k)\mu(\boldsymbol{y})\mathrm{d}s_{\boldsymbol{y}}, \quad \forall\, \boldsymbol{x} \in \Omega.$$

Proof Let

$$u_\mu(\boldsymbol{x}) = \int_\Gamma \frac{\partial}{\partial \boldsymbol{\nu}_y} E_n(\boldsymbol{x}, \boldsymbol{y}; k)\mu(\boldsymbol{y})\mathrm{d}s_{\boldsymbol{y}}, \quad \forall\, \boldsymbol{x} \in \mathbb{R}^n \backslash \Gamma.$$

On the domain Ω_e, $u_\mu(\boldsymbol{x})$ is the solution of exterior Neumann boundary value problem (5.1.4) with given function $h(\boldsymbol{x}) = 0, \forall \boldsymbol{x} \in \Gamma$.

The uniqueness of problem (5.1.4) yields

$$u_\mu(\boldsymbol{x}) \equiv 0, \quad \forall\, \boldsymbol{x} \in \Omega_e$$

and

$$u_\mu(\boldsymbol{x})\Big|_{\Gamma+0} = \frac{1}{2}\mu(\boldsymbol{x}) + \int_\Gamma \frac{\partial}{\partial \boldsymbol{\nu}_y} E_n(\boldsymbol{x}, \boldsymbol{y}; k)\mu(\boldsymbol{y})\mathrm{d}s_y = 0, \quad \forall\, \boldsymbol{x} \in \Gamma. \tag{5.4.16}$$

On the domain Ω, $u_\mu(\boldsymbol{x})$ satisfies

$$\left.\begin{aligned} -\Delta u_\mu(\boldsymbol{x}) - k^2 u_\mu(\boldsymbol{x}) &= 0, \ \text{ in } \Omega, \\ \frac{\partial u_\mu(\boldsymbol{x})}{\partial \boldsymbol{\nu}_x}\Big|_{\Gamma-0} &= 0, \ \ \forall\, \boldsymbol{x} \in \Gamma, \\ u_\mu(\boldsymbol{x})\Big|_{\Gamma-0} &= -\mu(\boldsymbol{x}). \end{aligned}\right\} \tag{5.4.17}$$

It means $k^2 > 0$ is an eigenvalue of interior Neumann eigenvalue problem (5.1.20) corresponding to the eigenfunction $u_\mu \in H^1(\Omega)$. Taking the real parts form Eqs. (5.4.15) and (5.4.16), we obtain an equivalent nonlinear boundary eigenvalue problem involved only $\mathrm{Re}\big\{E_n(\boldsymbol{x}, \boldsymbol{y}; k)\big\}$ of the problem (5.4.15)

$$\left.\begin{array}{l}\text{Find positive wave number } k \text{ and nontrivial real function } \mu \in H^{1/2}(\Gamma)\\ \text{such that}\\ \qquad \dfrac{\partial}{\partial \boldsymbol{\nu}_{\boldsymbol{x}}}\displaystyle\int_{\Gamma}\dfrac{\partial}{\partial \boldsymbol{\nu}_{\boldsymbol{y}}}\mathrm{Re}\{E_n(\boldsymbol{x},\boldsymbol{y};k)\}\mu(\boldsymbol{y})\mathrm{d}s_{\boldsymbol{y}} = 0, \quad \forall\, \boldsymbol{x}\in\Gamma\\ \text{and}\\ \qquad \dfrac{1}{2}\mu(\boldsymbol{x}) + \displaystyle\int_{\Gamma}\dfrac{\partial}{\boldsymbol{\nu}_{\boldsymbol{y}}}\mathrm{Re}\{E_n(\boldsymbol{x},\boldsymbol{y};k)\}\mu(\boldsymbol{y})\mathrm{d}s_{\boldsymbol{y}} = 0, \quad \forall\, \boldsymbol{x}\in\Gamma.\end{array}\right\} \tag{5.4.18}$$

□

Theorem 5.4.4 *Suppose that $k > 0$ is an eigenvalue of problem (5.4.18) corresponding eigenfunction $\mu \in H^{1/2}(\Gamma)$, then k^2 is an eigenvalue of interior Neumann eigenvalue problem (5.1.20) corresponding to eigenfunction*

$$u_{\mu}(\boldsymbol{x}) = \int_{\Gamma}\frac{\partial}{\partial \boldsymbol{\nu}_{\boldsymbol{y}}}\mathrm{Re}\{E_n(\boldsymbol{x},\boldsymbol{y};k)\}\mu(\boldsymbol{y})\mathrm{d}s_{\boldsymbol{y}}, \quad \forall\, \boldsymbol{x}\in\Omega.$$

Otherwise assume that $\lambda_j^N > 0$ $(2 \leqslant j < +\infty)$ is an eigenvalue of interior Neumann eigenvalue problem (5.1.20) corresponding real eigenfunction $u_j^N \in V_j^D$, then $k = \sqrt{\lambda_j^N} > 0$ is an eigenvalue of problem (5.4.18) corresponding to the eigenfunction

$$\mu(\boldsymbol{x}) = -u_j^N(\boldsymbol{x})\Big|_{\Gamma-0}, \quad \forall\, \boldsymbol{x}\in\Gamma.$$

The proof of this Theorem is omitted, which is similar to the proof of Theorem 5.4.3.

Furthermore by the Theorem 5.2.3 conclusion (iv), the problem (5.4.18) can be rewritten as

$$\left.\begin{array}{l}\text{Find positive wave number } k \text{ and nontrivial real function } \mu \in H^{1/2}(\Gamma)\\ \text{such that}\\ \qquad \displaystyle\sum_{1\leqslant i<j\leqslant n}\dfrac{\partial}{\partial \boldsymbol{\tau}_{\boldsymbol{x}}^{i,j}}\int_{\Gamma}\mathrm{Re}\{E_n(\boldsymbol{x},\boldsymbol{y};k)\}\dfrac{\partial}{\partial \boldsymbol{\tau}_{\boldsymbol{y}}^{i,j}}\mu(\boldsymbol{y})\mathrm{d}s_{\boldsymbol{y}} +\\ \qquad\qquad k^2\displaystyle\int_{\Gamma}\mathrm{Re}\{E_n(\boldsymbol{x},\boldsymbol{y};k)\}\big(\boldsymbol{\nu}(\boldsymbol{x})\cdot\boldsymbol{\nu}(\boldsymbol{y})\big)\mu(\boldsymbol{y})\mathrm{d}s_{\boldsymbol{y}} = 0, \quad \forall\, \boldsymbol{x}\in\Gamma\\ \text{and}\\ \qquad \dfrac{1}{2}\mu(\boldsymbol{x}) + \displaystyle\int_{\Gamma}\dfrac{\partial}{\boldsymbol{\nu}_{\boldsymbol{y}}}\mathrm{Re}\{E_n(\boldsymbol{x},\boldsymbol{y};k)\}\mu(\boldsymbol{y})\mathrm{d}s_{\boldsymbol{y}} = 0, \quad \forall\, \boldsymbol{x}\in\Gamma.\end{array}\right\} \tag{5.4.19}$$

For the Dirichlet-Neumann mixed eigenvalue problem of Laplace operator, a boundary integral algorithm was given by Akhmetgaliyev et al. (2015).

References

Akhmetgaliyev, E., Bruno, O. P., & Nigam, N. (2015). A boundary integral algorithm for the Laplace Dirichlet-Neumann mixed eigenvalue problem. *Journal of Computational Physics, 298*, 1–28.

Brakhage, H., & Werner, P. (1965). Über das Dirichletsche außenraumproblem für die helmholt z sche schwingungsgleichung. *Archiv der Mathematik, 16*, 325–329.

Colton D. & Kress R. (2013). *Integral equation methods in scattering theory*. SIAM.

Courant, R., & Hilbert, D. (2008). *Methods of mathematical physics: Partial differential equations*. Wiley.

Hsiao, G. C., & Wendland, W. L. (2008). *Boundary integral equations*. Springer.

Jones D. (1979). *Methods in electromagnetic wave propagation*. Clarendon Pr.

Kupradze V. D. (1967). *Potential methods in elasticity theory* (Technical Report).

Leis, R. (1958). Über das Neumannsche Randwertproblem für die Helmholtzsche schwingungsgleichung. *Archive for Rational Mechanics and Analysis, 2*, 101–113.

Leis, R. (1965). Zur Dirichletschen Randwertaufgabe des aussenraumes der schwingun- gsgleichung. *Mathematische Zeitschrift, 90*, 205–211.

Nédélec J. C. (2001). *Acoustic and electromagnetic equations: integral representations for harmonic problems* (Vol. 144). Springer.

Panich, O. (1965). On the question of the solvability of the exterior boundary-value problems for the wave equation and maxwells equations. *Usp. Mat. Nauk A, 20*, 221–226.

Stakgold, I., & Holst, M. J. (2011). *Green's functions and boundary value problems*. Wiley.

Steinbach, O., & Unger, G. (2009). A boundary element method for the Dirichlet eigenvalue problem of the Laplace operator. *Numerische Mathematik, 113*, 281–298.

Vekua I. N. (1967). *New methods for solving elliptic equations* (Vol. 1). North-Holland Publishing Company.

Chapter 6
Boundary Value Problems of the Navier Equations

The boundary reductions of boundary value problems of the Navier system are the main goal of this chapter. We first recall some basic boundary value problems of the Navier system, next discuss the single and double layer potentials corresponding to the Navier system, and then reduce the basic boundary value problems to boundary integral equations or boundary integro-differential equations.

6.1 Some Basic Boundary Value Problems

For the Navier system (2.4.10) with $\boldsymbol{f}(\boldsymbol{x}) = \boldsymbol{0}$, we have (Sokolnikoff et al., 1956)

$$-\mathcal{L}_n(D)\boldsymbol{u}(\boldsymbol{x}) = \boldsymbol{0}, \tag{6.1.1}$$

with

$$\begin{aligned} \mathcal{L}_n(D) = & \ (L_{ij}(D))_{n\times n}, \\ L_{ij}(D) = & \ \mu\,\delta_{ij}\,\Delta + (\lambda+\mu)\frac{\partial^2}{\partial x_i \partial x_j}, \quad 1 \leqslant i,\ j \leqslant n. \end{aligned}$$

The fundamental solution $\mathbb{E}(\boldsymbol{x}, \boldsymbol{y})$ of the Navier system is given by (2.4.19) ($n = 2$) and (2.4.26) ($n = 3$). It reads

$$\begin{aligned} \mathbb{E}(\boldsymbol{x}, \boldsymbol{y}) = & \ \frac{\lambda+3\mu}{4(n-1)\pi\mu(\lambda+2\mu)}\left[\gamma_n(\boldsymbol{x}, \boldsymbol{y})\,\boldsymbol{I} + \frac{\lambda+\mu}{\lambda+3\mu}\frac{(\boldsymbol{x}-\boldsymbol{y})(\boldsymbol{x}-\boldsymbol{y})^{\mathrm{T}}}{|\boldsymbol{x}-\boldsymbol{y}|^n}\right] \\ := & \ (E_{ij}(\boldsymbol{x}, \boldsymbol{y}))_{n\times n} \end{aligned} \tag{6.1.2}$$

with

H. Han and D. Yin, *Mathematical Foundation of the Boundary Integro-Differential Equation Method*, https://doi.org/10.1007/978-981-95-1088-7_6

$$\gamma_n(\boldsymbol{x}, \boldsymbol{y}) = \begin{cases} -\ln|\boldsymbol{x}-\boldsymbol{y}|, & \text{if } n=2, \\ \dfrac{1}{|\boldsymbol{x}-\boldsymbol{y}|}, & \text{if } n=3. \end{cases}$$

The Navier system can be rewritten in the vector form:

$$-\mu\, \Delta \boldsymbol{u}(\boldsymbol{x}) - (\lambda+\mu)\nabla(\nabla\cdot\boldsymbol{u}(\boldsymbol{x})) = \boldsymbol{0}, \tag{6.1.3}$$

or in another form in terms of the stress tensor $\boldsymbol{\sigma}$:

$$-\nabla\cdot\boldsymbol{\sigma}(\boldsymbol{u}(\boldsymbol{x})) = \boldsymbol{0} \tag{6.1.4}$$

with

$$\nabla\cdot\boldsymbol{\sigma}(\boldsymbol{u}) = \begin{pmatrix} \sum\limits_{k=1}^{n} \dfrac{\partial\sigma_{1k}(\boldsymbol{u})}{\partial x_k} \\ \sum\limits_{k=1}^{n} \dfrac{\partial\sigma_{2k}(\boldsymbol{u})}{\partial x_k} \\ \vdots \\ \sum\limits_{k=1}^{n} \dfrac{\partial\sigma_{nk}(\boldsymbol{u})}{\partial x_k} \end{pmatrix}_{n\times 1}.$$

For $\boldsymbol{x}\in\mathbb{R}^n$ $(n=2,\,3)$ and an arbitrary unit vector $\boldsymbol{\nu}(\boldsymbol{x}) = (\nu_1(\boldsymbol{x}),\, \nu_2(\boldsymbol{x}),\, \ldots,\, \nu_n(\boldsymbol{x}))^{\mathrm{T}}$, the matrix differential operator $\mathbb{T}(D_x, \boldsymbol{\nu}(\boldsymbol{x}))$ is defined by Kupradze (2012)

$$\begin{cases} \mathbb{T}(D_x, \boldsymbol{\nu}(\boldsymbol{x})) = \left(T_{ij}(D_x, \boldsymbol{\nu}(\boldsymbol{x}))\right)_{n\times n} \\ \text{with} \\ T_{ij}(D_x, \boldsymbol{\nu}(\boldsymbol{x})) = \lambda\,\nu_i(\boldsymbol{x})\dfrac{\partial}{\partial x_j} + \mu\,\nu_j(\boldsymbol{x})\dfrac{\partial}{\partial x_i} + \mu\,\delta_{ij}\dfrac{\partial}{\partial\boldsymbol{\nu}(\boldsymbol{x})}, \quad 1\leqslant i,\ j\leqslant n. \end{cases} \tag{6.1.5}$$

$\mathbb{T}(D_x, \boldsymbol{\nu}(\boldsymbol{x}))$ is called the stress operator. $\mathbb{T}(D_x, \boldsymbol{\nu}(\boldsymbol{x}))\boldsymbol{u}(\boldsymbol{x})$ denotes the stress at point $\boldsymbol{x}$ along the direction $\boldsymbol{\nu}(\boldsymbol{x})$ in terms of the displacement $\boldsymbol{u}(\boldsymbol{x}) = (u_1(\boldsymbol{x}),\, u_2(\boldsymbol{x}),\, \ldots,\, u_n(\boldsymbol{x}))^{\mathrm{T}}$. Namely,

$$\mathbb{T}(D_x,\, \boldsymbol{\nu}(\boldsymbol{x}))\boldsymbol{u}(\boldsymbol{x}) = (\boldsymbol{\sigma}(\boldsymbol{u}))\,\boldsymbol{\nu}(\boldsymbol{x}). \tag{6.1.6}$$

Furthermore, we introduce the differential operator (Günter, 1967)

$$\begin{cases} \mathbb{U}(D_x,\ \boldsymbol{\nu}(\boldsymbol{x})) = (U_{ij}(D_x, \boldsymbol{\nu}(\boldsymbol{x}))_{n\times n} \\ \text{with} \\ U_{ij}(D_x, \boldsymbol{\nu}(\boldsymbol{x})) = \nu_j(\boldsymbol{x})\dfrac{\partial}{\partial x_i} - \nu_i(\boldsymbol{x})\dfrac{\partial}{\partial x_j}, \qquad 1 \leqslant i,\ j \leqslant n. \end{cases} \tag{6.1.7}$$

Note that

$$\begin{cases} U_{ij}(D_x, \boldsymbol{\nu}(\boldsymbol{x})) = 0, \qquad 1 \leqslant i \leqslant n, & (6.1.8a) \\ U_{ij}(D_x, \boldsymbol{\nu}(\boldsymbol{x})) = -U_{ji}(D_x,\ \boldsymbol{\nu}(\boldsymbol{x})), \qquad 1 \leqslant i,\ j \leqslant n. & (6.1.8b) \end{cases}$$

Therefore $\mathbb{U}(D_x, \boldsymbol{\nu}(\boldsymbol{x}))$ is a skew-symmetric matrix differential operator. Specially for $n = 2$,

$$U_{21}(D_x, \boldsymbol{\nu}(\boldsymbol{x})) = -U_{12}(D_x, \boldsymbol{\nu}(\boldsymbol{x})) = \nu_1(\boldsymbol{x})\frac{\partial}{\partial x_2} - \nu_2(\boldsymbol{x})\frac{\partial}{\partial x_1} = \frac{\partial}{\partial \boldsymbol{\tau}(\boldsymbol{x})}$$

with the unit vector $\boldsymbol{\tau}(\boldsymbol{x}) = (-\nu_2(\boldsymbol{x}), \nu_1(\boldsymbol{x}))^{\mathrm{T}}$, perpendicular to the unit vector $\boldsymbol{\nu}(\boldsymbol{x})$. For the case $n = 3$, we have

$$\left.\begin{aligned} &U_{32}(D_x, \boldsymbol{\nu}(\boldsymbol{x})) = -\,U_{23}(D_x, \boldsymbol{\nu}(\boldsymbol{x})) = \nu_2(\boldsymbol{x})\frac{\partial}{\partial x_3} - \nu_3(\boldsymbol{x})\frac{\partial}{\partial x_2} = \frac{\partial}{\partial \boldsymbol{\tau}_1(\boldsymbol{x})}, \\ &U_{13}(D_x, \boldsymbol{\nu}(\boldsymbol{x})) = -\,U_{31}(D_x, \boldsymbol{\nu}(\boldsymbol{x})) = \nu_3(\boldsymbol{x})\frac{\partial}{\partial x_1} - \nu_1(\boldsymbol{x})\frac{\partial}{\partial x_3} = \frac{\partial}{\partial \boldsymbol{\tau}_2(\boldsymbol{x})}, \\ &U_{21}(D_x, \boldsymbol{\nu}(\boldsymbol{x})) = -\,U_{12}(D_x, \boldsymbol{\nu}(\boldsymbol{x})) = \nu_1(\boldsymbol{x})\frac{\partial}{\partial x_2} - \nu_2(\boldsymbol{x})\frac{\partial}{\partial x_1} = \frac{\partial}{\partial \boldsymbol{\tau}_3(\boldsymbol{x})}, \\ &\text{with} \quad \boldsymbol{\tau}_1(\boldsymbol{x}) = (0,\ -\nu_3(\boldsymbol{x}),\ \nu_2(\boldsymbol{x}))^{\mathrm{T}} := \boldsymbol{\tau}_{1,x}, \\ &\qquad\quad \boldsymbol{\tau}_2(\boldsymbol{x}) = (\nu_3(\boldsymbol{x}),\ 0,\ -\nu_1(\boldsymbol{x}))^{\mathrm{T}} := \boldsymbol{\tau}_{2,x}, \\ &\qquad\quad \boldsymbol{\tau}_3(\boldsymbol{x}) = (-\nu_2(\boldsymbol{x}),\ \nu_1(\boldsymbol{x}),\ 0)^{\mathrm{T}} := \boldsymbol{\tau}_{3,x}, \end{aligned}\right\} \tag{6.1.9}$$

which are perpendicular to the unit vector $\boldsymbol{\nu}(\boldsymbol{x})$.

Note that the strain tensor $\boldsymbol{\epsilon}(\boldsymbol{u})$ has a nontrivial kernel. Let

$$\mathbb{RM} = \left\{ \boldsymbol{v} \,\middle|\, \boldsymbol{\epsilon}(\boldsymbol{v}(\boldsymbol{x})) \equiv \boldsymbol{0}, \quad \forall \boldsymbol{x} \in \mathbb{R}^n \right\} \tag{6.1.10}$$

be the space of infinitesimal rigid motions. The dimension of the space $\mathbb{RM}$ is $3(n-1)$ and

$$\mathbb{RM} = \left\{ \boldsymbol{v} \,\middle|\, \boldsymbol{v}(\boldsymbol{x}) = \sum_{j=1}^{3(n-1)} m_j \boldsymbol{\omega}_j(\boldsymbol{x}) \right\} \tag{6.1.11}$$

with in the case $n = 2$,

$$\boldsymbol{\omega}_1(\boldsymbol{x}) = (1,\ 0)^{\mathrm{T}}, \quad \boldsymbol{\omega}_2(\boldsymbol{x}) = (0,\ 1)^{\mathrm{T}}, \quad \boldsymbol{\omega}_3(\boldsymbol{x}) = (-x_2,\ x_1)^{\mathrm{T}}; \tag{6.1.12}$$

and in the case $n = 3$,

$$\begin{cases} \omega_1(\boldsymbol{x}) = (1,\ 0,\ 0)^{\mathrm{T}}, \quad \omega_2(\boldsymbol{x}) = (0,\ 1,\ 0)^{\mathrm{T}}, \quad \omega_3(\boldsymbol{x}) = (0,\ 0,\ 1)^{\mathrm{T}}, \\ \omega_4(\boldsymbol{x}) = (0,\ -x_3,\ x_2)^{\mathrm{T}}, \quad \omega_5(\boldsymbol{x}) = (x_3,\ 0,\ -x_1)^{\mathrm{T}}, \quad \omega_6(\boldsymbol{x}) = (-x_2,\ x_1,\ 0)^{\mathrm{T}}. \end{cases} \tag{6.1.13}$$

Now we are ready to recount the following basic boundary value problems of the Navier system.

The interior displacement problem reads

$$\begin{cases} -\nabla \cdot \boldsymbol{\sigma}(\boldsymbol{u}(\boldsymbol{x})) = \boldsymbol{0}, \quad \forall \boldsymbol{x} \in \Omega, & (6.1.14a) \\ \boldsymbol{u}(\boldsymbol{x}) = g(\boldsymbol{x}), \quad \forall \boldsymbol{x} \in \Gamma, & (6.1.14b) \end{cases}$$

with given vector function $g(\boldsymbol{x})$ defined on Γ.

The interior traction problem reads

$$\begin{cases} -\nabla \cdot \boldsymbol{\sigma}(\boldsymbol{u}(\boldsymbol{x})) = \boldsymbol{0}, \quad \forall \boldsymbol{x} \in \Omega, & (6.1.15a) \\ \boldsymbol{\sigma}(\boldsymbol{u}(\boldsymbol{x}))\, \boldsymbol{\nu}(\boldsymbol{x}) = \boldsymbol{h}(\boldsymbol{x}), \quad \forall \boldsymbol{x} \in \Gamma, & (6.1.15b) \end{cases}$$

with $\boldsymbol{\nu}(\boldsymbol{x})$ be the unit outward normal vector of Ω at $\boldsymbol{x} \in \Gamma$, and the given vector function $\boldsymbol{h}(\boldsymbol{x})$ defined on Γ. It is the Neumann boundary value problem for the Navier system in Ω.

For the exterior problems in elasticity, the condition of the solution at infinity must be counted correctly, to guarantee the exterior problems that we considered are properly posed.

Let

$$\begin{cases} \Gamma_R = \left\{ \boldsymbol{x} \in \mathbb{R}^n \mid |\boldsymbol{x}| = R, \quad R \geqslant R_0 \right\} \subset \Omega_e, \\ \Omega_R = \left\{ \boldsymbol{x} \in \mathbb{R}^n \mid \boldsymbol{x} \in \Omega_e \ \text{ and } \ |\boldsymbol{x}| < R \right\}. \end{cases} \tag{6.1.16}$$

The growth condition of the solutions to exterior problems in elasticity was given by Hsiao and Wendland (1985, 2008)

$$\boldsymbol{u}(\boldsymbol{x}) = \mathbb{E}(\boldsymbol{x}, \boldsymbol{0})\boldsymbol{c}_0 + \sum_{j=1}^{3(n-1)} m_j \omega_j(\boldsymbol{x}) + \mathcal{O}(|\boldsymbol{x}|^{1-n}), \qquad \text{as } |\boldsymbol{x}| \to \infty, \tag{6.1.17}$$

with constant vector $\boldsymbol{c}_0 \in \mathbb{R}^n$ and $m_j \in \mathbb{R}$ ($j = 1,\ 2,\ \ldots, 3(n-1)$). It is a general growth condition. The constant vector $\boldsymbol{c}_0$ and constants $\{ m_j,\ 1 \leqslant j \leqslant 3(n-1) \}$ are related each other in general case.

Now we consider the following three basic exterior displacement problems. At first, taking $\boldsymbol{c}_0 = \boldsymbol{0}$, $m_j = 0$ ($n + 1 \leqslant j \leqslant 3(n-1)$) in (6.1.17), we have

$$\begin{cases} -\nabla\cdot\boldsymbol{\sigma}(\boldsymbol{u}(\boldsymbol{x}))=\mathbf{0}, \qquad \forall\boldsymbol{x}\in\Omega_e, & (6.1.18a)\\ \boldsymbol{u}(\boldsymbol{x})=g(\boldsymbol{x}), \qquad \forall\boldsymbol{x}\in\Gamma, & (6.1.18b)\\ \boldsymbol{u}(\boldsymbol{x})=\sum\limits_{j=1}^{n} m_j\boldsymbol{\omega}_j(\boldsymbol{x})+\mathcal{O}(|\boldsymbol{x}|^{1-n}), \quad \text{as } |\boldsymbol{x}|\to+\infty, & (6.1.18c)\\ \int_{\Gamma_R}\big(\boldsymbol{\sigma}(\boldsymbol{u}(\boldsymbol{x}))\boldsymbol{\nu}_R(\boldsymbol{x})\big)^{\mathrm{T}}\boldsymbol{\omega}_j(\boldsymbol{x})\,\mathrm{d}s_x=0, \quad \forall R\geqslant R_0, \quad j=1,2,\ldots,n. & (6.1.18d) \end{cases}$$

In the condition (6.1.18d), $\boldsymbol{\nu}_R(\boldsymbol{x})$ is the outward unit normal vector of Ω_R at $\boldsymbol{x}\in\Gamma_R$.

Taking $\boldsymbol{c}_0=\mathbf{0}$ in (6.1.17), we obtain

$$\begin{cases} -\nabla\cdot\boldsymbol{\sigma}(\boldsymbol{u}(\boldsymbol{x}))=\mathbf{0}, \qquad \forall\boldsymbol{x}\in\Omega_e, & (6.1.19a)\\ \boldsymbol{u}(\boldsymbol{x})=g(\boldsymbol{x}), \qquad \forall\boldsymbol{x}\in\Gamma, & (6.1.19b)\\ \boldsymbol{u}(\boldsymbol{x})=\sum\limits_{j=1}^{3(n-1)} m_j\boldsymbol{\omega}_j(\boldsymbol{x})+\mathcal{O}(|\boldsymbol{x}|^{1-n}), \qquad \text{as } |\boldsymbol{x}|\to+\infty, & (6.1.19c)\\ \int_{\Gamma_R}\big(\boldsymbol{\sigma}(\boldsymbol{u}(\boldsymbol{x}))\boldsymbol{\nu}_R(\boldsymbol{x})\big)^{\mathrm{T}}\boldsymbol{\omega}_j(\boldsymbol{x})\,\mathrm{d}s_x=0, \qquad \text{for } R\geqslant R_0, & \\ \qquad\qquad j=1,2,\ldots,3(n-1). & (6.1.19d) \end{cases}$$

Finally assuming $m_j=0$ for $j=1,2,\ldots,3(n-1)$, we arrive at

$$\begin{cases} -\nabla\cdot\boldsymbol{\sigma}(\boldsymbol{u}(\boldsymbol{x}))=\mathbf{0}, \qquad \forall\boldsymbol{x}\in\Omega_e, & (6.1.20a)\\ \boldsymbol{u}(\boldsymbol{x})=g(\boldsymbol{x}), \qquad \forall\boldsymbol{x}\in\Gamma, & (6.1.20b)\\ \boldsymbol{u}(\boldsymbol{x})=\mathbb{E}(\boldsymbol{x},\mathbf{0})\boldsymbol{c}_0+\mathcal{O}(|\boldsymbol{x}|^{1-n}), \qquad \text{as } |\boldsymbol{x}|\to+\infty. & (6.1.20c) \end{cases}$$

In order to ensure the uniqueness of the exterior traction problems, we must take $m_j=0$ for $j=1,2,\ldots,3(n-1)$ in the condition (6.1.19c). Naturally we have the exterior traction problem

$$\begin{cases} -\nabla\cdot\boldsymbol{\sigma}(\boldsymbol{u}(\boldsymbol{x}))=\mathbf{0}, \qquad \forall\boldsymbol{x}\in\Omega_e, & (6.1.21a)\\ \boldsymbol{\sigma}(\boldsymbol{u}(\boldsymbol{x}))\boldsymbol{\nu}(\boldsymbol{x})=\boldsymbol{h}(\boldsymbol{x}), \qquad \forall\boldsymbol{x}\in\Gamma, & (6.1.21b)\\ \boldsymbol{u}(\boldsymbol{x})=\mathbb{E}(\boldsymbol{x},\mathbf{0})\boldsymbol{c}_0+\mathcal{O}(|\boldsymbol{x}|^{1-n}), \quad \text{as } |\boldsymbol{x}|\to+\infty. & (6.1.21c) \end{cases}$$

Recall by the Betti-Somigliana representation formula, $\boldsymbol{u}(\boldsymbol{x})$, the solution to the interior problem of the homogeneous Navier system, is given by

$$\begin{aligned} \boldsymbol{u}(\boldsymbol{x})=&\int_{\Gamma}\mathbb{E}(\boldsymbol{x},\boldsymbol{y})\big(\mathbb{T}(D_y,\boldsymbol{\nu}(\boldsymbol{y}))\boldsymbol{u}(\boldsymbol{y})\big)\,\mathrm{d}s_y-\\ &\int_{\Gamma}\big(\mathbb{T}(D_y,\boldsymbol{\nu}(\boldsymbol{y}))\mathbb{E}(\boldsymbol{x},\boldsymbol{y})\big)^{\mathrm{T}}\boldsymbol{u}(\boldsymbol{y})\,\mathrm{d}s_y, \ \forall\boldsymbol{x}\in\Omega. \end{aligned} \tag{6.1.22}$$

It means that the solution $\boldsymbol{u}(\boldsymbol{x})$ to the interior problem of the homogeneous Navier system can be represented by two boundary integrals, which are the single layer potential and the double layer potential corresponding to the Navier system.

6.2 Single and Double Layer Potentials of the Navier System

For given n-dimensional vector-valued functions $\boldsymbol{\varphi}(\boldsymbol{x})$ and $\boldsymbol{\psi}(\boldsymbol{x})$ on Γ, the single and double layer potentials are defined correspondingly by

$$\mathbb{V}_I(\boldsymbol{x}) = \int_\Gamma \mathbb{E}(\boldsymbol{x}, \boldsymbol{y})\boldsymbol{\varphi}(\boldsymbol{y})\,\mathrm{d}s_y = \int_\Gamma \mathbb{K}_I(\boldsymbol{x}, \boldsymbol{y})\boldsymbol{\varphi}(\boldsymbol{y})\,\mathrm{d}s_y := (\mathbb{K}_I\boldsymbol{\varphi})(\boldsymbol{x}), \qquad \forall \boldsymbol{x} \in \mathbb{R}^n \tag{6.2.1}$$

with

$$\mathbb{K}_I(\boldsymbol{x}, \boldsymbol{y}) = \mathbb{E}(\boldsymbol{x}, \boldsymbol{y}), \qquad \forall \boldsymbol{x} \in \mathbb{R}^n,\ \boldsymbol{y} \in \Gamma, \tag{6.2.2}$$

and

$$\begin{aligned}\mathbb{V}_{II}(\boldsymbol{x}) &= \int_\Gamma \big(\mathbb{T}(D_y, \boldsymbol{\nu}(\boldsymbol{y}))\mathbb{E}(\boldsymbol{x}, \boldsymbol{y})\big)^{\mathrm{T}}\boldsymbol{\psi}(\boldsymbol{y})\,\mathrm{d}s_y \\ &= \int_\Gamma \mathbb{K}_{II}(\boldsymbol{x}, \boldsymbol{y})\boldsymbol{\psi}(\boldsymbol{y})\,\mathrm{d}s_y := (\mathbb{K}_{II}\boldsymbol{\psi})(\boldsymbol{x}), \qquad \forall \boldsymbol{x} \in \mathbb{R}^n,\end{aligned} \tag{6.2.3}$$

with

$$\mathbb{K}_{II}(\boldsymbol{x}, \boldsymbol{y}) = \big(\mathbb{T}(D_y, \boldsymbol{\nu}(\boldsymbol{y}))\mathbb{E}(\boldsymbol{x}, \boldsymbol{y})\big)^{\mathrm{T}}, \tag{6.2.4}$$

where $\boldsymbol{\nu}(\boldsymbol{x})$ denotes the unit outward normal vector on Γ of Ω at point $\boldsymbol{x} \in \Gamma$. $\mathbb{E}(\boldsymbol{x}, \boldsymbol{y})$ is the fundamental solution of the Navier system given by (6.1.2).

6.2.1 *Single Layer Potential* $\mathbb{V}_I(\boldsymbol{x})$

In this subsection, we discuss the properties of the single layer potential $\mathbb{V}_I(\boldsymbol{x})$ of the Navier system. Note that each element $E_{ij}(\boldsymbol{x}, \boldsymbol{y})$ $(1 \leqslant i,\ j \leqslant n)$ of the fundamental solution $\mathbb{E}(\boldsymbol{x}, \boldsymbol{y})$ is a weakly singular kernel and satisfies the condition (3.3.8). We have the following theorem.

Theorem 6.2.1 *Suppose that* $\Gamma \in C^2$ *and* $\boldsymbol{\varphi} \in (C(\Gamma))^n$.

(i) $\mathbb{V}_I(\boldsymbol{x})$ *is well defined on* $\mathbb{R}^n$. $\mathbb{V}_I(\boldsymbol{x})$ *is continuous on* $\mathbb{R}^n$ *and satisfies the Navier system (6.1.4) in* $\mathbb{R}^n\backslash\Gamma$.

(ii)

$$\mathbb{V}_I(\boldsymbol{x}) = \mathbb{E}(\boldsymbol{x}, \boldsymbol{0})\boldsymbol{c}_0 + \mathcal{O}(|\boldsymbol{x}|^{1-n}), \qquad as\ |\boldsymbol{x}| \to +\infty \tag{6.2.5}$$

with $\boldsymbol{c}_0 = \int_\Gamma \varphi(\boldsymbol{x})\, \mathrm{d}s_x$.

Proof The conclusion (i) is from Theorem 3.3.2 directly. The single layer potential $\mathbb{V}_I(\boldsymbol{x})$ can be rewritten as

$$\mathbb{V}_I(\boldsymbol{x}) = \int_\Gamma \big(\mathbb{E}(\boldsymbol{x}, \boldsymbol{y}) - \mathbb{E}(\boldsymbol{x}, \boldsymbol{0})\big)\varphi(\boldsymbol{y})\, \mathrm{d}s_y + \mathbb{E}(\boldsymbol{x}, \boldsymbol{0})\boldsymbol{c}_0.$$

It is obvious that

$$\int_\Gamma \big(\mathbb{E}(\boldsymbol{x}, \boldsymbol{y}) - \mathbb{E}(\boldsymbol{x}, \boldsymbol{0})\big)\, \varphi(\boldsymbol{y})\, \mathrm{d}s_y = \mathcal{O}(|\boldsymbol{x}|^{1-n}), \qquad \text{as } |\boldsymbol{x}| \to +\infty.$$

This yields (6.2.5). □

6.2.2 *Double Layer Potential* $\mathbb{V}_{II}(\boldsymbol{x})$

We study the properties of the double layer potential $\mathbb{V}_{II}(\boldsymbol{x})$ of the Navier system. By direct computation, we arrive at

Lemma 6.2.1 *The following equalities hold*

1.

$$\begin{aligned}&\mathbb{T}(D_y, \boldsymbol{\nu}(\boldsymbol{y}))\mathbb{E}(\boldsymbol{x}, \boldsymbol{y})\\ &= \frac{\mu}{2(n-1)\pi(\lambda+2\mu)}\left\{\left(\boldsymbol{I} + \frac{n(\lambda+\mu)}{\mu}\frac{(\boldsymbol{x}-\boldsymbol{y})(\boldsymbol{x}-\boldsymbol{y})^{\mathrm{T}}}{|\boldsymbol{x}-\boldsymbol{y}|^n}\right)\frac{\partial\gamma_n(\boldsymbol{x}, \boldsymbol{y})}{\partial\boldsymbol{\nu}(\boldsymbol{y})} + \right.\\ &\left.\frac{(\boldsymbol{x}-\boldsymbol{y})\boldsymbol{\nu}(\boldsymbol{y})^{\mathrm{T}} - \boldsymbol{\nu}(\boldsymbol{y})(\boldsymbol{x}-\boldsymbol{y})^{\mathrm{T}}}{|\boldsymbol{x}-\boldsymbol{y}|^n}\right\},\end{aligned} \tag{6.2.6}$$

$$\begin{aligned}&\mathbb{T}(D_x, \boldsymbol{\nu}(\boldsymbol{x}))\mathbb{E}(\boldsymbol{x}, \boldsymbol{y})\\ &= \frac{\mu}{2(n-1)\pi(\lambda+2\mu)}\left\{\left(\boldsymbol{I} + \frac{n(\lambda+\mu)}{\mu}\frac{(\boldsymbol{x}-\boldsymbol{y})(\boldsymbol{x}-\boldsymbol{y})^{\mathrm{T}}}{|\boldsymbol{x}-\boldsymbol{y}|^n}\right)\frac{\partial\gamma_n(\boldsymbol{x}, \boldsymbol{y})}{\partial\boldsymbol{\nu}(\boldsymbol{x})} - \right.\\ &\left.\frac{(\boldsymbol{x}-\boldsymbol{y})\boldsymbol{\nu}(\boldsymbol{x})^{\mathrm{T}} - \boldsymbol{\nu}(\boldsymbol{x})(\boldsymbol{x}-\boldsymbol{y})^{\mathrm{T}}}{|\boldsymbol{x}-\boldsymbol{y}|^n}\right\}.\end{aligned} \tag{6.2.7}$$

2.

$$\mathbb{U}(D_y,\ \boldsymbol{\nu}(\boldsymbol{y}))\gamma_n(\boldsymbol{x}, \boldsymbol{y}) = \frac{(\boldsymbol{x}-\boldsymbol{y})\boldsymbol{\nu}(\boldsymbol{y})^{\mathrm{T}} - \boldsymbol{\nu}(\boldsymbol{y})(\boldsymbol{x}-\boldsymbol{y})^{\mathrm{T}}}{|\boldsymbol{x}-\boldsymbol{y}|^n}, \tag{6.2.8}$$

$$\mathbb{U}(D_x,\ \boldsymbol{\nu}(\boldsymbol{x}))\gamma_n(\boldsymbol{x}, \boldsymbol{y}) = -\frac{(\boldsymbol{x}-\boldsymbol{y})\boldsymbol{\nu}(\boldsymbol{x})^{\mathrm{T}} - \boldsymbol{\nu}(\boldsymbol{x})(\boldsymbol{x}-\boldsymbol{y})^{\mathrm{T}}}{|\boldsymbol{x}-\boldsymbol{y}|^n}. \tag{6.2.9}$$

3.

$$\mathbb{U}(D_y,\ \boldsymbol{\nu}(\boldsymbol{y}))\left(\frac{(\boldsymbol{x}-\boldsymbol{y})(\boldsymbol{x}-\boldsymbol{y})^{\mathrm{T}}}{|\boldsymbol{x}-\boldsymbol{y}|^n}\right)=\frac{n(\boldsymbol{x}-\boldsymbol{y})(\boldsymbol{x}-\boldsymbol{y})^{\mathrm{T}}}{|\boldsymbol{x}-\boldsymbol{y}|^n}\frac{\partial\gamma_n(\boldsymbol{x},\boldsymbol{y})}{\partial\boldsymbol{\nu}(\boldsymbol{y})}-\frac{\partial\gamma_n(\boldsymbol{x},\boldsymbol{y})}{\partial\boldsymbol{\nu}(\boldsymbol{y})}\boldsymbol{I}, \tag{6.2.10}$$

$$\mathbb{U}(D_x,\ \boldsymbol{\nu}(\boldsymbol{x}))\left(\frac{(\boldsymbol{x}-\boldsymbol{y})(\boldsymbol{x}-\boldsymbol{y})^{\mathrm{T}}}{|\boldsymbol{x}-\boldsymbol{y}|^n}\right)=\frac{n(\boldsymbol{x}-\boldsymbol{y})(\boldsymbol{x}-\boldsymbol{y})^{\mathrm{T}}}{|\boldsymbol{x}-\boldsymbol{y}|^n}\frac{\partial\gamma_n(\boldsymbol{x},\boldsymbol{y})}{\partial\boldsymbol{\nu}(\boldsymbol{x})}-\frac{\partial\gamma_n(\boldsymbol{x},\boldsymbol{y})}{\partial\boldsymbol{\nu}(\boldsymbol{x})}\boldsymbol{I}. \tag{6.2.11}$$

Substituting (6.2.8), (6.2.10) into (6.2.6) yields

$$\mathbb{T}(D_y,\ \boldsymbol{\nu}(\boldsymbol{y}))\mathbb{E}(\boldsymbol{x},\boldsymbol{y})=\frac{1}{2(n-1)\pi}\left\{\frac{\partial\gamma_n(\boldsymbol{x},\boldsymbol{y})}{\partial\boldsymbol{\nu}(\boldsymbol{y})}\boldsymbol{I}-\mathbb{U}(D_y,\ \boldsymbol{\nu}(\boldsymbol{y}))\gamma_n(\boldsymbol{x},\boldsymbol{y})\right\}+2\mu\mathbb{U}(D_y,\ \boldsymbol{\nu}(\boldsymbol{x}))\mathbb{E}(\boldsymbol{x},\boldsymbol{y}) \tag{6.2.12}$$

and

$$\left(\mathbb{T}(D_y,\ \boldsymbol{\nu}(\boldsymbol{y}))\mathbb{E}(\boldsymbol{x},\boldsymbol{y})\right)^{\mathrm{T}}=\frac{1}{2(n-1)\pi}\left\{\frac{\partial\gamma_n(\boldsymbol{x},\boldsymbol{y})}{\partial\boldsymbol{\nu}(\boldsymbol{y})}\boldsymbol{I}+\mathbb{U}(D_y,\ \boldsymbol{\nu}(\boldsymbol{y}))\gamma_n(\boldsymbol{x},\boldsymbol{y})\right\}+2\mu\left(\mathbb{U}(D_y,\ \boldsymbol{\nu}(\boldsymbol{y}))\mathbb{E}(\boldsymbol{x},\boldsymbol{y})\right)^{\mathrm{T}}. \tag{6.2.13}$$

Similarly, we have

$$\mathbb{T}(D_x,\ \boldsymbol{\nu}(\boldsymbol{x}))\mathbb{E}(\boldsymbol{x},\boldsymbol{y})=\frac{1}{2(n-1)\pi}\left\{\frac{\partial\gamma_n(\boldsymbol{x},\boldsymbol{y})}{\partial\boldsymbol{\nu}(\boldsymbol{x})}\boldsymbol{I}-\mathbb{U}(D_x,\ \boldsymbol{\nu}(\boldsymbol{x}))\gamma_n(\boldsymbol{x},\boldsymbol{y})\right\}+2\mu\mathbb{U}(D_x,\ \boldsymbol{\nu}(\boldsymbol{x}))\mathbb{E}(\boldsymbol{x},\boldsymbol{y}). \tag{6.2.14}$$

The double layer potential $\mathbb{V}_{II}(\boldsymbol{x})$ can be rewritten as follows after integrating by parts in the form of Stokes theorem (Kupradze, 2012) in (6.2.13),

$$\mathbb{V}_{II}(\boldsymbol{x})=\frac{1}{2(n-1)\pi}\left\{\int_\Gamma\frac{\partial\gamma_n(\boldsymbol{x},\boldsymbol{y})}{\partial\boldsymbol{\nu}(\boldsymbol{y})}\boldsymbol{\psi}(\boldsymbol{y})\,\mathrm{d}s_y-\int_\Gamma\gamma_n(\boldsymbol{x},\boldsymbol{y})\mathbb{U}(D_y,\ \boldsymbol{\nu}(\boldsymbol{y}))\boldsymbol{\psi}(\boldsymbol{y})\,\mathrm{d}s_y\right\}+2\mu\int_\Gamma\mathbb{E}(\boldsymbol{x},\boldsymbol{y})\left(\mathbb{U}(D_y,\ \boldsymbol{\nu}(\boldsymbol{y}))\boldsymbol{\psi}(\boldsymbol{y})\right)\mathrm{d}s_y. \tag{6.2.15}$$

In (6.2.15), the double layer potential $\mathbb{V}_{II}(\boldsymbol{x})$ is expressed as a sum of the harmonic double layer potential, the harmonic single layer potential, and the single layer potential of the Navier system.

Based on the expression (6.2.15) and the properties of the harmonic single layer potential, harmonic double layer potential, and single layer potential of the Navier system, we have

Theorem 6.2.2 *Suppose that the boundary $\Gamma \in C^2$ and the given density $\psi(x) \in (C^1(\Gamma))^n$, then*

1. *The double layer potential $\mathbb{V}_{II}(x)$ of the Navier system is well-defined for all $x \in \mathbb{R}^n$. It satisfies the homogeneous Navier system on $\mathbb{R}^n \backslash \Gamma$, i.e.,*

$$-\nabla \cdot \sigma(\mathbb{V}_{II}(x)) = \mathbf{0}, \qquad \forall x \in \mathbb{R}^n \backslash \Gamma. \tag{6.2.16}$$

Especially for $x \in \Gamma$, we have

$$\begin{aligned} &\mathbb{V}_{II}(x) \\ = &\int_\Gamma \left(\mathbb{T}(D_y, \nu(y))\mathbb{E}(x, y)\right)^{\mathrm{T}} \psi(y)\, \mathrm{d}s_y \\ = &\frac{1}{2(n-1)\pi}\left\{\int_\Gamma \frac{\partial \gamma_n(x, y)}{\partial \nu(y)} \psi(y)\, \mathrm{d}s_y - \int_\Gamma \gamma_n(x, y)\mathbb{U}(D_y, \nu(y))\psi(y)\, \mathrm{d}s_y\right\} + \\ &2\mu \int_\Gamma \mathbb{E}(x, y)\left(\mathbb{U}(D_y, \nu(y))\right)\psi(y)\, \mathrm{d}s_y \\ := &(\mathbb{K}_{II}\psi)(x), \qquad \forall x \in \Gamma. \end{aligned} \tag{6.2.17}$$

2. *The restriction of $\mathbb{V}_{II}(x)$ in Ω (Ω_e) can be continuously extended to $\bar{\Omega}$ $(\bar{\Omega}_e)$ by*

$$\begin{aligned} \mathbb{V}_{II}(x)\Big|_{\Gamma-0} &:= \lim_{\substack{z \to x \in \Gamma \\ z \in \Omega}} \mathbb{V}_{II}(z) \\ &= -\frac{1}{2}\psi(x) + \int_\Gamma \left(\mathbb{T}(D_y, \nu_y)\mathbb{E}(x, y)\right)^{\mathrm{T}} \psi(y)\, \mathrm{d}s_y, \end{aligned} \tag{6.2.18}$$

$$\begin{aligned} \mathbb{V}_{II}(x)\Big|_{\Gamma+0} &:= \lim_{\substack{z \to x \in \Gamma_e \\ z \in \Omega_e}} \mathbb{V}_{II}(z) \\ &= \frac{1}{2}\psi(x) + \int_\Gamma \left(\mathbb{T}(D_y, \nu_y)\mathbb{E}(x, y)\right)^{\mathrm{T}} \psi(y)\, \mathrm{d}s_y. \end{aligned} \tag{6.2.19}$$

Let

$$\mathbb{V}_{II}^-(x) = \begin{cases} \mathbb{V}_{II}(x), & \forall x \in \Omega, \\ -\frac{1}{2}\psi(x) + \int_\Gamma \left(\mathbb{T}(D_y, \nu(y))\right)^{\mathrm{T}} \psi(y)\, \mathrm{d}s_y, & \forall x \in \Gamma \end{cases}$$

and

$$\mathbb{V}_{II}^{+}(x) = \begin{cases} \mathbb{V}_{II}(x), & \forall x \in \Omega_e, \\ \dfrac{1}{2}\psi(x) + \displaystyle\int_{\Gamma} \left(\mathbb{T}(D_y, \nu(y))\right)^{\mathrm{T}} \psi(y)\, \mathrm{d}s_y, & \forall x \in \Gamma. \end{cases}$$

These two functions $\mathbb{V}_{II}^{-}(x)$ and $\mathbb{V}_{II}^{+}(x)$ are two continuous vector-valued functions on $\bar{\Omega}$ and $\bar{\Omega}_e$. Moreover, we have

$$\left[\mathbb{V}_{II}(x)\right]_{\Gamma} := \mathbb{V}_{II}(x)\Big|_{\Gamma+0} - \mathbb{V}_{II}(x)\Big|_{\Gamma-0} = \psi(x), \qquad \forall x \in \Gamma.$$

3.

$$\mathbb{V}_{II}(x) = \mathcal{O}(|x|^{1-n}), \qquad as \ |x| \to \infty. \tag{6.2.20}$$

6.2.3 The Derivatives of the Single Layer Potential $\mathbb{V}_I(x)$

We first consider the following derivative of the single layer potential $\mathbb{V}_I(z)$ of the Navier system, for $z \in \mathbb{R}^n \backslash \Gamma$ given by (6.2.1)

$$\mathbb{T}(D_z, \nu(z))\mathbb{V}_I(z) = \int_{\Gamma} \mathbb{T}(D_z, \nu(z))\mathbb{E}(z, y)\varphi(y)\, \mathrm{d}s_y, \qquad \forall z \in \mathbb{R}^n \backslash \Gamma, \tag{6.2.21}$$

with $\nu(z)$ be a given unit vector.

Substituting (6.2.14) with $x = z$ into (6.2.21), we arrive at

$$\begin{aligned} \mathbb{T}(D_z, \nu(z))\mathbb{V}_I(z) = \ & \frac{1}{2(n-1)\pi} \left\{ \frac{\partial}{\partial \nu(z)} \int_{\Gamma} \gamma_n(z, y)\varphi(y)\, \mathrm{d}s_y - \right. \\ & \left. \mathbb{U}(D_z, \nu(z)) \int_{\Gamma} \gamma_n(z, y)\varphi(y)\, \mathrm{d}s_y \right\} + \\ & 2\mu \mathbb{U}(D_z, \nu(z)) \int_{\Gamma} \mathbb{E}(z, y)\varphi(y)\, \mathrm{d}s_y. \end{aligned} \tag{6.2.22}$$

Theorem 6.2.3 *Suppose that $\Gamma \in C^2$, $\varphi \in (C^1(\Gamma))^n$. For point $x \in \Gamma$, taking $\nu(z) = \nu(x)$ in (6.2.21) and (6.2.22) with $\nu(x)$ be the outward unit normal vector at $x \in \Gamma$ of Ω. We have*

1.

$$\begin{aligned}\mathbb{T}(D_x, \nu(x))\mathbb{V}_I(x)\Big|_{\Gamma-0} &= \lim_{\substack{z\to x\in\Gamma\\ z\in\Omega}} \mathbb{T}(D_z, \nu(x))\mathbb{V}_I(z)\\ &= \frac{1}{2}\varphi(x) + \frac{1}{2(n-1)\pi}\left\{\int_\Gamma \frac{\partial\gamma_n(x, y)}{\partial\nu(x)}\varphi(y)\,ds_y - \right.\\ &\left.\int_\Gamma \mathbb{U}(D_x, \nu(x))\gamma_n(x, y)\varphi(y)\,ds_y\right\} +\\ &2\mu\int_\Gamma \mathbb{U}(D_x, \nu(x))\mathbb{E}(x, y)\varphi(y)\,ds_y\\ &:= \frac{1}{2}\varphi(x) + \int_\Gamma \mathbb{T}(D_x, \nu(x))\mathbb{E}(x, y)\varphi(y)\,ds_y. \end{aligned} \tag{6.2.23}$$

2.

$$\begin{aligned}\mathbb{T}(D_x, \nu(x))\mathbb{V}_I(x)\Big|_{\Gamma+0} &= \lim_{\substack{z\to x\in\Gamma\\ z\in\Omega_e}} \mathbb{T}(D_z, \nu(x))\mathbb{V}_I(z)\\ &= -\frac{1}{2}\varphi(x) + \frac{1}{2(n-1)\pi}\left\{\int_\Gamma \frac{\partial\gamma_n(x, y)}{\partial\nu(x)}\varphi(y)\,ds_y - \right.\\ &\left.\int_\Gamma \mathbb{U}(D_x, \nu(x))\gamma_n(x, y)\varphi(y)\,ds_y\right\} +\\ &2\mu\int_\Gamma \mathbb{U}(D_x, \nu(x))\mathbb{E}(x, y)\varphi(y)\,ds_y\\ &:= -\frac{1}{2}\varphi(x) + \int_\Gamma \mathbb{T}(D_x, \nu(x))\mathbb{E}(x, y)\varphi(y)\,ds_y. \end{aligned} \tag{6.2.24}$$

3.

$$\begin{aligned}\Big[\mathbb{T}(D_x, \nu(x)\mathbb{V}_I(x)\Big]_\Gamma &= \mathbb{T}(D_x, \nu(x))\mathbb{V}_I(x)\Big|_{\Gamma+0} -\\ \mathbb{T}(D_x, \nu(x))\mathbb{V}_I(x)\Big|_{\Gamma-0} &= -\varphi(x),\ \forall x\in\Gamma. \end{aligned} \tag{6.2.25}$$

4.

$$\begin{aligned}&\int_{\Gamma_R} \left(\mathbb{T}(D_x, \nu_R(x))\mathbb{V}_I(x)\right)^{\mathrm{T}}\omega_j(x)\mathrm{d}s_x\\ &= -\int_\Gamma \varphi(x)^{\mathrm{T}}\omega_j(x)\,\mathrm{d}s_x, \qquad j = 1, 2, \ldots, 3(n-1), \end{aligned} \tag{6.2.26}$$

with Γ_R be given by (6.1.16) and $\nu_R(x)$ be the outward unit normal vector of Ω_R at point $x \in \Gamma_R$.

Proof Based on the representation Formula (6.2.22) and the properties of the single and double layer potentials of the Laplacian operator, the identities (6.2.23) and (6.2.24) follow directly. Subtracting (6.2.24) by (6.2.23) leads to (6.2.25). It is obvious that

$$\int_{\Gamma}\left(\mathbb{T}(D_x, \nu(x))\mathbb{V}_I(x)\Big|_{\Gamma-0}\right)^{\mathrm{T}}\omega_j(x)\,\mathrm{d}s_x = 0, \quad 1 \leqslant j \leqslant 3(n-1). \tag{6.2.27}$$

By the equality (6.2.25), for each j, $j = 1, 2, \ldots, 3(n-1)$, we have

$$\int_{\Gamma}\left(\mathbb{T}(D_x, \nu(x))\mathbb{V}_I(x)\Big|_{\Gamma+0}\right)^{\mathrm{T}}\omega_j(x)\,\mathrm{d}s_x = -\int_{\Gamma_R}\varphi(x)^{\mathrm{T}}\omega_j(x)\,\mathrm{d}s_x. \tag{6.2.28}$$

On the other hand, we know that

$$\int_{\Gamma}\left(\mathbb{T}(D_x, \nu(x))\mathbb{V}_I(x)\Big|_{\Gamma+0}\right)^{\mathrm{T}}\omega_j(x)\,\mathrm{d}s_x = \int_{\Gamma_R}(\mathbb{T}(D_x, \nu(x))\mathbb{V}_I(x))^{\mathrm{T}}\omega_j(x)\,\mathrm{d}s_x. \tag{6.2.29}$$

Combining (6.2.28) and (6.2.29) yields (6.2.26). □

Let

$$\begin{aligned}(\mathbb{K}_{II}^{*}\varphi)(x) = & \int_{\Gamma}\mathbb{T}(D_x, \nu(x))\mathbb{E}(x, y)\varphi(y)\,\mathrm{d}s_y \\ = & \frac{1}{2(n-1)\pi}\left\{\int_{\Gamma}\frac{\partial\gamma_n(x, y)}{\partial\nu(x)}\varphi(y)\,\mathrm{d}s_y - \right. \\ & \left.\int_{\Gamma}\mathbb{U}(D_x, \nu(x))\gamma_n(x, y)\varphi(y)\,\mathrm{d}s_y\right\} + \\ & 2\mu\int_{\Gamma}\mathbb{U}(D_x, \nu(x))\mathbb{E}(x, y)\varphi(y)\,\mathrm{d}s_y, \qquad \forall x \in \Gamma.\end{aligned} \tag{6.2.30}$$

6.2.4 The Derivatives of the Double Layer Potential $\mathbb{V}_{II}(x)$

In this subsection, we consider the following derivative of the double layer potential $\mathbb{V}_{II}(x)$ defined by (6.2.3). We first discuss the case $z \in \mathbb{R}^n\backslash\Gamma$,

$$\mathbb{T}(D_z, \nu(z))\mathbb{V}_{II}(z) = \mathbb{T}(D_z, \nu(z))\int_{\Gamma}\left[\mathbb{T}(D_y, \nu(y))\mathbb{E}(z, y)\right]^{\mathrm{T}}\psi(y)\,\mathrm{d}s_y \tag{6.2.31}$$

with $\nu(z)$ be a given unit vector.

Based on the representation (6.2.15), we compute $\mathbb{T}(D_z, \nu(z))\mathbb{V}_{II}(z)$ for $z \in \mathbb{R}^n\backslash\Gamma$ by

$$\begin{aligned}\mathbb{T}(D_z, \nu(z))\mathbb{V}_{II}(z) = & \frac{1}{2(n-1)\pi}\int_{\Gamma}\mathbb{T}(D_z, \nu(z))\frac{\partial\gamma_n(z, y)}{\partial\nu(y)}\psi(y)\,\mathrm{d}s_y + \\ & \int_{\Gamma}\mathbb{T}(D_z, \nu(z))\Big[2\mu\mathbb{E}(z, y) -\end{aligned}$$

$$\left.\frac{1}{2(n-1)\pi}\gamma_n(z,\, \boldsymbol{y})\boldsymbol{I}\right]\mathbb{U}(D_y,\, \boldsymbol{\nu}(\boldsymbol{y}))\boldsymbol{\psi}(\boldsymbol{y})\,\mathrm{d}s_y$$
$$:= I_0 + J. \tag{6.2.32}$$

By the definition of $\mathbb{T}(D_z,\, \boldsymbol{\nu}(z))$, we have

$$\mathbb{T}(D_z,\, \boldsymbol{\nu}(z)) = \mu\boldsymbol{I}\frac{\partial}{\partial\boldsymbol{\nu}(z)} + (\lambda+\mu)\boldsymbol{\nu}(z)(\mathrm{grad}_z)^{\mathrm{T}} + \mu\mathbb{U}(D_z,\, \boldsymbol{\nu}(z)). \tag{6.2.33}$$

Substituting it into I_0 yields

$$\begin{aligned} I_0 = {} & \frac{\mu}{2(n-1)\pi}\int_\Gamma \frac{\partial^2\gamma_n(z,\, \boldsymbol{y})}{\partial\boldsymbol{\nu}(z)\partial\boldsymbol{\nu}(\boldsymbol{y})}\boldsymbol{\psi}(\boldsymbol{y})\,\mathrm{d}s_y + \\ & \frac{\lambda+\mu}{2(n-1)\pi}\int_\Gamma \boldsymbol{\nu}(z)\left[\mathrm{grad}_z\frac{\partial\gamma_n(z,\, \boldsymbol{y})}{\partial\boldsymbol{\nu}(\boldsymbol{y})}\right]^{\mathrm{T}}\boldsymbol{\psi}(\boldsymbol{y})\,\mathrm{d}s_y + \\ & \frac{\mu}{2(n-1)\pi}\int_\Gamma \mathbb{U}(D_z,\, \boldsymbol{\nu}(z))\frac{\partial\gamma_n(z,\, \boldsymbol{y})}{\partial\boldsymbol{\nu}(\boldsymbol{y})}\boldsymbol{\psi}(\boldsymbol{y})\,\mathrm{d}s_y. \end{aligned} \tag{6.2.34}$$

Inserting (6.2.14) with $\boldsymbol{x} = z$ and (6.2.33) into J, we arrive at

$$\begin{aligned} J = {} & \int_\Gamma \mathbb{U}(D_z,\, \boldsymbol{\nu}(z))\left[4\mu^2\mathbb{E}(z,\, \boldsymbol{y}) - \frac{3\mu}{2(n-1)\pi}\gamma_n(z,\, \boldsymbol{y})\right]\mathbb{U}(D_y,\, \boldsymbol{\nu}(\boldsymbol{y}))\boldsymbol{\psi}(\boldsymbol{y})\,\mathrm{d}s_y - \\ & \frac{\lambda+\mu}{2(n-1)\pi}\int_\Gamma \boldsymbol{\nu}(z)[\mathrm{grad}_z\gamma_n(z,\, \boldsymbol{y})]^{\mathrm{T}}\mathbb{U}(D_y,\, \boldsymbol{\nu}(\boldsymbol{y}))\boldsymbol{\psi}(\boldsymbol{y})\,\mathrm{d}s_y + \\ & \frac{\mu}{2(n-1)\pi}\int_\Gamma \frac{\partial\gamma_n(z,\, \boldsymbol{y})}{\partial\boldsymbol{\nu}(z)}\mathbb{U}(D_y,\, \boldsymbol{\nu}(\boldsymbol{y}))\boldsymbol{\psi}(\boldsymbol{y})\,\mathrm{d}s_y. \end{aligned} \tag{6.2.35}$$

Combining (6.2.29), (6.2.34) and (6.2.35), we obtain

$$\begin{aligned} \mathbb{T}(D_z,\, \boldsymbol{\nu}(z))\mathbb{V}_{II}(z) = {} & \frac{\mu}{2(n-1)\pi}\int_\Gamma \frac{\partial^2\gamma_n(z,\, \boldsymbol{y})}{\partial\boldsymbol{\nu}(z)\partial\boldsymbol{\nu}(\boldsymbol{y})}\boldsymbol{\psi}(\boldsymbol{y})\,\mathrm{d}s_y + \\ & \int_\Gamma \mathbb{U}(D_z,\, \boldsymbol{\nu}(z))\left[4\mu^2\mathbb{E}(z,\, \boldsymbol{y}) - \frac{3\mu}{2(n-1)\pi}\gamma_n(z,\, \boldsymbol{y})\boldsymbol{I}\right]\cdot \\ & \mathbb{U}(D_y,\, \boldsymbol{\nu}(\boldsymbol{y}))\,\boldsymbol{\psi}(\boldsymbol{y})\,\mathrm{d}s_y + E_1 + E_2 \end{aligned} \tag{6.2.36}$$

with

$$\begin{aligned} E_1 = {} & \frac{\lambda+\mu}{2(n-1)\pi}\int_\Gamma \boldsymbol{\nu}(z)\left\{\left[\mathrm{grad}_z\frac{\partial\gamma_n(z,\, \boldsymbol{y})}{\partial\boldsymbol{\nu}(\boldsymbol{y})}\right]^{\mathrm{T}}\boldsymbol{\psi}(\boldsymbol{y}) - \right. \\ & \left.[\mathrm{grad}_z\gamma_n(z,\, \boldsymbol{y})]^{\mathrm{T}}\,\mathbb{U}(D_y,\, \boldsymbol{\nu}(\boldsymbol{y}))\boldsymbol{\psi}(\boldsymbol{y})\right\}\mathrm{d}s_y, \end{aligned} \tag{6.2.37}$$

$$E_2 = \frac{\mu}{2(n-1)\pi} \int_\Gamma \left\{ \mathbb{U}(D_z, \boldsymbol{\nu}(z)) \frac{\partial \gamma_n(z, \boldsymbol{y})}{\partial \boldsymbol{\nu}(\boldsymbol{y})} \boldsymbol{\psi}(\boldsymbol{y}) + \right. \\ \left. \frac{\partial \gamma_n(z, \boldsymbol{y})}{\partial \boldsymbol{\nu}(z)} \mathbb{U}(D_y, \boldsymbol{\nu}(\boldsymbol{y})) \boldsymbol{\psi}(\boldsymbol{y}) \right\} \mathrm{d}s_y. \tag{6.2.38}$$

In order to further simplify the Formula (6.2.36), we recall the following lemma (Han, 1994a).

Lemma 6.2.2 *The following equalities hold for $z \neq \boldsymbol{y}$,*

$$\mathrm{grad}_z \frac{\partial \gamma_n(z, \boldsymbol{y})}{\partial \boldsymbol{\nu}(\boldsymbol{y})} = \mathbb{U}(D_y, \boldsymbol{\nu}(\boldsymbol{y})) \mathrm{grad}_z \gamma_n(z, \boldsymbol{y}), \tag{6.2.39}$$

$$\mathbb{U}(D_z, \boldsymbol{\nu}(z)) \frac{\partial \gamma_n(z, \boldsymbol{y})}{\partial \boldsymbol{\nu}(\boldsymbol{y})} - \mathbb{U}(D_y, \boldsymbol{\nu}(\boldsymbol{y})) \frac{\partial \gamma_n(z, \boldsymbol{y})}{\partial \boldsymbol{\nu}(z)} \\ = \left\{ \mathbb{U}(D_y, \boldsymbol{\nu}(\boldsymbol{y})) \mathbb{U}(D_z, \boldsymbol{\nu}(z)) - \mathbb{U}(D_z, \boldsymbol{\nu}(z)) \mathbb{U}(D_y, \boldsymbol{\nu}(\boldsymbol{y})) \right\} \gamma_n(z, \boldsymbol{y}). \tag{6.2.40}$$

Applying (6.2.39), we have

$$\begin{aligned} &\int_\Gamma \left[\mathrm{grad}_z \frac{\partial \gamma_n(z, \boldsymbol{y})}{\partial \boldsymbol{\nu}(\boldsymbol{y})} \right]^{\mathrm{T}} \boldsymbol{\psi}(\boldsymbol{y}) \, \mathrm{d}s_y \\ = &\int_\Gamma \left[\mathbb{U}(D_y, \boldsymbol{\nu}(\boldsymbol{y})) \mathrm{grad}_z \gamma_n(z, \boldsymbol{y}) \right]^{\mathrm{T}} \boldsymbol{\psi}(\boldsymbol{y}) \, \mathrm{d}s_y \\ = &\int_\Gamma \sum_{i=1}^n \left[\sum_{j=1}^n U_{ij}(D_y, \boldsymbol{\nu}(\boldsymbol{y})) \frac{\partial \gamma_n(z, \boldsymbol{y})}{\partial z_j} \right] \psi_i(\boldsymbol{y}) \, \mathrm{d}s_y \\ = &-\int_\Gamma \sum_{j=1}^n \frac{\partial \gamma_n(z, \boldsymbol{y})}{\partial z_j} \left[\sum_{i=1}^n U_{ij}(D_y, \boldsymbol{\nu}(\boldsymbol{y})) \psi_i(\boldsymbol{y}) \right] \mathrm{d}s_y \\ = &\int_\Gamma \left[\mathrm{grad}_z \gamma_n(z, \boldsymbol{y}) \right]^{\mathrm{T}} \mathbb{U}(D_y, \boldsymbol{\nu}(\boldsymbol{y})) \boldsymbol{\psi}(\boldsymbol{y}) \, \mathrm{d}s_y. \end{aligned}$$

This indicates that

$$E_1 = 0. \tag{6.2.41}$$

Inserting (6.2.40) into (6.2.38) yields

$$E_2 = \frac{\mu}{2(n-1)\pi} \int_\Gamma \left\{ \left[\mathbb{U}(D_y, \boldsymbol{\nu}(\boldsymbol{y})) \mathbb{U}(D_z, \boldsymbol{\nu}(z)) - \right. \right. \\ \left. \left. \mathbb{U}(D_z, \boldsymbol{\nu}(z)) \mathbb{U}(D_y, \boldsymbol{\nu}(\boldsymbol{y})) \right] \gamma_n(z, \boldsymbol{y}) \right\} \boldsymbol{\psi}(\boldsymbol{y}) \, \mathrm{d}s_y$$

$$= -\frac{\mu}{2(n-1)\pi}\int_\Gamma \Big[\mathbb{U}(D_z,\ \boldsymbol{\nu}(z))\gamma_n(z,\ \boldsymbol{y})\mathbb{U}(D_y,\ \boldsymbol{\nu}(\boldsymbol{y}))\Big]^{\mathrm{T}}\boldsymbol{\psi}(\boldsymbol{y})\,\mathrm{d}s_y +$$
$$\frac{\mu}{2(n-1)\pi}\mathbb{U}(D_z,\ \boldsymbol{\nu}(z))\int_\Gamma \gamma_n(z,\ \boldsymbol{y})\mathbb{U}(D_y,\ \boldsymbol{\nu}(\boldsymbol{y}))\boldsymbol{\psi}(\boldsymbol{y})\,\mathrm{d}s_y. \qquad (6.2.42)$$

Combining the equalities (6.2.36), (6.2.41) and (6.2.42), we arrive at

$$\mathbb{T}(D_z,\ \boldsymbol{\nu}(z))\mathbb{V}_{II}(z)$$
$$= \frac{\mu}{2(n-1)\pi}\frac{\partial}{\partial\boldsymbol{\nu}(z)}\int_\Gamma \frac{\partial\gamma_n(z,\ \boldsymbol{y})}{\partial\boldsymbol{\nu}(\boldsymbol{y})}\boldsymbol{\psi}(\boldsymbol{y})\,\mathrm{d}s_y +$$
$$\mathbb{U}(D_z,\ \boldsymbol{\nu}(z))\int_\Gamma \left[4\mu^2\mathbb{E}(z,\ \boldsymbol{y}) - \frac{\mu}{(n-1)\pi}\gamma_n(z,\ \boldsymbol{y})\,\boldsymbol{I}\right]\cdot$$
$$\mathbb{U}(D_y,\ \boldsymbol{\nu}(\boldsymbol{y}))\boldsymbol{\psi}(\boldsymbol{y})\,\mathrm{d}s_y -$$
$$\frac{\mu}{2(n-1)\pi}\int_\Gamma \Big[\mathbb{U}(D_z,\ \boldsymbol{\nu}(z))\gamma_n(z,\ \boldsymbol{y})\mathbb{U}(D_y,\ \boldsymbol{\nu}(\boldsymbol{y}))\Big]^{\mathrm{T}}\boldsymbol{\psi}(\boldsymbol{y})\,\mathrm{d}s_y. \quad (6.2.43)$$

Finally $\mathbb{T}(D_z,\ \boldsymbol{\nu}(z))\mathbb{V}_{II}(z)$, the derivative of the double layer potential of the Navier system, is presented as a sum of a derivative of the double layer potential of the Laplacian operator and the tangential derivatives with respect to the unit vector $\boldsymbol{\nu}(z)$ of the single layer potentials of the Laplacian operator and the Navier system.

For any point $\boldsymbol{x} \in \Gamma$, we take $\boldsymbol{\nu}(z) = \boldsymbol{\nu}(\boldsymbol{x})$ in the Formula (6.2.43) with $\boldsymbol{\nu}(\boldsymbol{x})$ be the outward unit normal vector of Γ at $\boldsymbol{x} \in \Gamma$. Then we obtain

Theorem 6.2.4 *Suppose that $\Gamma \in C^2$ and $\boldsymbol{\psi} \in (C^2(\Gamma))^n$. The following results hold*

1.

$$\mathbb{T}(D_x,\ \boldsymbol{\nu}(\boldsymbol{x}))\mathbb{V}_{II}(\boldsymbol{x})\Big|_\Gamma = \lim_{\substack{z\to x\in\Gamma\\ z\in\mathbb{R}^n\setminus\Gamma}} \mathbb{T}(D_z,\ \boldsymbol{\nu}(\boldsymbol{x}))\mathbb{V}_{II}(z)$$
$$= \frac{1}{2(n-1)\pi}\sum_{1\leqslant k<j\leqslant n}\frac{\partial}{\partial\boldsymbol{\tau}_x^{k,j}}\int_\Gamma \gamma_n(\boldsymbol{x},\ \boldsymbol{y})\frac{\partial}{\partial\boldsymbol{\tau}_y^{k,j}}\boldsymbol{\psi}(\boldsymbol{y})\,\mathrm{d}s_y +$$
$$\mathbb{U}(D_x,\ \boldsymbol{\nu}(\boldsymbol{x}))\int_\Gamma \left[4\mu^2\mathbb{E}(\boldsymbol{x},\ \boldsymbol{y}) - \frac{\mu}{(n-1)\pi}\gamma_n(\boldsymbol{x},\ \boldsymbol{y})\,\boldsymbol{I}\right]\mathbb{U}(D_y,\ \boldsymbol{\nu}(\boldsymbol{y}))\boldsymbol{\psi}(\boldsymbol{y})\,\mathrm{d}s_y -$$
$$\frac{\mu}{2(n-1)\pi}\int_\Gamma \Big[\mathbb{U}(D_x,\ \boldsymbol{\nu}(\boldsymbol{x}))\gamma_n(\boldsymbol{x},\ \boldsymbol{y})\mathbb{U}(D_y,\ \boldsymbol{\nu}(\boldsymbol{y}))\Big]^{\mathrm{T}}\boldsymbol{\psi}(\boldsymbol{y})\,\mathrm{d}s_y$$
$$:= \int_\Gamma \mathbb{T}(D_x,\ \boldsymbol{\nu}(\boldsymbol{x}))\Big[\mathbb{T}(D_y,\ \boldsymbol{\nu}(\boldsymbol{y}))\mathbb{E}(\boldsymbol{x},\ \boldsymbol{y})\Big]^{\mathrm{T}}\boldsymbol{\psi}(\boldsymbol{y})\,\mathrm{d}s_y. \qquad (6.2.44)$$

The vectors $\{\boldsymbol{\tau}_x^{k,j},\ \boldsymbol{\tau}_y^{k,j},\ \ 1\leqslant k<j\leqslant n\}$ are given in Sect. 3.3.4. The last integral is hyper-singular, it exists as a Hardamard finite part integral.

2.

$$\int_{\Gamma_R}\Big[\mathbb{T}(D_x,\ \boldsymbol{\nu}_R(\boldsymbol{x}))\mathbb{V}_{II}(\boldsymbol{x})\Big]^{\mathrm{T}}\boldsymbol{\omega}_j(\boldsymbol{x})\,\mathrm{d}s_x = 0, \quad j = 1,\ 2,\ \ldots,\ 3(n-1), \qquad (6.2.45)$$

with Γ_R be given by (6.1.16) and $\boldsymbol{\nu}_R(\boldsymbol{x})$ be the unit outward normal vector of Ω_R at point $\boldsymbol{x} \in \Gamma_R$.

Proof 1. Based on the expression Formula (6.2.43), the conclusion (1) follows directly from the results in Theorems 3.3.5, 3.3.6 and 6.2.3.

2. Note that $\mathbb{V}_{II}(\boldsymbol{x})$ satisfies the homogeneous Navier system (6.1.4) in Ω. We have

$$\int_\Gamma \left[\mathbb{T}(D_x, \boldsymbol{\nu}(\boldsymbol{x}))\mathbb{V}_{II}(\boldsymbol{x})\Big|_{\Gamma-0}\right]^{\mathrm{T}} \boldsymbol{\omega}_j(\boldsymbol{x})\,\mathrm{d}s_x = 0, \quad 1 \leqslant j \leqslant 3(n-1).$$

By the conclusion (1), we obtain

$$\int_\Gamma \left[\mathbb{T}(D_x, \boldsymbol{\nu}(\boldsymbol{x}))\mathbb{V}_{II}(\boldsymbol{x})\Big|_{\Gamma+0}\right]^{\mathrm{T}} \boldsymbol{\omega}_j(\boldsymbol{x})\,\mathrm{d}s_x = 0, \quad 1 \leqslant j \leqslant 3(n-1).$$

Furthermore, $\mathbb{V}_{II}(\boldsymbol{x})$ satisfies the homogeneous Navier system in Ω_R. As a result, we get

$$\begin{aligned}
&\int_{\Gamma_R} \left[\mathbb{T}(D_x, \boldsymbol{\nu}_R(\boldsymbol{x}))\mathbb{V}_{II}(\boldsymbol{x})\right]^{\mathrm{T}} \boldsymbol{\omega}_j(\boldsymbol{x})\,\mathrm{d}s_x \\
= &\int_\Gamma \left[\mathbb{T}(D_x, \boldsymbol{\nu}(\boldsymbol{x}))\mathbb{V}_{II}(\boldsymbol{x})\Big|_{\Gamma+0}\right]^{\mathrm{T}} \boldsymbol{\omega}_j(\boldsymbol{x})\,\mathrm{d}s_x = 0, \quad 1 \leqslant j \leqslant 3(n-1).
\end{aligned}$$

This is the conclusion (2). □

Let

$$(\mathbb{K}_{III}\boldsymbol{\psi})(\boldsymbol{x}) = \int_\Gamma \mathbb{T}(D_x, \boldsymbol{\nu}(\boldsymbol{x}))\big[\mathbb{T}(D_y, \boldsymbol{\nu}(\boldsymbol{y}))\mathbb{E}(\boldsymbol{x}, \boldsymbol{y})\big]^{\mathrm{T}}\boldsymbol{\psi}(\boldsymbol{y})\,\mathrm{d}s_y, \quad \forall\, \boldsymbol{x} \in \Gamma. \tag{6.2.46}$$

The equality (6.2.44) shows that the hypersingular integral operator $\mathbb{K}_{III}$ can be regularized as a differential-integro-differential operator, in which the integral operators are weak singular. The equality (6.2.44) has been applied to the computational algorithm for Galerkin method in linear electrostatics by Of et al. (2004).

Remark 6.2.1 1. The content of Sect. 6.2.4 is mainly from the paper of Han (1994b). One may also refer to Bonnemay (1979), Nedelec (1982), Han (1993, 1994a), Hsiao and Wendland (2008).

2. The singule layer potential $\mathbb{V}_I(\boldsymbol{x})$ and the double layer potential $\mathbb{V}_{II}(\boldsymbol{x})$ with Hölder continuous densities $\boldsymbol{\varphi}(\boldsymbol{x})$ and $\boldsymbol{\psi}(\boldsymbol{x})$ were studied by many authors, such as Ahner and Hsiao (1976), Balaš et al. (2013), Kupradze (2012), Muskhelishvili (1953), Hsiao and Wendland (2008). By the results from them, Theorems 6.2.2 and 6.2.3 are still valid using the Hölder space $(C^\alpha(\Gamma))^n$ instead of $(C^1(\Gamma))^n$ and Theorem 6.2.4 is still correct using the Hölder space $(C^{1+\alpha}(\gamma))^n$ instead of $(C^2(\Gamma))^n$. In the coming subsection, we will discuss the layer potentials $\mathbb{V}_I(\boldsymbol{x})$ and $\mathbb{V}_{II}(\boldsymbol{x})$ with densities $\boldsymbol{\varphi} \in \mathbb{H}^{-1/2}(\Gamma) = \big(H^{-1/2}(\Gamma)\big)^n$ and $\boldsymbol{\psi} \in \mathbb{H}^{1/2}(\Gamma) = \big(H^{1/2}(\Gamma)\big)^n$.

6.2.5 The Layer Potentials $\mathbb{V}_I(x)$ and $\mathbb{V}_{II}(x)$ in Sobolev Spaces

In this subsection, we consider the single layer potential $\mathbb{V}_I(x)$ of the Navier system with density $\varphi(x) \in (C^1(\Gamma))^n$. We know that $\mathbb{V}_I(x)$ is a solution to the following transmission problem,

$$\begin{cases} -\nabla \cdot \sigma(\mathbb{V}(x)) = \mathbf{0}, \qquad \forall x \in \mathbb{R}^n \backslash \Gamma, & (6.2.47a) \\ \left[\mathbb{V}(x)\right]_\Gamma = \mathbf{0}, \qquad \forall x \in \Gamma, & (6.2.47b) \\ \left[\mathbb{T}(D_x, \nu(x))\mathbb{V}(x)\right]_\Gamma = -\varphi(x), \qquad \forall x \in \Gamma, & (6.2.47c) \\ \mathbb{V}(x) = \mathbb{E}(x, \mathbf{0})c_0 + \mathcal{O}(|x|^{1-n}), \qquad \text{as } |x| \to +\infty, & (6.2.47d) \end{cases}$$

with $c_0 = \displaystyle\int_\Gamma \varphi(x)\, ds_x$.

By Theorem 5.5.7 in Hsiao and Wendland (2008), we know that for given $\varphi \in \mathbb{H}^{-1/2}(\Gamma)$, the transmission problem (6.2.47) has a unique solution in the Sobolev space $\mathbb{H}^1_{\text{loc}}(\mathbb{R}^n) = (H^1_{\text{loc}}(\mathbb{R}^n))^n$, which is the so-called single layer potential $\mathbb{V}_I(x)$ of the Navier system with density $\varphi \in \mathbb{H}^{-1/2}(\Gamma)$. Furthermore, we have the following lemma (Hsiao and Wendland, 2008).

Lemma 6.2.3 *Given $\varphi \in \mathbb{H}^{-1/2}(\Gamma)$, the corresponding single layer potential $\mathbb{V}_I(x)$ of the Navier system satisfies*

1. $\mathbb{V}_I\Big|_\Omega \in \mathbb{H}^1(\Omega)$, $\mathbb{V}_I\Big|_{\Gamma-0} \in \mathbb{H}^{1/2}(\Gamma)$ *and* $\mathbb{T}(D_x, \nu(x))\mathbb{V}_I\Big|_{\Gamma-0} \in \mathbb{H}^{-1/2}(\Gamma)$,
   $\mathbb{V}_I\Big|_{\Omega_e} \in \mathbb{H}^1_{\text{loc}}(\Omega_e)$, $\mathbb{V}_I\Big|_{\Gamma+0} \in \mathbb{H}^{1/2}(\Gamma)$ *and* $\mathbb{T}(D_x, \nu(x))\mathbb{V}_I\Big|_{\Gamma+0} \in \mathbb{H}^{-1/2}(\Gamma)$.
2.
$$\mathbb{V}_I(x)\Big|_\Gamma = \mathbb{V}_I(x)\Big|_{\Gamma\pm0} = \int_\Gamma \mathbb{E}(x, y)\varphi(y)\, ds_y = (\mathbb{K}_I\varphi)(x), \quad \forall x \in \Gamma.$$

   The operator $\mathbb{K}_I$ is bounded from $\mathbb{H}^{-1/2}(\Gamma)$ to $\mathbb{H}^{1/2}(\Gamma)$. This implies that there exists a positive constant C such that

$$\left\| \mathbb{V}_I\Big|_\Gamma \right\|_{\mathbb{H}^{1/2}(\Gamma)} \leqslant C \left\| \varphi \right\|_{\mathbb{H}^{-1/2}(\Gamma)}. \tag{6.2.48}$$

3. *The equalities (6.2.23)–(6.2.24) hold in $\mathbb{H}^{-1/2}(\Gamma)$, i.e.,*

$$\mathbb{T}(D_x, \nu(x))\mathbb{V}_I(x)\Big|_{\Gamma-0} = \frac{1}{2}\varphi(x) + \int_\Gamma \mathbb{T}(D_x, \nu(x))\mathbb{E}(x, y)\varphi(y)\, ds_y,$$
$$\mathbb{T}(D_x, \nu(x))\mathbb{V}_I(x)\Big|_{\Gamma+0} = -\frac{1}{2}\varphi(x) + \int_\Gamma \mathbb{T}(D_x, \nu(x))\mathbb{E}(x, y)\varphi(y)\, ds_y.$$

   Furthermore, the operator $\mathbb{K}^*_{II}$ defined by

$$\int_{\Gamma} \mathbb{T}(D_x,\, \boldsymbol{\nu}(\boldsymbol{x}))\mathbb{E}(\boldsymbol{x},\, \boldsymbol{y})\boldsymbol{\varphi}(\boldsymbol{y})\, \mathrm{d}s_y := (\mathbb{K}_{II}^{*}\boldsymbol{\varphi})(\boldsymbol{x}), \quad \forall\, \boldsymbol{x} \in \Gamma,$$

is bounded from $\mathbb{H}^{-1/2}(\Gamma)$ *to* $\mathbb{H}^{-1/2}(\Gamma)$.

4. *For the case* $n = 3$ *and* $n = 2$ *with* $\int_{\Gamma} \boldsymbol{\varphi}(\boldsymbol{x})\, \mathrm{d}s_x = \mathbf{0}$, *the elastic strain energy* $J(\mathbb{V}_I(\boldsymbol{x}))$ *of* $\mathbb{V}_I(\boldsymbol{x})$ *is bounded, namely*

$$\begin{aligned} J(\mathbb{V}_I(\boldsymbol{x})) &= \frac{1}{2}\int_{\mathbb{R}^n} \Big\{\lambda\big[\nabla \cdot \mathbb{V}_I(\boldsymbol{x})\big]\big[\nabla \cdot \mathbb{V}_I(\boldsymbol{x})\big] + 2\mu\, \boldsymbol{\epsilon}(\mathbb{V}_I(\boldsymbol{x})) : \boldsymbol{\epsilon}(\mathbb{V}_I(\boldsymbol{x}))\Big\}\, \mathrm{d}\boldsymbol{x} \\ &:= \frac{1}{2} a(\mathbb{V}_I(\boldsymbol{x}),\, \mathbb{V}_I(\boldsymbol{x})) < +\infty \end{aligned} \tag{6.2.49}$$

with

$$\boldsymbol{\epsilon}(\mathbb{V}_I(\boldsymbol{x})) : \boldsymbol{\epsilon}(\mathbb{V}_I(\boldsymbol{x})) = \sum_{1 \leqslant i,\, j \leqslant n} \epsilon_{ij}(\mathbb{V}_I(\boldsymbol{x}))\epsilon_{ij}(\mathbb{V}_I(\boldsymbol{x})). \tag{6.2.50}$$

Moreover,

$$\Big\| \mathbb{T}(D_x,\, \boldsymbol{\nu})\mathbb{V}_I\Big|_{\Gamma\pm 0}\Big\|_{\mathbb{H}^{-1/2}(\Gamma)} \leqslant C\Big\{J(\mathbb{V}_I(\boldsymbol{x}))\Big\}^{1/2}. \tag{6.2.51}$$

The double layer potential $\mathbb{V}_{II}(\boldsymbol{x})$ with density $\boldsymbol{\psi} \in (C^2(\Gamma))^n$ is a solution to the following transmission problem,

$$\begin{cases} -\nabla \cdot \boldsymbol{\sigma}(\mathbb{V}(\boldsymbol{x})) = \mathbf{0}, & \forall\, \boldsymbol{x} \in \mathbb{R}^n \backslash \Gamma, & (6.2.52a) \\ \big[\mathbb{V}(\boldsymbol{x})\big]_{\Gamma} = \boldsymbol{\psi}(\boldsymbol{x}), & \forall\, \boldsymbol{x} \in \Gamma, & (6.2.52b) \\ \big[\mathbb{T}(D_x,\, \boldsymbol{\nu}(\boldsymbol{x}))\mathbb{V}(\boldsymbol{x})\big]_{\Gamma} = \mathbf{0}, & \forall\, \boldsymbol{x} \in \Gamma, & (6.2.52c) \\ \mathbb{V}(\boldsymbol{x}) = \mathcal{O}(|\boldsymbol{x}|^{1-n}), & \text{as } |\boldsymbol{x}| \to +\infty & (6.2.52d) \end{cases}$$

subject to the condition

$$\int_{\Gamma_R} \Big[\mathbb{T}(D_x,\, \boldsymbol{\nu}_R(\boldsymbol{x}))\mathbb{V}(\boldsymbol{x})\Big]^{\mathrm{T}} \boldsymbol{\omega}_j(\boldsymbol{x})\, \mathrm{d}s_x = 0, \quad j = 1, 2, \ldots, 3(n-1), \text{ with } R \geqslant R_0. \tag{6.2.53}$$

Similarly, given $\boldsymbol{\psi} \in \mathbb{H}^{1/2}(\Gamma)$, the transmission problem (6.2.52)–(6.2.53) has a unique solution, which is the so-called double layer potential $\mathbb{V}_{II}(\boldsymbol{x})$ of the Navier system with density $\boldsymbol{\psi} \in \mathbb{H}^{1/2}(\Gamma)$.

Lemma 6.2.4 (Han, 1994b). *The double layer potential of the Navier system has the following properties*

1. $\mathbb{V}_{II} \in \mathbb{H}^1(\Omega)$, *and its trace on* Γ,

$$\mathbb{V}_{II}\Big|_{\Gamma-0} \in \mathbb{H}^{1/2}(\Gamma), \qquad \mathbb{T}(D_x,\, \boldsymbol{\nu})\mathbb{V}_{II}\Big|_{\Gamma-0} \in \mathbb{H}^{-1/2}(\Gamma).$$

$\mathbb{V}_{II} \in \mathbb{H}^1_{\rm loc}(\Omega_e)$, *and its trace on* Γ,

$$\mathbb{V}_{II}\Big|_{\Gamma+0} \in \mathbb{H}^{1/2}(\Gamma), \qquad \mathbb{T}(D_x, \nu)\mathbb{V}_{II}\Big|_{\Gamma+0} \in \mathbb{H}^{-1/2}(\Gamma).$$

2.

$$\mathbb{V}_{II}(x)\Big|_{\Gamma-0} = -\frac{1}{2}\psi(x) + \int_\Gamma \Big[\mathbb{T}(D_x, \nu(x))\mathbb{E}(x, y)\Big]^{\rm T} \psi(y)\, {\rm d}s_y, \quad \textit{in } \mathbb{H}^{1/2}(\Gamma),$$

$$\mathbb{V}_{II}(x)\Big|_{\Gamma+0} = \frac{1}{2}\psi(x) + \int_\Gamma \Big[\mathbb{T}(D_x, \nu(x))\mathbb{E}(x, y)\Big]^{\rm T} \psi(y)\, {\rm d}s_y, \quad \textit{in } \mathbb{H}^{1/2}(\Gamma).$$

The operator $\mathbb{K}_{II}$ *defined by*

$$(\mathbb{K}_{II}\psi)(x) = \int_\Gamma \Big[\mathbb{T}(D_x, \nu(x))\mathbb{E}(x, y)\Big]^{\rm T} \psi(y)\, {\rm d}s_y, \quad \forall\ x \in \Gamma,$$

is bounded from $\mathbb{H}^{1/2}(\Gamma)$ *to* $\mathbb{H}^{1/2}(\Gamma)$.

3.

$$\mathbb{T}(D_x, \nu(x))\mathbb{V}_{II}(x)\Big|_{\Gamma-0} = \mathbb{T}(D_x, \nu(x))\mathbb{V}_{II}(x)\Big|_{\Gamma+0}, \qquad \textit{in } \mathbb{H}^{-1/2}(\Gamma).$$

Furthermore,

$$\Big\|\mathbb{T}(D_x, \nu)\mathbb{V}_{II}\Big|_\Gamma\Big\|_{\mathbb{H}^{-1/2}(\Gamma)} \leqslant C\|\psi\|_{\mathbb{H}^{1/2}(\Gamma)}, \qquad \forall\, \psi \in \mathbb{H}^{1/2}(\Gamma). \tag{6.2.54}$$

4. *For the case* $\psi \in \mathbb{H}^{1/2}(\Gamma)$ *and* $\int_\Gamma \psi(x) \cdot \omega_j(x)\, {\rm d}s_x = 0$ *for* $j = 1, 2, \ldots, 3(n-1)$, *we have*

$$\Big\|\mathbb{V}_{II}\Big|_{\Gamma\pm0}\Big\|_{\mathbb{H}^{1/2}(\Gamma)} \leqslant C\Big\{J(\mathbb{V}_{II})\Big\}^{1/2}. \tag{6.2.55}$$

6.3 Boundary Reduction for the Boundary Value Problems of the Navier System

We are now ready to discuss the boundary reduction for the boundary value problems of the Navier system to boundary integral (or differential-integro-differential) equations.

6.3.1 First Kind Integral (Differential-Integro-Differential) Equations of the Boundary Value Problems of the Navier System

We first introduce some Sobolev spaces

$$\begin{cases} \mathbb{H}_*^{-1/2}(\Gamma)=\left\{\varphi \,\middle|\, \varphi\in\mathbb{H}^{-1/2}(\Gamma) \text{ and } \int_\Gamma \varphi(x)\cdot\omega_j(x)\,ds_x=0,\ j=1,2,\ldots,n\right\}, \\ \mathbb{H}_{**}^{-1/2}(\Gamma)=\left\{\varphi \,\middle|\, \varphi\in\mathbb{H}^{-1/2}(\Gamma) \text{ and } \int_\Gamma \varphi\cdot\omega_j(x)\,ds_x=0,\ j=1,2,\ldots,3(n-1)\right\}, \end{cases} \tag{6.3.1}$$

$$\begin{cases} \mathbb{H}_*^{1/2}(\Gamma)=\left\{\varphi \,\middle|\, \varphi\in\mathbb{H}^{1/2}(\Gamma) \text{ and } \int_\Gamma \varphi(x)\cdot\omega_j(x)\,ds_x=0,\ j=1,2,\ldots,n\right\}, \\ \mathbb{H}_{**}^{1/2}(\Gamma)=\left\{\varphi \,\middle|\, \varphi\in\mathbb{H}^{1/2}(\Gamma) \text{ and } \int_\Gamma \varphi(x)\cdot\omega_j(x)\,ds_x=0,\ j=1,2,\ldots,3(n-1)\right\}. \end{cases} \tag{6.3.2}$$

We know that for any $\psi\in\mathbb{H}^{1/2}(\Gamma)$, the tangential derivatives of ψ, $\partial\psi/\partial\tau_x^{k,j}\in\mathbb{H}_*^{-1/2}(\Gamma)$ $(1\leqslant k<j\leqslant n)$ and the semi-norm of $\psi\in\mathbb{H}^{1/2}(\Gamma)$ is given by

$$\left|\psi\right|_{\mathbb{H}^{1/2}(\Gamma)}=\left\{\sum_{1\leqslant k<j\leqslant n}\left\|\frac{\partial\psi}{\partial\tau_x^{k,j}}\right\|^2_{\mathbb{H}^{-1/2}(\Gamma)}\right\}^{1/2}, \tag{6.3.3}$$

which is an equivalent norm of the space $\mathbb{H}_*^{1/2}(\Gamma)$.

Let $\mathbb{V}_I(x)+\sum_{j=1}^{n} m_j\omega_j(x)$ be the solution to the interior displacement problem (6.1.14) on $\bar\Omega$ and the exterior displacement problem (6.1.18) on $\bar\Omega_e$ with the unknown density $\varphi\in\mathbb{H}_*^{-1/2}(\Gamma)$ and constants $\{m_j : j=1,2,\ldots,n\}$. Since the density $\varphi\in\mathbb{H}_*^{1/2}(\Gamma)$, $\mathbb{V}_I(x)+\sum_{j=1}^{n} m_j\omega_j(x)$ satisfies the conditions (6.1.18c)–(6.1.18d) in the exterior displacement (6.1.18). By the displacement condition on Γ, the unknown vector-valued function $\varphi\in\mathbb{H}_*^{-1/2}(\Gamma)$ and the unknown constants $\{m_j : j=1,2,\ldots,n\}$ satisfy the problem:

$$\begin{cases} \text{Find } \varphi\in\mathbb{H}_*^{-1/2}(\Gamma) \text{ and constants } \{m_j : j=1,2,\ldots,n\}, \text{ such that} \\ \int_\Gamma \mathbb{E}(x,y)\varphi(y)\,ds_y+\sum_{j=1}^{n} m_j\omega_j(x)=g(x), \qquad \forall x\in\Gamma \end{cases} \tag{6.3.4}$$

with given $g\in\mathbb{H}^{1/2}(\Gamma)$.

Similarly, let $\mathbb{V}_I(x)+\sum_{j=1}^{3(n-1)} m_j\omega_j(x)$ with the unknown density $\varphi\in\mathbb{H}_{**}^{-1/2}(\Gamma)$ and the unknown constants $\{m_j : j=1,2,\ldots,3(n-1)\}$ be the solution to the interior displacement problem (6.1.14) on $\bar\Omega$ and the exterior displacement problem

(6.1.19) on $\bar{\Omega}_e$. Since $\varphi \in \mathbb{H}_{**}^{-1/2}(\Gamma)$, the conditions (6.1.19c)–(6.1.19d) are satisfied by $\mathbb{V}_I(\boldsymbol{x}) + \sum_{j=1}^{3(n-1)} m_j \boldsymbol{\omega}_j(\boldsymbol{x})$. Using the displacement condition on Γ, we obtain the problem:

$$\begin{cases} \text{Find} \varphi \in \mathbb{H}_{**}^{-1/2}(\Gamma) \text{ and constants } \{ m_j \ : \ j = 1, 2, \ldots, 3(n-1) \}, \text{ such that} \\ \displaystyle\int_\Gamma \mathbb{E}(\boldsymbol{x}, \boldsymbol{y})\varphi(\boldsymbol{y})\, \mathrm{d}s_y + \sum_{j=1}^{3(n-1)} m_j \boldsymbol{\omega}_j(\boldsymbol{x}) = g(\boldsymbol{x}), \qquad \forall \boldsymbol{x} \in \Gamma, \end{cases} \tag{6.3.5}$$

with given $g \in \mathbb{H}^{1/2}(\Gamma)$.

Furthermore, let the single layer potential $\mathbb{V}_I(\boldsymbol{x})$ with unknown density $\varphi \in \mathbb{H}^{-1/2}(\Gamma)$ be the solution to the interior displacement problem (6.1.14) on $\bar{\Omega}$ and the exterior displacement problem (6.1.20). In this case $\mathbb{V}_I(\boldsymbol{x})$ satisfies the condition (6.1.20c) by Theorem 6.2.1. From the displacement condition on Γ, we have

$$\begin{cases} \text{Find } \varphi \in \mathbb{H}^{-1/2}(\Gamma), \ \text{such that} \\ \displaystyle\int_\Gamma \mathbb{E}(\boldsymbol{x}, \boldsymbol{y})\varphi(\boldsymbol{y})\, \mathrm{d}s_y = g(\boldsymbol{x}), \qquad \forall \boldsymbol{x} \in \Gamma, \end{cases} \tag{6.3.6}$$

with given $g \in \mathbb{H}^{1/2}(\Gamma)$.

We now return to the traction problems (6.1.15) and (6.1.21). For the interior traction problem (6.1.15), we know that given $\boldsymbol{h} \in \mathbb{H}^{-1/2}(\Gamma)$, the problem has a solution if and only if

$$\int_\Gamma \left[\boldsymbol{h}(\boldsymbol{x})\right]^{\mathrm{T}} \boldsymbol{\omega}_j(\boldsymbol{x})\, \mathrm{d}s_x = 0, \qquad j = 1, 2, \ldots, 3(n-1).$$

For the exterior traction problem (6.1.21) with given $\boldsymbol{h} \in \mathbb{H}^{-1/2}(\Gamma)$, let $\mathbb{U}(\boldsymbol{x})$ be the solution to (6.1.21), and

$$\mathbb{U}_I(\boldsymbol{x}) = \int_\Gamma \mathbb{E}(\boldsymbol{x}, \boldsymbol{y})\boldsymbol{h}(\boldsymbol{y})\, \mathrm{d}s_y, \qquad \forall \boldsymbol{x} \in \Omega_e.$$

Let

$$\mathbb{U}_*(\boldsymbol{x}) = \mathbb{U}(\boldsymbol{x}) + \mathbb{U}_I(\boldsymbol{x}).$$

It is obvious that $\mathbb{U}_*(\boldsymbol{x})$ satisfies

$$\begin{cases} -\nabla \cdot \boldsymbol{\sigma}(\mathbb{U}_*(\boldsymbol{x})) = \boldsymbol{0}, \qquad \text{in } \Omega_e, \\ \mathbb{T}(D_x, \boldsymbol{\nu}(\boldsymbol{x}))\mathbb{U}_*(\boldsymbol{x})\Big|_{\Gamma+0} = \boldsymbol{h}(\boldsymbol{x}) + \mathbb{T}(D_x, \boldsymbol{\nu}(\boldsymbol{x}))\mathbb{U}_I(\boldsymbol{x})\Big|_{\Gamma+0} := \boldsymbol{h}^*(\boldsymbol{x}), \\ \mathbb{U}_*(\boldsymbol{x}) = \mathbb{E}(\boldsymbol{x}, \boldsymbol{0})\boldsymbol{c}_0 + \boldsymbol{o}(|\boldsymbol{x}|^{1-n}), \qquad \text{as } |\boldsymbol{x}| \to +\infty. \end{cases}$$

The equality (6.2.26) leads to

$$\int_\Gamma \left[\boldsymbol{h}^*(\boldsymbol{x})\right]^{\mathrm{T}} \boldsymbol{\omega}_j(\boldsymbol{x})\,\mathrm{d}s_x = 0, \qquad j = 1, 2, \ldots, 3(n-1). \tag{6.3.7}$$

This indicates that the exterior traction problem (6.1.21) with $\boldsymbol{h} \in \mathbb{H}^{-1/2}(\Gamma)$ is reduced to the same problem with $\boldsymbol{h} \in \mathbb{H}_{**}^{-1/2}(\Gamma)$. In the following we only consider the traction problem with $\boldsymbol{h} \in \mathbb{H}_{**}^{-1/2}(\Gamma)$.

Let the double layer potential $\mathbb{V}_{II}(\boldsymbol{x})$ with unknown density $\boldsymbol{\psi} \in \mathbb{H}^{1/2}(\Gamma)$ be the solution to the interior traction problem (6.1.15) on $\bar{\Omega}$ and the exterior traction problem (6.1.21) on Ω_e with given $\boldsymbol{h} \in \mathbb{H}_{**}^{-1/2}(\Gamma)$. Clearly $\mathbb{V}_{II}(\boldsymbol{x})$ satisfies the homogeneous Navier system in $\mathbb{R}^n \backslash \Gamma$ and the condition (6.1.21c). By the traction condition on Γ and the equality (6.2.44), we have the following differential-integro-differential equations to determine the unknown density $\boldsymbol{\psi}$ (Han, 1994a),

$$\begin{aligned}
& -\frac{1}{2(n-1)\pi} \sum_{1\leqslant k<j\leqslant n} \frac{\partial}{\partial \tau_x^{k,j}} \int_\Gamma \gamma_n(\boldsymbol{x}, \boldsymbol{y}) \frac{\partial}{\partial \tau_y^{k,j}} \boldsymbol{\psi}(\boldsymbol{y})\,\mathrm{d}s_y - \\
& \mathbb{U}(D_x, \boldsymbol{\nu}(\boldsymbol{x})) \int_\Gamma \left[4\mu \mathbb{E}(\boldsymbol{x}, \boldsymbol{y}) - \frac{\mu}{(n-1)\pi} \gamma_n(\boldsymbol{x}, \boldsymbol{y})\, \boldsymbol{I}\right] \mathbb{U}(D_y, \boldsymbol{\nu}(\boldsymbol{y})) \boldsymbol{\psi}(\boldsymbol{y})\,\mathrm{d}s_y + \\
& \frac{\mu}{2(n-1)\pi} \int_\Gamma \left[\mathbb{U}(D_x, \boldsymbol{\nu}(\boldsymbol{x})) \gamma_n(\boldsymbol{x}, \boldsymbol{y}) \mathbb{U}(D_y, \boldsymbol{\nu}(\boldsymbol{y}))\right]^{\mathrm{T}} \boldsymbol{\psi}(\boldsymbol{y})\,\mathrm{d}s_y \\
= \;& -\boldsymbol{h}(\boldsymbol{x}), \qquad \forall \boldsymbol{x} \in \Gamma,
\end{aligned} \tag{6.3.8}$$

which is equivalent to the hypersingular integral equations

$$-\int_\Gamma \mathbb{T}(D_x, \boldsymbol{\nu}(\boldsymbol{x})) \left[\mathbb{T}(D_y, \boldsymbol{\nu}(\boldsymbol{y})) \mathbb{E}(\boldsymbol{x}, \boldsymbol{y})\right]^{\mathrm{T}} \boldsymbol{\psi}(\boldsymbol{y})\,\mathrm{d}s_y = -\boldsymbol{h}(\boldsymbol{x}), \qquad \forall \boldsymbol{x} \in \Gamma. \tag{6.3.9}$$

In the next, we will discuss the solvability of the problems (6.3.4)–(6.3.6) and (6.3.8) by the variational approach.

We first introduce the bilinear forms $\mathbb{A}(\boldsymbol{\varphi}, \boldsymbol{\rho})$ on $\mathbb{H}^{-1/2}(\Gamma) \times \mathbb{H}^{-1/2}(\Gamma)$ and $\mathbb{B}(\boldsymbol{\psi}, \boldsymbol{\mu})$ on $\mathbb{H}^{1/2}(\Gamma) \times \mathbb{H}^{1/2}(\Gamma)$, which are defined by Han (1994a)

$$\mathbb{A}(\boldsymbol{\varphi}, \boldsymbol{\rho}) = \int_\Gamma \int_\Gamma \left[\mathbb{E}(\boldsymbol{x}, \boldsymbol{y}) \boldsymbol{\varphi}(\boldsymbol{y})\right] \cdot \boldsymbol{\rho}(\boldsymbol{x})\,\mathrm{d}s_y\,\mathrm{d}s_x, \quad \forall \boldsymbol{\varphi},\ \boldsymbol{\rho} \in \mathbb{H}^{-1/2}(\Gamma), \tag{6.3.10}$$

$$\begin{aligned}
\mathbb{B}(\boldsymbol{\psi}, \boldsymbol{\mu}) = \;& \frac{1}{2(n-1)\pi} \sum_{1\leqslant k<j\leqslant n} \int_\Gamma \int_\Gamma \gamma_n(\boldsymbol{x}, \boldsymbol{y}) \frac{\partial \boldsymbol{\psi}(\boldsymbol{y})}{\partial \tau_y^{k,j}} \cdot \frac{\partial \boldsymbol{\mu}(\boldsymbol{x})}{\partial \tau_x^{k,j}}\,\mathrm{d}s_y\,\mathrm{d}s_x + \\
& \int_\Gamma \int_\Gamma \left[\mathbb{U}(D_y, \boldsymbol{\nu}(\boldsymbol{y})) \boldsymbol{\psi}(\boldsymbol{y})\right]^{\mathrm{T}} \left[4\mu \mathbb{E}(\boldsymbol{x}, \boldsymbol{y}) - \frac{\mu}{(n-1)\pi} \gamma_n(\boldsymbol{x}, \boldsymbol{y})\, \boldsymbol{I}\right] \cdot \\
& \mathbb{U}(D_x, \boldsymbol{\nu}(\boldsymbol{x})) \boldsymbol{\mu}(\boldsymbol{x})\mathrm{d}s_y\mathrm{d}s_x +
\end{aligned}$$

$$\frac{\mu}{2(n-1)\pi}\int_\Gamma\int_\Gamma\sum_{i,j,k=1}^{n} U_{k,i}(D_y, \boldsymbol{\nu}(\boldsymbol{y}))\varphi_j(\boldsymbol{y})\gamma_n(\boldsymbol{x}, \boldsymbol{y})\cdot$$
$$U_{k,j}(D_x, \boldsymbol{\nu}(\boldsymbol{x}))\mu_i(\boldsymbol{x})\,\mathrm{d}s_y\,\mathrm{d}s_x,$$
$$\forall\,\boldsymbol{\psi},\ \boldsymbol{\mu}\in\mathbb{H}^{1/2}(\Gamma). \tag{6.3.11}$$

We directly obtain

Lemma 6.3.1 *1. The system (6.3.4) of integral equations is equivalent to the following variational problem:*

$$\begin{cases}\text{Find } \boldsymbol{\varphi}\in\mathbb{H}_*^{-1/2}(\Gamma), \text{ such that}\\ \mathbb{A}(\boldsymbol{\varphi}, \boldsymbol{\rho}) = \displaystyle\int_\Gamma g(\boldsymbol{x})\cdot\boldsymbol{\rho}(\boldsymbol{x})\,\mathrm{d}s_x, \qquad \forall\,\boldsymbol{\rho}\in\mathbb{H}_*^{-1/2}(\Gamma)\end{cases} \tag{6.3.12}$$

and

$$m_j = \frac{1}{\mathrm{meas}(\Gamma)}\left\{\int_\Gamma g(\boldsymbol{x})\cdot\boldsymbol{\omega}_j(\boldsymbol{x})\mathrm{d}s_x - \int_\Gamma\int_\Gamma \mathbb{E}(\boldsymbol{x}, \boldsymbol{y})\boldsymbol{\varphi}(\boldsymbol{y})\cdot\boldsymbol{\omega}_j(\boldsymbol{x})\,\mathrm{d}s_y\,\mathrm{d}s_x\right\}, \qquad j = 1, 2, \ldots, n. \tag{6.3.13}$$

2. The integral equations (6.3.5) is equivalent to the variational problem:

$$\begin{cases}\text{Find } \boldsymbol{\varphi}\in\mathbb{H}_{**}^{-1/2}(\Gamma), \text{ such that}\\ \mathbb{A}(\boldsymbol{\varphi}, \boldsymbol{\rho}) = \displaystyle\int_\Gamma g(\boldsymbol{x})\cdot\boldsymbol{\rho}(\boldsymbol{x})\,\mathrm{d}s_x, \qquad \forall\,\boldsymbol{\rho}\in\mathbb{H}_{**}^{-1/2}(\Gamma),\end{cases} \tag{6.3.14}$$

and the constants $\{m_j,\ j = 1, 2, \ldots, 3(n-1)\}$ *are determined by*

$$\sum_{j=1}^{3(n-1)} m_j\int_\Gamma \boldsymbol{\omega}_j(\boldsymbol{x})\cdot\boldsymbol{\omega}_k(\boldsymbol{x})\,\mathrm{d}s_x$$
$$= \int_\Gamma g(\boldsymbol{x})\cdot\boldsymbol{\omega}_k(\boldsymbol{x})\,\mathrm{d}s_x - \int_\Gamma\int_\Gamma\left[\mathbb{E}(\boldsymbol{x}, \boldsymbol{y})\boldsymbol{\varphi}(\boldsymbol{y})\right]\cdot\boldsymbol{\omega}_k(\boldsymbol{x})\,\mathrm{d}s_y\mathrm{d}s_x,$$
$$k = 1, 2, \ldots, 3(n-1). \tag{6.3.15}$$

3. The equivalent variational problem to the integral equations (6.3.6) is given by

$$\begin{cases}\text{Find } \boldsymbol{\varphi}\in\mathbb{H}^{-1/2}(\Gamma), \text{ such that}\\ \mathbb{A}(\boldsymbol{\varphi}, \boldsymbol{\rho}) = \displaystyle\int_\Gamma g(\boldsymbol{x})\cdot\boldsymbol{\rho}(\boldsymbol{x})\,\mathrm{d}s_x, \qquad \forall\,\boldsymbol{\rho}\in\mathbb{H}^{-1/2}(\Gamma).\end{cases} \tag{6.3.16}$$

4. *The equivalent variational problem of the differential-integro-differential equations (6.3.8) is given by*

$$\begin{cases} \text{Find } \psi \in \mathbb{H}_{**}^{-1/2}(\Gamma), \text{ such that} \\ \mathbb{B}(\psi, \mu) = -\displaystyle\int_\Gamma \boldsymbol{h}(\boldsymbol{x}) \cdot \boldsymbol{\mu}(\boldsymbol{x})\, \mathrm{d}s_{\boldsymbol{x}}, \qquad \forall\, \mu \in \mathbb{H}_{**}^{-1/2}(\Gamma). \end{cases} \tag{6.3.17}$$

For the bilinear forms $\mathbb{A}(\varphi, \rho)$ and $\mathbb{B}(\psi, \mu)$, we have the following lemma

Lemma 6.3.2 *1.* $\mathbb{A}(\varphi, \rho)$ *is a symmetric and bounded bilinear form on* $\mathbb{H}^{-1/2}(\Gamma) \times \mathbb{H}^{-1/2}(\Gamma)$*, namely,*

$$\mathbb{A}(\varphi, \rho) = \mathbb{A}(\rho, \varphi), \qquad \forall\, \varphi, \rho \in \mathbb{H}^{-1/2}(\Gamma), \tag{6.3.18}$$

and there exists a constant $M > 0$ *such that*

$$|\mathbb{A}(\varphi, \rho)| \leqslant M \|\varphi\|_{\mathbb{H}^{-1/2}(\Gamma)} \|\rho\|_{\mathbb{H}^{-1/2}(\Gamma)}, \qquad \forall\, \varphi, \rho \in \mathbb{H}^{-1/2}(\Gamma). \tag{6.3.19}$$

2. *There is a constant* $\alpha > 0$ *such that*

$$\mathbb{A}(\varphi, \varphi) \geqslant \alpha \|\varphi\|^2_{\mathbb{H}^{-1/2}(\Gamma)}, \qquad \forall\, \varphi \in \mathbb{H}_*^{-1/2}(\Gamma). \tag{6.3.20}$$

This implies that $\mathbb{A}(\varphi, \rho)$ *is coercive on* $\mathbb{H}_*^{-1/2}(\Gamma) \times \mathbb{H}_*^{-1/2}(\Gamma)$.

3. *For the case* $n = 3$*, there exists a constant* $\alpha > 0$ *such that*

$$\mathbb{A}(\varphi, \varphi) \geqslant \alpha \|\varphi\|^2_{\mathbb{H}^{-1/2}(\Gamma)}, \qquad \forall\, \varphi \in \mathbb{H}^{-1/2}(\Gamma). \tag{6.3.21}$$

This says that in the case $n = 3$*,* $\mathbb{A}(\varphi, \varphi)$ *is also coercive on* $\mathbb{H}^{-1/2}(\Gamma) \times \mathbb{H}^{-1/2}(\Gamma)$.

4. $\mathbb{B}(\psi, \mu)$ *is a symmetric and bounded bilinear form on* $\mathbb{H}^{1/2}(\Gamma) \times \mathbb{H}^{1/2}(\Gamma)$*, i.e.,*

$$\mathbb{B}(\psi, \mu) = \mathbb{B}(\mu, \psi), \qquad \forall\, \psi, \mu \in \mathbb{H}^{1/2}(\Gamma) \tag{6.3.22}$$

and there exists a constant $M > 0$ *such that*

$$|\mathbb{B}(\psi, \mu)| \leqslant M \|\psi\|_{\mathbb{H}^{1/2}(\Gamma)} \|\mu\|_{\mathbb{H}^{1/2}(\Gamma)}, \qquad \forall\, \psi, \mu \in \mathbb{H}^{1/2}(\Gamma). \tag{6.3.23}$$

5. $\mathbb{B}(\psi, \mu)$ *is coercive on* $\mathbb{H}_{**}^{1/2}(\Gamma) \times \mathbb{H}_{**}^{1/2}(\Gamma)$*. Namely there is a constant* $\beta > 0$ *such that*

$$\mathbb{B}(\psi, \psi) \geqslant \beta \|\psi\|^2_{\mathbb{H}^{1/2}(\Gamma)}, \qquad \forall\, \psi \in \mathbb{H}_{**}^{1/2}(\Gamma). \tag{6.3.24}$$

Proof 1. From the definition of $\mathbb{A}(\varphi, \rho)$ and the inequality (6.2.48), the conclusions (6.3.18)–(6.3.19) follow directly.

2. For $\varphi \in \mathbb{H}_*^{-1/2}(\Gamma)$, let

$$\mathbb{V}_{I,\varphi}(\boldsymbol{x}) = \int_\Gamma \mathbb{E}(\boldsymbol{x}, \boldsymbol{y})\varphi(\boldsymbol{y})\, \mathrm{d}s_{\boldsymbol{y}}.$$

Then $\mathbb{V}_{I,\varphi}(\boldsymbol{x})$ is the unique solution to the transmission problem (6.2.47) with $\boldsymbol{c}_0 = \boldsymbol{0}$, i.e.,

$$\begin{cases} -\nabla \cdot \sigma(\mathbb{V}_{I,\varphi}(\boldsymbol{x})) = \boldsymbol{0}, \quad \forall \boldsymbol{x} \in \mathbb{R}^n \backslash \Gamma, & (6.3.25a) \\ \left[\mathbb{V}_{I,\varphi}(\boldsymbol{x})\right]_\Gamma = \boldsymbol{0}, \quad \forall \boldsymbol{x} \in \Gamma, & (6.3.25b) \\ \left[\mathbb{T}(D_x, \boldsymbol{\nu}(\boldsymbol{x}))\mathbb{V}_{I,\varphi}(\boldsymbol{x})\right]_\Gamma = -\varphi(\boldsymbol{x}), \quad \forall \boldsymbol{x} \in \Gamma, & (6.3.25c) \\ \mathbb{V}_{I,\varphi}(\boldsymbol{x}) = \mathcal{O}(|\boldsymbol{x}|^{1-n}), \quad \text{as } |\boldsymbol{x}| \to \infty. & (6.3.25d) \end{cases}$$

By the problem (6.3.25), we obtain

$$J\big(\mathbb{V}_{I,\varphi}(\boldsymbol{x})\big) = \mathbb{A}(\varphi, \varphi), \quad \forall\, \varphi \in \mathbb{H}_*^{-1/2}(\Gamma).$$

Furthermore

$$\begin{aligned} \left\|\varphi\right\|_{-1/2,\Gamma} &= \left\|\left[\mathbb{T}(D_x, \boldsymbol{\nu}_x \mathbb{V}_{I,\varphi}\right]_\Gamma\right\|^2_{-1/2,\Gamma} \\ &\leqslant C J\big(\mathbb{V}_{I,\varphi}(\boldsymbol{x})\big) \\ &= C\mathbb{A}(\varphi, \varphi), \end{aligned}$$

with a constant $C > 0$. Then the equality (6.3.21) follows directly.
3. The inequality (6.3.21) for $n = 3$ can be proved by the same process as in (2). Note that in general the inequality (6.3.21) does not hold for the case $n = 2$.
4. The proofs for inequalities (6.3.23)–(6.3.24) were given by Han (1994b) Lemma 3.1. □

We are now ready to discuss the variational problems (6.3.12), (6.3.14), (6.3.16), (6.3.17).

Applying the Lax-Milgram theorem and the results given in Lemma 6.3.2, we obtain.

Theorem 6.3.1 *1. For given $g \in \mathbb{H}^{1/2}(\Gamma)$, the variational problem (6.3.12) has a unique $\varphi \in \mathbb{H}_*^{-1/2}(\Gamma)$.*
2. For given $g \in \mathbb{H}^{1/2}(\Gamma)$, there is a unique solution $\varphi \in \mathbb{H}_{**}^{-1/2}(\Gamma)$ of the variational problem (6.3.14).
3. For the case $n = 3$ and given $g \in \mathbb{H}^{1/2}(\Gamma)$, the variational problem (6.3.16) has a unique solution $\varphi \in \mathbb{H}^{-1/2}(\Gamma)$.
4. For given $\boldsymbol{h} \in \mathbb{H}_{**}^{-1/2}(\Gamma)$, there is a unique solution $\psi \in \mathbb{H}_{**}^{1/2}(\Gamma)$ of the variational problem (6.3.17).

By the results given in Theorem 6.3.1, we immediately have the following theorem

Theorem 6.3.2 *1. For given $g \in \mathbb{H}^{1/2}(\Gamma)$, the boundary integral equations (6.3.4) has a unique solution $\{\varphi \in \mathbb{H}_*^{-1/2}(\Gamma),\ m_j,\ j = 1, 2, \ldots, n\}$, in which the function $\varphi \in \mathbb{H}_*^{-1/2}(\Gamma)$ is the unique solution to the variational problem (6.3.12) and the constants $\{m_j,\ j = 1, 2, \ldots, n\}$ are given by (6.3.13).*

2. *For given $g \in \mathbb{H}^{1/2}(\Gamma)$, the boundary integral equations (6.3.5) has a unique solution $\{\varphi \in \mathbb{H}_{**}^{-1/2}(\Gamma),\ m_j,\ j = 1, 2, \ldots, 3(n-1)\}$, in which $\varphi \in \mathbb{H}_{**}^{-1/2}(\Gamma)$ is the unique solution to the variational problem (6.3.14) and the constants $\{m_j,\ j = 1, 2, \ldots, 3(n-1)\}$ are determined by (6.3.15).*
3. *For given $g \in \mathbb{H}^{1/2}(\Gamma)$ and $n = 3$, the boundary integral equations (6.3.6) has a unique solution $\varphi \in \mathbb{H}^{-1/2}(\Gamma)$, which is also the unique solution to the variational problem (6.3.16).*
4. *For given $\boldsymbol{h} \in \mathbb{H}_{**}^{-1/2}(\Gamma)$, the general solution to the boundary differential-integro-differential equation (6.3.8) is given by*

$$\psi(\boldsymbol{x}) = \psi_{**}(\boldsymbol{x}) + \sum_{j=1}^{3(n-1)} c_j \boldsymbol{\omega}_j(\boldsymbol{x}), \tag{6.3.26}$$

with $\psi_{**} \in \mathbb{H}_{**}^{-1/2}(\Gamma)$, the unique solution of the variational problem (6.3.17) and constants $\{c_j \in \mathbb{R},\ j = 1, 2, \ldots, 3(n-1)\}$.

Combining the results given in Theorem 6.3.2 and the solution uniqueness of the displacement and traction problems of the Navier system, which are considered in Sect. 6.1 (Kupradze, 2012; Hsiao & Wendland, 2008), we arrive at

Corollary 6.3.1 *Suppose that $\Gamma \in C^2$. Then*

1. *For given $g \in \mathbb{H}^{1/2}(\Gamma)$, the interior displacement problem* (6.1.14) *has a unique solution $\boldsymbol{u} \in \mathbb{H}^1(\Omega)$.*
2. *The exterior displacement problem (6.1.18) (problem (6.1.19) and problem (6.1.20) with $n = 3$) has a unique solution $\boldsymbol{u} \in \mathbb{H}^1_{\text{loc}}(\Omega_e)$ with given $g \in \mathbb{H}^{1/2}(\Gamma)$.*
3. *For given $\boldsymbol{h} \in \mathbb{H}_{**}^{-1/2}(\Gamma)$, the interior traction problem (6.1.15) has a unique solution $\boldsymbol{u}_{**} \in \mathbb{H}^1_{**}(\Omega)$ and the general solution to problem (6.1.15) is given by*

$$\boldsymbol{u}(\boldsymbol{x}) = \boldsymbol{u}_{**}(\boldsymbol{x}) + \sum_{j=1}^{3(n-1)} c_j \boldsymbol{\omega}_j(\boldsymbol{x}), \qquad \forall \boldsymbol{x} \in \Omega. \tag{6.3.27}$$

Here $\mathbb{H}^1_{**}(\Omega)$ is a subspace of $\mathbb{H}^1(\Omega)$ and

$$\mathbb{H}^1_{**}(\Omega) = \left\{ \boldsymbol{u} \in \mathbb{H}^1(\Omega) \text{ and } \int_\Omega \boldsymbol{u}(\boldsymbol{x}) \cdot \boldsymbol{\omega}_j(\boldsymbol{x})\, \mathrm{d}\boldsymbol{x} = 0,\ j = 1, 2, \ldots, 3(n-1) \right\}. \tag{6.3.28}$$

Solvability of the exterior displacement problem (6.1.20) with $n = 2$ will be discussed in the coming section.

6.3.2 *Solvability of the First Kind Integral Equations with n = 2 and the Degenerate Scales*

We first rewrite the first kind integral equations (6.3.6) with $n = 2$ as

$$\begin{cases} \text{Find } \varphi \in \mathbb{H}^{-1/2}(\Gamma), \text{ such that} \\ \displaystyle\int_\Gamma \mathbb{E}_2(\boldsymbol{x}, \boldsymbol{y})\varphi(\boldsymbol{y})\,\mathrm{d}s_y = g(\boldsymbol{x}), \qquad \forall \boldsymbol{x} \in \Gamma, \end{cases} \tag{6.3.29}$$

with given $g \in \mathbb{H}^{1/2}(\Gamma)$.

The first kind integral equations (6.3.29) have similar properties of the first kind integral equation (3.4.20) for the Laplacian. The degenerate scales also exist in the problem (6.3.29) (He et al., 1996; Chen et al., 2017).

Let $\nu = \dfrac{\lambda}{2(\lambda + \mu)}$, which is the Poisson ratio, and $\kappa = 3 - 4\nu$.

For the two special cases (Chen et al., 2017):

1. If Γ is the boundary of a circle with radius R, the degenerate scale of problem (6.3.29) occurs at
$$R = \mathrm{e}^{1/(2\kappa)}. \tag{6.3.30}$$

2. If Γ is the boundary of an ellipse with major radius a and minor radius b, the degenerate scales of problem (6.3.29) occur at
$$\kappa \ln\left(\frac{a+b}{2}\right) = \frac{a}{a+b} \quad \text{or} \quad \kappa \ln\left(\frac{a+b}{2}\right) = \frac{b}{a+b}. \tag{6.3.31}$$

In the following, we discuss how to judge that the given boundary Γ is a degenerate scale of problem (6.3.29) or not.

Definition 6.3.1 Suppose that $\Gamma \in C^2$, which is a simple closed curve in the plane if

1. The homogeneous integral equations
$$\int_\Gamma \mathbb{E}_2(\boldsymbol{x}, \boldsymbol{y})\varphi(\boldsymbol{y})\,\mathrm{d}s_y = \mathbf{0}, \qquad \forall \boldsymbol{x} \in \Gamma \tag{6.3.32}$$
has a nontrivial solution.
2. There is a nonzero vector $(\alpha_1, \alpha_2, \alpha_3)^{\mathrm{T}} = \boldsymbol{\alpha} \in \mathbb{R}^3$ such that the integral equations
$$\int_\Gamma \mathbb{E}_2(\boldsymbol{x}, \boldsymbol{y})\varphi(\boldsymbol{y})\,\mathrm{d}s_y = \sum_{j=1}^{3} \alpha_j \boldsymbol{\omega}_j(\boldsymbol{x}), \qquad \forall \boldsymbol{x} \in \Gamma \tag{6.3.33}$$
do not have any solution.

The curve Γ is called a *degenerate scale* of the first kind integral equations (6.3.29).

Obviously in the case that Γ is a degenerate scale of (6.3.29), the first kind integral equations (6.3.29) are improperly posed. How to judge whether the given boundary Γ is a degenerate scale of integral equations (6.3.29) or not? A computable index γ_Γ will be introduced to answer this question.

Let

$$g_j^*(\boldsymbol{x}) = \int_\Gamma \mathbb{E}_2(\boldsymbol{x}, \boldsymbol{y})\boldsymbol{\omega}_j(\boldsymbol{y})\, \mathrm{d}s_y, \qquad \forall \boldsymbol{x} \in \Gamma, \quad j = 1, 2, 3. \tag{6.3.34}$$

We know that $g_j^* \in \mathbb{H}^{1/2}(\Gamma)$ (for $j = 1, 2, 3$).

Now we consider the following three auxiliary problems for $j = 1, 2, 3$:

$$\begin{cases} \text{Find } \varphi_j^{**} \in \mathbb{H}_{**}^{-1/2}(\Gamma) \text{ such that} \\ \displaystyle\int_\Gamma \mathbb{E}_2(\boldsymbol{x}, \boldsymbol{y})\varphi_j^{**}(\boldsymbol{y})\, \mathrm{d}s_y = g_j^*(\boldsymbol{x}) + \sum_{i=1}^{3} m_{ij}\boldsymbol{\omega}_i(\boldsymbol{x}), \qquad \forall \boldsymbol{x} \in \Gamma. \end{cases} \tag{6.3.35}$$

For each $j = 1, 2, 3$, Theorem 6.3.2 shows that the auxiliary problem (6.3.35) has a unique solution $\{\varphi_j^{**}(\boldsymbol{x});\ m_{1j}, m_{2j}, m_{3j}\}$. Let $\boldsymbol{M}$ be the 3×3 matrix,

$$\boldsymbol{M} = \begin{pmatrix} m_{11} & m_{12} & m_{13} \\ m_{21} & m_{22} & m_{23} \\ m_{31} & m_{32} & m_{33} \end{pmatrix}, \qquad \gamma_\Gamma = \det \boldsymbol{M}. \tag{6.3.36}$$

The index γ_Γ is computable. We have

Theorem 6.3.3 *Suppose that* $\Gamma \in C^2$.

1. *The boundary* Γ *is a degenerate scale of the first kind integral equations (6.3.29) if*
$$\gamma_\Gamma = 0. \tag{6.3.37}$$
2. *If* $\gamma_\Gamma \neq 0$, *then the first kind integral equations (6.3.29) has a unique solution* $\varphi \in \mathbb{H}^{-1/2}(\Gamma)$ *for given* $g \in \mathbb{H}^{1/2}(\Gamma)$.

Proof 1. By the condition $\gamma_\Gamma = 0$, we know that there is a non-zero vector $\boldsymbol{\beta} = (\beta_1, \beta_2, \beta_3)^{\mathrm{T}} \in \mathbb{R}^3$ such that

$$\begin{pmatrix} m_{11} & m_{12} & m_{13} \\ m_{21} & m_{22} & m_{23} \\ m_{31} & m_{32} & m_{33} \end{pmatrix} \begin{pmatrix} \beta_1 \\ \beta_2 \\ \beta_3 \end{pmatrix} = \begin{pmatrix} 0 \\ 0 \\ 0 \end{pmatrix}. \tag{6.3.38}$$

Multiplying (6.3.35) with β_j for each j and adding up all of them ($j = 1, 2, 3$) yields

$$\int_\Gamma \mathbb{E}_2(\boldsymbol{x}, \boldsymbol{y})\Big(\sum_{j=1}^{3} \beta_j \boldsymbol{\varphi}_j^{**}(\boldsymbol{y})\Big)\,\mathrm{d}s_y = \int_\Gamma \mathbb{E}_2(\boldsymbol{x}, \boldsymbol{y})\Big(\sum_{j=1}^{3} \beta_j \boldsymbol{\omega}_j(\boldsymbol{y})\Big)\,\mathrm{d}s_y + \sum_{i=1}^{3}\Big(\sum_{j=1}^{3} m_{ij}\beta_j\Big)\boldsymbol{\omega}_i(\boldsymbol{x}), \qquad \forall \boldsymbol{x} \in \Gamma.$$

By the equality (6.3.38), we arrive at

$$\int_\Gamma \mathbb{E}_2(\boldsymbol{x}, \boldsymbol{y})\Big(\sum_{j=1}^{3} \beta_j \boldsymbol{\varphi}_j^{**}(\boldsymbol{y}) - \sum_{j=1}^{3} \beta_j \boldsymbol{\omega}_j(\boldsymbol{y})\Big)\,\mathrm{d}s_y = \mathbf{0}.$$

This means that

$$\boldsymbol{\varphi}_\beta(\boldsymbol{x}) := \sum_{j=1}^{3} \beta_j \boldsymbol{\varphi}_j^{**}(\boldsymbol{y}) - \sum_{j=1}^{3} \beta_j \boldsymbol{\omega}_j(\boldsymbol{y}) \tag{6.3.39}$$

is a non-trivial solution to the homogeneous integral equations (6.3.32). Moreover, since $\gamma_\Gamma = 0$, there exists a non-zero vector $\boldsymbol{\alpha} = (\alpha_1, \alpha_2, \alpha_3)^{\mathrm{T}} \in \mathbb{R}^3$ such that the following linear algebraic equations

$$\begin{pmatrix} m_{11} & m_{12} & m_{13} \\ m_{21} & m_{22} & m_{23} \\ m_{31} & m_{32} & m_{33} \end{pmatrix}\begin{pmatrix} x_1 \\ x_2 \\ x_3 \end{pmatrix} = \begin{pmatrix} \alpha_1 \\ \alpha_2 \\ \alpha_3 \end{pmatrix} \tag{6.3.40}$$

do not have any solution.

We consider the integral equations below

$$\int_\Gamma \mathbb{E}_2(\boldsymbol{x}, \boldsymbol{y})\boldsymbol{\varphi}(\boldsymbol{y})\,\mathrm{d}s_y = \sum_{j=1}^{3} \alpha_j \boldsymbol{\omega}_j(\boldsymbol{x}), \qquad \forall \boldsymbol{x} \in \Gamma. \tag{6.3.41}$$

The problem (6.3.41) does not have a solution, which will be proved by contradiction in the next. Suppose that $\boldsymbol{\varphi} \in \mathbb{H}^{-1/2}(\Gamma)$ is a solution to the problem (6.3.41), then $\boldsymbol{\varphi}(\boldsymbol{x})$ can be written as

$$\boldsymbol{\varphi}(\boldsymbol{x}) = \boldsymbol{\varphi}^{**}(\boldsymbol{x}) + \sum_{j=1}^{3} \xi_j \boldsymbol{\omega}_j(\boldsymbol{x}), \qquad \forall \boldsymbol{x} \in \Gamma, \tag{6.3.42}$$

with $\boldsymbol{\varphi}^{**} \in \mathbb{H}_{**}^{-1/2}(\boldsymbol{x})$ and $\boldsymbol{\xi} = (\xi_1, \xi_2, \xi_3)^{\mathrm{T}} \in \mathbb{R}^3$. Substituting (6.3.42) into (6.3.41) yields

$$\int_\Gamma \mathbb{E}_2(\boldsymbol{x}, \boldsymbol{y})\boldsymbol{\varphi}^{**}(\boldsymbol{y})\,\mathrm{d}s_y = \sum_{j=1}^{3} \alpha_j \boldsymbol{\omega}_j(\boldsymbol{x}) - \sum_{j=1}^{3} \xi_j \int_\Gamma \mathbb{E}_2(\boldsymbol{x}, \boldsymbol{y})\boldsymbol{\omega}_j(\boldsymbol{y})\,\mathrm{d}s_y. \tag{6.3.43}$$

By (6.3.34)–(6.3.35), we have

$$\int_\Gamma \mathbb{E}_2(\boldsymbol{x}, \boldsymbol{y})\boldsymbol{\omega}_j(\boldsymbol{y})\,\mathrm{d}s_y = \int_\Gamma \mathbb{E}_2(\boldsymbol{x}, \boldsymbol{y})\boldsymbol{\varphi}_j^{**}(\boldsymbol{y})\,\mathrm{d}s_y - \sum_{i=1}^{3} m_{ij}\boldsymbol{\omega}_i(\boldsymbol{x}), \quad j = 1,\ 2,\ 3. \tag{6.3.44}$$

Substituting (6.3.44) into (6.3.43), we arrive at

$$\int_\Gamma \mathbb{E}_2(\boldsymbol{x}, \boldsymbol{y})\left[\boldsymbol{\varphi}^{**}(\boldsymbol{y}) + \sum_{j=1}^{3}\xi_j\boldsymbol{\varphi}_j^{**}(\boldsymbol{y})\right]\mathrm{d}s_y = \sum_{j=1}^{3}\alpha_j\boldsymbol{\omega}_j(\boldsymbol{x}) + \sum_{i,\ j=1}^{3} m_{ij}\xi_j\boldsymbol{\omega}_i(\boldsymbol{x}),$$

namely

$$\begin{aligned}&\int_\Gamma \mathbb{E}_2(\boldsymbol{x}, \boldsymbol{y})\left[\boldsymbol{\varphi}^{**}(\boldsymbol{y}) + \sum_{j=1}^{3}\xi_j\boldsymbol{\varphi}_j^{**}(\boldsymbol{y})\right]\mathrm{d}s_y \\ &= \left[\boldsymbol{\alpha} - \boldsymbol{M}(-\boldsymbol{\xi})\right]\cdot(\boldsymbol{\omega}_1(\boldsymbol{x}),\ \boldsymbol{\omega}_2(\boldsymbol{x}),\ \boldsymbol{\omega}_3(\boldsymbol{x}))^{\mathrm{T}}.\end{aligned} \tag{6.3.45}$$

We see that $\left\{\boldsymbol{\varphi}^{**}(\boldsymbol{x}) + \sum\limits_{j=1}^{3}\xi_j\boldsymbol{\varphi}_j^{**}(\boldsymbol{x}),\ (\boldsymbol{\alpha} - \boldsymbol{M}(-\boldsymbol{\xi})) \in \mathbb{R}^3\right\}$ is the unique solution to the problem (6.3.5) with $\boldsymbol{g}(\boldsymbol{x}) = \boldsymbol{0}$, which is the zero-solution. Then we have

$$\boldsymbol{M}(-\boldsymbol{\xi}) = \boldsymbol{\alpha}.$$

This means that $-\boldsymbol{\xi} = -(\xi_1,\ \xi_2,\ \xi_3)^{\mathrm{T}}$ is a solution to the problem (6.3.40), which contradicts with that problem (6.3.40) does not have any solution.

2. In the case that $\gamma_\Gamma \neq 0$, the matrix $\boldsymbol{M}$ is invertible.
   Let
$$M^{-1} = \left(m_{ij}^{-}\right)_{3\times 3},$$
   then we have
$$\sum_{j=1}^{3} m_{ij}m_{jk}^{-} = \delta_{i,k}, \quad 1 \leqslant i,\ k \leqslant 3.$$

The equalities (6.3.44) can be rewritten as

$$\sum_{i=1}^{3} m_{ij}\boldsymbol{\omega}_i(\boldsymbol{x}) = \int_\Gamma \mathbb{E}_2(\boldsymbol{x}, \boldsymbol{y})\big(\boldsymbol{\varphi}_j^{**}(\boldsymbol{y}) - \boldsymbol{\omega}_j(\boldsymbol{y})\big)\mathrm{d}s_y, \quad j = 1, 2, 3, \tag{6.3.46}$$

furthermore we obtain

$$\sum_{i=1}^{3}\left(\sum_{j=1}^{3} m_{il}m_{jk}^{-}\right)\omega_i(\boldsymbol{x}) = \int_\Gamma \mathbb{E}_2(\boldsymbol{x},\boldsymbol{y})\sum_{j=1}^{3} m_{jk}^{-}\big(\varphi_j^{**}(\boldsymbol{y}) - \omega_j(\boldsymbol{y})\big)\mathrm{d}s_{\boldsymbol{y}},\ k=1,2,3.$$

Finally we get

$$\omega_k(\boldsymbol{x}) = \int_\Gamma \mathbb{E}_2(\boldsymbol{x},\boldsymbol{y})\varphi_k^{\omega}(\boldsymbol{y})\mathrm{d}s_{\boldsymbol{y}},\quad k=1,2,3, \tag{6.3.47}$$

with

$$\varphi_k^{\omega}(\boldsymbol{x}) = \sum_{j=1}^{3} m_{jk}^{-}\big(\varphi_j^{**}(\boldsymbol{y}) - \omega_j(\boldsymbol{y})\big). \tag{6.3.48}$$

For given $g \in \mathbb{H}^{1/2}(\Gamma)$, we know that problem (6.3.5) with $n=2$ has a unique solution $\left\{\varphi_g^{**} \in \mathbb{H}_{**}^{-1/2}(\Gamma) \text{ and constants } \beta_1,\ \beta_2,\ \beta_3\right\}$ such that

$$\int_\Gamma \mathbb{E}_2(\boldsymbol{x},\boldsymbol{y})\varphi_g^{**}(\boldsymbol{y})\,\mathrm{d}s_y + \sum_{j=1}^{3}\beta_j\omega_j(\boldsymbol{x}) = g(\boldsymbol{x}),\qquad \forall \boldsymbol{x}\in\Gamma. \tag{6.3.49}$$

Substituting (6.3.48) into (6.3.49), we arrive at

$$\int_\Gamma \mathbb{E}_2(\boldsymbol{x},\boldsymbol{y})\left\{\varphi_g^{**}(\boldsymbol{y}) + \sum_{j=1}^{3}\beta_j\varphi_j^{\omega}(\boldsymbol{y})\right\}\mathrm{d}s_y = g(\boldsymbol{x}),\qquad \forall \boldsymbol{x}\in\Gamma.$$

This means that for given $g \in \mathbb{H}^{1/2}(\Gamma)$, the first kind equations (6.3.6) with $n=2$ has a solution in $\mathbb{H}^{-1/2}(\Gamma)$ provided by $\gamma_\Gamma \neq 0$. Furthermore, we will show the solution uniqueness of problem (6.3.6) with $n=2$ when $\gamma_\Gamma \neq 0$.

Suppose that $\varphi \in \mathbb{H}^{-1/2}(\Gamma)$ is a solution to the homogeneous integral equations (6.3.32). Then $\varphi(\boldsymbol{x})$ can be decomposed as

$$\varphi(\boldsymbol{x}) = \varphi^{**}(\boldsymbol{x}) + \sum_{j=1}^{3}\xi_j\omega_j(\boldsymbol{x})$$

with $\varphi^{**}(\boldsymbol{x}) \in \mathbb{H}_{**}^{-1/2}(\Gamma)$. We have

$$\int_\Gamma \mathbb{E}_2(\boldsymbol{x},\boldsymbol{y})\varphi^{**}(\boldsymbol{y})\,\mathrm{d}s_y + \sum_{j=1}^{3}\xi_j\int_\Gamma \mathbb{E}_2(\boldsymbol{x},\boldsymbol{y})\omega_j(\boldsymbol{y})\,\mathrm{d}s_y = \mathbf{0}. \tag{6.3.50}$$

Substituting (6.3.44) into (6.3.50), we arrive at

$$\int_{\Gamma} \mathbb{E}_2(\boldsymbol{x}, \boldsymbol{y})\Big\{\boldsymbol{\varphi}^{**}(\boldsymbol{y}) + \sum_{j=1}^{3} \xi_j \boldsymbol{\varphi}_j^{**}(\boldsymbol{y})\Big\}\, \mathrm{d}s_y - \sum_{i=1}^{3}\Big(\sum_{j=1}^{3} m_{ij}\xi_j\Big)\boldsymbol{\omega}_i(\boldsymbol{x}) = \mathbf{0},\ \forall \boldsymbol{x} \in \Gamma.$$

The uniqueness of the first integral equations (6.3.6) leads to

$$\boldsymbol{M}\,\boldsymbol{\xi} = \mathbf{0}, \quad \boldsymbol{\varphi}^{**}(\boldsymbol{x}) + \sum_{j=1}^{3} \xi_j \boldsymbol{\varphi}_j^{**}(\boldsymbol{x}) \equiv \mathbf{0}, \qquad \forall \boldsymbol{x} \in \Gamma.$$

In the case that $\gamma_\Gamma \neq 0$, the matrix $\boldsymbol{M}$ is invertible. Directly we obtain

$$\boldsymbol{\xi} = (\xi_1,\ \xi_2,\ \xi_3)^{\mathrm{T}} = (0,\ 0,\ 0)^{\mathrm{T}},$$
$$\boldsymbol{\varphi}^{**}(\boldsymbol{x}) \equiv \mathbf{0}, \qquad \forall \boldsymbol{x} \in \Gamma.$$

Namely,

$$\boldsymbol{\varphi}(\boldsymbol{x}) \equiv \mathbf{0}, \qquad \forall \boldsymbol{x} \in \Gamma.$$

The uniqueness of problem (6.3.6) with $n = 2$ provided by $\gamma_\Gamma \neq 0$ is thus proved completely. □

Corollary 6.3.2 *The boundary* Γ *is a degenerate scale of the first kind integral equations (6.3.29) if and only if*

$$\gamma_\Gamma = 0.$$

Corollary 6.3.3 *For given* $g \in \mathbb{H}^{1/2}(\Gamma)$*, we have*

1. *The exterior displacement problem (6.1.20) with* $n = 2$ *and* $\gamma_\Gamma = 0$ *is improperly posed.*
2. *The exterior displacement problem (6.1.20) with* $n = 2$ *and* $\gamma_\Gamma \neq 0$ *has a unique solution.*

6.3.3 *The Second Kind Integral Equations of the Boundary Value Problems of The Navier System*

First, let the double layer potential $\mathbb{V}_{II}(\boldsymbol{x})$ with density $\boldsymbol{\psi}(\boldsymbol{x})$ be the solution to the interior displacement problem (6.1.14). The displacement boundary condition (6.1.14b) leads to the following second kind integral equations to determine the unknown density $\boldsymbol{\psi}(\boldsymbol{x})$,

$$-\frac{1}{2}\boldsymbol{\psi}(\boldsymbol{x}) + (\mathbb{K}_{II}\boldsymbol{\psi})(\boldsymbol{x}) = g(\boldsymbol{x}), \qquad \forall \boldsymbol{x} \in \Gamma. \tag{6.3.51}$$

By the equality (6.2.20), the property of the double layer potential $\mathbb{V}_{II}(\boldsymbol{x})$ as $|\boldsymbol{x}| \to \infty$, we know that $\mathbb{V}_{II}(\boldsymbol{x})$ itself can not represent the solutions to those three basic exterior displacement problems given in Sect. 6.1. Thus, we look for a function in the form,

$$\mathbb{U}(\boldsymbol{x}) = \mathbb{V}_{II}(\boldsymbol{x}) + \sum_{j=1}^{3(n-1)} m_j \boldsymbol{\omega}_j(\boldsymbol{x}), \tag{6.3.52}$$

as the solution to the exterior displacement problem (6.1.18). Here, the double layer potential $\mathbb{V}_{II}(\boldsymbol{x})$ with unknown density $\boldsymbol{\psi}(\boldsymbol{x})$ and $\{m_j,\ j = 1, 2, \ldots, 3(n-1)\}$ are unknown constants. It is obvious that $\mathbb{U}(\boldsymbol{x})$ given by (6.3.52) satisfies the homogeneous Navier system in Ω_e and the conditions (6.1.18c)–(6.1.18d). By the displacement boundary condition (6.1.18b), we obtain the following second integral equations to determine the unknown density $\boldsymbol{\psi}(\boldsymbol{x})$ and constants $\{m_j,\ j = 1, 2, \ldots, 3(n-1)\}$,

$$\frac{1}{2}\boldsymbol{\psi}(\boldsymbol{x}) + (\mathbb{K}_{II}\boldsymbol{\psi})(\boldsymbol{x}) = g(\boldsymbol{x}) - \sum_{j=1}^{3(n-1)} m_j \boldsymbol{\omega}_j(\boldsymbol{x}) := g^*(\boldsymbol{x}), \qquad \forall \boldsymbol{x} \in \Gamma. \tag{6.3.53}$$

Let the single layer potential $\mathbb{V}_I(\boldsymbol{x})$ with unknown density $\boldsymbol{\varphi}(\boldsymbol{x})$ be the solution to the interior traction problem (6.1.15) on $\bar{\Omega}$ and the solution to the exterior traction problem (6.1.21) on $\bar{\Omega}_e$.

For the interior traction problem, applying the traction boundary condition (6.1.15b) yields

$$\frac{1}{2}\boldsymbol{\varphi}(\boldsymbol{x}) + (\mathbb{K}_{II}^*\boldsymbol{\varphi})(\boldsymbol{x}) = \boldsymbol{h}(\boldsymbol{x}), \qquad \forall \boldsymbol{x} \in \Gamma. \tag{6.3.54}$$

For the exterior traction problem, applying the traction boundary condition (6.1.21b) yields

$$-\frac{1}{2}\boldsymbol{\varphi}(\boldsymbol{x}) + (\mathbb{K}_{II}^*\boldsymbol{\varphi})(\boldsymbol{x}) = \boldsymbol{h}(\boldsymbol{x}), \qquad \forall \boldsymbol{x} \in \Gamma. \tag{6.3.55}$$

We now obtain the four systems of second kind integral equations (6.3.51), (6.3.53)–(6.3.55) involve operators $\mathbb{K}_{II}$ and $\mathbb{K}_{II}^*$. Let

$$\mathbb{T}_- = -\frac{1}{2}\boldsymbol{I} + \mathbb{K}_{II}, \qquad \mathbb{T}_+ = \frac{1}{2}\boldsymbol{I} + \mathbb{K}_{II},$$

$$\mathbb{T}_-^* = -\frac{1}{2}\boldsymbol{I} + \mathbb{K}_{II}^*, \qquad \mathbb{T}_+^* = \frac{1}{2}\boldsymbol{I} + \mathbb{K}_{II}^*.$$

The four second kind integral equations (6.3.51), (6.3.53)–(6.3.55) can be rewritten as

$$\mathbb{T}_-\boldsymbol{\psi}(\boldsymbol{x}) = g(\boldsymbol{x}), \quad \forall \boldsymbol{x} \in \Gamma, \tag{6.3.56}$$

$$\mathbb{T}_+\boldsymbol{\psi}(\boldsymbol{x}) = g^*(\boldsymbol{x}), \quad \forall \boldsymbol{x} \in \Gamma, \tag{6.3.57}$$

$$\mathbb{T}_+^*\boldsymbol{\varphi}(\boldsymbol{x}) = \boldsymbol{h}(\boldsymbol{x}), \quad \forall \boldsymbol{x} \in \Gamma, \tag{6.3.58}$$

$$\mathbb{T}_-^*\varphi(\boldsymbol{x}) = \boldsymbol{h}(\boldsymbol{x}), \quad \forall\, \boldsymbol{x} \in \Gamma. \tag{6.3.59}$$

We now recall some properties of the operators $\mathbb{K}_{II}$, $\mathbb{K}_{II}^*$, $\mathbb{T}_-$, $\mathbb{T}_+$, $\mathbb{T}_-^*$ and $\mathbb{T}_+^*$ (Hsiao & Wendland, 2008; Mikhlin, 1938; Mikhlin & Prössdorf, 1987; Kupradze, 2012; Costabel 1988).

Lemma 6.3.3 *Suppose that* $\Gamma \in C^2(\Gamma)$.

1. *The operators* $\mathbb{K}_{II}$ *and* $\mathbb{K}_{II}^*$ *are continuous and compact mappings from* $\mathbb{C}^\alpha(\Gamma)$ *to* $\mathbb{C}^\alpha(\Gamma)$ *with* $0 < \alpha < 1$, *i.e.,*

$$\mathbb{K}_{II} \;\; \textit{and} \;\; \mathbb{K}_{II}^* : \mathbb{C}^\alpha(\Gamma) \to \mathbb{C}^\alpha(\Gamma).$$

2. *For given* g, g^*, *and* $\boldsymbol{h} \in \mathbb{C}^\alpha(\Gamma)$, *the Fredholm alternative remains valid for the second kind integral equation (6.3.56)–(6.3.59).*
3. *The operator* $\mathbb{K}_{II}^*$ *is a continuous mapping from* $\mathbb{H}^{-1/2}(\Gamma)$ *to* $\mathbb{H}^{-1/2}(\Gamma)$*; the operator* $\mathbb{K}_{II}$ *is a continuous mapping from* $\mathbb{H}^{1/2}(\Gamma)$ *to* $\mathbb{H}^{1/2}(\Gamma)$. *That is,*

$$\begin{aligned} \mathbb{K}_{II}^* &: \mathbb{H}^{-1/2}(\Gamma) \to \mathbb{H}^{-1/2}(\Gamma), \\ \mathbb{K}_{II} &: \mathbb{H}^{1/2}(\Gamma) \to \mathbb{H}^{1/2}(\Gamma). \end{aligned}$$

4. *Furthermore, we can obtain that for given* $g, g^* \in \mathbb{H}^{1/2}(\Gamma)$ *and* $\boldsymbol{h} \in \mathbb{H}^{-1/2}(\Gamma)$, *the Fredholm alternative remains valid for the second integral equation (6.3.56)–(6.3.59) by the approach used in Chap. 5.*

The solvability of the second kind integral equation (6.3.56)–(6.3.59) for given g, g^* *and* $\in \mathbb{C}^\alpha(\Gamma)$ *with* $0 < \alpha < 1$ *can be found in the book by Hsiao and Wendland (2008). In the following we discuss the solvability of (6.3.56)–(6.3.59) for given* $g, g^* \in \mathbb{H}^{1/2}(\Gamma)$ *and* $\boldsymbol{h} \in \mathbb{H}^{-1/2}(\Gamma)$.

Let $\mathbb{N}(\mathbb{T}_-)$, $\mathbb{N}(\mathbb{T}_+)$ and $\mathbb{N}(\mathbb{T}_-^*)$, $\mathbb{N}(\mathbb{T}_+^*)$ be the null-spaces of the operators $\mathbb{T}_-$, $\mathbb{T}_+$ and $\mathbb{T}_-^*$, $\mathbb{T}_+^*$ in $\mathbb{H}^{1/2}(\Gamma)$ and $\mathbb{H}^{-1/2}(\Gamma)$ respectively. Then

$$\begin{aligned} \mathbb{H}^{1/2}(\Gamma) &= \mathbb{N}(\mathbb{L}) \oplus \mathbb{N}(\mathbb{L})^\perp, \qquad \text{for } \mathbb{L} = \mathbb{T}_-, \text{ and } \mathbb{T}_+, \\ \mathbb{H}^{-1/2}(\Gamma) &= \mathbb{N}(\mathbb{L}) \oplus \mathbb{N}(\mathbb{L})^\perp, \qquad \text{for } \mathbb{L} = \mathbb{T}_-^* \text{ and } \mathbb{T}_+^*. \end{aligned}$$

Furthermore, we have

Lemma 6.3.4 *1.* $\mathbb{N}(\mathbb{T}_-) = \mathbb{N}(\mathbb{T}_-^*) = \{\mathbf{0}\}$.
2. $\dim\mathbb{N}(\mathbb{T}_+) = \dim\mathbb{N}(\mathbb{T}_+^*) = 3(n-1)$,

$$\mathbb{N}(\mathbb{T}_+) = \;\{\psi : \psi(\boldsymbol{x}) = \sum_{j=1}^{3(n-1)} c_j\omega_j(\boldsymbol{x}), \quad c_j \in \mathbb{R},\; j = 1, 2, \ldots, 3(n-1)\,\},$$

$$\mathbb{N}(\mathbb{T}_+^*) = \;\{\varphi : \varphi(\boldsymbol{x}) = \sum_{j=1}^{3(n-1)} c_j\varphi_j(\boldsymbol{x}), \quad c_j \in \mathbb{R},\; j = 1, 2, \ldots, 3(n-1)\,\}.$$

The functions $\{\varphi_j(\boldsymbol{x}),\ j = 1,\ 2,\ \ldots,\ 3(n-1)\}$ *will be given in the proof below.*

Proof 1. For any $\boldsymbol{\psi} \in \mathbb{N}(\mathbb{T}_-)$, i.e.,

$$-\frac{1}{2}\boldsymbol{\psi}(\boldsymbol{x}) + \big(\mathbb{K}_{II}\boldsymbol{\psi}\big)(\boldsymbol{x}) = \mathbf{0}, \qquad \forall \boldsymbol{x} \in \Gamma.$$

Let

$$\boldsymbol{u}(\boldsymbol{x}) = \int_\Gamma \Big[\mathbb{T}(D_y,\ \boldsymbol{\nu}(\boldsymbol{y}))\mathbb{E}(\boldsymbol{x},\boldsymbol{y})\Big]^{\mathrm{T}}\boldsymbol{\psi}(\boldsymbol{y})\mathrm{d}s_y,$$

then we have

$$\boldsymbol{u}(\boldsymbol{x}) = \int_\Gamma \big[\mathbb{T}(D_y,\ \boldsymbol{\nu}(\boldsymbol{y}))\mathbb{E}(\boldsymbol{x},\boldsymbol{y})\big]^{\mathrm{T}}\boldsymbol{\psi}(\boldsymbol{y})\,\mathrm{d}s_y = \mathbf{0}, \qquad \forall \boldsymbol{x} \in \Omega,$$

and consequently

$$\mathbb{T}(D_x,\ \boldsymbol{\nu}(\boldsymbol{x}))\boldsymbol{u}(\boldsymbol{x})\Big|_{\Gamma+0} = \mathbb{T}(D_x,\ \boldsymbol{\nu}(\boldsymbol{x}))\boldsymbol{u}(\boldsymbol{x})\Big|_{\Gamma-0} = \mathbf{0}, \qquad \forall \boldsymbol{x} \in \Gamma.$$

On the domain Ω_e, $\boldsymbol{u}(\boldsymbol{x})$ satisfies

$$\begin{cases} -\nabla\cdot\boldsymbol{\sigma}(\boldsymbol{u}(\boldsymbol{x})) = \mathbf{0}, & \forall \boldsymbol{x} \in \Omega_e, \\ \boldsymbol{\sigma}(\boldsymbol{u}(\boldsymbol{x}))\boldsymbol{\nu}(\boldsymbol{x})\Big|_\Gamma = \mathbf{0}, & \forall \boldsymbol{x} \in \Gamma, \\ \boldsymbol{u}(\boldsymbol{x}) = \mathcal{O}(|\boldsymbol{x}|^{1-n}), & \text{as } |\boldsymbol{x}| \to \infty. \end{cases}$$

By the uniqueness of the exterior traction problem (6.1.21), we have $\boldsymbol{u}(\boldsymbol{x}) = 0$, $\forall \boldsymbol{x} \in \Omega_e$. Therefore, we have

$$\boldsymbol{\psi}(\boldsymbol{x}) = \big[\boldsymbol{u}(\boldsymbol{x})\big]_\Gamma = \mathbf{0}, \qquad \forall \boldsymbol{x} \in \Gamma.$$

As a result we get

$$\mathbb{N}(\mathbb{T}_-) = \{\mathbf{0}\}.$$

We have known $\langle\mathbb{H}^{1/2}(\Gamma), \mathbb{H}^{-1/2}(\Gamma)\rangle$ is a dual system and $\mathbb{K}_{II}$, $\mathbb{K}_{II}^*$ are adjoint operator, the we obtain and

$$\mathbb{N}(\mathbb{T}_-^*) = \{\mathbf{0}\}.$$

2. For $\boldsymbol{\psi} \in \mathbb{N}(\mathbb{T}_+)$, i.e.,

$$\frac{1}{2}\boldsymbol{\psi}(\boldsymbol{x}) + \big(\mathbb{K}_{II}\boldsymbol{\psi}\big)(\boldsymbol{x}) = \mathbf{0}, \qquad \forall \boldsymbol{x} \in \Gamma,$$

Let

$$u(x) = \int_{\Gamma} \left[\mathbb{T}(D_y, \nu(y))\mathbb{E}(x, y) \right]^{\mathrm{T}} \psi(y) \mathrm{d}s_y,$$

we have

$$u(x) = \int_{\Gamma} \left[\mathbb{T}(D_y, \nu(y))\mathbb{E}(x, y) \right]^{\mathrm{T}} \psi(y) \, \mathrm{d}s_y = \mathbf{0}, \qquad \forall x \in \Omega_e,$$

and consequently

$$\mathbb{T}(D_x, \nu(x))u(x)\Big|_{\Gamma-0} = \mathbb{T}(D_x, \nu(x))u(x)\Big|_{\Gamma+0} = \mathbf{0}, \qquad \forall x \in \Gamma.$$

On the domain Ω, $u(x)$ satisfies

$$\begin{cases} -\nabla \cdot \sigma(u(x)) = \mathbf{0}, & \forall x \in \Omega, \\ \sigma(u(x))\nu(x)\Big|_{\Gamma} = \mathbf{0}, & \forall x \in \Gamma. \end{cases}$$

This is an interior traction problem with given $h(x) = \mathbf{0}$. The solution to this problem is given by

$$u(x) = -\sum_{j=1}^{3(n-1)} c_j \omega_j(x), \qquad \forall x \in \Omega.$$

Namely,

$$u(x) = \begin{cases} -\sum\limits_{j=1}^{3(n-1)} c_j \omega_j(x), & \forall x \in \Omega, \\ -\frac{1}{2} \sum\limits_{j=1}^{3(n-1)} c_j \omega_j(x), & \forall x \in \Gamma, \\ \mathbf{0}, & \forall x \in \Omega_e. \end{cases}$$

Therefore $u(x)$ is the unique solution to the transmission problem (6.2.52)–(6.2.53) with

$$\psi(x) = \sum_{j=1}^{3(n-1)} c_j \omega_j(x).$$

We now arrive at

$$\mathbb{N}(\mathbb{T}_+) = \left\{ \psi : \psi(x) = \sum_{j=1}^{3(n-1)} c_j \, \omega_j(x), \quad c_j \in \mathbb{R}, \; j = 1, 2, \ldots, 3(n-1) \right\}$$

and

$$\dim \mathbb{N}(\mathbb{T}_+) = 3(n-1).$$

Similarly, for $\varphi \in \mathbb{N}(\mathbb{T}_+^*)$, i.e.,

$$\frac{1}{2}\varphi(\boldsymbol{x}) + \big(\mathbb{K}_{II}^*\varphi\big)(\boldsymbol{x}) = \boldsymbol{0}, \qquad \forall \boldsymbol{x} \in \Gamma.$$

Let

$$\boldsymbol{v}(\boldsymbol{x}) = \int_\Gamma \mathbb{E}(\boldsymbol{x}, \boldsymbol{y})\varphi(\boldsymbol{y})\, \mathrm{d}s_y,$$

we have

$$\mathbb{T}(D_x, \boldsymbol{\nu}(\boldsymbol{x}))\boldsymbol{v}(\boldsymbol{x})\Big|_{\Gamma-0} = \boldsymbol{0}, \qquad \forall \boldsymbol{x} \in \Gamma.$$

On the domain Ω, $\boldsymbol{v}(\boldsymbol{x})$ satisfies

$$\begin{cases} -\nabla\cdot\boldsymbol{\sigma}(\boldsymbol{v}(\boldsymbol{x})) = \boldsymbol{0}, & \forall \boldsymbol{x} \in \Omega, \\ \mathbb{T}(D_x, \boldsymbol{\nu}(\boldsymbol{x}))\boldsymbol{v}(\boldsymbol{x}) = \boldsymbol{0}, & \forall \boldsymbol{x} \in \Gamma, \end{cases}$$

which is an interior traction problem with given $\boldsymbol{h}(\boldsymbol{x}) = \boldsymbol{0}$ and $\boldsymbol{v}(\boldsymbol{x})$ is given by

$$\boldsymbol{v}(\boldsymbol{x}) = \sum_{j=1}^{3(n-1)} c_j\boldsymbol{\omega}_j(\boldsymbol{x}), \quad c_j \in \mathbb{R},\ j = 1, 2, \ldots, 3(n-1), \qquad \forall \boldsymbol{x} \in \Omega.$$

For each $\boldsymbol{\omega}_j(\boldsymbol{x})$ $(1 \leqslant j \leqslant 3(n-1))$, we consider the following first kind integral equations

$$\int_\Gamma \mathbb{E}(\boldsymbol{x}, \boldsymbol{y})\varphi_j(\boldsymbol{y})\, \mathrm{d}s_y = \boldsymbol{\omega}_j(\boldsymbol{x}), \qquad \forall \boldsymbol{x} \in \Gamma. \tag{6.3.60}$$

In the cases $n > 2$ and $n = 2$ with $\gamma_\Gamma \neq 0$, Theorems 6.3.1 and 6.3.3 indicate that the problem (6.3.60) has a unique solution $\varphi_j(\boldsymbol{x})$. Let

$$\boldsymbol{v}_j(\boldsymbol{x}) = \int_\Gamma \mathbb{E}(\boldsymbol{x}, \boldsymbol{y})\varphi_j(\boldsymbol{y})\, \mathrm{d}s_y, \qquad \forall \boldsymbol{x} \in \mathbb{R}^n.$$

Then we see

$$\boldsymbol{v}_j(\boldsymbol{x}) = \boldsymbol{\omega}_j(\boldsymbol{x}), \qquad \forall \boldsymbol{x} \in \bar{\Omega}$$

and

$$\mathbb{T}(D_x, \boldsymbol{\nu}(\boldsymbol{x}))\boldsymbol{v}_j(\boldsymbol{x}) = \frac{1}{2}\varphi_j(\boldsymbol{x}) + \big(\mathbb{K}_{II}^*\varphi_j\big)(\boldsymbol{x}) = \boldsymbol{0}, \qquad \forall \boldsymbol{x} \in \Gamma.$$

Namely $\varphi_j \in \mathbb{N}(\mathbb{T}_+^*)$ for $j = 1, 2, \ldots, 3(n-1)$ and

$$\mathbb{N}(\mathbb{T}_+^*) = \left\{ \varphi \,\middle|\, \varphi(\boldsymbol{x}) = \sum_{j=1}^{3(n-1)} c_j\varphi_j(\boldsymbol{x}), \quad c_j \in \mathbb{R},\ j = 1, 2, \ldots, 3(n-1) \right\}$$

since $\left\{ \varphi_j(\boldsymbol{x}),\ j = 1, 2, \ldots, 3(n-1) \right\}$ are linearly independent.

Here we do not discuss the case $n = 2$ with $\gamma_\Gamma = 0$ and would like to leave it to interested readers as an exercise. □

Applying the Fredholm Alternative Theorem, we can directly obtain the following results.

Theorem 6.3.4 *Suppose that $\Gamma \in C^2(\Gamma)$ and $g \in \mathbb{H}^{1/2}(\Gamma)$, $\boldsymbol{h} \in \mathbb{H}^{-1/2}(\Gamma)$.*

1. *The second kind integral equations (6.3.51) has a unique solution $\psi \in \mathbb{H}^{1/2}(\Gamma)$ and*

$$\boldsymbol{u}(\boldsymbol{x}) = \int_\Gamma \left[\mathbb{T}(D_y, \boldsymbol{\nu}(\boldsymbol{y}))\mathbb{E}(\boldsymbol{x}, \boldsymbol{y}) \right]^{\mathrm{T}} \psi(\boldsymbol{y})\, \mathrm{d}s_y, \qquad \forall \boldsymbol{x} \in \Omega$$

is the unique solution to the interior displacement problem (6.1.14).

2. *The second kind integral equations (6.3.55) has a unique solution $\varphi \in \mathbb{H}^{1/2}(\Gamma)$ and*

$$\boldsymbol{v}(\boldsymbol{x}) = \int_\Gamma \mathbb{E}(\boldsymbol{x}, \boldsymbol{y})\varphi(\boldsymbol{y})\, \mathrm{d}s_y, \qquad \forall \boldsymbol{x} \in \Omega_e$$

is the unique solution to the exterior traction problem (6.1.21).

3. *The second kind integral equations (6.3.53) has general solutions $\{ m_j,\ j = 1, 2, \ldots, 3(n-1) \}$ and*

$$\psi(\boldsymbol{x}) = \psi^{**}(\boldsymbol{x}) + \sum_{j=1}^{3(n-1)} m_j \omega_j(\boldsymbol{x}).$$

The constants $\{ m_j,\ j = 1, 2, \ldots, 3(n-1) \}$ are uniquely determined by

$$\sum_{j=1}^{3(n-1)} m_j \int_\Gamma \omega_j(\boldsymbol{x}) \cdot \omega_i(\boldsymbol{x})\, \mathrm{d}s_x = \int_\Gamma g(\boldsymbol{x}) \cdot \omega_i(\boldsymbol{x})\, \mathrm{d}s_x, \quad i = 1, 2, \ldots, 3(n-1)$$

and $\psi^{**} \in \mathbb{N}(\mathbb{T}_+)^{\mathrm{T}}$ is uniquely determined. Then

$$\boldsymbol{u}(\boldsymbol{x}) = \int_\Gamma \left[\mathbb{T}(D_y, \boldsymbol{\nu}(\boldsymbol{y}))\mathbb{E}(\boldsymbol{x}, \boldsymbol{y}) \right]^{\mathrm{T}} \psi^{**}(\boldsymbol{y})\, \mathrm{d}s_y + \sum_{j=1}^{3(n-1)} m_j \omega_j(\boldsymbol{x}), \qquad \forall \boldsymbol{x} \in \Omega_e$$

is the unique solution to the exterior displacement problem (6.1.18).

Proof Assume that $\boldsymbol{h} \in \mathbb{N}(\mathbb{T}_+^*)^\perp$, namely

$$\int_\Gamma \boldsymbol{h}(\boldsymbol{x}) \cdot \varphi_j(\boldsymbol{x})\, \mathrm{d}s_x = 0, \qquad \forall j = 1, 2, \ldots, 3(n-1).$$

Then the second kind integral equations (6.3.54) has a solution. The general solution takes the form

$$\varphi(\boldsymbol{x}) = \varphi^{**}(\boldsymbol{x}) + \sum_{j=1}^{3(n-1)} c_j \varphi_j(\boldsymbol{x}), \qquad \forall \boldsymbol{x} \in \Gamma,$$

with $\varphi^{**} \in \mathbb{N}(\mathbb{T}_+^*)^\perp$, which is uniquely determined. Then

$$\boldsymbol{v}(\boldsymbol{x}) = \int_\Gamma \mathbb{E}(\boldsymbol{x}, \boldsymbol{y})\varphi^{**}(\boldsymbol{y})\, \mathrm{d}s_y + \sum_{j=1}^{3(n-1)} c_j \boldsymbol{\omega}_j(\boldsymbol{x}), \qquad \forall \boldsymbol{x} \in \Omega$$

is the general solution to the interior traction problem (6.1.15). This means that

$$\int_\Gamma \boldsymbol{h}(\boldsymbol{x}) \cdot \boldsymbol{\omega}_j(\boldsymbol{x})\, \mathrm{d}s_x = 0, \qquad j = 1, 2, \ldots, 3(n-1).$$

Namely if $\boldsymbol{h} \in \mathbb{N}(\mathbb{T}_+^*)^\perp$, then $\boldsymbol{h} \in \mathbb{N}(\mathbb{T}_+)^\perp$ and

$$\mathbb{N}(\mathbb{T}_+^*)^\perp \subset \mathbb{N}(\mathbb{T}_+)^\perp, \qquad \mathbb{N}(\mathbb{T}_+) \subset \mathbb{N}(\mathbb{T}_+^*).$$

Since $\dim \mathbb{N}(\mathbb{T}_+) = \dim \mathbb{N}(\mathbb{T}_+^*) = 3(n-1)$, we have

$$\mathbb{N}(\mathbb{T}_+^*) = \mathbb{N}(\mathbb{T}_+).$$

The proof of Theorem 6.3.4 is thus completed. □

Corollary 6.3.4

$$\mathbb{N}(\mathbb{T}_+^*) = \left\{ \varphi : \varphi(\boldsymbol{x}) = \sum_{j=1}^{3(n-1)} c_j \boldsymbol{\omega}_j(\boldsymbol{x}), \quad c_j \in \mathbb{R}, \ j = 1, 2, \ldots, 3(n-1) \right\}.$$

Remark 6.3.1 The boundary reduction of the boundary value problems for harmonic Navier equations (2.4.26) has been discussed by Han (1995). In this paper, the hypersingular integral operator corresponding to the derivative of the double layer potential for harmonic Navier operator has been regularized as a differential-integro-differential operator, in which the integral operators are weak singular. Therefore the traction problem of harmonic Navier equations can be reduced to the system of boundary integro-differential equations.

References

Ahner, J. F., & Hsiao, G. C. (1976). On the two-dimensional exterior boundary-value problems of elasticity. *SIAM Journal on Applied Mathematics, 31*(4), 677–685.

Balaš, J., Sládek, J., & Sládek, V. (2013). *Stress analysis by boundary element methods*. Elsevier.

Bonnemay, P. (1979). *Equations intégrales pour l'élasticité plane*. Unpublished doctoral dissertation, éditeur non identifié.

Chen, J. T., Huang, W. S., Lee, Y. T., et al. (2017). Revisit of degenerate scales in the BIEM/BEM for 2d elasticity problems. *Mechanics of Advanced Materials and Structures, 24*(1), 1–15.

Costabel, M. (1988). Boundary integral operators on Lipschitz domains: Elementary results. *SIAM Journal on Mathematical Analysis, 19*(3), 613–626.

Günter, N. M. (1967). *Potential theory and its applications to basic problems of mathematical physics*. Ungar.

Han, H. (1993). The boundary integro-differential equation of boundary value problems in linear elasticity. *Journal of Tsinghua University, 33*, 15–25.

Han, H. (1994). A boundary element procedure for the Signorini problem in three-dimensional elasticity. *Numerische Mathematik, 3*, 104–117.

Han, H. (1994). The boundary integro-differential equations of three-dimensional Neumann problem in linear elasticity. *Numerische Mathematik, 68*(2), 269–281.

Han, H. (1995). A system of boundary integro-differential equations for harmonic elastic waves in $\mathbb{R}^3$. *Numerische Mathematik, 4*, 203–621.

He, W., Ding, H., & Hu, H. (1996). Non-equivalence of the conventional boundary integral formulation and its elimination for plane elasticity problems. *Computers & structures, 59*(6), 1059–1062.

Hsiao, G. C., & Wendland, W. L. (1985). *On a boundary integral method for some exterior problems in elasticity*. Tbilisi: Tbilisi University Press.

Hsiao, G. C., & Wendland, W. L. (2008). *Boundary integral equations*. Springer.

Kupradze, V. D. (2012). *Three-dimensional problems of elasticity and thermoelasticity*. Elsevier.

Mikhlin, S. G. (1938). The problem of equivalence in the theory of singular integral equations. *Sbornik: Mathematics, 3*(45), 121–141.

Mikhlin, S. G., & Prössdorf, S. (1987). *Singular integral operators* (Vol. 68). Springer Science & Business Media.

Muskhelishvili N. I. (1953). *Some basic problems of the mathematical theory of elasticity* (Vol. 15). Noordhoff.

Nedelec, J. (1982). Integral equations with non integrable kernels. *Integral Equations and Operator Theory, 5*, 562–572.

Of, G., Steinbach, O., & Wendland, W. L. (2004). *Applications of a fast multipole Galerkin boundary element method in linear elastostatics*. Springer.

Sokolnikoff I. S., Specht R. D., et al. (1956). *Mathematical theory of elasticity* (Vol. 83). McGraw-Hill.

Chapter 7
Boundary Value Problems of the Stokes Equations

The boundary integro-differential method for solving the boundary value problems of Stokes equations is discussed in this chapter.

7.1 Principal Boundary Value Problems of Stokes equations

The Stokes equations are the basic partial differential equations to describe the viscous incompressible fluid with low Reynold numbers (Happel & Brenner, 1983). In this case, the unknown function $\boldsymbol{u} = (u_1, u_2 \ldots, u_n)^{\mathrm{T}}$ denotes the fluid velocity and p denotes the fluid pressure. The Stokes equations are also the basic partial differential equations describing the linear incompressible elastic body (Sokolnikoff et al., 1956), in which the unknown function $\boldsymbol{u} = (u_1, u_2, \ldots, u_n)^{\mathrm{T}}$ denotes the displacement and p denotes the pressure of the elastic body. The Stokes equations are given by

$$\begin{cases} -\mu\Delta \boldsymbol{u}(\boldsymbol{x}) + \nabla p(\boldsymbol{x}) = 0, & (7.1.1\text{a}) \\ \nabla \cdot \boldsymbol{u}(\boldsymbol{x}) = 0, & (7.1.1\text{b}) \end{cases}$$

with Lamé constant $\mu > 0$.

The fundamental solution $\{\mathbb{E}_n(\boldsymbol{x}, \boldsymbol{y}), \mathbb{Q}_n(\boldsymbol{x}, \boldsymbol{y})\}$ of Stokes operator is given by (2.4.67)–(2.4.68), which is

$$\begin{aligned} \mathbb{E}_n(\boldsymbol{x}, \boldsymbol{y}) &= \frac{1}{4(n-1)\pi\mu}\left\{\gamma_n(\boldsymbol{x}, \boldsymbol{y})\mathbb{I} + \frac{(\boldsymbol{x}-\boldsymbol{y})(\boldsymbol{x}-\boldsymbol{y})^{\mathrm{T}}}{|\boldsymbol{x}-\boldsymbol{y}|^n}\right\} \\ &:= \left(\mathbb{E}_n^1(\boldsymbol{x}, \boldsymbol{y}), \mathbb{E}_n^2(\boldsymbol{x}, \boldsymbol{y}), \ldots, \mathbb{E}_n^n(\boldsymbol{x}, \boldsymbol{y})\right), \qquad (7.1.2) \\ \mathbb{Q}_n(\boldsymbol{x}, \boldsymbol{y}) &= \frac{1}{2(n-1)\pi}\left(-\nabla_x\gamma_n(\boldsymbol{x}, \boldsymbol{y})\right)^{\mathrm{T}} = \frac{1}{2(n-1)\pi}\left(\nabla_y\gamma_n(\boldsymbol{x}, \boldsymbol{y})\right)^{\mathrm{T}} \\ &:= \left(Q_n^1(\boldsymbol{x}, \boldsymbol{y}), Q_n^2(\boldsymbol{x}, \boldsymbol{y}), \ldots, Q_n^n(\boldsymbol{x}, \boldsymbol{y})\right), \qquad (7.1.3) \end{aligned}$$

H. Han and D. Yin, *Mathematical Foundation of the Boundary Integro-Differential Equation Method*, https://doi.org/10.1007/978-981-95-1088-7_7

with

$$\gamma_n(\boldsymbol{x}, \boldsymbol{y}) = \begin{cases} -\ln|\boldsymbol{x} - \boldsymbol{y}|, & n = 2, \\ \dfrac{1}{|\boldsymbol{x} - \boldsymbol{y}|}, & n = 3. \end{cases} \tag{7.1.4}$$

For the given functions $\boldsymbol{u}(\boldsymbol{x}) = (u_1(\boldsymbol{x}), u_2(\boldsymbol{x}), \ldots, u_n(\boldsymbol{x}))^{\mathrm{T}}$ and $p(\boldsymbol{x})$, let

$$\epsilon(\boldsymbol{u}) = \Big(\epsilon_{ij}(\boldsymbol{u})\Big)_{n\times n}, \tag{7.1.5}$$

with

$$\epsilon_{ij}(\boldsymbol{u}) = \frac{1}{2}\left(\frac{\partial u_i}{\partial x_j} + \frac{\partial u_j}{\partial x_i}\right), \quad 1 \leqslant i, j \leqslant n, \tag{7.1.6}$$

denotes the strain tensor and

$$\boldsymbol{\sigma}(\boldsymbol{u}, p) = \Big(\sigma_{ij}(\boldsymbol{u}, p)\Big)_{n\times n}, \tag{7.1.7}$$

with

$$\sigma_{ij}(\boldsymbol{u}, p) = -p\delta_{ij} + 2\mu\epsilon_{ij}(\boldsymbol{u}), \quad 1 \leqslant i, j \leqslant n \tag{7.1.8}$$

denotes the stress tensor. Namely

$$\boldsymbol{\sigma}(\boldsymbol{u}, p) = -p\mathbb{I} + \mu\left((\nabla\boldsymbol{u})^{\mathrm{T}} + \nabla\boldsymbol{u}^{\mathrm{T}}\right). \tag{7.1.9}$$

For $\boldsymbol{x} \in \mathbb{R}^n$ and arbitrary unit vector $\boldsymbol{\nu}(\boldsymbol{x}) = (\nu_1(\boldsymbol{x}), \nu_2(\boldsymbol{x}), \ldots, \nu_n(\boldsymbol{x}))^{\mathrm{T}}$, the stress operator $\mathbb{T}(D_x, \boldsymbol{\nu}(\boldsymbol{x}))$ is defined by

$$\begin{aligned} \mathbb{T}(D_x, \boldsymbol{\nu}(\boldsymbol{x}))(\boldsymbol{u}(\boldsymbol{x}), p(\boldsymbol{x})) &= \boldsymbol{\sigma}(\boldsymbol{u}(\boldsymbol{x}), p(\boldsymbol{x}))\boldsymbol{\nu}(\boldsymbol{x}) \\ &= -p(\boldsymbol{x})\boldsymbol{\nu}(\boldsymbol{x}) + \mu\Big((\nabla\boldsymbol{u})^{\mathrm{T}} + \nabla\boldsymbol{u}^{\mathrm{T}}\Big)\boldsymbol{\nu}(\boldsymbol{x}) \\ &= -p(\boldsymbol{x})\boldsymbol{\nu}(\boldsymbol{x}) + \mathbb{T}^*\big(D_x, \boldsymbol{\nu}(\boldsymbol{x})\big)\boldsymbol{u}(\boldsymbol{x}), \end{aligned} \tag{7.1.10}$$

with

$$\mathbb{T}^*\big(D_x, \boldsymbol{\nu}(\boldsymbol{x})\big) = \left(\mu\nu_j(\boldsymbol{x})\frac{\partial}{\partial x_i} + \mu\delta_{ij}\frac{\partial}{\partial\boldsymbol{\nu}(\boldsymbol{x})}\right)_{n\times n}. \tag{7.1.11}$$

Furthermore the stress operator $\mathbb{T}'(D_x, \boldsymbol{\nu}(\boldsymbol{x}))$ is defined by

$$\begin{aligned} \mathbb{T}'\big(D_x, \boldsymbol{\nu}(\boldsymbol{x})\big)\big(\boldsymbol{u}(\boldsymbol{x}), p(\boldsymbol{x})\big) &= \boldsymbol{\sigma}\big(\boldsymbol{u}(\boldsymbol{x}), -p(\boldsymbol{x})\big)\boldsymbol{\nu}(\boldsymbol{x}) \\ &= p(\boldsymbol{x})\boldsymbol{\nu}(\boldsymbol{x}) + \mathbb{T}^*\big(D_x, \boldsymbol{\nu}(\boldsymbol{x})\big)\boldsymbol{u}(\boldsymbol{x}). \end{aligned} \tag{7.1.12}$$

It is ready to discuss the following principal boundary value problems of Stokes equations.

Interior Dirichlet problem:

$$\begin{cases} -\mu\Delta \boldsymbol{u}(\boldsymbol{x}) + \nabla p(\boldsymbol{x}) = \boldsymbol{0}, \quad \forall \boldsymbol{x} \in \Omega, & (7.1.13a)\\ \nabla \cdot \boldsymbol{u}(\boldsymbol{x}) = 0, \quad \forall\, \boldsymbol{x} \in \Omega, & (7.1.13b)\\ \boldsymbol{u}(\boldsymbol{x})\Big|_{\Gamma} = g(\boldsymbol{x}), \quad \forall \boldsymbol{x} \in \Gamma, & (7.1.13c)\\ \displaystyle\int_{\Omega} p(\boldsymbol{x})\mathrm{d}\boldsymbol{x} = 0, & (7.1.13d)\end{cases}$$

with given function $g(\boldsymbol{x})$.

Exterior Dirichlet problem (I):

$$\begin{cases} -\mu\Delta \boldsymbol{u}(\boldsymbol{x}) + \nabla p(\boldsymbol{x}) = \boldsymbol{0}, \quad \forall \boldsymbol{x} \in \Omega_e, & (7.1.14a)\\ \nabla \cdot \boldsymbol{u}(\boldsymbol{x}) = 0, \quad \forall\, \boldsymbol{x} \in \Omega_e, & (7.1.14b)\\ \boldsymbol{u}(\boldsymbol{x})\Big|_{\Gamma} = g(\boldsymbol{x}), \quad \forall \boldsymbol{x} \in \Gamma, & (7.1.14c)\\ \boldsymbol{u}(\boldsymbol{x}) = \displaystyle\sum_{j=1}^{n} m_j \omega_j(\boldsymbol{x}) + \mathcal{O}(|\boldsymbol{x}|^{1-n}), \quad \text{as } |\boldsymbol{x}| \to +\infty, & (7.1.14d)\\ p(\boldsymbol{x}) = \mathcal{O}(|\boldsymbol{x}|^{-n}), \quad \text{as } |\boldsymbol{x}| \to +\infty, & (7.1.14e)\end{cases}$$

with unknown constants $\{m_1, m_2, \ldots, m_n\}$, the functions $\{\omega_j(\boldsymbol{x}),\, j=1, 2, \ldots, n\}$ are defined by (6.1.12)–(6.1.13).

Exterior Dirichlet problem (II):

$$\begin{cases} -\mu\Delta \boldsymbol{u}(\boldsymbol{x}) + \nabla p(\boldsymbol{x}) = \boldsymbol{0}, \quad \forall \boldsymbol{x} \in \Omega_e, & (7.1.15a)\\ \nabla \cdot \boldsymbol{u}(\boldsymbol{x}) = 0, \quad \forall\, \boldsymbol{x} \in \Omega_e, & (7.1.15b)\\ \boldsymbol{u}(\boldsymbol{x})\Big|_{\Gamma} = g(\boldsymbol{x}), \quad \forall \boldsymbol{x} \in \Gamma, & (7.1.15c)\\ \boldsymbol{u}(\boldsymbol{x}) = \mathbb{E}_n(\boldsymbol{x}, \boldsymbol{0})\mathbb{C} + \mathcal{O}(|\boldsymbol{x}|^{1-n}), \quad \text{as } |\boldsymbol{x}| \to +\infty, & (7.1.15d)\\ p(\boldsymbol{x}) = \mathcal{O}(|\boldsymbol{x}|^{1-n}), \quad \text{as } |\boldsymbol{x}| \to +\infty, & (7.1.15e)\end{cases}$$

with unknown vector $\mathbb{C} \in \mathbb{R}^n$.

Interior traction problem:

$$\begin{cases} -\mu\Delta \boldsymbol{u}(\boldsymbol{x}) + \nabla p(\boldsymbol{x}) = \boldsymbol{0}, \quad \forall \boldsymbol{x} \in \Omega_e, & (7.1.16a)\\ \nabla \cdot \boldsymbol{u}(\boldsymbol{x}) = 0, \quad \forall\, \boldsymbol{x} \in \Omega_e, & (7.1.16b)\\ \mathbb{T}\big(D_x, \boldsymbol{\nu}(\boldsymbol{x})\big)\big(\boldsymbol{u}(\boldsymbol{x}), p(\boldsymbol{x})\big)\Big|_{\Gamma} = \boldsymbol{h}(\boldsymbol{x}), \quad \forall \boldsymbol{x} \in \Gamma. & (7.1.16c)\end{cases}$$

Exterior traction problem (I):

$$\begin{cases} -\mu\Delta\boldsymbol{u}(\boldsymbol{x}) + \nabla p(\boldsymbol{x}) = \mathbf{0}, \quad \forall \boldsymbol{x} \in \Omega_e, & (7.1.17a) \\ \nabla \cdot \boldsymbol{u}(\boldsymbol{x}) = 0, \quad \forall\, \boldsymbol{x} \in \Omega_e, & (7.1.17b) \\ \mathbb{T}\big(D_x, \boldsymbol{\nu}(\boldsymbol{x})\big)\big(\boldsymbol{u}(\boldsymbol{x}), p(\boldsymbol{x})\big)\Big|_\Gamma = \boldsymbol{h}(\boldsymbol{x}), \quad \forall \boldsymbol{x} \in \Gamma, & (7.1.17c) \\ \boldsymbol{u}(\boldsymbol{x}) = \mathcal{O}(|\boldsymbol{x}|^{1-n}), \quad \text{as } |\boldsymbol{x}| \to +\infty, & (7.1.17d) \\ p(\boldsymbol{x}) = \mathcal{O}(|\boldsymbol{x}|^{-n}), \quad \text{as } |\boldsymbol{x}| \to +\infty. & (7.1.17e) \end{cases}$$

Exterior traction problem (II):

$$\begin{cases} -\mu\Delta\boldsymbol{u}(\boldsymbol{x}) + \nabla p(\boldsymbol{x}) = \mathbf{0}, \quad \forall \boldsymbol{x} \in \Omega_e, & (7.1.18a) \\ \nabla \cdot \boldsymbol{u}(\boldsymbol{x}) = 0, \quad \forall\, \boldsymbol{x} \in \Omega_e, & (7.1.18b) \\ \mathbb{T}\big(D_x, \boldsymbol{\nu}(\boldsymbol{x})\big)\big(\boldsymbol{u}(\boldsymbol{x}), p(\boldsymbol{x})\big)\Big|_\Gamma = \boldsymbol{h}(\boldsymbol{x}), \quad \forall \boldsymbol{x} \in \Gamma, & (7.1.18c) \\ \boldsymbol{u}(\boldsymbol{x}) = \mathbb{E}_n(\boldsymbol{x}, \mathbf{0})\mathbb{C} + \mathcal{O}(|\boldsymbol{x}|^{1-n}), \quad \text{as } |\boldsymbol{x}| \to +\infty, & (7.1.18d) \\ p(\boldsymbol{x}) = \mathcal{O}(|\boldsymbol{x}|^{1-n}), \quad \text{as } |\boldsymbol{x}| \to +\infty. & (7.1.18e) \end{cases}$$

For the interior Dirichlet problem and the interior traction problem, the given data $g(\boldsymbol{x})$ and $\boldsymbol{h}(\boldsymbol{x})$ need to satisfy respectively the compatibility conditions

$$\int_\Gamma g(\boldsymbol{x}) \cdot \boldsymbol{\nu}(\boldsymbol{x}) \mathrm{d}s_x = 0 \tag{7.1.19}$$

and

$$\int_\Gamma \boldsymbol{h}(\boldsymbol{x}) \cdot \boldsymbol{\omega}_j(\boldsymbol{x}) \mathrm{d}s_x = 0, \quad \forall \boldsymbol{\omega}_j \in \mathbb{RM}. \tag{7.1.20}$$

The $3(n-1)$ dimensional space $\mathbb{RM}$ is defined by (6.1.11).

In fact compatibility condition (7.1.19) is from the incompressible equation (7.1.1b), and compatibility condition (7.1.20) is from Green formula directly.

In the following we only give the definitions of the generalized solutions of the boundary value problems of Stokes equations, the definitions of classical solution of which are omitted here. Introduce the following notations, let

$$\begin{aligned} \mathbb{H}^1(\Omega) &= \big(H^1(\Omega)\big)^n, \\ \mathbb{H}^1_g(\Omega) &= \Big\{\boldsymbol{v}\Big|\ \boldsymbol{v} \in \mathbb{H}^1(\Omega) \text{ and } \boldsymbol{v}(\boldsymbol{x})\Big|_\Gamma = g(\boldsymbol{x})\Big\}, \\ \mathbb{H}^1_0(\Omega) &= \Big\{\boldsymbol{v}\Big|\boldsymbol{v} \in \mathbb{H}^1(\Omega) \text{ and } \boldsymbol{v}(\boldsymbol{x})\Big|_\Gamma = \mathbf{0}\Big\}, \\ L_{2,*}(\Omega) &= \Big\{q\Big|q \in L_2(\Omega) \text{ and } \int_\Omega q(\boldsymbol{x})\mathrm{d}\boldsymbol{x} = 0\Big\}, \\ \mathbb{H}^1_{\mathrm{loc}}(\Omega_e) &= \Big\{\boldsymbol{v}\Big|\boldsymbol{v} \in \mathbb{H}^1(G), \quad \text{for any bounded domain } G \subset \Omega_e\Big\}, \\ \mathbb{H}^1_{\mathrm{comp}}(\Omega_e) &= \big\{\boldsymbol{v} \in \mathbb{H}^1_{\mathrm{loc}}(\Omega_e) \text{ and the support of } \boldsymbol{v}(\boldsymbol{x}) \text{ is compact}\big\}, \end{aligned}$$

$$\mathbb{H}^1_{\mathrm{loc},g}(\Omega_e) = \left\{ \boldsymbol{v} \middle| \boldsymbol{v} \in \mathbb{H}^1_{\mathrm{loc}}(\Omega_e) \text{ and } \boldsymbol{v}(\boldsymbol{x})\Big|_{\Gamma} = g(\boldsymbol{x}) \right\},$$

$$\mathbb{H}^1_{\mathrm{comp},0}(\Omega_e) = \left\{ \boldsymbol{v} \middle| \boldsymbol{v} \in \mathbb{H}^1_{\mathrm{comp}}(\Omega_e) \text{ and } \boldsymbol{v}(\boldsymbol{x})\Big|_{\Gamma} = \boldsymbol{0} \right\},$$

$$L_{2,\mathrm{loc}}(\Omega_e) = \left\{ q \middle| q \in L_2(G), \text{ for any bounded domain } G \subset \Omega_e \right\},$$

$$L_{2,\mathrm{comp}}(\Omega_e) = \left\{ q \middle| q \in L_{2,\mathrm{loc}}(G), \text{ and the support of } g(\boldsymbol{x}) \text{ is compact} \right\}.$$

The Stokes equations (7.1.1) can be rewritten in the following equivalent form

$$\begin{cases} -2\mu\nabla \cdot \epsilon(\boldsymbol{u}) + \nabla p = \boldsymbol{0}, & (7.1.21\text{a}) \\ \nabla \cdot \boldsymbol{u} = 0. & (7.1.21\text{b}) \end{cases}$$

Furthermore introduce the bilinear forms $\mathbb{A}(\boldsymbol{u}, \boldsymbol{v})$, $\mathbb{B}(p, \boldsymbol{v})$, $\mathbb{A}_e(\boldsymbol{u}, \boldsymbol{v})$ and $\mathbb{B}_e(p, \boldsymbol{v})$:

$$\mathbb{A}(\boldsymbol{u}, \boldsymbol{v}) = 2\mu \int_{\Omega} \sum_{i,j=1}^{n} \epsilon_{ij}(\boldsymbol{u})\epsilon_{ij}(\boldsymbol{v})\mathrm{d}\boldsymbol{x} := 2\mu \int_{\Omega} \epsilon(\boldsymbol{u}) : \epsilon(\boldsymbol{v})\mathrm{d}\boldsymbol{x},$$

$$\mathbb{B}(p, \boldsymbol{v}) = -\int_{\Omega} p\nabla \cdot \boldsymbol{v}\mathrm{d}\boldsymbol{x},$$

$$\mathbb{A}_e(\boldsymbol{u}, \boldsymbol{v}) = 2\mu \int_{\Omega_e} \epsilon(\boldsymbol{u}) : \epsilon(\boldsymbol{v})\mathrm{d}\boldsymbol{x},$$

$$\mathbb{B}_e(p, \boldsymbol{v}) = -\int_{\Omega_e} p\nabla \cdot \boldsymbol{v}\mathrm{d}\boldsymbol{x}.$$

Definition 7.1.1 Suppose that $g \in \mathbb{H}^{1/2}(\Gamma)$ and $\boldsymbol{h} \in \mathbb{H}^{-1/2}(\Gamma)$, which satisfy the compatibility condition (7.1.19) and (7.1.20) respectively.

1. Find $(\boldsymbol{u}, p) \in \mathbb{H}^1_g(\Omega) \times L_{2,*}(\Omega)$, such that

$$\begin{cases} A(\boldsymbol{u}, \boldsymbol{v}) + \mathbb{B}(p, \boldsymbol{v}) = 0, \quad \forall \boldsymbol{v} \in \mathbb{H}^1_0(\Omega), & (7.1.22\text{a}) \\ \mathbb{B}(q, \boldsymbol{u}) = 0, \quad \forall q \in L_2(\Omega). & (7.1.22\text{b}) \end{cases}$$

   Then $(\boldsymbol{u}, p)$, the solution of variational problem (7.1.22), is called generalized solution of the interior Dirichlet problem (7.1.13).

2. Find $(\boldsymbol{u}, p) \in \mathbb{H}^1_{\mathrm{loc},g}(\Omega_e) \times L_{2,\mathrm{loc}}(\Omega_e)$, such that

$$\begin{cases} A_e(\boldsymbol{u}, \boldsymbol{v}) + \mathbb{B}_e(p, \boldsymbol{v}) = 0, \quad \forall \boldsymbol{v} \in \mathbb{H}^1_{\mathrm{comp},0}(\Omega_e), & (7.1.23\text{a}) \\ \mathbb{B}_e(q, \boldsymbol{u}) = 0, \quad \forall q \in L_{2,\mathrm{comp}}(\Omega_e), & (7.1.23\text{b}) \\ \boldsymbol{u}(\boldsymbol{x}) = \mathbb{C} + \mathcal{O}(|\boldsymbol{x}|^{1-n}), \quad \text{as } |\boldsymbol{x}| \to +\infty, & (7.1.23\text{c}) \\ p(\boldsymbol{x}) = \mathcal{O}(|\boldsymbol{x}|^{-n}), \quad \text{as } |\boldsymbol{x}| \to +\infty. & (7.1.23\text{d}) \end{cases}$$

   Then $(\boldsymbol{u}, p)$, the solution of variational problem (7.1.23), is called generalized solution of the exterior Dirichlet problem (I) (7.1.14).

3. Find $\big(\boldsymbol{u}, p\big) \in \mathbb{H}^1_{\text{loc},g}(\Omega_e) \times L_{2,\text{loc}}(\Omega_e)$, such that

$$\begin{cases} A_e(\boldsymbol{u}, \boldsymbol{v}) + \mathbb{B}_e(p, \boldsymbol{v}) = 0, \quad \forall \boldsymbol{v} \in \mathbb{H}^1_{\text{comp},0}(\Omega_e), & (7.1.24\text{a}) \\ \mathbb{B}_e(q, \boldsymbol{u}) = 0, \quad \forall q \in L_{2,\text{comp}}(\Omega_e), & (7.1.24\text{b}) \\ \boldsymbol{u}(\boldsymbol{x}) = \mathbb{E}_n(\boldsymbol{x}, \boldsymbol{0})\mathbb{C} + \mathcal{O}(|\boldsymbol{x}|^{1-n}), \quad \text{as } |\boldsymbol{x}| \to +\infty, & (7.1.24\text{c}) \\ p(\boldsymbol{x}) = \mathcal{O}(|\boldsymbol{x}|^{1-n}), \quad \text{as } |\boldsymbol{x}| \to +\infty. & (7.1.24\text{d}) \end{cases}$$

Then $(\boldsymbol{u}, p)$, the solution of variational problem (7.1.24), is called generalized solution of the exterior Dirichlet problem (Ⅱ) (7.1.15).

4. Find $\big(\boldsymbol{u}, p\big) \in \mathbb{H}^1(\Omega) \times L_{2,*}(\Omega)$, such that

$$\begin{cases} A(\boldsymbol{u}, \boldsymbol{v}) + \mathbb{B}(p, \boldsymbol{v}) = \displaystyle\int_\Gamma \boldsymbol{h} \cdot \boldsymbol{v} \mathrm{d}\boldsymbol{x}, \quad \forall \boldsymbol{v} \in \mathbb{H}^1(\Omega), & (7.1.25\text{a}) \\ \mathbb{B}(q, \boldsymbol{u}) = 0, \quad \forall q \in L_2(\Omega). & (7.1.25\text{b}) \end{cases}$$

Then $(\boldsymbol{u}(\boldsymbol{x}), p(\boldsymbol{x}))$, the solution of variational problem (7.1.25), is called generalized solution of the interior traction problem.

5. Find $\big(\boldsymbol{u}, p\big) \in \mathbb{H}^1_{\text{loc}}(\Omega_e) \times L_{2,\text{loc}}(\Omega_e)$, such that

$$\begin{cases} A_e(\boldsymbol{u}, \boldsymbol{v}) + \mathbb{B}_e(p, \boldsymbol{v}) = -\displaystyle\int_\Gamma \boldsymbol{h} \cdot \boldsymbol{v} \mathrm{d}\boldsymbol{x} \quad \forall \boldsymbol{v} \in \mathbb{H}^1_{\text{comp}}(\Omega_e), & (7.1.26\text{a}) \\ \mathbb{B}_e(q, \boldsymbol{u}) = 0, \quad \forall q \in L_{2,\text{comp}}(\Omega_e), & (7.1.26\text{b}) \\ \boldsymbol{u}(\boldsymbol{x}) = \mathcal{O}(|\boldsymbol{x}|^{1-n}), \quad \text{as } |\boldsymbol{x}| \to +\infty, & (7.1.26\text{c}) \\ p(\boldsymbol{x}) = \mathcal{O}(|\boldsymbol{x}|^{-n}), \quad \text{as } |\boldsymbol{x}| \to +\infty. & (7.1.26\text{d}) \end{cases}$$

Then $(\boldsymbol{u}(\boldsymbol{x}), p(\boldsymbol{x}))$, the solution of variational problem (7.1.26), is called generalized solution of the exterior traction problem (Ⅰ).

6. Find $\big(\boldsymbol{u}, p\big) \in \mathbb{H}^1_{\text{loc}}(\Omega_e) \times L_{2,\text{loc}}(\Omega_e)$, such that

$$\begin{cases} A_e(\boldsymbol{u}, \boldsymbol{v}) + \mathbb{B}_e(p, \boldsymbol{v}) = -\displaystyle\int_\Gamma \boldsymbol{h} \cdot \boldsymbol{v} \mathrm{d}\boldsymbol{x}, \quad \forall \boldsymbol{v} \in \mathbb{H}^1_{\text{comp}}(\Omega_e), & (7.1.27\text{a}) \\ \mathbb{B}_e(q, \boldsymbol{u}) = 0, \quad \forall q \in L_{2,\text{comp}}(\Omega_e), & (7.1.27\text{b}) \\ \boldsymbol{u}(\boldsymbol{x}) = \mathbb{E}_n(\boldsymbol{x}, \boldsymbol{0})\mathbb{C} + \mathcal{O}(|\boldsymbol{x}|^{1-n}), \quad \text{as } |\boldsymbol{x}| \to +\infty, & (7.1.27\text{c}) \\ p(\boldsymbol{x}) = \mathcal{O}(|\boldsymbol{x}|^{1-n}), \quad \text{as } |\boldsymbol{x}| \to +\infty. & (7.1.27\text{d}) \end{cases}$$

Then $(\boldsymbol{u}(\boldsymbol{x}), p(\boldsymbol{x}))$, the solution of variational problem (7.1.27), is called generalized solution of the exterior traction problem (Ⅱ).

Suppose that $\big(\boldsymbol{u}(\boldsymbol{x}), p(\boldsymbol{x})\big)$ is a solution of Stokes equations (7.1.1) on the domain Ω, in the following we discuss the representation of $\big(\boldsymbol{u}(\boldsymbol{x}), p(\boldsymbol{x})\big)$ by the potential integrals (Hsiao & Wendland, 2008). For the smooth $\big(\boldsymbol{u}(\boldsymbol{x}), p(\boldsymbol{x})\big)$ and pair function $\big(\boldsymbol{v}(\boldsymbol{x}), q(\boldsymbol{x})\big)$ on Ω, we have the second Green formula

$$\int_\Omega \boldsymbol{u}(\boldsymbol{y})\cdot\big(-\mu\Delta\boldsymbol{v}(\boldsymbol{y})-\nabla q(\boldsymbol{y})\big)\mathrm{d}\boldsymbol{y}-\int_\Omega\big(-\mu\Delta\boldsymbol{u}(\boldsymbol{y})+\nabla p(\boldsymbol{y})\big)\cdot\boldsymbol{v}(\boldsymbol{y})\mathrm{d}\boldsymbol{y}$$
$$=\int_\Gamma\big\{\mathbb{T}(D_y,\boldsymbol{\nu}(\boldsymbol{y}))(\boldsymbol{u}(\boldsymbol{y}),p(\boldsymbol{y}))\cdot\boldsymbol{v}(\boldsymbol{y})-\boldsymbol{u}(\boldsymbol{y})\cdot\mathbb{T}'(D_y,\boldsymbol{\nu}(\boldsymbol{y}))(\boldsymbol{v}(\boldsymbol{y}),q(\boldsymbol{y}))\big\}\,\mathrm{d}s_y. \tag{7.1.28}$$

Replacing $\big(\boldsymbol{v}(\boldsymbol{y}),q(\boldsymbol{y})\big)$ by $\big(\mathbb{E}_n^k(\boldsymbol{x},\boldsymbol{y}),Q_n^k(\boldsymbol{x},\boldsymbol{y})\big)$ in (7.1.28), then $u_k(\boldsymbol{x})$, the component of $\boldsymbol{u}(\boldsymbol{x})$, has the following representation

$$u_k(\boldsymbol{x})=\int_\Gamma \mathbb{E}_n^k(\boldsymbol{x},\boldsymbol{y})\cdot\mathbb{T}(D_y,\boldsymbol{\nu}(\boldsymbol{y}))(\boldsymbol{u}(\boldsymbol{y}),p(\boldsymbol{y}))\mathrm{d}s_y-$$
$$\int_\Gamma \mathbb{T}'(D_y,\boldsymbol{\nu}(\boldsymbol{y}))(\mathbb{E}_n^k(\boldsymbol{x},\boldsymbol{y}),Q_n^k(\boldsymbol{x},\boldsymbol{y}))\cdot\boldsymbol{u}(\boldsymbol{y})\mathrm{d}s_y,\quad \forall\boldsymbol{x}\in\Omega,\ \ 1\leqslant k\leqslant n.$$

Namely, we obtain

$$\boldsymbol{u}(\boldsymbol{x})=\int_\Gamma \mathbb{E}_n(\boldsymbol{x},\boldsymbol{y})\mathbb{T}(D_y,\boldsymbol{\nu}(\boldsymbol{y}))\big(\boldsymbol{u}(\boldsymbol{y}),p(\boldsymbol{y})\big)\mathrm{d}s_y-$$
$$\int_\Gamma\big\{\mathbb{T}'(D_y,\boldsymbol{\nu}(\boldsymbol{y}))\big(\mathbb{E}_n(\boldsymbol{x},\boldsymbol{y}),\mathbb{Q}_n(\boldsymbol{x},\boldsymbol{y})\big)\big\}^{\mathrm{T}}\boldsymbol{u}(\boldsymbol{y})\mathrm{d}s_y,\ \forall\boldsymbol{x}\in\Omega. \tag{7.1.29}$$

In which

$$\mathbb{T}'(D_y,\boldsymbol{\nu}(\boldsymbol{y}))\big(\mathbb{E}_n(\boldsymbol{x},\boldsymbol{y}),\mathbb{Q}_n(\boldsymbol{x},\boldsymbol{y})\big)$$
$$=\Big(\mathbb{T}'(D_y,\boldsymbol{\nu}(\boldsymbol{y}))\big(\mathbb{E}_n^1(\boldsymbol{x},\boldsymbol{y}),Q_n^1(\boldsymbol{x},\boldsymbol{y})\big),\ldots,\mathbb{T}'(D_y,\boldsymbol{\nu}(\boldsymbol{y}))\big(\mathbb{E}_n^n(\boldsymbol{x},\boldsymbol{y}),Q_n^n(\boldsymbol{x},\boldsymbol{y})\big)\Big)$$
$$=\Big(\boldsymbol{\sigma}(\mathbb{E}_n^1(\boldsymbol{x},\boldsymbol{y}),-Q_n^1(\boldsymbol{x},\boldsymbol{y})\boldsymbol{\nu}(\boldsymbol{y}),\ldots,\boldsymbol{\sigma}(\mathbb{E}_n^n(\boldsymbol{x},\boldsymbol{y}),-Q_n^n(\boldsymbol{x},\boldsymbol{y})\boldsymbol{\nu}(\boldsymbol{y})\Big)$$
$$:=\boldsymbol{\sigma}\Big(\mathbb{E}_n(\boldsymbol{x},\boldsymbol{y}),-Q_n(\boldsymbol{x},\boldsymbol{y})\Big)\boldsymbol{\nu}(\boldsymbol{y}).$$

For obtaining the representation of the pressure $p(\boldsymbol{x})$ on Ω, recall the equation (7.1.1a), namely

$$\nabla_x p(\boldsymbol{x})=\mu\Delta\boldsymbol{u}(\boldsymbol{x}),\quad \forall\boldsymbol{x}\in\Omega.$$

Substituting (7.1.29) into the above equation, it yields

$$\nabla_x p(\boldsymbol{x})=\nabla_x\Big\{\int_\Gamma \mathbb{Q}_n(\boldsymbol{x},\boldsymbol{y})\cdot\mathbb{T}(D_y,\boldsymbol{\nu}(\boldsymbol{y}))(\boldsymbol{u}(\boldsymbol{y}),p(\boldsymbol{y}))\mathrm{d}s_y-$$
$$2\mu\int_\Gamma\frac{\partial}{\partial\boldsymbol{\nu}_y}\mathbb{Q}_n(\boldsymbol{x},\boldsymbol{y})\cdot\boldsymbol{u}(\boldsymbol{y})\mathrm{d}s_y\Big\},\quad\forall\boldsymbol{x}\in\Omega,$$

then we obtain

$$p(\boldsymbol{x}) = \int_\Gamma \mathbb{Q}_n(\boldsymbol{x}, \boldsymbol{y}) \cdot \mathbb{T}(D_y, \boldsymbol{\nu}(\boldsymbol{y}))(\boldsymbol{u}(\boldsymbol{y}), p(\boldsymbol{y}))\mathrm{d}s_y - 2\mu \int_\Gamma \frac{\partial}{\partial \boldsymbol{\nu}_y} \mathbb{Q}_n(\boldsymbol{x}, \boldsymbol{y}) \cdot \boldsymbol{\nu}(\boldsymbol{y})\mathrm{d}s_y + C, \forall \boldsymbol{x} \in \Omega. \tag{7.1.30}$$

It means that the pressure $p(\boldsymbol{x})$ is uniquely determined only up to an additive constant C by $\boldsymbol{u}(\boldsymbol{x})$.

In the coming section, the single and double potentials of Stokes operator appearing in the representation (7.1.29)–(7.1.30) will be discussed further.

7.2 Single Layer Potential and Double Layer Potential of Stokes Operator

For given vector value functions $\varphi(\boldsymbol{x})$ and $\psi(\boldsymbol{x})$ on Γ, let

$$\mathbb{V}_I(\boldsymbol{x}) = \int_\Gamma \mathbb{E}_n(\boldsymbol{x}, \boldsymbol{y})\varphi(\boldsymbol{y})\mathrm{d}s_y, \quad \forall \boldsymbol{x} \in \mathbb{R}^n, \tag{7.2.1}$$

$$q_I(\boldsymbol{x}) = \int_\Gamma \mathbb{Q}_n(\boldsymbol{x}, \boldsymbol{y}) \cdot \varphi(\boldsymbol{y})\mathrm{d}s_y, \quad \forall \boldsymbol{x} \in \mathbb{R}^n, \tag{7.2.2}$$

and

$$\mathbb{V}_{II}(\boldsymbol{x}) = \int_\Gamma \left\{\mathbb{T}'(D_y, \boldsymbol{\nu}(\boldsymbol{y}))\Big(\mathbb{E}_n(\boldsymbol{x}, \boldsymbol{y}), \mathbb{Q}_n(\boldsymbol{x}, \boldsymbol{y})\Big)\right\}^{\mathrm{T}} \psi(\boldsymbol{y})\mathrm{d}s_y, \quad \forall \boldsymbol{x} \in \mathbb{R}^n, \tag{7.2.3}$$

$$q_{II}(\boldsymbol{x}) = 2\mu \int_\Gamma \frac{\partial}{\partial \boldsymbol{\nu}_y} \mathbb{Q}_n(\boldsymbol{x}, \boldsymbol{y}) \cdot \psi(\boldsymbol{y})\mathrm{d}s_y, \quad \forall \boldsymbol{x} \in \mathbb{R}^n. \tag{7.2.4}$$

$\big(\mathbb{V}_I(\boldsymbol{x}), q_I(\boldsymbol{x})\big)$ and $\big(\mathbb{V}_{II}(\boldsymbol{x}), q_{II}(\boldsymbol{x})\big)$ are defined as the single and double layer potentials of Stokes operator with given density functions $\varphi(\boldsymbol{x})$ and $\psi(\boldsymbol{x})$ correspondingly.

It is known that Stokes equations are closely related to Navier system, therefore the layer potentials of Stokes operator are also closely related to the potentials of Navier system. In the following, we discuss the relations of them.

Let

$$E_n^L(\boldsymbol{x}, \boldsymbol{y}) = \frac{1}{2(n-1)\pi}\gamma_n(\boldsymbol{x}, \boldsymbol{y}), \tag{7.2.5}$$

which is the fundamental solution of Laplace operator given by (3.3.3), and

$$\mathbb{E}_n^N(\boldsymbol{x}, \boldsymbol{y}) = \frac{\lambda + 3\mu}{4(n-1)\pi\mu(\lambda + 2u)} \left\{\gamma_n(\boldsymbol{x}, \boldsymbol{y})\mathbb{I} + \frac{\lambda + \mu}{\lambda + 3\mu} \frac{(\boldsymbol{x} - \boldsymbol{y})(\boldsymbol{x} - \boldsymbol{y})^{\mathrm{T}}}{|\boldsymbol{x} - \boldsymbol{y}|^n}\right\}, \tag{7.2.6}$$

which is the fundamental solution of Navier system given by (6.1.2), the stress operator corresponding Navier system

$$\begin{cases} \mathbb{T}^N(D_x, \boldsymbol{\nu}(\boldsymbol{x})) = \left(T_{ij}^N(D_x, \boldsymbol{\nu}(\boldsymbol{x}))\right)_{n\times n} \\ \text{with } T_{ij}^N(D_x, \boldsymbol{\nu}(\boldsymbol{x})) = \lambda \nu_i(\boldsymbol{x})\dfrac{\partial}{\partial x_j} + \mu\nu_j(\boldsymbol{x})\dfrac{\partial}{\partial x_i} + \mu\delta_{ij}\dfrac{\partial}{\partial \boldsymbol{\nu}(\boldsymbol{x})}, \quad 1 \leqslant i, j \leqslant n, \end{cases} \tag{7.2.7}$$

which is given by (6.1.5).

Let

$$\mathbb{T}^{\Delta}(D_x, \boldsymbol{\nu}(\boldsymbol{x})) = \lambda\left(\nu_j(\boldsymbol{x})\frac{\partial}{\partial x_j}\right)_{n\times n}, \tag{7.2.8}$$

$$\mathbb{T}^*(D_x, \boldsymbol{\nu}(\boldsymbol{x})) = \mathbb{T}^N(D_x, \boldsymbol{\nu}(\boldsymbol{x})) - \mathbb{T}^{\Delta}(D_x, \boldsymbol{\nu}(\boldsymbol{x})). \tag{7.2.9}$$

Namely

$$\mathbb{T}^N(D_x, \boldsymbol{\nu}(\boldsymbol{x})) = \mathbb{T}^*(D_x, \boldsymbol{\nu}(\boldsymbol{x})) + \mathbb{T}^{\Delta}(D_x, \boldsymbol{\nu}(\boldsymbol{x})). \tag{7.2.10}$$

Furthermore let

$$\mathbb{Q}_n^N(\boldsymbol{x}, \boldsymbol{y}) = -(\lambda+\mu)\nabla_x \cdot \mathbb{E}_n^N(\boldsymbol{x}, \boldsymbol{y}), \tag{7.2.11}$$

a direct computation shows that

$$\begin{aligned} \mathbb{Q}_n^N(\boldsymbol{x}, \boldsymbol{y}) &= -\frac{\lambda+\mu}{\lambda+2\mu}\frac{1}{2(n-1)\pi}(\nabla_x \gamma_n(\boldsymbol{x}, \boldsymbol{y}))^{\mathrm{T}} \\ &= \frac{\lambda+\mu}{\lambda+2\mu}\mathbb{Q}_n(\boldsymbol{x}, \boldsymbol{y}). \end{aligned} \tag{7.2.12}$$

Obviously

$$\left.\begin{aligned} \lim_{\lambda\to+\infty} \mathbb{Q}_n^N(\boldsymbol{x}, \boldsymbol{y}) &= \mathbb{Q}_n(\boldsymbol{x}, \boldsymbol{y}), \\ \lim_{\lambda\to+\infty} \mathbb{E}_n^N(\boldsymbol{x}, \boldsymbol{y}) &= \mathbb{E}_n(\boldsymbol{x}, \boldsymbol{y}). \end{aligned}\right\} \tag{7.2.13}$$

For the fundamental solutions $\left(\mathbb{E}_n(\boldsymbol{x}, \boldsymbol{y}), \mathbb{Q}_n(\boldsymbol{x}, \boldsymbol{y})\right)$, $\mathbb{E}_n^N(\boldsymbol{x}, \boldsymbol{y})$ and $E_n^L(\boldsymbol{x}, \boldsymbol{y})$ of Stokes equations, Navier operator and Laplace operator, the following results hold.

Lemma 7.2.1 *The following equations hold*

1.

$$\mathbb{E}_n(\boldsymbol{x}, \boldsymbol{y}) = \frac{\lambda+2\mu}{\lambda+\mu}\mathbb{E}_n^N(\boldsymbol{x}, \boldsymbol{y}) - \frac{1}{\lambda+\mu}E_n^L(\boldsymbol{x}, \boldsymbol{y})\mathbb{I}, \tag{7.2.14}$$

$$\mathbb{Q}_n(\boldsymbol{x}, \boldsymbol{y}) = -(\nabla_x E_n^L(\boldsymbol{x}, \boldsymbol{y}))^{\mathrm{T}} = (\nabla_y E_n^L(\boldsymbol{x}, \boldsymbol{y}))^{\mathrm{T}}. \tag{7.2.15}$$

2. *For $\boldsymbol{y} \in \Gamma$ and $\boldsymbol{\nu}(\boldsymbol{y})$, the unit outward normal at $\boldsymbol{y} \in \Gamma$ of Ω, the following equations hold*

$$
\begin{aligned}
\mathbb{W}_n^{II}(x, y) &:= \Big\{\mathbb{T}'(D_y, \nu(y))(\mathbb{E}_n^N(x, y), \mathbb{Q}_n(x, y))\Big\}^{\mathrm{T}} \\
&= \frac{\lambda+2\mu}{\lambda+\mu}\Big\{\mathbb{T}^N(D_y, \nu(y))\mathbb{E}_n^N(x, y)\Big\}^{\mathrm{T}} + \\
&\quad \frac{\mu}{\lambda+\mu}\left(\frac{\partial}{\partial \tau_y^{i,j}}\right)_{n\times n} E_n^L(x, y) - \frac{\mu}{\lambda+\mu}\frac{\partial}{\partial \nu(y)}E_n^L(x, y)\mathbb{I}, \quad (7.2.16) \\
\mathbb{Q}_n^{II}(x, y) &:= 2\mu\frac{\partial}{\partial \nu(y)}\mathbb{Q}_n(x, y) = -2\mu\Big(\nabla_x\big(\frac{\partial}{\partial \nu(y)}E_n^L(x, y)\big)\Big)^{\mathrm{T}}. \quad (7.2.17)
\end{aligned}
$$

Proof 1. The definitions of the three fundamental solutions $\big(\mathbb{E}_n(x, y), \mathbb{Q}_n(x, y)\big)$, $\mathbb{E}_n^N(x, y)$ and $E_n^L(x, y)$ yield that

$$
\begin{aligned}
\mathbb{E}_n^N(x, y) &= \frac{\lambda+2\mu}{4(n-1)\pi\mu(\lambda+2\mu)}\left\{\gamma_n(x, y)\mathbb{I} + \frac{\lambda+\mu}{\lambda+3\mu}\frac{(x-y)(x-y)^{\mathrm{T}}}{|x-y|^n}\right\} \\
&= \frac{1}{4(n-1)\pi\mu}\frac{\lambda+\mu}{\lambda+2\mu}\left\{\gamma_n(x, y)\mathbb{I} + \frac{(x-y)(x-y)^{\mathrm{T}}}{|x-y|^n}\right\} + \\
&\quad \frac{1}{2(n-1)\pi}\frac{1}{\lambda+2\mu}\gamma_n(x, y)\mathbb{I} \\
&= \frac{\lambda+\mu}{\lambda+2\mu}\mathbb{E}_n(x, y) + \frac{1}{\lambda+2\mu}E_n^L(x, y)\mathbb{I}.
\end{aligned}
$$

Then the equations (7.2.14) and (7.2.15) are proved directly.

2. By the definition of the operator $\mathbb{T}'(D_y, \nu(y))$, we obtain

$$
\begin{aligned}
&\mathbb{T}'(D_y, \nu(y))\big(\mathbb{E}_n(x, y), \mathbb{Q}_n(x, y)\big) \\
&= \mathbb{T}^*(D_y, \nu(y))\mathbb{E}_n(x, y) + \nu(y)\mathbb{Q}_n(x, y) \\
&= \mathbb{T}^*(D_y, \nu(y))\left\{\frac{\lambda+2\mu}{\lambda+\mu}\mathbb{E}_n^N(x, y) - \frac{1}{\lambda+\mu}E_n^L(x, y)\mathbb{I}\right\} + \nu(y)\mathbb{Q}_n(x, y) \\
&= \frac{\lambda+2\mu}{\lambda+\mu}\mathbb{T}^N(D_y, \nu(y))\mathbb{E}_n^N(x, y) - \frac{\lambda+2\mu}{\lambda+\mu}\mathbb{T}^\Delta(D_y, \nu(y))\mathbb{E}_n^N(x, y) - \\
&\quad \frac{1}{\lambda+\mu}\mathbb{T}^*(D_y, \nu(y))E_n^L(x, y) + \nu(y)\mathbb{Q}_n(x, y). \quad (7.2.18)
\end{aligned}
$$

Furthermore we have

$$
\begin{aligned}
\mathbb{T}^\Delta(D_y, \nu(y))\mathbb{E}_n^N(x, y) &= \lambda\left(\nu_i(y)\frac{\partial}{\partial y_j}\right)_{n\times n}\mathbb{E}_n^N(x, y) \\
&= \lambda\nu(y)\nabla_y\cdot\mathbb{E}_n^N(x, y)
\end{aligned}
$$

$$
\begin{aligned}
&= \frac{\lambda}{\lambda+2\mu}\boldsymbol{\nu}(\boldsymbol{y})\mathbb{Q}_n(\boldsymbol{x},\boldsymbol{y}),\\
\mathbb{T}^*(D_y,\boldsymbol{\nu}(\boldsymbol{y}))E_n^L(\boldsymbol{x},\boldsymbol{y}) &= \left(\mu\boldsymbol{\nu}_j(\boldsymbol{y})\frac{\partial}{\partial y_i}+\mu\delta_{ij}\frac{\partial}{\partial\boldsymbol{\nu}(\boldsymbol{y})}\right)_{n\times n}E_n^L(\boldsymbol{x},\boldsymbol{y})\\
&= \mu\left(\nu_j(\boldsymbol{y})\frac{\partial}{\partial y_i}\right)_{n\times n}E_n^L(\boldsymbol{x},\boldsymbol{y})+\mu\frac{\partial}{\partial\boldsymbol{\nu}(\boldsymbol{y})}E_n^L(\boldsymbol{x},\boldsymbol{y})\mathbb{I},\\
\boldsymbol{\nu}(\boldsymbol{y})\mathbb{Q}_n(\boldsymbol{x},\boldsymbol{y}) &= \left(\nu_j(\boldsymbol{y})Q_n^j(\boldsymbol{x},\boldsymbol{y})\right)_{n\times n}\\
&= \left(\nu_j(\boldsymbol{y})\frac{\partial}{\partial y_j}\right)_{n\times n}E_n^L(\boldsymbol{x},\boldsymbol{y}).
\end{aligned}
$$

The above three equalities yield

$$
\begin{aligned}
&-\frac{\lambda+2\mu}{\lambda+\mu}\mathbb{T}^{\Delta}(D_y,\boldsymbol{\nu}(\boldsymbol{y}))\mathbb{E}_n^N(\boldsymbol{x},\boldsymbol{y})-\frac{1}{\lambda+\mu}\mathbb{T}^*(D_y,\boldsymbol{\nu}(\boldsymbol{y}))E_n^L(\boldsymbol{x},\boldsymbol{y})+\boldsymbol{\nu}(\boldsymbol{y})\mathbb{Q}_n(\boldsymbol{x},\boldsymbol{y})\\
&= \frac{\mu}{\lambda+\mu}\left(\nu_i(\boldsymbol{y})\frac{\partial}{\partial y_j}\right)_{n\times n}E_n^L(\boldsymbol{x},\boldsymbol{y})-\frac{\mu}{\lambda+\mu}\left(\nu_j(\boldsymbol{y})\frac{\partial}{\partial y_i}\right)_{n\times n}E_n^L(\boldsymbol{x},\boldsymbol{y})-\\
&\quad\frac{\mu}{\lambda+\mu}\frac{\partial}{\partial\boldsymbol{\nu}(\boldsymbol{y})}E_n^L(\boldsymbol{x},\boldsymbol{y})\mathbb{I}\\
&= -\frac{\mu}{\lambda+\mu}\left(\frac{\partial}{\partial\boldsymbol{\tau}_{\boldsymbol{y}}^{i,j}}\right)_{n\times n}E_n^L(\boldsymbol{x},\boldsymbol{y})-\frac{\mu}{\lambda+\mu}\frac{\partial}{\partial\boldsymbol{\nu}(\boldsymbol{y})}E_n^L(\boldsymbol{x},\boldsymbol{y})\mathbb{I},
\end{aligned}
\tag{7.2.19}
$$

in which the vectors $\left\{\boldsymbol{\tau}_{\boldsymbol{y}}^{i,j},\ 1\leqslant i,j\leqslant n\right\}$ are defined in Subsect. 3.3.4 and $\boldsymbol{\tau}_{\boldsymbol{y}}^{i,j}$ $(1\leqslant i,j\leqslant n)$ is orthogonal with the unit vector $\boldsymbol{\nu}(\boldsymbol{y})$. Substituting (7.2.19) into (7.2.18) it yields

$$
\begin{aligned}
&\mathbb{T}'(D_y,\boldsymbol{\nu}(\boldsymbol{y}))\big(\mathbb{E}_n(\boldsymbol{x},\boldsymbol{y}),\mathbb{Q}_n(\boldsymbol{x},\boldsymbol{y})\big)\\
&= \frac{\lambda+2\mu}{\lambda+\mu}\mathbb{T}^N(D_y,\boldsymbol{\nu}(\boldsymbol{y}))\mathbb{E}_n^N(\boldsymbol{x},\boldsymbol{y})-\frac{\mu}{\lambda+\mu}\left(\frac{\partial}{\partial\boldsymbol{\tau}_{\boldsymbol{y}}^{i,j}}\right)_{n\times n}E_n^L(\boldsymbol{x},\boldsymbol{y})-\\
&\quad\frac{\mu}{\lambda+\mu}\frac{\partial}{\partial\boldsymbol{\nu}(\boldsymbol{y})}E_n^L(\boldsymbol{x},\boldsymbol{y})\mathbb{I},
\end{aligned}
$$

then the equalities (7.2.16) and (7.2.17) follows immediately. □

Now it is ready to discuss the properties of the single layer and double layer potentials of Stokes equations. Substituting (7.2.14) and (7.2.15) into (7.2.1) and (7.2.2) respectively, then the single layer potential $\big(\mathbb{V}_I(\boldsymbol{x}),q_I(\boldsymbol{x})\big)$ is given by

$$
\begin{aligned}
\mathbb{V}_I(\boldsymbol{x}) = &\frac{\lambda+2\mu}{\lambda+\mu}\int_\Gamma\mathbb{E}_n^N(\boldsymbol{x},\boldsymbol{y})\boldsymbol{\varphi}(\boldsymbol{y})\mathrm{d}s_y-\\
&\frac{1}{\lambda+\mu}\int_\Gamma E_n^L(\boldsymbol{x},\boldsymbol{y})\boldsymbol{\varphi}(\boldsymbol{y})\mathrm{d}s_y,\quad\forall\boldsymbol{x}\in\mathbb{R}^n,
\end{aligned}
\tag{7.2.20}
$$

$$q_I(\boldsymbol{x}) = -\sum_{k=1}^{n} \frac{\partial}{\partial x_k} \int_{\Gamma} E_n^L(\boldsymbol{x}, \boldsymbol{y})\varphi_k(\boldsymbol{y})\mathrm{d}s_y, \quad \forall \boldsymbol{x} \in \mathbb{R}^n. \tag{7.2.21}$$

It means that $\mathbb{V}_I(\boldsymbol{x})$ is represented by the sum of single layer potential of Navier operator and single layer potentials of Laplace operator, and $q_I(\boldsymbol{x})$ is represented by the sum of the first derivatives of single layer potentials of Laplace operator. By the definition of the fundamental solution $\{\mathbb{E}_n(\boldsymbol{x}, \boldsymbol{y}), \mathbb{Q}_n(\boldsymbol{x}, \boldsymbol{y})\}$ and the properties of single layer potentials of Laplace operator and Navier operator given in Chap. 3 (Theorems 3.3.4–3.3.5, Lemma 3.3.6) and Chap. 6 (Theorem 6.2.3, Lemma 6.2.3) directly we obtain

Theorem 7.2.1 *For given $\varphi \in \mathbb{H}^{-1/2}(\Gamma)$, the corresponding single layer potential $\{\mathbb{V}_I(\boldsymbol{x}), q_I(\boldsymbol{x})\}$ of Stokes equations satisfies:*

1. *The single layer potential $\{\mathbb{V}_I(\boldsymbol{x}), q_I(\boldsymbol{x})\}$ is well difined on $\mathbb{R}^n$, and $\mathbb{V}_I \in \mathbb{H}^1_{\text{loc}}(\mathbb{R}^n)$ and $q_I \in L_{2,\text{loc}}(\mathbb{R}^2)$ in which*

$$\mathbb{H}^1_{\text{loc}}(\mathbb{R}^n) = \left\{\boldsymbol{v}\middle|\boldsymbol{v} \in \mathbb{H}^1(G), \textit{ for any bounded domain } G \subset \mathbb{R}^n\right\},$$

$$L_{2,\text{loc}}(\mathbb{R}^n) = \left\{q\middle|q \in L_2(G), \textit{ for any bounded domain } G \subset \mathbb{R}^n\right\}.$$

2. *On $\mathbb{R}^n\backslash\Gamma$, $\{\mathbb{V}_I(\boldsymbol{x}), q_I(\boldsymbol{x})\}$ is smooth and satisfies the Stokes equations (7.1.1).*
3.

$$\mathbb{V}_I(\boldsymbol{x}) = \mathbb{E}_n(\boldsymbol{x}, \boldsymbol{0})\mathbb{C}_0 + \mathcal{O}\big(|\boldsymbol{x}|^{1-n}\big), \quad \textit{as } |\boldsymbol{x}| \to +\infty,$$

with

$$\mathbb{C}_0 = \int_{\Gamma} \varphi(\boldsymbol{x})\mathrm{d}s_x,$$

$$q_I(\boldsymbol{x}) = \begin{cases} \mathcal{O}\big(|\boldsymbol{x}|^{1-n}\big), & \mathbb{C}_0 \neq 0, \\ \mathcal{O}\big(|\boldsymbol{x}|^{-n}\big), & \mathbb{C}_0 = 0. \end{cases}$$

4. *Let $\mathbb{V}_I(\boldsymbol{x})\Big|_{\Gamma} := \big(\mathbb{K}_I\varphi\big)(\boldsymbol{x})$, then $\mathbb{K}_I : \mathbb{H}^{1/2}(\Gamma) \to \mathbb{H}^{1/2}(\Gamma)$ is bounded.*

$$\begin{aligned}
&\mathbb{T}(D_x, \boldsymbol{\nu}(\boldsymbol{x}))\big\{\mathbb{V}_I(\boldsymbol{x}), q_I(\boldsymbol{x})\big\}\Big|_{\Gamma-0} \\
&= \frac{1}{2}\varphi(\boldsymbol{x}) + \int_{\Gamma} \mathbb{T}(D_x, \boldsymbol{\nu}(\boldsymbol{x}))\{\mathbb{E}_n(\boldsymbol{x}, \boldsymbol{y}), \mathbb{Q}_n(\boldsymbol{x}, \boldsymbol{y})\}\varphi(\boldsymbol{y})\mathrm{d}s_y \\
&:= \frac{1}{2}\varphi(\boldsymbol{x}) + \Big(\mathbb{K}^*_{II}\varphi\Big)(\boldsymbol{x}), \\
&\mathbb{T}(D_x, \boldsymbol{\nu}(\boldsymbol{x}))\big\{\mathbb{V}_I(\boldsymbol{x}), q_I(\boldsymbol{x})\big\}\Big|_{\Gamma+0} \\
&= -\frac{1}{2}\varphi(\boldsymbol{x}) + \int_{\Gamma} \mathbb{T}(D_x, \boldsymbol{\nu}(\boldsymbol{x}))\{\mathbb{E}_n(\boldsymbol{x}, \boldsymbol{y}), \mathbb{Q}_n(\boldsymbol{x}, \boldsymbol{y})\}\varphi(\boldsymbol{y})\mathrm{d}s_y
\end{aligned}$$

$$:= -\frac{1}{2}\varphi(x) + \left(\mathbb{K}_{II}^{*}\varphi\right)(x).$$

The operator $\mathbb{K}_{II}^{*} : \mathbb{H}^{-1/2}(\Gamma) \to \mathbb{H}^{-1/2}(\Gamma)$ *is bounded.*

5.

$$\left.\begin{aligned}
&\int_{\Gamma} \mathbb{T}(D_x, \boldsymbol{\nu}(x))\{\mathbb{V}_I(x), q_I(x)\}\Big|_{\Gamma-0} \cdot \boldsymbol{\omega}_j(x)\mathrm{d}s_x = 0,\\
&\int_{\Gamma} \mathbb{T}(D_x, \boldsymbol{\nu}(x))\{\mathbb{V}_I(x), q_I(x)\}\Big|_{\Gamma+0} \cdot \boldsymbol{\omega}_j(x)\mathrm{d}s_x = -\int_{\Gamma} \varphi(x)\cdot \boldsymbol{\omega}_j \mathrm{d}s_x,\\
&\qquad\qquad \forall \boldsymbol{\omega}_j \in \mathbb{R}\mathrm{M},\ j = 1, 2, \ldots, 3(n-1).
\end{aligned}\right\}$$

We now return to the double layer potential $\{\mathbb{V}_{II}(x), q_{II}(x)\}$ of Stokes operator with given density $\psi \in \mathbb{H}^{1/2}(\Gamma)$. Substituting (7.2.16) and (7.2.17) into (7.2.3) and (7.2.4) respectively, it arrives at

$$\begin{aligned}
\mathbb{V}_{II}(x) = {} & \frac{\lambda + 2\mu}{\lambda + \mu}\int_{\Gamma} \left\{\mathbb{T}^{N}(D_y, \boldsymbol{\nu}(y))\mathbb{E}_n^{N}(x, y)\right\}^{\mathrm{T}} \psi(y)\mathrm{d}s_y - \\
& \frac{\mu}{\lambda + \mu}\int_{\Gamma} E_n^{L}(x, y)\left(\frac{\partial}{\partial \mathbb{S}_y^{i,j}}\right)_{n\times n} \psi(y)\mathrm{d}s_y - \\
& \frac{\mu}{\lambda + \mu}\int_{\Gamma} \frac{\partial}{\partial \boldsymbol{\nu}(y)} E_n^{L}(x, y)\psi(y)\mathrm{d}s_y, \qquad (7.2.22)
\end{aligned}$$

$$q_{II}(x) = -2\mu\nabla_x \cdot \int_{\Gamma} \frac{\partial}{\partial \boldsymbol{\nu}(y)} E_n^{L}(x, y)\psi(y)\mathrm{d}s_y. \qquad (7.2.23)$$

The equality (7.2.22) shows that the double layer potential $\mathbb{V}_{II}(x)$ is represented as a double layer potential of Navier operator, single and double layer potentials of Laplace operator. Furthermore, $q_{II}(x)$ is represented as the derivatives of the double layer potential of Laplace operator. Therefore the properties of the single and double layer potentials of Laplace operator given in Chap. 3 and the properties of the double layer potentials of Navier operator given in Chap. 6 yield the following results.

Theorem 7.2.2 *For given density* $\psi \in \mathbb{H}^{1/2}(\Gamma)$, *the corresponding double layer potential* $\{\mathbb{V}_{II}(x), q_{II}(x)\}$ *of Stokes equations satisfies:*

1. *The double layer potential* $\{\mathbb{V}_{II}(x), q_{II}(x)\}$ *is well defined on* $\mathbb{R}^n$ *and*

$$\mathbb{V}_{II} \in \mathbb{H}^{1}_{\mathrm{loc}}(\mathbb{R}^n\backslash\Gamma), \quad q_{II} \in L_{2,\mathrm{loc}}(\mathbb{R}^n).$$

2. *On the domain* $\mathbb{R}^n\backslash\Gamma, \{\mathbb{V}_{II}(x), q_{II}(x)\}$ *is smooth and satisfies the Stoked equations (7.1.1).*
3.

$$\mathbb{V}_{II}(x) = \mathcal{O}\left(|x|^{1-n}\right), \quad \textit{as } |x| \to +\infty,$$

$$q_{II}(x) = \mathcal{O}(|x|^{-n}), \quad as \ |x| \to +\infty.$$

4. *The trace* $\mathbb{V}_{II}(x)\Big|_{\Gamma-0}$ *and* $\mathbb{V}_{II}(x)\Big|_{\Gamma+0}$ *are given by the following equalities:*

$$\begin{aligned}\mathbb{V}_{II}(x)\Big|_{\Gamma-0} &= -\frac{1}{2}\psi(x) + \int_\Gamma \mathbb{T}(D_y, \nu(y))\{\mathbb{E}_n(x, y), \mathbb{Q}_n(x, y)\}\psi(y)ds_y \\ &:= -\frac{1}{2}\psi(x) + (\mathbb{K}_{II}\psi)(x), \quad \forall x \in \Gamma, \qquad (7.2.24)\\ \mathbb{V}_{II}(x)\Big|_{\Gamma+0} &= \frac{1}{2}\psi(x) + (\mathbb{K}_{II}\psi)(x), \quad \forall x \in \Gamma. \qquad (7.2.25)\end{aligned}$$

The operator $\mathbb{K}_{II} : \mathbb{H}^{1/2}(\Gamma) \to \mathbb{H}^{1/2}(\Gamma)$ *is bounded.*

5.

$$\begin{aligned}&\mathbb{T}(D_x, \nu(x))\Big\{\mathbb{V}_{II}(x), q_{II}(x)\Big\}\Big|_{\Gamma-0} = \mathbb{T}(D_x, \nu(x))\Big\{\mathbb{V}_{II}(x), q_{II}(x)\Big\}\Big|_{\Gamma+0}\\ &= \mathbb{T}(D_x, \nu(x))\Big\{\mathbb{V}_{II}(x), q_{II}(x)\Big\}\Big|_{\Gamma}\\ &:= (\mathbb{K}_{III}\psi)(x).\end{aligned}$$

The operator $\mathbb{K}_{III} : \mathbb{H}^{1/2}(\Gamma) \to \mathbb{H}^{-1/2}(\Gamma)$ *is bounded, which is a hypersingular integral operator. Furthermore the operator* $\mathbb{K}_{III}$ *can be regularized as a differential-integro operator with weakly singular kernel, namely*

$$\begin{aligned}(\mathbb{K}_{III}\psi)(x) = &\frac{1}{2(n-1)\pi}\sum_{1\leqslant k,j\leqslant n}\frac{\partial}{\partial\tau_x^{k,j}}\int_\Gamma \gamma_n(x, y)\frac{\partial}{\partial\tau_y^{k,j}}\psi(y)ds_y + \\ &\mathbb{U}(D_x, \nu(x))\int_\Gamma\left[4\mu^2\mathbb{E}_n(x, y) - \frac{\mu}{(n-1)\pi}\gamma_n(x, y)\mathbb{I}\right]\cdot\\ &\mathbb{U}(D_y, \nu(y))\psi(y)ds_y - \\ &\frac{\mu}{2(n-1)\pi}\int_\Gamma\Big[\mathbb{U}(D_x, \nu(x))\gamma_n(x, y)\mathbb{U}(D_y, \nu(y))\Big]^{\mathrm{T}}\psi(y)ds_y, \qquad (7.2.26)\end{aligned}$$

in which the matrix differential operator $\mathbb{U}(D_x, \nu(x))$ *is defined in (6.1.7).*

Remark 7.2.1 (The null-spaces of operator $\mathbb{K}_{III}$ and $\mathbb{K}_I$). Let $\psi \in \mathbb{H}^{1/2}(\Gamma)$ and

$$(\mathbb{K}_{III}\psi)(x) = 0, \quad \forall x \in \Gamma. \qquad (7.2.27)$$

The corresponding double layer potential

$$\begin{aligned}\mathbb{V}_{II,\psi}(x) &= \int_\Gamma \{\mathbb{T}'(D_y, \nu(y))(\mathbb{E}_n(x, y), \mathbb{Q}_n(x, y))\}^{\mathrm{T}}\psi(y)ds_y, \quad \forall x \in \mathbb{R}^n,\\ q_{II,\psi}(x) &= 2\mu\int_\Gamma \frac{\partial}{\partial\nu(y)}\mathbb{Q}_n(x, y)\cdot\psi(y)ds_y, \quad \forall x \in \mathbb{R}^n,\end{aligned}$$

then

$$\mathbb{V}_{\mathit{II},\psi}(\boldsymbol{x}) = \begin{cases} -\sum_{j=1}^{3(n-1)} m_j \boldsymbol{\omega}_j(\boldsymbol{x}), & \forall \boldsymbol{x} \in \Omega, \\ 0, & \forall \boldsymbol{x} \in \Omega_e, \end{cases}$$

$$q_{\mathit{II},\psi}(\boldsymbol{x}) = 0, \quad \forall \boldsymbol{x} \in \mathbb{R}^n,$$

with constants $\{m_j \in \mathbb{R},\ j = 1, 2, \ldots, 3(n-1)\}$. Therefore

$$\left[\mathbb{V}_{\mathit{II},\psi}(\boldsymbol{x})\right]_\Gamma = -\mathbb{V}_{\mathit{II}}(\boldsymbol{x})\Big|_{\Gamma-0} = \sum_{j=1}^{3(n-1)} m_j \boldsymbol{\omega}_j(\boldsymbol{x}) = \boldsymbol{\psi}(\boldsymbol{x}).$$

It means that $\mathbb{N}(\mathbb{K}_{\mathit{III}})$, the null-space of operator $\mathbb{K}_{\mathit{III}}$, is given by

$$\mathbb{N}(\mathbb{K}_{\mathit{III}}) = \left\{ \boldsymbol{\psi} \middle| \boldsymbol{\psi}(\boldsymbol{x}) = \sum_{j=1}^{3(n-1)} m_j \boldsymbol{\omega}_j(\boldsymbol{x}),\ \ m_j \in \mathbb{R}, j = 1, 2, \ldots, 3(n-1) \right\}. \tag{7.2.28}$$

Let

$$\mathbb{V}_{I,\nu}(\boldsymbol{x}) = \int_\Gamma \mathbb{E}_n(\boldsymbol{x}, \boldsymbol{y})\boldsymbol{\nu}(\boldsymbol{y})\mathrm{d}s_y, \quad \forall \boldsymbol{x} \in \mathbb{R}^n,$$

$$q_{I,\nu}(\boldsymbol{x}) = \int_\Gamma \mathbb{Q}_n(\boldsymbol{x}, \boldsymbol{y})\boldsymbol{\nu}(\boldsymbol{y})\mathrm{d}s_y, \quad \forall \boldsymbol{x} \in \mathbb{R}^n.$$

in which $\boldsymbol{\nu}(\boldsymbol{x})$ is the unit outward normal at $\boldsymbol{x} \in \Gamma$ of Ω. Note that

$$\int_\Gamma \boldsymbol{\nu}(\boldsymbol{y})\mathrm{d}s_y = 0,$$

which yields

$$\mathbb{V}_{I,\nu}(\boldsymbol{x}) + \mathcal{O}\big(|\boldsymbol{x}|^{1-n}\big), \quad \text{as } |\boldsymbol{x}| \to +\infty.$$

For any $\boldsymbol{\rho} \in \mathbb{H}^{-1/2}(\Gamma)$,

$$\begin{aligned}\int_\Gamma \mathbb{V}_{I,\nu}(\boldsymbol{x}) \cdot \boldsymbol{\rho}(\boldsymbol{x})\mathrm{d}s_x &= \int_\Gamma \int_\Gamma \boldsymbol{\nu}(\boldsymbol{y})^{\mathrm{T}}\mathbb{E}_n(\boldsymbol{x}, \boldsymbol{y})\boldsymbol{\rho}(\boldsymbol{x})\mathrm{d}s_y\mathrm{d}s_x \\ &= \int_\Gamma \boldsymbol{\nu}(\boldsymbol{y})^{\mathrm{T}}\mathbb{V}_{I,\rho}(\boldsymbol{y})\mathrm{d}s_y.\end{aligned}$$

Since

$$\nabla \cdot \mathbb{V}_{I,\rho}(\boldsymbol{x}) = 0, \quad \forall \boldsymbol{x} \in \mathbb{R}^n \backslash \Gamma,$$

therefore we obtain

$$\int_{\Gamma} \mathbb{V}_{I,\nu}(\boldsymbol{x}) \cdot \boldsymbol{\rho}(\boldsymbol{x}) \mathrm{d}s_x = 0, \quad \forall \boldsymbol{\rho} \in \mathbb{H}^{-1/2}(\Gamma).$$

Namely

$$\mathbb{V}_{I,\nu}(\boldsymbol{x})\Big|_{\Gamma} = 0.$$

It means that

$$\boldsymbol{\nu} \in \mathbb{N}(\mathbb{K}_I). \tag{7.2.29}$$

Furthermore

$$\mathbb{V}_{I,\nu}(\boldsymbol{x}) = 0, \quad \forall \boldsymbol{x} \in \mathbb{R}^n,$$

$$q_{I,\nu}(\boldsymbol{x}) = \frac{1}{2(n-1)\pi} \int_{\Gamma} \frac{\partial}{\partial \boldsymbol{\nu}(\boldsymbol{y})} E_n^L(\boldsymbol{x}, \boldsymbol{y}) \mathrm{d}s_y = \begin{cases} -1, & \boldsymbol{x} \in \Omega, \\ -1/2, & \boldsymbol{x} \in \Gamma, \\ 0, & \boldsymbol{x} \in \Omega_e. \end{cases}$$

7.3 Boundary Reduction of the Boundary Value Problems of Stokes Equations

We are ready to reduce the boundary value problems of Stokes equations to boundary integral equations or boundary differential-integro equations.

Let

$$\mathbb{H}_*^{-1/2}(\Gamma) = \left\{\boldsymbol{\varphi} \,\middle|\, \boldsymbol{\varphi} \in \mathbb{H}^{-1/2}(\Gamma) \text{ and } \int_{\Gamma} \boldsymbol{\varphi}(\boldsymbol{x}) \cdot \boldsymbol{\omega}_j(\boldsymbol{x}) \mathrm{d}s_x = 0, \quad j = 1, 2, \ldots, n.\right\}$$

$$\mathbb{H}_{*,\Delta}^{-1/2}(\Gamma) = \left\{\boldsymbol{\varphi} \,\middle|\, \boldsymbol{\varphi} \in \mathbb{H}_*^{-1/2}(\Gamma) \text{ and } \int_{\Gamma} \boldsymbol{\varphi}(\boldsymbol{x}) \cdot \boldsymbol{\nu}(\boldsymbol{x}) \mathrm{d}s_x = 0.\right\}$$

$$\mathbb{H}_{*,*}^{-1/2}(\Gamma) = \left\{\boldsymbol{\varphi} \,\middle|\, \boldsymbol{\varphi} \in \mathbb{H}^{-1/2}(\Gamma) \text{ and } \int_{\Gamma} \boldsymbol{\varphi}(\boldsymbol{x}) \cdot \boldsymbol{\omega}_j(\boldsymbol{x}) \mathrm{d}s_x = 0, \ j = 1, 2, \ldots, 3(n-1)\right\}$$

$$\mathbb{H}_{*,*}^{1/2}(\Gamma) = \left\{\boldsymbol{\psi} \,\middle|\, \boldsymbol{\psi} \in \mathbb{H}^{1/2}(\Gamma) \text{ and } \int_{\Gamma} \boldsymbol{\psi}(\boldsymbol{x}) \cdot \boldsymbol{\omega}_j(\boldsymbol{x}) \mathrm{d}s_x = 0, \ j = 1, 2, \ldots, 3(n-1)\right\}$$

and

$$\mathbb{V}_{I,\varphi}(\boldsymbol{x}) = \int_{\Gamma} \mathbb{E}_n(\boldsymbol{x}, \boldsymbol{y}) \boldsymbol{\varphi}(\boldsymbol{y}) \mathrm{d}s_y, \quad \forall \boldsymbol{x} \in \mathbb{R}^n, \tag{7.3.1}$$

$$q_{I,\varphi}(\boldsymbol{x}) = \int_{\Gamma} \mathbb{Q}_n(\boldsymbol{x}, \boldsymbol{y}) \cdot \boldsymbol{\varphi}(\boldsymbol{x}) \mathrm{d}s_y, \quad \forall \boldsymbol{x} \in \mathbb{R}^n. \tag{7.3.2}$$

Let $\left\{\mathbb{V}_{I,\varphi}(\boldsymbol{x}) + \sum_{j=1}^{n} m_j \boldsymbol{\omega}_j(\boldsymbol{x}), q_{I,\varphi}(\boldsymbol{x}) + C\right\}$ and $\left\{\mathbb{V}_{I,\varphi} + \sum_{j=1}^{n} m_j \boldsymbol{\omega}_j(\boldsymbol{x}), q_{I,\varphi}(\boldsymbol{x})\right\}$ denote the solution of interior Dirichlet problem (7.1.13) and the solution of exterior Dirichlet problem (7.1.14) separately with $\varphi \in \mathbb{H}^{-1/2}_{*,\Delta}(\Gamma)$ and constants $\{C, m_j, j = 1, 2, \ldots, n\}$.

The Dirichlet condition yields the following boundary integral equation of first kind to determine the density $\varphi \in \mathbb{H}^{-1/2}_{*,\Delta}(\Gamma)$ and constants $\{m_j, j = 1, 2, \ldots, n\}$.

$$\begin{cases} \text{Find } \varphi \in \mathbb{H}^{-1/2}_{*,\Delta}(\Gamma) \text{ and constants } \{m_j, j=1, 2, \ldots, n\}, \quad \text{such that} \\ \displaystyle\int_{\Gamma} \mathbb{E}(\boldsymbol{x}, \boldsymbol{y})\varphi(\boldsymbol{y})\,\mathrm{d}s_y + \sum_{j=1}^{n} m_j \boldsymbol{\omega}_j(\boldsymbol{x}) = g(\boldsymbol{x}), \qquad \forall \boldsymbol{x} \in \Gamma, \end{cases} \tag{7.3.3}$$

with given $g \in \mathbb{H}^{1/2}(\Gamma)$ satisfies the compatibility condition (7.1.20). The problem (7.3.3) is equivalent to the following variational problem

$$\begin{cases} \text{Find } \varphi \in \mathbb{H}^{-1/2}_{*,\Delta}(\Gamma) \text{ such that} \\ \mathbb{A}_{\Gamma}(\varphi, \boldsymbol{\rho}) := \displaystyle\int_{\Gamma}\int_{\Gamma} \mathbb{E}_n(\boldsymbol{x}, \boldsymbol{y}) \cdot \boldsymbol{\rho}(\boldsymbol{x})\mathrm{d}s_y\mathrm{d}s_x = \int_{\Gamma} g(\boldsymbol{x}) \cdot \boldsymbol{\rho}(\boldsymbol{x})\mathrm{d}s_y, \forall \boldsymbol{\rho} \in \mathbb{H}^{-1/2}_{*,\Delta}(\Gamma). \end{cases} \tag{7.3.4}$$

The constants $\{m_j, j = 1, 2, \ldots, n\}$ are determined by $\varphi(\boldsymbol{x})$, the solution of variational problem (7.3.4),

$$m_j = \frac{1}{\text{meas}(\Gamma)}\left\{-\int_{\Gamma}\int_{\Gamma} \mathbb{E}_n(\boldsymbol{x}, \boldsymbol{y}) \cdot \boldsymbol{\omega}_j(\boldsymbol{x})\mathrm{d}s_y\mathrm{d}s_x + \int_{\Gamma} g(\boldsymbol{x}) \cdot \boldsymbol{\omega}(\boldsymbol{x})\mathrm{d}s_y\right\}, \quad j = 1, 2, \ldots, n. \tag{7.3.5}$$

For the bilinear form $\mathbb{A}_{\Gamma}(\varphi, \boldsymbol{\rho})$ defined in (7.3.4), we have:

(1) $\mathbb{A}_{\Gamma}(\varphi, \boldsymbol{\rho})$ is a bounded bilinear form on $\mathbb{H}^{-1/2}(\Gamma) \times \mathbb{H}^{-1/2}(\Gamma)$.
(2) For $\varphi, \boldsymbol{\rho} \in \mathbb{H}^{-1/2}_{*,\Delta}(\Gamma)$, we obtain

$$\mathbb{A}_{\Gamma}(\varphi, \boldsymbol{\rho}) = 2\mu \int_{\mathbb{R}^n} \epsilon(\mathbb{V}_{I,\varphi}) : \epsilon(\mathbb{V}_{I,\rho})\mathrm{d}\boldsymbol{x}. \tag{7.3.6}$$

It means that

$$\mathbb{A}_{\Gamma}(\varphi, \varphi) \geqslant 0, \quad \forall \varphi \in \mathbb{H}^{-1/2}_{*,\Delta}(\Gamma). \tag{7.3.7}$$

(3) Suppose that $\varphi \in \mathbb{H}^{-1/2}_{*,\Delta}(\Gamma)$ and $\mathbb{A}_{\Gamma}(\varphi, \varphi) = 0$, then $\varphi(\boldsymbol{x}) \equiv 0$. It yields the uniqueness of variational problem (7.3.4).
(4) We have the following conjecture, the bilinear form $\mathbb{A}_{\Gamma}(\varphi, \varphi)$ is coercive on $\mathbb{H}^{-1/2}_{*,\Delta}(\Gamma) \times \mathbb{H}^{-1/2}_{*,\Delta}(\Gamma)$.

After obtained $\varphi \in \mathbb{H}^{-1/2}_{*,\Delta}(\Gamma)$ and constants $\{m_j,\ j = 1, 2, \ldots, n\}$, the solution of boundary integral equations (7.3.3), let

$$u(x) = \int_\Gamma \mathbb{E}_n(x, y)\varphi(x)\mathrm{d}s_y + \sum_{j=1}^{n} m_j\omega_j(x), \quad \forall x \in \mathbb{R}^n,$$

$$p(x) = \begin{cases} \displaystyle\int_\Gamma \mathbb{Q}_n(x, y)\cdot\varphi(y)\mathrm{d}s_y + C, & \forall x \in \Omega, \\ \displaystyle\int_\Gamma \mathbb{Q}_n(x, y)\cdot\varphi(y)\mathrm{d}s_y, & \forall x \in \Omega_e, \end{cases}$$

with

$$C = -\frac{1}{\mathrm{meas}(\Gamma)}\int_\Gamma\int_\Gamma \mathbb{Q}_n(x, y)\cdot\varphi(y)\mathrm{d}s_y\mathrm{d}s_x.$$

Obviously $\{u(x), p(x)\}$ is the generalized solution of the interior Dirichlet problem (7.1.13) on domain Ω and the generalized solution of the exterior Dirichelt problem (7.1.14) on Ω_e.

Let $\{\mathbb{V}_{\mathit{II},\psi}(x), q_{\mathit{II},\psi}(x)\}$, the double layer potential of Stokes operator with unknown density $\psi \in \mathbb{H}^{1/2}_{*,*}(\Gamma)$, be a solution of interior traction problem (7.1.16) in domain Ω and the solution of exterior traction problem (7.1.17). The traction boundary condition on Γ yields the following boundary integro-differential equations of first kind to determine the unknown density $\psi \in \mathbb{H}^{1/2}_{*,*}(\Gamma)$ with given $h \in \mathbb{H}^{-1/2}_{*,*}(\Gamma)$, namely

$$\begin{cases} \text{Find } \psi \in \mathbb{H}^{1/2}_{*,*}(\Gamma) \text{ such that} \\ -(\mathbb{K}_{\mathit{III}}\psi)(x) = -\dfrac{1}{2(n-1)\pi}\displaystyle\sum_{1\leqslant k,j\leqslant n}\frac{\partial}{\partial\tau_x^{k,j}}\int_\Gamma \gamma_n(x, y)\frac{\partial}{\partial\tau_y^{k,j}}\psi(y)\mathrm{d}s_y - \\ \mathbb{U}(D_x, \nu(x))\displaystyle\int_\Gamma\left[4\mu^2\mathbb{E}_n(x, y) - \frac{\mu}{(n-1)\pi}\gamma_n(x, y)\mathbb{I}\right]\mathbb{U}(D_y, \nu(y))\psi(y)\mathrm{d}s_y + \\ \dfrac{\mu}{2(n-1)\pi}\displaystyle\int_\Gamma\left[\mathbb{U}(D_x, \nu(x))\gamma_n(x, y)\mathbb{U}(D_y, \nu(y))\right]^{\mathrm{T}}\psi(y)\mathrm{d}s_y = -h(x), \quad \forall x \in \Gamma. \end{cases} \tag{7.3.8}$$

The boundary integro-differential equations (7.3.8) is equivalent to the following variational problem:

$$\begin{cases} \text{Find } \psi \in \mathbb{H}^{1/2}_{*,*}(\Gamma) \text{ such that} \\ \mathbb{B}_\Gamma(\psi, \mu) = \dfrac{1}{2(n-1)\pi} \displaystyle\sum_{1 \leqslant k,j \leqslant n} \int_\Gamma \gamma_n(x, y) \frac{\partial \psi(y)}{\partial \tau_y^{k,j}} \cdot \frac{\partial \mu(x)}{\partial \tau_x^{k,j}} \mathrm{d}s_y \mathrm{d}s_x + \\ \displaystyle\int_\Gamma \int_\Gamma \left[\mathbb{U}(D_y, \nu(y)) \psi(y)\right]^{\mathrm{T}} \left[4\mu^2 \mathbb{E}_n(x, y) - \frac{\mu \gamma_n(x, y)}{(n-1)\pi} \mathbb{I}\right] \mathbb{U}(D_x, \nu(x)) \mu(x) \mathrm{d}s_y \mathrm{d}s_x + \\ \dfrac{\mu}{2(n-1)\pi} \displaystyle\int_\Gamma \int_\Gamma \psi(y)^{\mathrm{T}} \left[\mathbb{U}(D_x, \nu(x)) \gamma_n(x, y) \mathbb{U}(D_y, \nu(y))\right]^{\mathrm{T}} \mu(x) \mathrm{d}s_y \mathrm{d}s_x, \\ = -h(x), \quad \forall \mu \in \mathbb{H}^{1/2}_{*,*}(\Gamma). \end{cases} \tag{7.3.9}$$

For the bilinear form $\mathbb{B}_\Gamma(\psi, \mu)$ we have:

(1) $\mathbb{B}_\Gamma(\psi, \mu)$ is a bounded bilinear form on $\mathbb{H}^{1/2}(\Gamma) \times \mathbb{H}^{1/2}(\Gamma)$.
(2) For $\psi, \mu \in \mathbb{H}^{1/2}_{*,*}(\Gamma)$, then

$$\mathbb{B}_\Gamma(\psi, \mu) = 2\mu \int_{\mathbb{R}^n} \epsilon(\mathbb{V}_{II,\psi}) : \epsilon(V_{II,\mu}) \mathrm{d}x.$$

It yields that

$$\mathbb{B}_\Gamma(\psi, \psi) \geqslant 0, \quad \forall \psi \in \mathbb{H}^{1/2}(\Gamma).$$

(3) Assume that $\psi \in \mathbb{H}^{1/2}_{*,*}(\Gamma)$ and $\mathbb{B}_\Gamma(\psi, \psi) = 0$, then we obtain

$$\psi(x) = 0, \quad \forall x \in \Gamma.$$

The uniqueness of variational problem (7.3.9) is directly available from this.
(4) Conjecture: The bilinear form $\mathbb{B}_\Gamma(\psi, \mu)$ is coercive on $\mathbb{H}^{1/2}_{*,*}(\Gamma) \times \mathbb{H}^{1/2}_{*,*}(\Gamma)$.

After solving the variational problem (7.3.9), the density $\psi \in \mathbb{H}^{1/2}_{*,*}(\Gamma)$ is obtained, then the double layer potential $\{\mathbb{V}_{II,\psi}(x), q_{II,\psi}(x)\}$ is a generalized solution of the interior traction problem (7.1.16) on domain Ω and the generalized solution of the exterior traction problem (7.1.17).

The boundary value problems of Stokes equations can also reduced into the boundary integral equations of second kind. Let $\left\{\mathbb{V}_{II,\psi}(x), q_{II,\psi}(x) + C\right\}$ be the solution of interior Dirichlet problem (7.1.13) with unknown density $\psi \in \mathbb{H}^{1/2}(\Gamma)$. The Dirichlet boundary condition (7.1.13c) yields the following integral equations to determine the unknown density $\psi \in \mathbb{H}^{1/2}(\Gamma)$:

$$-\frac{1}{2}\psi(x) + (\mathbb{K}_{II}\psi)(x) = g(x), \quad \forall x \in \Gamma. \tag{7.3.10}$$

Next let $\left\{\mathbb{V}_{II,\psi}(x) + \displaystyle\sum_{j=1}^{n} m_j \omega_j(x), q_{II,\psi}(x)\right\}$ be the solution of exterior Dirichlet problem(I) (7.1.14) with unknown density $\psi \in \mathbb{H}^{1/2}(\Gamma)$ and unknown constants $\{m_j, \ j = 1, 2, \ldots, n\}$. The Dirichlet boundary condition (7.1.14c) provides the boundary integral equations of the second kind to find the unknown density $\psi \in \mathbb{H}^{1/2}(\Gamma)$ and constants $\{m_j,\ j = 1, 2, \ldots, n\}$:

$$\frac{1}{2}\psi(\boldsymbol{x}) + (\mathbb{K}_{II}\psi)(\boldsymbol{x}) = g(\boldsymbol{x}) - \sum_{j=1}^{n} m_j \omega_j(\boldsymbol{x}), \quad \forall \boldsymbol{x} \in \Gamma. \tag{7.3.11}$$

For the traction problems, let $\{\mathbb{V}_{I,\varphi}(\boldsymbol{x}), q_{I,\varphi}(\boldsymbol{x})\}$ with unknown density $\varphi \in \mathbb{H}^{-1/2}(\Gamma)$ be a solution of interior traction problem (7.1.16), the traction boundary condition yields that

$$\frac{1}{2}\varphi(\boldsymbol{x}) + (\mathbb{K}_{II}^{*}\varphi)(\boldsymbol{x}) = \boldsymbol{h}(\boldsymbol{x}), \quad \forall \boldsymbol{x} \in \Gamma. \tag{7.3.12}$$

Furthermore let $\{\mathbb{V}_{I,\varphi}(\boldsymbol{x}), q_{I,\varphi}(\boldsymbol{x})\}$ with unknown density $\varphi \in \mathbb{H}_{*}^{-1/2}(\Gamma)$ be the solution of exterior traction problem (7.1.17), the traction boundary condition (7.1.17c) provides the following boundary integral equations to find the unknown density $\varphi \in \mathbb{H}_{*}^{-1/2}(\Gamma)$:

$$-\frac{1}{2}\varphi(\boldsymbol{x}) + (\mathbb{K}_{II}^{*}\varphi)(\boldsymbol{x}) = \boldsymbol{h}(\boldsymbol{x}), \quad \forall \boldsymbol{x} \in \Gamma. \tag{7.3.13}$$

The exterior Dirichlet problem (II) and exterior traction problem (II) are omitted, which leave to readers. For more results on this subject the readers can refer to the book by Hsiao and Wendland (2008).

References

Happel, J., & Brenner, H. (1983). *Low reynolds number hydrodynamics: With special applications to particulate media* (Vol. 1). Springer Science & Business Media.

Hsiao, G. C., & Wendland, W. L. (2008). *Boundary integral equations*. Springer.

Sokolnikoff, I. S., Specht, R. D., et al. (1956). *Mathematical theory of elasticity* (Vol. 83). McGraw-Hill.

Chapter 8
Some Nonlinear Problems

In this chapter, we discuss the boundary reduction of some nonlinear problems including heat radiation problems, Signoroni problems, contact problems and Steklov eigenvalue problems. A common characteristic of these nonlinear problems we consider is that their solutions satisfy linear partial differential equations with constant coefficients in the open domain. It is natural to reduce these nonlinear problems to nonlinear boundary integro-differential problems by boundary reduction.

8.1 Heat Radiation Problems

In this section we consider the steady-state heat transfer with radiation, which is mathematically described by the following nonlinear boundary value problem of Laplace equation.

$$\begin{cases} \Delta u = 0, \qquad \text{in} \quad \Omega, & (8.1.1a) \\ \dfrac{\partial u}{\partial \boldsymbol{\nu}_x} + \gamma |u|^3 u = f(x), \quad \text{on} \ \ \Gamma. & (8.1.1b) \end{cases}$$

Here u is the unknown temperature, $\Omega \in \mathbb{R}^n (n = 2, 3)$ is the given bounded domain with smooth boundary Γ, γ is a given positive constant associated with the body's emittance, and $f(x)$ is a given function on Γ, $\boldsymbol{\nu}_x = (\nu_1(x), \nu_2(x), \dots, \nu_n(x))^{\mathrm{T}}$ is the unit outward normal vector of Ω at $x \in \Gamma$ (Sparrow, 2018; Han, 1995). Ruotsalainen and Wendland (1988), Atkinson and Chandler (1990) discussed the boundary element method for Laplace equation with more general nonlinear boundary condition:

H. Han and D. Yin, *Mathematical Foundation of the Boundary Integro-Differential Equation Method*, https://doi.org/10.1007/978-981-95-1088-7_8

$$\begin{cases} \Delta u = 0, \quad \text{in} \quad \Omega, & (8.1.2a) \\ \dfrac{\partial u}{\partial \boldsymbol{\nu}} + g(x, u) = f(x), \quad \text{on} \quad \Gamma. & (8.1.2b) \end{cases}$$

Here the function $g(x, u)$ is given, which is assumed $g : \Gamma \times \mathbb{R} \to \mathbb{R}$ is Caratheodory-function, namely, $g(\cdot, u)$ is measurable for $u \in \mathbb{R}$, and $g(x, \cdot)$ is continuous for almost all $x \in \Gamma$, and $(\partial/\partial u)g(x, u)$ is Borel measurable satisfying

$$0 < l \leqslant \frac{\partial}{\partial u} g(x, u) \leqslant L, \quad \forall x \in \Gamma, \ u \in \mathbb{R}. \tag{8.1.3}$$

Obviously the nonlinear boundary condition (8.1.1b) does not satisfy the condition (8.1.3). We focus on the problem (8.1.1) mainly in this section.

8.1.1 Boundary Condition of Nonlinear Boundary Problem (8.1.1)

Suppose that $u(x)$ is a solution of problem (8.1.1) and $u \in H^1(\Omega)$. Let

$$p(x) = \frac{\partial u(x)}{\partial \boldsymbol{\nu}_x}\bigg|_{\Gamma}, \tag{8.1.4}$$

then $p \in H^{-\frac{1}{2}}(\Gamma)$ by the trace theorem of Sobolev space $H^1(\Omega)$. Furthermore we know

$$p \in H_*^{-\frac{1}{2}}(\Gamma) = \left\{ q \in H^{-\frac{1}{2}}(\Gamma) \text{ and } \int_\Gamma q(x) \mathrm{d}s_x = 0 \right\}.$$

By Green's formula we obtain

$$u(x) = -\int_\Gamma \frac{\partial E_n(x, y)}{\partial \boldsymbol{\nu}_y} u(u) \mathrm{d}s_y + \int_\Gamma E_n(x, y) p(y) \mathrm{d}s_y, \quad \forall x \in \Omega, \tag{8.1.5}$$

where $E_n(x, y)$ is the fundamental solution of Laplace operator given by (3.3.3) It means that $u(x)$, the solution of problem (8.1.1), is represented as the sum of a double layer potential and a single layer potential. By the properties of double (single) layer potential and formula (8.1.4), let $x \in \Omega$ approaches the boundary Γ, we obtain the following two relationships between $p(x)$ and $u(x)\Big|_\Gamma$ (Han, 1988; Nedelec, 1982):

$$\frac{1}{2} u(x) = -\int_\Gamma \frac{\partial E_n(x, y)}{\partial \boldsymbol{\nu}_y} u(y) \mathrm{d}s_y + \int_\Gamma E_n(x, y) p(y) \mathrm{d}s_y, \quad \forall x \in \Gamma, \tag{8.1.6}$$

$$\frac{1}{2}p(x) = -\int_{\Gamma} \frac{\partial^2 E_n(x, y)}{\partial \boldsymbol{\nu}_x \partial \boldsymbol{\nu}_y} u(y) \mathrm{d}s_y + \int_{\Gamma} \frac{\partial E_n(x, y)}{\partial \boldsymbol{\nu}_x} p(y) \mathrm{d}s_y, \quad \forall x \in \Gamma. \tag{8.1.7}$$

The derivative of double layers potential in (8.1.7) is given by (3.3.37) ($n = 2$) and (3.3.38) ($n = 3$), namely

$$\int_{\Gamma} \frac{\partial^2 E_n(x, y)}{\partial \boldsymbol{\nu}_x \partial \boldsymbol{\nu}_y} u(y) \mathrm{d}s_y = \begin{cases} \dfrac{\partial}{\partial \boldsymbol{\tau}_x} \displaystyle\int_{\Gamma} E_2(x, y)(\frac{\partial}{\partial \boldsymbol{\tau}_y} u(y)) \mathrm{d}s_y, & n = 2, \\ \displaystyle\sum_{k=1}^{3} \frac{\partial}{\partial \boldsymbol{\tau}_{k,x}} \int_{\Gamma} E_3(x, y)(\frac{\partial}{\partial \boldsymbol{\tau}_{k,y}} u(y)) \mathrm{d}s_y, & n = 3. \end{cases} \tag{8.1.8}$$

Substituting (8.1.7) into the nonlinear boundary condition (8.1.1b) and combining (8.1.6), we obtain the system of the following nonlinear boundary integro-differential equations to determine the unknown functions $u(x)\Big|_{\Gamma}$ and $p(x)$:

$$\frac{1}{2}p(x) - \int_{\Gamma} \frac{\partial^2 E_n(x, y)}{\partial \boldsymbol{\nu}_x \partial \boldsymbol{\nu}_y} u(y) \mathrm{d}s_y + \int_{\Gamma} \frac{\partial E_n(x, y)}{\partial \boldsymbol{\nu}_x} p(y) \mathrm{d}s_y + \gamma \left|u(x)\right|^3 u(x) = f(x), \forall x \in \Gamma, \tag{8.1.9}$$

$$-\frac{1}{2}u(x) - \int_{\Gamma} \frac{\partial E_n(x, y)}{\partial \boldsymbol{\nu}_y} u(y) \mathrm{d}s_y + \int_{\Gamma} E_n(x, y) p(y) \mathrm{d}s_y = 0, \quad \forall x \in \Gamma \tag{8.1.10}$$

The nonlinear boundary system (8.1.9)–(8.1.10) is equivalent to the nonlinear boundary value problem of problem (8.1.1). Similarly nonlinear boundary value problem (8.1.2) is reduced to the following nonlinear boundary system

$$\frac{1}{2}p(x) - \int_{\Gamma} \frac{\partial^2 E_n(x, y)}{\partial \boldsymbol{\nu}_x \partial \boldsymbol{\nu}_y} u(y) \mathrm{d}s_y + \int_{\Gamma} \frac{\partial E_n(x, y)}{\partial \boldsymbol{\nu}_x} p(y) \mathrm{d}s_y + g(x, u) = f(x), \quad \forall x \in \Gamma, \tag{8.1.11}$$

$$-\frac{1}{2}u(x) - \int_{\Gamma} \frac{\partial E_n(x, y)}{\partial \boldsymbol{\nu}_y} u(y) \mathrm{d}s_y + \int_{\Gamma} E_n(x, y) p(y) \mathrm{d}s_y = 0, \quad \forall x \in \Gamma. \tag{8.1.12}$$

8.1.2 Equivalent Formula of Problem (8.1.1)

Suppose that $u \in H^1(\Omega) \cap L_5(\Gamma)$ is the weak solution of nonlinear boundary value problem (8.1.1), then we know that

$$u\Big|_{\Gamma} \in H^{\frac{1}{2}}(\Gamma) \cap L_5(\Gamma), \quad p \in H_*^{-\frac{1}{2}}(\Gamma).$$

Naturally we choose space $(H^{\frac{1}{2}}(\Gamma) \cap L_5(\Gamma)) \times H_*^{-\frac{1}{2}}(\Gamma)$ as the solution space and the test function space of the coming variational problem, which is equivalent to nonlinear boundary value problem (8.1.9)–(8.1.10).

Multiplying (8.1.9) by a function $v \in H^{\frac{1}{2}}(\Gamma) \cap L_5(\Gamma)$ and integrating on Γ, multiplying (8.1.10) by a function $q \in H_*^{-\frac{1}{2}}(\Gamma)$ and integrating on Γ, we obtain:

$$\begin{cases} \text{Find } u \in H^{\frac{1}{2}}(\Gamma) \cap L_5(\Gamma), \quad p \in H_*^{-\frac{1}{2}}(\Gamma), \quad \text{such that} \\ \frac{1}{2}\int_\Gamma p(x)v(x)\mathrm{d}s_x - \int_\Gamma\int_\Gamma \frac{\partial^2 E_n(x,y)}{\partial \boldsymbol{\nu}_x \partial \boldsymbol{\nu}_y} u(y)\mathrm{d}s_y\mathrm{d}s_x + \\ \int_\Gamma\int_\Gamma \frac{\partial E_n(x,y)}{\partial \boldsymbol{\nu}_x} p(y)v(x)\mathrm{d}s_y\mathrm{d}s_x + \\ \gamma \int_\Gamma |u(x)|^3 u(x)v(x)\mathrm{d}s_x = \int_\Gamma f(x)v(x)\mathrm{d}s_x, \quad \forall v \in H^{\frac{1}{2}}(\Gamma) \cap L_5(\Gamma), \quad (8.1.13a) \\ -\frac{1}{2}\int_\Gamma q(x)u(x)\mathrm{d}s_x - \int_\Gamma\int_\Gamma \frac{\partial E_n(x,y)}{\partial \boldsymbol{\nu}_y} q(x)u(y)\mathrm{d}s_y\mathrm{d}s_x + \\ \int_\Gamma\int_\Gamma E_n(x,y)p(y)q(x)\mathrm{d}s_y\mathrm{d}s_x = 0, \quad \forall q \in H_*^{-\frac{1}{2}}(\Gamma). \quad (8.1.13b) \end{cases}$$

Recall the bilinear forms $A(\varphi, \mu)$ and $B(\psi, \rho)$ defined by (3.4.16) and (3.4.17). Furthermore let

$$C(q,v) = \frac{1}{2}\int_\Gamma q(x)v(x)\mathrm{d}s_x + \int_\Gamma\int_\Gamma \frac{\partial E_n(x,y)}{\partial \boldsymbol{\nu}_y} q(x)v(y)\mathrm{d}s_y\mathrm{d}s_x,$$

$$\forall q \in H^{-\frac{1}{2}}(\Gamma), \ v \in H^{\frac{1}{2}}(\Gamma), \tag{8.1.14}$$

$$A_1(u,v) = \gamma \int_\Gamma |u(x)|^3 u(x)v(x)\mathrm{d}s_x, \quad \forall u, v \in L_5(\Gamma), \tag{8.1.15}$$

$$(f, v)_\Gamma = \int_\Gamma f v \mathrm{d}s. \tag{8.1.16}$$

Therefore the variational problem (8.1.13) can be rewritten as

$$\begin{cases} \text{Find } u(x) \in H^{\frac{1}{2}}(\Gamma) \cap L_5(\Gamma), \ p(x) \in H_*^{-\frac{1}{2}}(\Gamma), \quad \text{such that} \\ B(u,v) + C(p,v) + A_1(u,v) = \int_\Gamma f v \mathrm{d}s_x, \quad \forall v \in H^{\frac{1}{2}}(\Gamma) \cap L_5(\Gamma), \quad (8.1.17a) \\ -C(q,u) + A(q,p) = 0, \quad \forall q \in H_*^{-\frac{1}{2}}(\Gamma). \quad (8.1.17b) \end{cases}$$

For the bilinear form $C(q, v)$ we have

Lemma 8.1.1 *(i) $C(q, v)$ is bounded on $H^{-\frac{1}{2}}(\Gamma) \times H^{\frac{1}{2}}(\Gamma)$, namely, there is a constant $M > 0$ such that*

$$|C(q,v)| \leqslant M\|q\|_{H^{-\frac{1}{2}}(\Gamma)}\|v\|_{H^{\frac{1}{2}}(\Gamma)}, \quad \forall q \in H^{-\frac{1}{2}}(\Gamma),\ v \in H^{\frac{1}{2}}(\Gamma). \tag{8.1.18}$$

(ii) Furthermore there exists a constant $M > 0$, such that

$$\left|C(q,v)\right| \leqslant M\|q\|_{H^{-\frac{1}{2}}(\Gamma)}\left|v\right|_{H^{\frac{1}{2}}(\Gamma)}, \quad \forall q \in H_*^{-\frac{1}{2}}(\Gamma),\ \ v \in H^{\frac{1}{2}}(\Gamma). \tag{8.1.19}$$

Proof (i) Since that the operator $\mathbb{K}_{II}$ defined by (3.3.22) is bounded from $H^{\frac{1}{2}}(\Gamma)$ to $H^{\frac{1}{2}}(\Gamma)$, the estimate (8.1.18) follows directly.

(ii) For any $v(x) \in H^{\frac{1}{2}}(\Gamma)$, let

$$v_0 = \frac{1}{\text{meas}(\Gamma)}\int_\Gamma v(x)\mathrm{d}s_x,$$

then $u(x) = v_0$ is a solution of $\Delta u = 0$ and $p_0 = \frac{\partial u(x)}{\partial \nu_x}\Big|_\Gamma \equiv 0$, v_0, p_0 satisfy

$$-C(q,v_0) + A(q,p_0) = 0, \quad \forall q \in H_*^{-\frac{1}{2}}(\Gamma),$$

namely

$$C(q,v_0) = 0, \qquad \forall q \in H_*^{-\frac{1}{2}}(\Gamma).$$

Furthermore for any $q(x) \in H_*^{-\frac{1}{2}}(\Gamma)$, $v(x) \in H^{\frac{1}{2}}(\Gamma)$, we have

$$\begin{aligned}\left|C(q,v)\right| = \left|C(q,v-v_0)\right| &\leqslant M\|q\|_{H^{-\frac{1}{2}}(\Gamma)}\|v-v_0\|_{H^{\frac{1}{2}}(\Gamma)}\\ &\leqslant M\|q\|_{H^{-\frac{1}{2}}(\Gamma)}\left|v\right|_{H^{\frac{1}{2}}(\Gamma)}.\end{aligned}$$

(Noting the positive constant M can have different values in different locations). Estimate (8.1.19) is proved completely. □

$A_1(u,v)$ is well defined on $L_5(\Gamma)\times L_5(\Gamma)$, and we have

Lemma 8.1.2 *(i) Functional $A_1(v,v)$ is strictly convex, and*

$$A_1(v,v) = \gamma\|v\|_{L_5(\Gamma)}^5, \qquad \forall v \in L_5(\Gamma). \tag{8.1.20}$$

(ii) The following inequalities hold

$$\left|A_1(u,w) - A_1(v,w)\right| \leqslant 4\gamma\int_\Gamma (|u|^3+|w|^3)|u-v||w|\mathrm{d}s, \quad \forall u,v,w \in L_5(\Gamma), \tag{8.1.21}$$

$$A_1(v,v-w) - A_1(w,v-w) \geqslant \frac{\gamma}{4}\int_\Gamma (|v|^3+|w|^3)(v-w)^2\mathrm{d}s,$$

$$\forall u, v, w \in L_5(\Gamma). \tag{8.1.22}$$

Proof (i) The conclusion follows by using the strict convexity of mapping $t \in \mathbb{R} \to |t|^5$.

(ii) Since $C^0(\Gamma)$, the continuous function space, is dense in $L_5(\Gamma)$, we only need to prove the inequalities (8.1.21)–(8.1.22) for functions $u, v, w \in C^0(\Gamma)$.

By the definition of $A_1(u, v)$, we obtain

$$A_1(u, w) - A_1(v, w) = \int_\Gamma (|u|^3 u - |v|^3 v) w \mathrm{d}s,$$

$$A_1(u, v - w) - A_1(v, v - w) = \int_\Gamma (|u|^3 u - |v|^3 v)(v - w) \mathrm{d}s.$$

Moreover we get

$$\left|(|u|^3 u - |v|^3 v) w\right| = \left|4w \int_v^u |\xi|^3 \mathrm{d}\xi\right| \leqslant 4(|u|^3 + |v|^3)|u - v||w|.$$

On the other hand for $vw \geqslant 0$, we get

$$\begin{aligned}(|v|^3 v - |w|^3 w)(v - w) &= (|v|^4 - |w|^4)(|v| - |w|)\\ &\geqslant (|v|^3 + |w|^3)(|v| - |w|)^2\\ &= (|v|^3 + |w|^3)(v - w)^2\end{aligned}$$

and for $vw < 0$, we have

$$\begin{aligned}(|v|^3 v - |w|^3 w)(v - w) &= (|v|^4 + |w|^4)(|v| + |w|)\\ &\geqslant \frac{1}{2}(|v|^2 + |w|^2)^2(|v| + |w|)\\ &\geqslant \frac{1}{4}(|v|^2 + |w|^2)(|v| + |w|)^3\\ &\geqslant \frac{1}{4}(|v|^3 + |w|^3)(|v| + |w|)^2\\ &\geqslant \frac{1}{4}(|v|^3 + |w|^3)(v - w)^2.\end{aligned}$$

Then the inequalities (8.1.21)–(8.1.22) are proved directly. □

For the variational problem (8.1.17), we have

Theorem 8.1.1 *Suppose that $f \in L_{\frac{5}{4}}(\Gamma)$, then problem (8.1.17) exists a unique solution $(u, p) \in (H^{\frac{1}{2}}(\Gamma) \cap L_5(\Gamma)) \times H_*^{-\frac{1}{2}}(\Gamma)$. Furthermore the following estimate holds*

$$|u|^2_{H^{\frac{1}{2}}(\Gamma)} + \|p\|^2_{H^{-\frac{1}{2}}(\Gamma)} + \|u\|^5_{L_5(\Gamma)} \leqslant C\left(\|f\|_{L_{\frac{5}{4}}(\Gamma)}\right)^{\frac{5}{4}} \tag{8.1.23}$$

with positive constant C.

Proof For given $v \in H^{\frac{1}{2}}(\Gamma)$ consider the following problem:

$$\begin{cases} \text{Find} \quad t \in H_*^{-\frac{1}{2}}(\Gamma), \quad \text{such that} \\ A(q,t) = C(q,v), \qquad \forall q \in H_*^{-\frac{1}{2}}(\Gamma). \end{cases} \tag{8.1.24}$$

By Lemma 3.4.1 we know that $A(p,q)$ is bounded on $H^{-\frac{1}{2}}(\Gamma) \times H^{-\frac{1}{2}}(\Gamma)$ and coercive on $H_*^{-\frac{1}{2}}(\Gamma)$, then for any given $v \in H^{\frac{1}{2}}(\Gamma)$, problem (8.1.24) has a unique solution $t \in H_*^{-\frac{1}{2}}(\Gamma)$, which denoted by $t = Kv$. K is a bounded operator from $H^{\frac{1}{2}}(\Gamma)$ to $H_*^{-\frac{1}{2}}(\Gamma)$. If (u,p) is a solution of problem (8.1.17), then $p = Ku$ and

$$C(p,v) = A(p,Kv) = A(Ku,Kv), \quad \forall v \in H^{\frac{1}{2}}(\Gamma). \tag{8.1.25}$$

Substituting (8.1.25) into (8.1.17a), then the variational problem (8.1.17) is reduced as

$$\begin{cases} \text{Find} \quad u \in H^{\frac{1}{2}}(\Gamma) \cap L_5(\Gamma), \quad \text{such that} \\ B(u,v) + A(Ku,Kv) + A_1(u,v) = (f,v)_\Gamma, \quad \forall v \in H^{\frac{1}{2}}(\Gamma) \cap L_5(\Gamma). \end{cases} \tag{8.1.26}$$

Let

$$I(v) = \frac{1}{2}(B(v,v) + A(Kv,Kv)) + \frac{1}{5}A_1(v,v) - (f,v)_\Gamma.$$

It is direct to find that variational problem (8.1.26) is equivalent to following minimization problem

$$\begin{cases} \text{Find} \quad u \in H^{\frac{1}{2}}(\Gamma) \cap L_5(\Gamma), \quad \text{such that} \\ I(u) = \inf\limits_{v \in H^{\frac{1}{2}}(\Gamma) \cap L_5(\gamma)} I(v). \end{cases} \tag{8.1.27}$$

Let

$$\|v\|_* = |v|_{H^{\frac{1}{2}}(\Gamma)} + \|v\|_{L_5(\Gamma)},$$

then $\|\cdot\|_*$ is a norm of space $H^{\frac{1}{2}} \cap L_5(\Gamma)$. Applying Lemma 3.4.1 we obtain

$$I(v) \geqslant \frac{\delta}{2}|v|^2_{H^{\frac{1}{2}}(\Gamma)} + \gamma\|v\|^5_{L_5(\Gamma)} - \|f\|_{L_{\frac{5}{4}}(\Gamma)}\|v\|_{L_5(\Gamma)}, \quad \forall v \in H^{\frac{1}{2}}(\Gamma) \cap L_5(\Gamma),$$

which yields

$$\lim_{\|v\|_* \to +\infty} I(v) = +\infty.$$

By the results in Lemma 8.1.2, the strict convexity and the differentiability of functional $I(v)$ on $H^{\frac{1}{2}}(\Gamma) \cap L_5(\Gamma)$ are obtained. By the theorem of minimization of convex functionals we deduce that the minimization problem (8.1.27) exists a unique solution $u \in H^{\frac{1}{2}}(\Gamma) \cap L_5(\Gamma)$, which is the unique solution of problem (8.1.26). Let $p = Ku$, then $(u, p) \in (H^{\frac{1}{2}}(\Gamma) \cap L_5(\Gamma)) \times H_*^{-\frac{1}{2}}(\Gamma)$ is the unique solution of problem (8.1.17). Furthermore for the unique solution (u, p) of problem (8.1.17) taking $v = u$ in (8.1.17a), $q = p$ in (8.1.17b), we arrive at

$$B(u, u) + A_1(u, u) + A(p, p) = (f, u)_\Gamma.$$

Since $B(u, u) \geqslant 0$, $A(p, p) \geqslant 0$, we obtain

$$\gamma \|u\|^5_{L_5(\Gamma)} \leqslant (f, u)_\Gamma \leqslant \|f\|_{L_{\frac{5}{4}}(\Gamma)} \|u\|_{L_5(\Gamma)},$$

namely

$$\|u\|_{L_5(\Gamma)} \leqslant \gamma^{-1/4} \left(\|f\|_{L_{\frac{5}{4}}(\Gamma)} \right)^{1/4}, \tag{8.1.28}$$

and

$$(f, u)_\Gamma \leqslant \gamma^{-1/4} \left(\|f\|_{L_{\frac{5}{4}}(\Gamma)} \right)^{5/4}.$$

By properties of $B(u, u)$ and $A(p, p)$, we obtain the following inequality with a constant $\delta > 0$

$$\delta |u|^2_{H^{\frac{1}{2}}(\Gamma)} + \delta \|p\|_{H^{-\frac{1}{2}}(\Gamma)} + \gamma \|u\|^5_{L_5(\Gamma)} \leqslant \gamma^{-1/4} \left(\|f\|_{L_{\frac{5}{4}}(\Gamma)} \right)^{5/4}.$$

We have

$$\|u\|^2_{H^{\frac{1}{2}}(\Gamma)} + \|p\|^2_{H^{-\frac{1}{2}}(\Gamma)} + \|u\|^5_{L_5(\Gamma)} \leqslant C \left(\|f\|_{L_{\frac{5}{4}}(\Gamma)} \right)^{5/4},$$

with positive constant $C = \gamma^{-1/4} / \min(\delta, \gamma)$. □

8.1.3 Equivalent Saddle-Point Problem

In this subsection the equivalent saddle-point problem of problem (8.1.17) is given. Let

$$L(v, q) = \frac{1}{2} B(v, v) + \frac{1}{5} A_1(v, v) + C(q, v) - \frac{1}{2} A(q, q) - (f, v)_\Gamma,$$

$$\forall v \in H^{\frac{1}{2}}(\Gamma) \cap L_5(\Gamma), \quad q \in H_*^{-\frac{1}{2}}(\Gamma). \quad (8.1.29)$$

Theorem 8.1.2 *The variational problem (8.1.17) is equivalent to the following saddle-point problem:*

$$\begin{cases} \textit{Find} \quad u \in H^{\frac{1}{2}}(\Gamma) \cap L_5(\Gamma), \quad p \in H_*^{-\frac{1}{2}}(\Gamma), \quad \textit{such that} \\ L(u, q) \leqslant L(u, p) \leqslant L(v, p), \quad \forall v \in H^{\frac{1}{2}}(\Gamma) \cap L_5(\Gamma), q \in H_*^{-\frac{1}{2}}(\Gamma). \end{cases} \tag{8.1.30}$$

Proof Assume that (u, p) is a solution of saddle-point problem (8.1.30), then for any $q \in H_*^{-\frac{1}{2}}(\Gamma)$ and $\lambda \in \mathbb{R}$, we have

$$L(u, p + \lambda q) \leqslant L(u, p),$$

that is

$$\lambda\{C(q, u) - A(q, p)\} - \frac{\lambda^2}{2} A(q, q) \leqslant 0, \quad \forall q \in H_*^{-\frac{1}{2}}(\Gamma), \quad \lambda \in \mathbb{R}.$$

Since $A(q, q) \geqslant 0$, it arrives at

$$C(q, u) - A(q, p) = 0, \quad \forall q \in H_*^{-\frac{1}{2}}(\Gamma).$$

Furthermore for any $v \in H^{\frac{1}{2}}(\Gamma) \cap L_5(\Gamma)$, $\lambda \in \mathbb{R}$, we get

$$L(u, p) \leqslant L(u + \lambda v, p), \quad \forall v \in H^{\frac{1}{2}} \cap L_5(\Gamma), \quad \lambda \in \mathbb{R},$$

namely

$$\begin{aligned} L(u + \lambda v, p) - L(u, p) = \lambda\{B(u, v) + C(p, v) - (f, v)_\Gamma\} + \frac{\lambda^2}{2} B(v, v) + \\ \frac{1}{5}(A_1(u + \lambda v, u + \lambda v) - A_1(u, v)) \geqslant 0. \end{aligned} \tag{8.1.31}$$

Noting

$$J(\lambda) = \frac{1}{5} A_1(u + \lambda v, u + \lambda v),$$

then we obtain

$$\begin{aligned} J(0) &= \frac{1}{5} A_1(u, u), \\ \frac{\mathrm{d}J(\lambda)}{\mathrm{d}\lambda} &= \gamma \int_\Gamma |u + \lambda v|^3 (u + \lambda v) v \mathrm{d}s, \\ \frac{\mathrm{d}J(\lambda)}{\mathrm{d}\lambda}\bigg|_{\lambda=0} &= \gamma \int_\Gamma |u|^3 u v \mathrm{d}s = A_1(u, v), \end{aligned}$$

$$\frac{\mathrm{d}^2 J}{\mathrm{d}\lambda^2} = 4\gamma \int_\Gamma |u + \lambda v|^3 v^2 \mathrm{d}s \geqslant 0, \quad \forall \lambda \in \mathbb{R}, \quad u, v \in L_5(\Gamma).$$

It yields

$$\frac{1}{5}(A_1(u + \lambda v, u + \lambda v) - A_1(u, v)) = J(\lambda) - J(0) = \lambda A_1(u, v) + \frac{\lambda^2}{2} \frac{\mathrm{d}^2 J}{\mathrm{d}\lambda^2}\bigg|_{\lambda=\tilde{\lambda}}, \tag{8.1.32}$$

where $\tilde{\lambda}$ lies between 0 and λ. Substituting (8.1.32) into (8.1.31), it arrives at

$$\begin{aligned} &\lambda \{B(u, v) + A_1(u, v) + C(p, v) - (f, v)_\Gamma\} + \\ &\frac{\lambda^2}{2} \left\{ B(v, v) + \frac{\mathrm{d}^2 J}{\mathrm{d}\lambda^2}\bigg|_{\lambda=\tilde{\lambda}} \right\} \geqslant 0, \qquad \forall v \in H^{\frac{1}{2}}(\Gamma) \cap L_5(\Gamma), \quad \lambda \in \mathbb{R}. \end{aligned}$$

Since λ is an arbitrary real number, we obtain

$$B(u, v) + A_1(u, v) + C(p, v) = (f, v)_\Gamma, \qquad \forall v \in H^{\frac{1}{2}}(\Gamma) \cap L_5(\Gamma).$$

It means that $(u, p) \in (H^{\frac{1}{2}}(\Gamma) \cap L_5(\Gamma)) \times H_*^{-\frac{1}{2}}(\Gamma)$ is the solution of variational problem (8.1.17). Each of the above steps is reversible, therefor the proof is completed. □

8.1.4 The Numerical Solutions of Nonlinear Boundary Variational Problem (8.1.17)

We consider the numerical solutions of problem (8.1.17) by boundary element method. Suppose that V_h and M_h are finite dimensional subspace of $H^{\frac{1}{2}}(\Gamma) \cap L_5(\Gamma)$ and $H_*^{-\frac{1}{2}}(\Gamma)$ respectively. In problem (8.1.17) replace space $H^{\frac{1}{2}}(\Gamma) \cap L_5(\Gamma)$ and $H_*^{-\frac{1}{2}}(\Gamma)$ with V_h and M_h, we obtain the discrete problem of problem (8.1.17).

$$\begin{cases} \text{Find} \quad (u_h, p_h) \in V_h \times M_h, \quad \text{such that} \\ B(u_h, v_h) + A_1(u_h, v_h) + C(p_h, v_h) = (f, v_h)_\Gamma, \quad \forall v_h \in V_h, & (8.1.33\text{a}) \\ -C(q_h, u_h) + A(q_h, p_h) = 0, \quad \forall q_h \in M_h. & (8.1.33\text{b}) \end{cases}$$

Theorem 8.1.3 *Suppose that $f \in L_{5/4}(\Gamma)$, then problem (8.1.33) exists a unique solution $(u_h, p_h) \in V_h \times M_h$.*

Furthermore the following estimate holds

$$|u_h|^2_{H^{\frac{1}{2}}(\Gamma)} + \|p_h\|^2_{H^{-\frac{1}{2}}(\Gamma)} + \|u_h\|^5_{L_5(\Gamma)} \leqslant C\left(\|f\|_{L^{5/4}(\Gamma)}\right)^{5/4}, \tag{8.1.34}$$

with constant $C > 0$.

The proof of Theorem 8.1.3 is similar to the proof of Theorem 8.1.2, which is omitted here.

We now consider the error estimate of the numerical solution (u_h, p_h) given by problem (8.1.33).

Let

$$e_h = u - u_h, \quad \epsilon_h = p - p_h,$$

here (u, p) is the solution of problem (8.1.17).

Theorem 8.1.4 *Suppose that $f \in L_{5/4}(\Gamma)$, then the following error estimate holds*

$$\begin{aligned}
&|e_h|^2_{H^{\frac{1}{2}}(\Gamma)} + \|\epsilon_h\|^2_{H^{-\frac{1}{2}}(\Gamma)} + \int_\Gamma (|u|^3 + |u_h|^3) e_h^2 \mathrm{d}s \\
&\leqslant \Bigg\{ \inf_{\forall v_h \in V_h} |v - v_h|^2_{H^{\frac{1}{2}}(\Gamma)} + \inf_{\forall q_h \in M_h} \|p - q_h\|_{H^{-\frac{1}{2}}(\Gamma)} + \\
&\left(\|f\|_{L^{5/4}(\Gamma)}\right)^{3/4} \inf_{\forall v_h \in V_h} \|u - v_h\|^2_{L_5(\Gamma)} \Bigg\}
\end{aligned} \tag{8.1.35}$$

with positive constant C independent of h.

Proof By the variational equalities (8.1.17) and (8.1.33), we obtain

$$B(e_h, v_h) + A_1(u, v_h) - A_1(u_h, v_h) + C(\epsilon_h, v_h) = 0, \quad \forall v_h \in V_h, \tag{8.1.36}$$

$$- C(q_h, e_h) + A(q_h, \epsilon_h) = 0, \quad \forall q_h \in M_h. \tag{8.1.37}$$

In (8.1.36) using $v_h - u_h \in V_h$ instead of v_h, it arrives

$$B(e_h, v_h - u_h) + A_1(u, v_h - u_h) - A_1(u_h, v_h - u_h) + C(\epsilon_h, v_h - u_h) = 0, \tag{8.1.38}$$

and $q_h - p_h \in M_h$ instead of q_h in (8.1.37) yields

$$- C(q_h - p_h, e_h) + A(q_h - p_h, \epsilon_h) = 0. \tag{8.1.39}$$

A computation shows that

$$\begin{aligned}
&B(e_h, e_h) + A_1(u, e_h) - A_1(u_h, e_h) + A(\epsilon_h, \epsilon_h) \\
&= B(e_h, u - v_h) + A_1(u, u - v_h) - A_1(u_h, u - v_h) + A(p - q_h, \epsilon_h) + \\
&B(e_h, v_h - u_h) + A_1(u, v_h - u_h) - A_1(u_h, v_h - u_h) + A(q_h - p_h, \epsilon_h) \\
&= B(e_h, u - v_h) + A_1(u, u - v_h) - A_1(u_h, u - v_h) + A(p - q_h, \epsilon_h) - \\
&C(p - q_h, e_h) + C(\epsilon_h, u - v_h), \quad \forall v_h \in V_h, \ q_h \in M_h.
\end{aligned}$$

The last equality is from (8.1.36)–(8.1.37).

Combining the results given in Lemma 3.4.1, Lemmas 8.1.1–8.1.2, we obtain

$$
\begin{aligned}
&\delta\left(|e_h|^2_{H^{\frac{1}{2}}(\Gamma)} + \|\epsilon_h\|^2_{H^{-\frac{1}{2}}(\Gamma)}\right) + \frac{\gamma}{4}\int_\Gamma (|u|^3 + |u_h|^3)e_h^2 \mathrm{d}s \\
&\leqslant M\left\{|e_h|_{H^{\frac{1}{2}}(\Gamma)}|u - v|_{H^{\frac{1}{2}}(\Gamma)} + \|\epsilon_h\|_{H^{-\frac{1}{2}}(\Gamma)}\|p - q_h\|_{H^{-\frac{1}{2}}(\Gamma)} + \right.\\
&\left.|e_h|_{H^{\frac{1}{2}}(\Gamma)}\|p - q_h\|_{H^{-\frac{1}{2}}(\Gamma)} + \|\epsilon_h\|_{H^{-\frac{1}{2}}(\Gamma)}|u - v_h|_{H^{\frac{1}{2}}(\Gamma)}\right\} + \\
&4\gamma\int_\Gamma (|u|^3 + |u_h|^3)e_h(u - v_h)\mathrm{d}s. \qquad (8.1.40)
\end{aligned}
$$

Furthermore by the inequalities (8.1.28) and (8.1.34) we have

$$
4\gamma\int_\Gamma (|u|^3 + |u_h|^3)e_h(u - u_h)\mathrm{d}s \leqslant C\left(\|f\|_{L^{5/4}(\Gamma)}\right)^{3/4}\|u - v_h\|^2_{L_5(\Gamma)}, \qquad (8.1.41)
$$

with positive constant C.

The inequalities (8.1.40)–(8.1.41) yields the error estimate (8.1.35) directly. □

Remark 8.1.1 In the case $n = 2$ by the Sobolev imbedding theorem (Adams & Fournier, 2003), we know that for any $v \in H^{\frac{1}{2}}(\Gamma)$, then $v \in L_5(\Gamma)$ and

$$
\|v\|_{L_5(\Gamma)} \leqslant C\|v\|_{H^{\frac{1}{2}}(\Gamma)}, \quad \forall v \in H^{\frac{1}{2}}(\Gamma).
$$

Therefore the variational problem (8.1.17) can be rewritten as Han (1995):

$$
\begin{cases}
\text{Find} \quad u \in H^{\frac{1}{2}}(\Gamma), \quad p \in H_*^{-\frac{1}{2}}(\Gamma), \quad \text{suchthat} \\
B(u, v) + C(p, v) + A_1(u, v) = (f, v)_\Gamma, \quad \forall v \in H^{\frac{1}{2}}(\Gamma), \\
-C(q, u) + A(q, p) = 0, \quad \forall q \in H_*^{-\frac{1}{2}}(\Gamma).
\end{cases}
$$

8.2 Variational Inequality (I)-Laplace Equation with Unilateral Boundary Conditions

The variational inequalities arise in many fields of applied sciences (Duvaut & Lions, 1976), which attract a lot of scholars working on. For the existence, uniqueness and regularity results of variational inequalities, the readers can refer to Glowinski et al. (1981), and Friedman (1982). The numerical solutions of variational inequalities by finite element method are treated, for instance, by Glowinski et al. (1981), Glowinski (2013), Kikuchi and Oden (1988). We know that there is a large class of variational inequalities, whose solutions satisfy linear partial differential equations in the given domain, even the variational inequalities are nonlinear problems. The typical examples are fluid mechanics in media with semipermeable boundary, the electropaint process, with unilateral boundary conditions (Goeleven & Motreanu, 2003; Goeleven et al., 2004; Han 1990, 1991, 1994a, 1994b, 1994c; Han, Guan, & Yu, 1994a). In this and coming section, we discuss the boundary integro-differential method for these problems, which are also called Signorini problems (Signorini, 1933; Kikuchi & Oden, 1988). Laplace equation with unilateral boundary conditions, which is a model of the Signorini problem, is discussed in this section.

$$
\begin{cases}
\Delta u = 0, \quad \text{in} \quad \Omega, & (8.2.1a) \\
u = 0, \quad \text{on} \quad \Gamma_0, & (8.2.1b) \\
u \geqslant 0, \quad \dfrac{\partial u}{\partial \boldsymbol{v}_x} \geqslant g, \quad \text{and} \quad u\left(\dfrac{\partial u}{\partial \boldsymbol{v}_x} - g\right) = 0, \quad \text{on} \quad \Gamma_1, & (8.2.1c)
\end{cases}
$$

where $\Omega \in R^n$ $(n = 2, 3)$ is a bounded domain with boundary $\Gamma = \Gamma_0 \cup \Gamma_1$ and g is given on Γ_1.

Let

$$
K = \left\{ v \middle| v \in H^1(\Omega), v\Big|_{\Gamma_0} = 0, \ v|_{\Gamma_1} \geqslant 0 \right\},
$$

K is a nonempty convex set of Sobolev space $H^1(\Omega)$, then the problem (8.2.1) is characterized by the following variational inequality:

$$
\begin{cases}
\text{Find} \quad u \in K, \quad \text{such that} \\
a(u, v - u) \geqslant \int_{\Gamma_1} g(v - u)\mathrm{d}s, \quad \forall v \in K,
\end{cases}
\tag{8.2.2}
$$

with

$$
a(u, v) = \int_\Omega \nabla u \cdot \nabla v \mathrm{d}x.
$$

The existence and uniqueness of problem (8.2.2) can be found in Glowinski (2013).

8.2.1 *Equivalent Boundary Variational Inequality of Problem (8.2.2)*

Let $u \in K$ is the solution of variational inequality (8.2.2), and

$$
p(x) = \left.\frac{\partial u(x)}{\partial \boldsymbol{v}_x}\right|_\Gamma . \tag{8.2.3}
$$

Since $u \in H^1(\Omega)$ and $\Delta u = 0$ in Ω, then we know that $p \in H_*^{-\frac{1}{2}}(\Gamma)$. Furthermore $u(x)$, the solution of variational inequality (8.2.2), is given by (8.1.5) and $p(x), u(x)\Big|_\Gamma$ satisfy the two equalities (8.1.6) and (8.1.7). For any $v \in H^1(\Omega)$ and $u(x)$, the solution of variational inequality (8.2.2), we have

$$a(u, v) = \int_\Omega \nabla u \cdot \nabla v \mathrm{d}x = \int_\Gamma p(x)v(x)\mathrm{d}s. \tag{8.2.4}$$

Substituting (8.1.7) into (8.2.4) it yields

$$\begin{aligned} a(u, v) = &-\int_\Gamma \int_\Gamma \frac{\partial^2 E_n(x, y)}{\partial \boldsymbol{v}_x \partial \boldsymbol{v}_y} u(y)v(x)\mathrm{d}s_y \mathrm{d}s_x + \\ &\int_\Gamma \int_\Gamma \frac{\partial E(x, y)}{\partial \boldsymbol{v}_x} p(y)v(x)\mathrm{d}s_y \mathrm{d}s_x + \frac{1}{2}\int_\Gamma p(x)v(x)\mathrm{d}s_x \\ &= B(u, v) + C(p, v), \end{aligned} \tag{8.2.5}$$

where the bilinear form $B(u, v)$ and $C(p, v)$ are defined by (3.4.17) and (8.1.14). Introduce

$$K_\Gamma = \left\{ v \middle| v \in H^{\frac{1}{2}}(\Gamma),\ v\big|_{\Gamma_0} = 0,\ \text{and}\ v\big|_{\Gamma_1} \geqslant 0 \right\}, \tag{8.2.6}$$

K_Γ is a convex set of $H^{\frac{1}{2}}(\Gamma)$, then the variational inequality (8.2.2) is reduced to the following boundary variational inequality and equality:

$$\begin{cases} \text{Find}\quad (u, p) \in K_\Gamma \times H_*^{-\frac{1}{2}}(\Gamma), \quad \text{such that} \\ B(u, v-u) + C(p, v-u) \geqslant \displaystyle\int_{\Gamma_1} g(v-u)\mathrm{d}s, \quad \forall v \in K_\Gamma, & (8.2.7\text{a}) \\ -C(q, u) + A(q, p) = 0, \quad \forall q \in H_*^{-\frac{1}{2}}(\Gamma), & (8.2.7\text{b}) \end{cases}$$

the variational equality (8.2.7b) is from the equality (8.1.6).
Let

$$\begin{aligned} H_\Delta^{\frac{1}{2}} &= \left\{ v \mid v \in H^{\frac{1}{2}}(\Gamma),\ v\big|_{\Gamma_0} = 0 \right\}, \\ \mathbb{V} &= H_\Delta^{\frac{1}{2}}(\Gamma) \times H_*^{-\frac{1}{2}}(\Gamma), \quad \text{with norm} \quad \|(u, p)\|_{\mathbb{V}} = \left\{ \|u\|^2_{\frac{1}{2},\Gamma} + \|p\|^2_{-\frac{1}{2},\Gamma} \right\}^{\frac{1}{2}}, \\ \mathbb{W}_\Gamma &= K_\Gamma \times H_*^{-\frac{1}{2}}(\Gamma), \end{aligned}$$

where $\mathbb{W}_\Gamma$ is a convex set of $\mathbb{V}$. Introduce

$$D(u, p; v, q) = B(u, v) + C(p, v) - C(q, u) + A(q, p), \quad \forall (u, p), (v, q) \in \mathbb{V}. \tag{8.2.8}$$

By the properties of the bilinear forms $A(q, p)$, $B(u, v)$ and $C(p, v)$, we have

Lemma 8.2.1 *(i) $D(u, p; v, q)$ is a bounded bilinear form on $\mathbb{V} \times \mathbb{V}$, namely there exists a positive constant M, such that*

$$|D(u, p; v, q)| \leqslant M \|(u, p)\|_{\mathbb{V}} \|(v, q)\|_{\mathbb{V}}, \quad \forall (u, p), (v, q) \in \mathbb{V}.$$

(ii) There is a positive constant β, such that

$$D(v, q; v, q) \geqslant \beta \|(v, q)\|_{\mathbb{V}}, \quad \forall (v, q) \in \mathbb{V}.$$

Now the boundary variational inequality and equality (8.2.7) can be rewritten as the coming boundary variational inequality.

$$\begin{cases} \text{Find} \quad (u, p) \in \mathbb{W}_\Gamma, \quad \text{such that} \\ D(u, p; v - u, q) \geqslant \int_{\Gamma_1} g(v - u) \mathrm{d}s, \quad \forall (v, q) \in \mathbb{W}_\Gamma. \end{cases} \tag{8.2.9}$$

The Lemma 8.2.1 and the result on the existence and uniqueness of variational inequality in the book by Glowinski (2013) yield the following results.

Theorem 8.2.1 *Suppose that $g \in H^{-\frac{1}{2}}(\Gamma_1)$, the boundary variational problem (8.2.9) exists a unique solution $(u, p) \in \mathbb{W}_\Gamma$.*

8.2.2 Abstract Error Estimate of the Numerical Solution of Boundary Variational Inequality (8.2.9)

Let V_h and M_h be finite dimensional subspace of $H^{\frac{1}{2}}(\Gamma)$ and $H_*^{-\frac{1}{2}}(\Gamma)$, and

$$\begin{aligned} \mathbb{K}_\Gamma^h &= \left\{ v_h \middle| v_h \in V_h, \ v_h\big|_{\Gamma_0} = 0, \ v_h\big|_{\Gamma_1} \geqslant 0 \right\}, \\ \mathbb{W}_\Gamma^h &= K_\Gamma^h \times M_h. \end{aligned}$$

Obviously $\mathbb{W}_\Gamma^h \subset \mathbb{W}_\Gamma$ is a convex set, for example, in the case $n = 2$, the subspace V_h, M_h of $H^{\frac{1}{2}}(\Gamma)$, $H_*^{-\frac{1}{2}}(\Gamma)$ and the convex set $\mathbb{K}_\Gamma^h$ were given in Han (1990). Using $\mathbb{W}_\Gamma^h$ instead of $\mathbb{W}_\Gamma$ in inequality (8.2.9), it yields the following discrete problem:

$$\begin{cases} \text{Find} \quad (u_h, p_h) \in \mathbb{W}_\Gamma^h, \quad \text{such that} \\ D(u_h, p_h; v_h - u_h, q_h) \geqslant \int_{\Gamma_1} g(v_h - u_h) \mathrm{d}s, \quad \forall (v_h, q_h) \in \mathbb{W}_\Gamma^h. \end{cases} \tag{8.2.10}$$

It is straightforward to have

Theorem 8.2.2 *The discrete variational inequality (8.2.10) exists a unique solution $(u_h, p_h) \in \mathbb{W}_\Gamma^h$.*

Moreover the following abstract error estimate holds.

Theorem 8.2.3 *Suppose that (u, p), the solution of boundary variational inequality (8.2.9), satisfies $(u, p) \in \mathbb{K}_\Gamma$, and $u \in H^{\frac{1}{2}+\alpha}(\Gamma)$, $p \in H^{-\frac{1}{2}+\alpha}(\Gamma)$ and $g \in H^{-\frac{1}{2}+\alpha}(\Gamma_1)$ with $0 < \alpha \leqslant 1$; then the following error estimate holds:*

$$\|(u-u_h, p-p_h)\|_{\mathbb{V}} \leqslant$$
$$C_\alpha \inf_{(v_h,q_h)\in \mathbb{W}_\Gamma^h} \Big\{ \|(u-v_h, p-p_h)\|_{\mathbb{V}} + \|u-v_h\|_{\frac{1}{2}-\alpha,\Gamma} +$$
$$\|p-q_h\|_{-\frac{1}{2}-\alpha,\Gamma} \Big\}, \quad (8.2.11)$$

where C_α is a constant independent of h.

Proof By Lemma 8.2.1 we have

$$\begin{aligned}\big\|(u-u_h, p-p_h)\big\|_{\mathbb{V}}^2 &\leqslant \frac{1}{\beta} D(u-u_h, p-p_h; u-u_h, p-p_h)\\ &= \frac{1}{\beta}\Big\{ D(u-u_h, p-p_h; u-v_h, p-q_h) +\\ &\quad D(u, p; v_h-u, q_h-p) + D(u, p; u-u_h, p-p_h) -\\ &\quad D(u_h, p_h; v_h-u_h, q_h-p_h)\Big\}, \quad \forall (v_h, q_h) \in \mathbb{W}_\Gamma^h.\end{aligned} \quad (8.2.12)$$

In the inequality (8.2.9) taking $(v, q) = (u_h, p_h - p) \in \mathbb{V}_\Gamma$, it arrives at

$$D(u, p; u_h - u, p_h - p) \geqslant \int_{\Gamma_1} g(u_h - u)\mathrm{d}s,$$

namely

$$D(u, p; u - u_h, p - p_h) \leqslant \int_{\Gamma_1} g(u - u_h)\mathrm{d}s, \quad (8.2.13)$$

Similarly using $q_h - p_h \in M_h$ instead of q_h in the inequality (8.2.10) it yields

$$D(u_h, p_h; v_h - u_h, q_h - p_h) \geqslant \int_{\Gamma_1} g(v_h - u_h)\mathrm{d}s,$$

that is

$$-D(u_h, p_h; v_h - u_h, q_h - p_h) \leqslant \int_{\Gamma_1} g(u_h - v_h)\mathrm{d}s. \quad (8.2.14)$$

Combining the inequalities (8.2.13)–(8.2.14), we obtain

$$\begin{aligned}&D(u, p : u_h - u, p - p_h) - D(u_h, p_h; v_h - u_h, q_h - p_h)\\ &\leqslant \int_{\Gamma_1} g(u - v_h)\mathrm{d}s \leqslant \|g\|_{-\frac{1}{2}+\alpha,\Gamma_1} \|u - v_h\|_{\frac{1}{2}-\alpha,\Gamma}.\end{aligned} \quad (8.2.15)$$

Furthermore we have

$$\begin{aligned}D(u, p; v_h - u, q_h - p) \leqslant M_\alpha &\Big\{ \Big(\|u\|^2_{\frac{1}{2}+\alpha,\Gamma} + \|p\|^2_{-\frac{1}{2}+\alpha,\Gamma} \Big) \cdot\\ &\Big(\|u - v_h\|^2_{\frac{1}{2}-\alpha,\Gamma} + \|p - q_h\|^2_{-\frac{1}{2}-\alpha,\Gamma} \Big) \Big\}^{\frac{1}{2}},\end{aligned} \quad (8.2.16)$$

with constant $M_\alpha > 0$.

By the estimates (8.2.12), (8.2.15)–(8.2.16) and the boundness of the bilinear form $D(u, p; v, q)$ given in Lemma 8.2.1, the abstract error estimate (8.2.11) is proved directly.

□

8.3 Variational Inequality (II)-Signorini Problems in Linear Elasticity

8.3.1 Signorini Problems in Linear Elasticity

In this subsection we return to Signorini problems in linear elasticity which are discussed by Kinderlehrer (1981), Kikuchi and Oden (1988). An elastic body is given in its natural configuration occupying a bounded domain $\Omega \in \mathbb{R}^n$ $(n = 2, 3)$ with boundary Γ. The boundary Γ contains three smooth open $n-1$ dimensional manifold $\Gamma_0, \Gamma_1, \Gamma_2$ such that $\Gamma = \bar{\Gamma}_0 \cup \bar{\Gamma}_1 \cup \bar{\Gamma}_2$ and $\Gamma_i \cap \Gamma_j = \varnothing$, for $i \neq j,\ i, j = 0, 1, 2$. The elastic body is subjected to assigned surface force $\boldsymbol{g}(x)$ on the partial boundary Γ_2, the displacement $\boldsymbol{u} = (u_1, u_2, \ldots, u_n)^{\mathrm{T}}$ is zero vector on Γ_0 and the elastic body is in contact with a functionless rigid foundation Γ_1.

The notations strain and stress tensors $\boldsymbol{\epsilon}(\boldsymbol{u}), \boldsymbol{\sigma}(\boldsymbol{u})$ and other notations used in Chap. 6 are used in this section without explanations. Introduce the bilinear form $a(\boldsymbol{u}, \boldsymbol{v})$ on $\mathbb{H}^1(\Omega) \times \mathbb{H}^1(\Omega)$,

$$a(\boldsymbol{u}, \boldsymbol{v}) = \int_\Omega \sum\nolimits_{i,j=1}^{n} \sigma_{ij}(\boldsymbol{u}(\boldsymbol{x}))\epsilon_{ij}(\boldsymbol{v}(\boldsymbol{x}))\mathrm{d}\boldsymbol{x}, \quad \forall \boldsymbol{u}, \boldsymbol{v} \in \mathbb{H}^1(\Omega) \tag{8.3.1}$$

and the linear function

$$(\boldsymbol{g}, \boldsymbol{v})_{\Gamma_2} = \int_{\Gamma_2} \boldsymbol{g}(\boldsymbol{x}) \cdot \boldsymbol{v}(\boldsymbol{x})\mathrm{d}s, \quad \forall \boldsymbol{v} \in \mathbb{H}^{\frac{1}{2}}(\Gamma), \tag{8.3.2}$$

with $\boldsymbol{g} \in \mathbb{H}^{-\frac{1}{2}}(\Gamma_2)$.

Let

$$\begin{aligned}
\mathbb{H}^1_\triangle(\Omega) &= \left\{\boldsymbol{v} \in \mathbb{H}^1(\Omega) \text{ and } \boldsymbol{v}\big|_{\Gamma_0} = 0\right\}, \\
\mathbb{H}^{\frac{1}{2}}_\triangle(\Gamma) &= \left\{\text{the trance of } \boldsymbol{v} \text{ on } \Gamma,\ \forall \boldsymbol{v} \in \mathbb{H}^1_\triangle(\Omega)\right\} \\
\mathbb{K} &= \left\{\boldsymbol{v} \in \mathbb{H}^1_\triangle(\Omega) \text{ and } \boldsymbol{v}(\boldsymbol{x}) \cdot \boldsymbol{\nu}(\boldsymbol{x}) \leqslant 0, \text{ on } \Gamma_0\right\}, \\
\mathbb{K}_\Gamma &= \{\text{the trace of } \boldsymbol{v}(\boldsymbol{x}) \text{ on } \Gamma,\ \forall \boldsymbol{v} \in \mathbb{K}\},
\end{aligned}$$

where $\boldsymbol{\nu}(\boldsymbol{x})$ is the unit outward vector of Ω on boundary Γ. Obviously $\mathbb{K}$ is a nonempty convex set of $\mathbb{H}^1(\Omega)$, and $\mathbb{K}_\Gamma$ is a nonempty convex set of $\mathbb{H}^{\frac{1}{2}}(\Gamma)$.

The Signorini problem consider in this subsection is the following variational inequality (Kinderlehrer, 1981; Kikuchi & Oden, 1988):

$$\begin{cases} \text{Find } \boldsymbol{u} \in \mathbb{K}, \quad \text{such that} \\ a(\boldsymbol{u}, \boldsymbol{v} - \boldsymbol{u}) \geqslant (\boldsymbol{g}, \boldsymbol{v} - \boldsymbol{u})_{\Gamma_2}, \quad \forall \boldsymbol{v} \in \mathbb{K}. \end{cases} \tag{8.3.3}$$

Suppose the partial boundary Γ_0 is not empty. For the variational inequality (8.3.3), is characterized by

$$-\begin{cases} \mathcal{L}_n(D)\boldsymbol{u}(\boldsymbol{x}) = 0, \quad \boldsymbol{x} \in \Omega, & \text{(8.3.4a)} \\ \boldsymbol{u}(\boldsymbol{x})\big|_{\Gamma_0} = 0, & \text{(8.3.4b)} \\ \left.\begin{aligned} & -\boldsymbol{u}(\boldsymbol{x}) \cdot \boldsymbol{\nu}(\boldsymbol{x}) \geqslant 0, \quad \text{on} \quad \Gamma_1, \\ & -\boldsymbol{\nu}(\boldsymbol{x})^{\mathrm{T}} \boldsymbol{\sigma}(\boldsymbol{u}(\boldsymbol{x}))\boldsymbol{\nu}(\boldsymbol{x}) \geqslant 0 \quad \text{on} \quad \Gamma_1, \\ & \big(\boldsymbol{u}(\boldsymbol{x}) \cdot \boldsymbol{\nu}(\boldsymbol{x})\big)\big(\boldsymbol{\nu}(\boldsymbol{x})^{\mathrm{T}} \boldsymbol{\sigma}(\boldsymbol{u}(\boldsymbol{x}))\boldsymbol{\nu}(\boldsymbol{x})\big) = 0, \quad \text{on} \quad \Gamma_1, \\ & \boldsymbol{\tau}(\boldsymbol{x})^{\mathrm{T}} \boldsymbol{\sigma}(\boldsymbol{u}(\boldsymbol{x}))\boldsymbol{\nu}(\boldsymbol{x}) = 0, \ \forall \boldsymbol{\tau}(\boldsymbol{x}) \text{ with } \boldsymbol{\tau}(\boldsymbol{x}) \cdot \boldsymbol{\nu}(\boldsymbol{x}) = 0, \ \text{on } \Gamma_1. \end{aligned}\right\} & \text{(8.3.4c)} \\ \mathbb{T}(D_x, \boldsymbol{\nu}(\boldsymbol{x}))\boldsymbol{u}(\boldsymbol{x}) = \boldsymbol{g}(\boldsymbol{x}), \quad \text{on} \quad \Gamma_2. & \text{(8.3.4d)} \end{cases}$$

Obviously the variational inequality (8.3.3) describes an elastic contact problem with the functionless rigid foundation Γ_1.

8.3.2 *An Equivalent Boundary Variational Inequality of Problem (8.3.3)*

Suppose that $\boldsymbol{u} \in \mathbb{K}$ is the unique solution of the variational inequality (8.3.3), then $\boldsymbol{u}(\boldsymbol{x})$ satisfies the homogeneous Navier system (8.3.4) in Ω. From the Betti-Sonigliana representation Formula (6.1.22) the solution $\boldsymbol{u} \in \mathbb{K}$ is given by

$$\begin{aligned} \boldsymbol{u}(\boldsymbol{x}) = & \int_\Gamma \mathbb{E}_n(\boldsymbol{x}, \boldsymbol{y})\mathbb{T}(D_y, \boldsymbol{\nu}(\boldsymbol{y}))\boldsymbol{u}(\boldsymbol{y})\mathrm{d}s_y - \\ & \int_\Gamma \Big(\mathbb{T}(D_x, \boldsymbol{\nu}(\boldsymbol{x}))\mathbb{E}_n(\boldsymbol{x}, \boldsymbol{y})\Big)^{\mathrm{T}} \boldsymbol{u}(\boldsymbol{y})\mathrm{d}s_y, \quad \forall \boldsymbol{x} \in \Omega. \end{aligned}$$

It means that $\boldsymbol{u}(\boldsymbol{x})$ is represented by a simple layer potential and a double layer potential corresponding to Navier system. By the properties of the simple layer potential and double layer potential given in Theorems 6.2.1 and 6.2.2, directly we obtain the following two integral relations on the boundary Γ for $\boldsymbol{u}(\boldsymbol{x})\big|_\Gamma$ and $\boldsymbol{h} := \mathbb{T}(D_x, \boldsymbol{\nu})\boldsymbol{u}\big|_\Gamma \in \mathbb{H}_{**}^{-\frac{1}{2}}(\Gamma)$. They are

$$\frac{1}{2}\boldsymbol{u}(\boldsymbol{x}) = \int_\Gamma \mathbb{E}_n(\boldsymbol{x}, \boldsymbol{y})\boldsymbol{h}(\boldsymbol{y})\mathrm{d}s_y - \int_\Gamma \left(\mathbb{T}(D_y, \boldsymbol{\nu}(\boldsymbol{y}))\mathbb{E}_n(\boldsymbol{x}, \boldsymbol{y})\right)^{\mathrm{T}} \boldsymbol{u}(\boldsymbol{y})\mathrm{d}s_y, \quad \forall \boldsymbol{x} \in \Gamma, \tag{8.3.5}$$

$$\frac{1}{2}\boldsymbol{h}(\boldsymbol{x}) = \int_\Gamma \mathbb{T}(D_x, \boldsymbol{\nu}(\boldsymbol{x}))\mathbb{E}_n(\boldsymbol{x}, \boldsymbol{y})\boldsymbol{h}(\boldsymbol{y})\mathrm{d}s_y - \int_\Gamma \mathbb{T}(D_x, \boldsymbol{\nu}(\boldsymbol{x}))(\mathbb{T}(D_y, \boldsymbol{\nu}(\boldsymbol{y}))\mathbb{E}_n(\boldsymbol{x}, \boldsymbol{y}))^{\mathrm{T}}\boldsymbol{u}(\boldsymbol{x})\mathrm{d}s_y, \quad \forall \boldsymbol{x} \in \Gamma. \tag{8.3.6}$$

Substituting (6.2.44) into (8.3.6), it arrives at

$$\begin{aligned}
&\frac{1}{2}\boldsymbol{h}(\boldsymbol{x}) \\
&= \int_\Gamma \mathbb{T}(D_x, \boldsymbol{\nu}(\boldsymbol{x}))\mathbb{E}_n(\boldsymbol{x}, \boldsymbol{y})\boldsymbol{h}(\boldsymbol{y})\mathrm{d}s_y - \\
&\frac{1}{2(n-1)\pi} \sum\nolimits_{1 \leqslant k < j \leqslant n} \frac{\partial}{\partial \boldsymbol{\tau}_x^{k,j}} \int_\Gamma \boldsymbol{\gamma}_n(\boldsymbol{x}, \boldsymbol{y}) \frac{\partial}{\partial \boldsymbol{\tau}_y^{k,j}} \boldsymbol{u}(\boldsymbol{y})\mathrm{d}s_y - \\
&\mathbb{U}(D_x, \boldsymbol{\nu}(\boldsymbol{x})) \int_\Gamma \left(4\mu^2 \mathbb{E}_n(\boldsymbol{x}, \boldsymbol{y}) - \frac{\mu}{(n-1)\pi}\gamma_n(\boldsymbol{x}, \boldsymbol{y})\boldsymbol{I}\right) \cdot \\
&\mathbb{U}(D_y, \boldsymbol{\nu}(\boldsymbol{y}))\boldsymbol{u}(\boldsymbol{y})\mathrm{d}s_y + \\
&\frac{\mu}{2(n-1)\pi} \int_\Gamma \left(\mathbb{U}(D_x, \boldsymbol{\nu}(\boldsymbol{x}))\boldsymbol{\gamma}_n(\boldsymbol{x}, \boldsymbol{y})\mathbb{U}(D_y, \boldsymbol{\nu}(\boldsymbol{y}))\right)^{\mathrm{T}} \boldsymbol{u}(\boldsymbol{y})\mathrm{d}s_y, \quad \forall \boldsymbol{x} \in \Gamma.
\end{aligned} \tag{8.3.7}$$

For $\boldsymbol{u}(\boldsymbol{x})$, the solution of variational inequality (8.3.3), and $\boldsymbol{v} \in \mathbb{H}^1(\Omega)$, we have

$$\begin{aligned}
a(\boldsymbol{u}, \boldsymbol{v}) &= \int_\Omega \sum\nolimits_{i,j=1}^n \sigma_{ij}(\boldsymbol{u}(\boldsymbol{x}))\epsilon_{ij}(\boldsymbol{v}(\boldsymbol{x}))\mathrm{d}\boldsymbol{x} \\
&= \int_\Gamma \mathbb{T}(D_x, \boldsymbol{\nu}(\boldsymbol{x}))\boldsymbol{u}(\boldsymbol{x}) \cdot \boldsymbol{v}(\boldsymbol{x})\mathrm{d}s_x = \int_\Gamma \boldsymbol{h}(\boldsymbol{x}) \cdot \boldsymbol{v}(\boldsymbol{x})\mathrm{d}s_y \\
&= \frac{1}{2}\int_\Gamma \boldsymbol{h}(\boldsymbol{x}) \cdot \boldsymbol{v}(\boldsymbol{x})\mathrm{d}s_x + \frac{1}{2}\int_\Gamma \boldsymbol{h}(\boldsymbol{x}) \cdot \boldsymbol{v}(\boldsymbol{x})\mathrm{d}s_x.
\end{aligned} \tag{8.3.8}$$

Substituting (8.3.7) into (8.3.8) yields that

$$a(\boldsymbol{u}, \boldsymbol{v}) = \mathbb{B}(\boldsymbol{u}, \boldsymbol{v}) + \mathbb{C}(\boldsymbol{h}, \boldsymbol{v}), \tag{8.3.9}$$

where the bilinear form $\mathbb{B}(\boldsymbol{u}, \boldsymbol{v})$ is defined by (6.3.11), and the bilinear form $\mathbb{C}(\boldsymbol{h}, \boldsymbol{v})$ is given by

$$\mathbb{C}(\boldsymbol{h}, \boldsymbol{v}) = \frac{1}{2} \int_{\Gamma} \boldsymbol{h}(x)\boldsymbol{v}(x)\mathrm{d}s_y + \int_{\Gamma}\int_{\Gamma} \boldsymbol{v}(x)^{\mathrm{T}}\Big(\mathbb{T}(D_x, \boldsymbol{\nu}(x))E_n(x, y)\Big)\boldsymbol{h}(y)\mathrm{d}s_y\mathrm{d}s_x. \tag{8.3.10}$$

In the equality (8.3.9) $\boldsymbol{u}(\boldsymbol{x})\big|_{\Gamma}$ and $\boldsymbol{h}(\boldsymbol{x}) = \mathbb{T}(D_x, \boldsymbol{\nu}(\boldsymbol{x}))\boldsymbol{u}(\boldsymbol{x})\big|_{\Gamma}$ are two unknown vector-value functions on Γ, but they are not independent, which satisfy the integral equation (8.3.5), which is equivalent to

$$\mathbb{C}(\boldsymbol{p}, \boldsymbol{u}) - \mathbb{A}(\boldsymbol{p}, \boldsymbol{h}) = 0, \quad \forall \boldsymbol{p} \in \mathbb{H}_{**}^{-\frac{1}{2}}(\Gamma),$$

the bilinear form $\mathbb{A}(\boldsymbol{p}, \boldsymbol{h})$ is defined in (6.3.10). Inserting (8.3.9) into (8.3.3) and coupling the above equality it yields the following boundary variational inequality, which is equivalent to the variational inequality (8.3.3).

$$\begin{cases} \text{Find} \quad (\boldsymbol{u}, \boldsymbol{h}) \in \mathbb{K}_{\Gamma} \times \mathbb{H}_{**}^{-\frac{1}{2}}(\Gamma), \quad \text{such that} \\ \mathbb{B}(\boldsymbol{u}, \boldsymbol{v} - \boldsymbol{u}) + \mathbb{C}(\boldsymbol{h}, \boldsymbol{v} - \boldsymbol{u}) \geqslant (\boldsymbol{g}, \boldsymbol{v} - \boldsymbol{u})_{\Gamma_2}, \quad \forall \boldsymbol{v} \in \mathbb{K}_{\Gamma}, & (8.3.11\text{a}) \\ \mathbb{C}(\boldsymbol{p}, \boldsymbol{u}) - \mathbb{A}(\boldsymbol{p}, \boldsymbol{h}) = 0, \quad \forall \boldsymbol{p} \in \mathbb{H}_{**}^{-\frac{1}{2}}(\Gamma). & (8.3.11\text{b}) \end{cases}$$

The problem (8.3.11) can be rewritten as the following equivalent form:

$$\begin{cases} \text{Find} \quad (\boldsymbol{u}, \boldsymbol{h}) \in \mathbb{K}_{\Gamma} \times \mathbb{H}_{**}^{-\frac{1}{2}}(\Gamma), \quad \text{such that} \\ \mathbb{B}(\boldsymbol{u}, \boldsymbol{v} - \boldsymbol{u}) + \mathbb{C}(\boldsymbol{h}, \boldsymbol{v} - \boldsymbol{u}) \geqslant (\boldsymbol{g}, \boldsymbol{v} - \boldsymbol{u})_{\Gamma_2}, \quad \forall \boldsymbol{v} \in \mathbb{K}_{\Gamma}, & (8.3.12\text{a}) \\ -\mathbb{C}(\boldsymbol{p}, \boldsymbol{u}) + \mathbb{A}(\boldsymbol{p}, \boldsymbol{h}) = 0, \quad \forall \boldsymbol{p} \in \mathbb{H}_{**}^{-\frac{1}{2}}(\Gamma). & (8.3.12\text{b}) \end{cases}$$

Let

$$\mathbb{D}(\boldsymbol{u}, \boldsymbol{h}; \boldsymbol{v}, \boldsymbol{p}) = \mathbb{B}(\boldsymbol{u}, \boldsymbol{v}) - \mathbb{C}(\boldsymbol{h}, \boldsymbol{v}) - \mathbb{C}(\boldsymbol{p}, \boldsymbol{u}) + \mathbb{A}(\boldsymbol{p}, \boldsymbol{h}). \tag{8.3.13}$$

Finally it yields the following boundary variational inequality:

$$\begin{cases} \text{Find} \quad (\boldsymbol{u}, \boldsymbol{h}) \in \mathbb{K}_{\Gamma} \times \mathbb{H}_{**}^{-\frac{1}{2}}(\Gamma), \quad \text{such that} \\ \mathbb{D}(\boldsymbol{u}, \boldsymbol{h}; \boldsymbol{v} - \boldsymbol{u}, \boldsymbol{p}) \geqslant (\boldsymbol{g}, \boldsymbol{v} - \boldsymbol{u})_{\Gamma_2}, \quad \forall (\boldsymbol{v}, \boldsymbol{p}) \in \mathbb{K}_{\Gamma} \times \mathbb{H}_{**}^{-\frac{1}{2}}(\Gamma), \end{cases} \tag{8.3.14}$$

which is equivalent to the variational problem (8.3.3).

The bilinear form $\mathbb{D}(\boldsymbol{u}, \boldsymbol{h}; \boldsymbol{v}, \boldsymbol{p})$ is defined on $(\mathbb{H}^{\frac{1}{2}}(\Gamma) \times \mathbb{H}^{-\frac{1}{2}}(\Gamma)) \times (\mathbb{H}^{\frac{1}{2}}(\Gamma) \times \mathbb{H}^{-\frac{1}{2}}(\Gamma))$, which satisfies (see Han, 1992, 1994c):

Lemma 8.3.1 *(i) $\mathbb{D}(\boldsymbol{u}, \boldsymbol{h}; \boldsymbol{v}, \boldsymbol{p})$ is a bounded bilinear form, namely there exists a constant $M > 0$, such that*

$$|\mathbb{D}(\boldsymbol{u}, \boldsymbol{h}; \boldsymbol{v}, \boldsymbol{p})| \leqslant M \left(\|\boldsymbol{u}\|_{\frac{1}{2},\Gamma}^2 + \|\boldsymbol{h}\|_{-\frac{1}{2},\Gamma}^2\right)^{\frac{1}{2}} \left(\|\boldsymbol{v}\|_{\frac{1}{2},\Gamma}^2 + \|\boldsymbol{p}\|_{-\frac{1}{2},\Gamma}^2\right)^{\frac{1}{2}},$$
$$\forall (\boldsymbol{u}, \boldsymbol{h}), (\boldsymbol{v}, \boldsymbol{p}) \in \mathbb{H}^{\frac{1}{2}}(\Gamma) \times \mathbb{H}^{-\frac{1}{2}}(\Gamma). \tag{8.3.15}$$

(ii) $\mathbb{D}(\boldsymbol{u},\boldsymbol{h};\boldsymbol{v},\boldsymbol{p})$ *is coercive on* $\mathbb{H}^{\frac{1}{2}}_{\Delta}(\Gamma)\times\mathbb{H}^{-\frac{1}{2}}_{**}(\Gamma)$, *namely there is a constant* $\alpha>0$, *such that*

$$\mathbb{D}(\boldsymbol{v},\boldsymbol{p};\boldsymbol{v},\boldsymbol{p})\geqslant\alpha\left(\|\boldsymbol{v}\|^2_{\frac{1}{2},\Gamma}+\|\boldsymbol{p}\|^2_{-\frac{1}{2},\Gamma}\right),\quad\forall(\boldsymbol{v},\boldsymbol{p})\in\mathbb{H}^{\frac{1}{2}}_{\Delta}(\Gamma)\times\mathbb{H}^{-\frac{1}{2}}_{**}(\Gamma).\tag{8.3.16}$$

Furthermore we have Han (1992, 1994c).

Theorem 8.3.1 *For any given* $\boldsymbol{g}\in\mathbb{H}^{-\frac{1}{2}}(\Gamma_2)$, *then the boundary variational inequality (8.3.14) exists a unique solution* $\boldsymbol{u}\in\mathbb{K}_\Gamma$.

The discussion of the numerical solutions and its error estimates of problem (8.3.14) is omitted here, which can be found in Han (1992, 1994c) for interested readers.

8.4 Steklov Eigenvalue Problems

Steklov eigenvalue problems appear in many fields of physics and mathematics (Bergman & Schiffer 2005; Conca et al., 1995).

In this section we consider the following Steklov eigenvalue problem:

$$\begin{cases}\text{Find non-zero function } u(x) \text{ and number } \lambda, \text{ such that}\\ \Delta u=0,\quad \text{in } \Omega, & (8.4.1\text{a})\\ \dfrac{\partial u}{\partial \nu}=\lambda u,\quad \text{on } \Gamma, & (8.4.1\text{b})\end{cases}$$

where $\Omega\in\mathbb{R}^n$ $(n=2,3)$ is a bounded domain with smooth boundary Γ. The equivalent variational problem is given by

$$\begin{cases}\text{Find non-zero } u(x)\in H^1(\Omega) \text{ and number } \lambda, \text{ such that}\\ \int_\Omega \nabla u\cdot\nabla v\mathrm{d}x=\lambda\int_\Gamma uv\mathrm{d}s,\quad \forall v\in H^1(\Omega).\end{cases}\tag{8.4.2}$$

For the Steklov eigenvalue problem (8.4.1) we know that

(i) Problem (8.4.1) exists infinitely many eigenvalues, which are nonnegative real numbers.

(ii) Suppose that $u_n(x)$, $u_m(x)$ are two eigenfunctions of (8.4.1) corresponding to two different eigenvalues λ_n, λ_m, then

$$\int_\Gamma u_n(x)u_m(x)\mathrm{d}s=0,$$

namely the trace of $u_n(x)$ and $u_m(x)$ on Γ are orthogonal on the space $L_2(\Gamma)$.

Based on the variation problem (8.4.2), Bramble and Osborn (1972) developed finite element method of Steklov eigenvalue problem and the optimal error estimate was given. After that, a lot of work on this subject appears. The isoparametric finite element approximation was applied by Andreev and Todorov (2004). The nonconforming finite element approximation was considered by Yang et al. (2009), Li et al. (2013). A type of multilevel method was discussed by Xie (2014). A virtual element method was proposed by Mora et al. (2015). The extrapolation and superconvergence of the Steklov eigenvalue problem were considered in Li et al. (2010). Also, a posterior error estimate for Steklov eigenvalue problem was given by Armentano and Padra (2008).

In fact, using the boundary integro-differential equation method to solve the Steklov eigenvalue problem numerically has special advantages, which reduce the dimension of the problem by one.

In the case $n = 2$, Courant and Hilbert (2008) consider the Laplace equation with boundary condition with parameter, which was reduced to an equivalent integral form on the boundary Γ, but the Green function corresponding to Neumann boundary condition used in the book is not corrected, the correct Green function corresponding to Neumann boundary condition should be:

$$\begin{cases} -\Delta_y G_n(x, y) = \delta(x - y), \quad \forall x \in \Omega, & (8.4.3a) \\ \dfrac{\partial G_n(x, y)}{\partial \boldsymbol{\nu}_y}\Big|_\Gamma = -\dfrac{1}{\text{meas}(\Gamma)}. & (8.4.3b) \end{cases}$$

For any solution (u, λ) of problem (8.4.1), $u(x)$ is given by the Green formula

$$u(x) = -\int_\Gamma \frac{\partial}{\partial \boldsymbol{\nu}_y} G_n(x, y) u(y) \mathrm{d}s_y + \int_\Gamma G_n(x, y) \frac{\partial u(y)}{\partial \boldsymbol{\nu}_y}\Big|_\Gamma \mathrm{d}s_y, \quad \forall x \in \Omega. \quad (8.4.4)$$

Substituting (8.4.3b) and (8.4.1b) into (8.4.4), let $x \in \Omega$ goes to boundary Γ, then the Steklov eigenvalue problem (8.4.1) is reduced to an eigenvalue problem of a integral operator on the boundary Γ. That is

$$\begin{cases} \text{Find non-zero } u \in H^{\frac{1}{2}}(\Gamma) \text{ and number } \lambda, \text{ such that} \\ u(x) = \frac{1}{\text{meas}(\Gamma)} \int_\Gamma u(y) \mathrm{d}s_y + \lambda \int_\Gamma G_n(x, y) u(y) \mathrm{d}s_y, \quad \forall x \in \Gamma. \end{cases} \quad (8.4.5)$$

It means that the Steklov eigenvalue problem (8.4.1) is simplified on the boundary Γ, and the dimension of the given problem is reduced by one. However, it is impossible to obtain the numerical solution of the Steklov eigenvalue problem based on the boundary integral form (8.4.5) in general. The reason is that for a given bounded Ω, the Green function $G_n(x, y)$ defined by (8.4.3) is unknown except for only a few special domains. Han, Guan, and He (1991, 1994a), Han and Guan (1992), and Tang et al. (1998) developed the boundary integro-differential equations method for the Steklov eigenvalue problems. Huang and Lü (2004) proposed a boundary integral equation method to solve Steklov eigenvalue problem, which does not preserve the symmetry of the original problems. The advantage of this formula is that there is only an unknown function $u\big|_\Gamma$ in the given boundary integral equation. In the following, we discuss the numerical solution of Steklov eigenvalue problems by the boundary integro-differential equations method.

8.4.1 The Boundary Reduction of Steklov Eigenvalue Problem

Suppose that $u \in H^1(\Omega)$ and number λ is a solution of the Steklov eigenvalue problem (8.4.2). Let $\frac{\partial u(x)}{\partial \boldsymbol{v}_x}\big|_\Gamma = p(x)$, then $p \in H_*^{-\frac{1}{2}}(\Gamma)$. Since $u(x)$ satisfies Laplace equation in Ω, $u(x)\big|_\Gamma$ and $p(x)$ satisfy the two boundary integro-differential equation (8.1.6) and (8.1.7). Substituting the boundary condition (8.4.1b) into (8.1.6) and (8.1.7) respectively, we obtain two equivalent problems of Steklov eigenvalue problem (8.4.2) on the boundary Γ. That are

$$\begin{cases} \text{Find non-zero } u \in H^{\frac{1}{2}}(\Gamma) \text{ and number } \lambda, \text{ such that} \\ \frac{1}{2}u(x) + \int_\Gamma \frac{\partial E_n(x,y)}{\partial \boldsymbol{v}_y} u(y)\mathrm{d}s_y = \lambda \int_\Gamma E_n(x,y)u(y)\mathrm{d}s_y, \quad \forall x \in \Gamma. \end{cases} \tag{8.4.6}$$

and

$$\begin{cases} \text{Find non-zero } u \in H^{\frac{1}{2}}(\Gamma) \text{ and number } \lambda, \text{ such that} \\ \int_\Gamma \frac{\partial^2 E_n(x,y)}{\partial \boldsymbol{v}_x \partial \boldsymbol{v}_y} \mathrm{d}s_y = \lambda\Big(-\frac{1}{2}u(x) + \int_\Gamma \frac{\partial E_n(x,y)}{\partial \boldsymbol{v}_x} u(y)\mathrm{d}s_y \Big), \quad \forall x \in \Gamma. \end{cases} \tag{8.4.7}$$

The boundary integral form (8.4.6) was given by Huang and Lü (2004). Both of problem (8.4.6) and (8.4.7) do not maintain the symmetry of the original problem.

The equality (8.1.7) can be rewritten as

$$p(x) = \frac{1}{2}p(x) + \int_\Gamma \frac{\partial E_n(x,y)}{\partial \boldsymbol{v}_x} p(y)\mathrm{d}s_y - \int_\Gamma \frac{\partial^2 E_n(x,y)}{\partial \boldsymbol{v}_x \partial \boldsymbol{v}_y} u(y)\mathrm{d}s_y, \quad \forall x \in \Gamma. \tag{8.4.8}$$

Substituting the boundary condition (8.4.1b) into (8.4.8) and combining the equality (8.1.6), the Steklov eigenvalue problem is reduced to

$$
\begin{cases}
\text{Find non-zero } (u, p) \in H^{\frac{1}{2}}(\Gamma) \times H_*^{-\frac{1}{2}}(\Gamma), \\
\quad \text{and number } \lambda, \text{ such that} \\
\frac{1}{2}p(x) + \int_\Gamma \frac{\partial E_n(x, y)}{\partial \boldsymbol{v}_x} p(y)\mathrm{d}s_y - \\
\quad \int_\Gamma \frac{\partial^2 E_n(x, y)}{\partial \boldsymbol{v}_x \partial \boldsymbol{v}_y} u(y)\mathrm{d}s_y = \lambda u(x), \quad \forall x \in \Gamma, & (8.4.9a) \\
-\frac{1}{2}u(x) - \int_\Gamma \frac{\partial E_n(x, y)}{\partial \boldsymbol{v}_y} u(y)s\mathrm{d}s_y + \\
\quad \int_\Gamma E_n(x, y)p(y)\mathrm{d}s_y = 0, \quad \forall x \in \Gamma. & (8.4.9b)
\end{cases}
$$

The derivative of double layer potential in (8.4.9a) is given by (8.1.8) for $n = 2, 3$. It means that the problem (8.4.9) is an eigenvalue problem of boundary integro-differential operators. Furthermore the problem (8.4.9) can be rewritten as the following equivalent symmetry form:

$$
\begin{cases}
\text{Find non-zero } (u, p) \in H^{\frac{1}{2}}(\Gamma) \times H_*^{-\frac{1}{2}}(\Gamma), \\
\quad \text{and number } \lambda, \text{ such that} \\
\frac{1}{2}p(x) + \int_\Gamma \frac{\partial E_n(x, y)}{\partial \boldsymbol{v}_x} p(y)\mathrm{d}s_y - \\
\quad \int_\Gamma \frac{\partial^2 E_n(x, y)}{\partial \boldsymbol{v}_x \partial \boldsymbol{v}_y} u(y)\mathrm{d}s_y = \lambda u(x), \ \forall x \in \Gamma, & (8.4.10a) \\
\frac{1}{2}u(x) + \int_\Gamma \frac{\partial E_n(x, y)}{\partial \boldsymbol{v}_y} u(y)s\mathrm{d}s_y - \\
\quad \int_\Gamma E_n(x, y)p(y)\mathrm{d}s_y = 0, \ \forall x \in \Gamma. & (8.4.10b)
\end{cases}
$$

Multiplying (8.4.9a) and (8.4.9b) by a function $v \in H^{\frac{1}{2}}(\Gamma)$ and a function $q \in H_*^{-\frac{1}{2}}(\Gamma)$, integrating on the boundary Γ respectively, then we obtain the following variational form of problem (8.4.9).

$$
\begin{cases}
\text{Find non-zero } (u, p) \in H^{\frac{1}{2}}(\Gamma) \times H_*^{-\frac{1}{2}}(\Gamma) \text{ and number } \lambda, \text{ such that} \\
B(u, v) + C(p, v) = \lambda \int_\Gamma u(x)v(x)\mathrm{d}s_x, \quad \forall u \in H^{\frac{1}{2}}(\Gamma), & (8.4.11a) \\
-C(q, u) + A(q, p) = 0, \quad \forall q \in H_*^{-\frac{1}{2}}(\Gamma), & (8.4.11b)
\end{cases}
$$

where the bilinear forms $A(q, p)$, $B(u, v)$ and $C(q, u)$ are defined by (3.4.16), (3.4.17) and (8.1.14).

Similarly the symmetry problem (8.4.10) is equivalent to the following variational problem

$$\begin{cases} \text{Find non-zero } (u,p) \in H^{\frac{1}{2}}(\Gamma) \times H_*^{-\frac{1}{2}}(\Gamma) \text{ and number } \lambda, \text{ such that} \\ B(u,v) + C(p,v) = \lambda \int_\Gamma u(x)v(x)\mathrm{d}s_x, \quad \forall u \in H^{\frac{1}{2}}(\Gamma), & (8.4.12a) \\ C(q,u) - A(q,p) = 0, \quad \forall q \in H_*^{-\frac{1}{2}}(\Gamma). & (8.4.12b) \end{cases}$$

Let

$$V = H^{\frac{1}{2}}(\Gamma) \times H_*^{-\frac{1}{2}}(\Gamma), \quad \text{with } \|(u,p)\|_V = \left\{\|u\|^2_{\frac{1}{2},\Gamma} + \|p\|^2_{-\frac{1}{2},\Gamma}\right\}^{\frac{1}{2}},$$

$$(u,v)_\Gamma = \int_\Gamma u(x)v(x)\mathrm{d}s_x,$$

$$D(u,p;v,q) = B(u,v) + C(p,v) - C(q,u) + A(q,p),$$

$$D_S(u,p;v,q) = B(u,v) + C(p,v) + C(q,u) - A(q,p),$$

then the problem (8.4.11) and problem (8.4.12) can be rewritten as

$$\begin{cases} \text{Find non-zero } (u,p) \in V \text{ and number } \lambda, \text{ such that} \\ D(u,p;v,q) = \lambda(u,v)_\Gamma, \quad \forall (v,q) \in V, \end{cases} \tag{8.4.13}$$

and

$$\begin{cases} \text{Find non-zero } (u,p) \in V \text{ and number } \lambda, \text{ such that} \\ D_S(u,p;v,q) = \lambda(u,v)_\Gamma, \quad \forall (v,q) \in V. \end{cases} \tag{8.4.14}$$

The properties of bilinear forms $A(q,p)$, $B(u,v)$ and $C(p,v)$ yield that

Lemma 8.4.1 *(i) $D(u,v;p,q)$ and $D_S(u,v;p,q)$ are two bounded bilinear form on $V \times V$.*

(ii) $D_S(u,v;p,q)$ is symmetric.

(iii) There exists a constant $\alpha > 0$, such that

$$D(v,q;v,q) \geqslant \alpha \left\{|v|^2_{\frac{1}{2},\Gamma} + \|q\|^2_{-\frac{1}{2},\Gamma}\right\}^{\frac{1}{2}}, \quad \forall (v,q) \in V.$$

8.4.2 The Numerical Solutions of Steklov Eigenvalue Problem Based on the Variational Form (8.4.13)

Suppose that S_h and M_h are two finite dimensional subspaces of $H^{\frac{1}{2}}(\Gamma)$ and $H_*^{-\frac{1}{2}}(\Gamma)$. Furthermore assume that S_h and M_h are two regular finite element spaces in the sense of Babuska and Aziz (1972), which satisfy the following approximation properties:

$$\inf_{v_h \in S_h} \|u - u_h\|_{\frac{1}{2},\Gamma} \leqslant Ch^t |u|_{\frac{1}{2}+t,\Gamma}, \quad \forall u \in H^{\frac{1}{2}+t}(\Gamma), \tag{8.4.15}$$

$$\inf_{q_h \in M_h} \|q - q_h\|_{-\frac{1}{2},\Gamma} \leqslant Ch^t |p|_{-\frac{1}{2}+t,\Gamma}, \quad \forall p \in H_*^{-\frac{1}{2}+t}(\Gamma), \tag{8.4.16}$$

with $0 < t \leqslant 1$.

Let

$$V_h = S_h \times M_h.$$

We consider the discrete problem of (8.4.13).

$$\begin{cases} \text{Find non-zero } (u_h, p_h) \in V_h \text{ and number } \lambda_h, \text{ such that} \\ D(u_h, p_h; v_h, q_h) = \lambda_h (u_h, v_h)_\Gamma, \quad \forall (v_h, q_h) \in V_h. \end{cases} \tag{8.4.17}$$

It's equivalent form is given

$$\begin{cases} \text{Find non-zero } (u_h, p_h) \in V_h \text{ and number } \lambda_h, \text{ such that} \\ D_S(u_h, p_h; v_h, q_h) = \lambda_h (u_h, v_h)_\Gamma, \quad \forall (v_h, q_h) \in V_h. \end{cases} \tag{8.4.18}$$

Assume that $\{S_k(x), k = 1, 2, \ldots, K\}$ are the base functions of space S_h and $\{m_l(x), l = 1, 2, \ldots, L\}$ are the base functions of space M_h, then the problem (8.4.17) and (8.4.18) can be rewritten in an eigenvalue problem of $K + L$ order matrices. By solving the matrix eigenvalue problem, finally we obtain the numerical approximations of the original Steklov eigenvalue problem (8.4.1).

8.4.3 The Error Estimate of Numerical Solution of Steklov Eigenvalue Problem

In this subsection we discuss the error estimate of numerical solution given by (8.4.17) using the general approximation theory of eigenvalue problems for a certain class of compact operators (Babuska & Aziz, 1972). Therefore we identify the Steklov-eigenvalue problem (8.4.13) (namely problem (8.4.12)) is equivalent to the eigenvalue of a certain compact operator.

Let $\mu = \lambda + 1$, then the Steklov eigenvalue problem (8.4.13) can be rewritten as

$$\begin{cases} \text{Find non-zero } (u, p) \in V \text{ and number } \mu, \text{ such that} \\ B(u, v) + C(p, v) + (u, v)_\Gamma = \mu(u, v)_\Gamma, \quad \forall v \in H^{\frac{1}{2}}(\Gamma), \quad (8.4.19\text{a}) \\ C(q, u) - A(q, p) = 0, \quad \forall q \in H_*^{-\frac{1}{2}}(\Gamma). \quad (8.4.19\text{b}) \end{cases}$$

For any given $u \in H^{\frac{1}{2}}(\Gamma)$, $C(q, u)$ is a bounded linear functional on $H_*^{-\frac{1}{2}}(\Gamma)$. The existence and uniqueness of problem (3.4.8) yield that for given $u \in H^{\frac{1}{2}}(\Gamma)$, the following variational problem

$$\begin{cases} \text{Find } p \in H_*^{-\frac{1}{2}}(\Gamma), \text{ such that} \\ A(q, p) = C(q, u), \quad \forall q \in H_*^{-\frac{1}{2}}(\Gamma), \end{cases} \tag{8.4.20}$$

has a unique solution p.

Let $p = \mathcal{P}u$, then we obtain a linear operator $\mathcal{P}$ from $H^{\frac{1}{2}}(\Gamma)$ to $H_*^{-\frac{1}{2}}(\Gamma)$ and operator $\mathcal{P}$ is bounded, namely there is a positive constant C, such that

$$\|\mathcal{P}u\|_{-\frac{1}{2},\Gamma} \leqslant C\|u\|_{\frac{1}{2},\Gamma}, \quad \forall u \in H^{\frac{1}{2}}(\Gamma). \tag{8.4.21}$$

In fact for given $u \in H^{\frac{1}{2}}(\Gamma)$, it can be extended on Ω and $u \in H^1(\Omega)$ is the solution of Laplace equation $\Delta u = 0$, then $\mathcal{P}u = \frac{\partial u}{\partial \nu}\big|_\Gamma$. Therefore the operator $\mathcal{P}$ can be extended to

$$\mathcal{P} : H^{s+1}(\Gamma) \to H^s(\Gamma), \quad \frac{1}{2} \geqslant s \geqslant -\frac{1}{2},$$

and

$$\|\mathcal{P}u\|_{s,\Gamma} \leqslant C\|u\|_{s+1,\Gamma}, \quad \forall u \in H^{s+1}(\Gamma).$$

Let $p = \mathcal{P}u$, the unknown function p can be eliminated in the problem (8.4.19).

Note

$$C(q, u) = A(q, \mathcal{P}u), \quad \forall q \in H_*^{-\frac{1}{2}}(\Gamma),$$

taking $q = \mathcal{P}v$ for $v \in H^{-\frac{1}{2}}(\Gamma)$, we obtain

$$C(\mathcal{P}v, u) = A(\mathcal{P}v, \mathcal{P}u) = A(\mathcal{P}u, \mathcal{P}v) = C(\mathcal{P}u, v) = C(p, v),$$

namely

$$C(p, v) = A(\mathcal{P}u, \mathcal{P}v). \tag{8.4.22}$$

Therefore the eigenvalue problem (8.4.19) is reduced to

$$\begin{cases} \text{Find non-zero } u \in H^{\frac{1}{2}}(\Gamma) \text{ and number } \lambda, \text{ such that} \\ E(u, v) = \mu(u, v)_\Gamma, \quad \forall v \in H^{\frac{1}{2}}(\Gamma), \end{cases} \tag{8.4.23}$$

with

$$E(u, v) = B(u, v) + A(\mathcal{P}u, \mathcal{P}v) + (u, v)_\Gamma.$$

From Lemma 8.4.1 directly we obtain

Lemma 8.4.2 *$E(u, v)$ is a bounded bilinear form on $H^{\frac{1}{2}}(\Gamma) \times H^{\frac{1}{2}}(\Gamma)$ and coercive, namely there exist two constant $M > 0$ and $\alpha > 0$, such that*

$$|E(u, v)| \leqslant M\|u\|_{\frac{1}{2},\Gamma}\|v\|_{\frac{1}{2},\Gamma}, \quad \forall u, v \in H^{\frac{1}{2}}(\Gamma),$$

$$E(v, v) \geqslant \alpha \|v\|^2_{\frac{1}{2},\Gamma}, \quad \forall v \in H^{\frac{1}{2}}(\Gamma).$$

For any given $g \in H^{-\frac{1}{2}}(\Gamma)$, we consider the following variational problem

$$\begin{cases} \text{Find } u \in H^{\frac{1}{2}}(\Gamma), \text{ such that} \\ E(u, v) = (g, v)_\Gamma, \quad \forall v \in H^{\frac{1}{2}}(\Gamma). \end{cases} \tag{8.4.24}$$

Lax-Milgram Theorem yields that the problem (8.4.24) has a unique solution $u \in H^{\frac{1}{2}}(\Gamma)$. Let $Tg = u$, then operator T is a bounded operator from $H^{-\frac{1}{2}}(\Gamma) \to H^{\frac{1}{2}}(\Gamma)$. In fact for given $g \in H^{-\frac{1}{2}}(\Gamma)$, $Tg = u$ is the restriction on Γ of the solution of the following boundary value problem

$$\begin{cases} \Delta u = 0, \quad \text{in } \Omega, & \text{(8.4.25a)} \\ \dfrac{\partial u}{\partial \boldsymbol{\nu}} + u = g, \quad \text{on } \Gamma. & \text{(8.4.25b)} \end{cases}$$

By the regularity of the solution of problem (8.4.25), we know that T is a bounded operator form $H^s(\Gamma) \to H^{s+1}(\Gamma)$ with $-\frac{1}{2} \leqslant s \leqslant \frac{1}{2}$, that is

$$T : H^s(\Gamma) \to H^{s+1}(\Gamma), \quad -\frac{1}{2} \leqslant s \leqslant \frac{1}{2},$$
$$\|Tg\|_{s+1,\Gamma} \leqslant C\|g\|_{s,\Gamma}, \quad \forall g \in H^s(\Gamma), \quad -\frac{1}{2} \leqslant s \leqslant \frac{1}{2}.$$

Obviously T is a compact operator on $H^{\frac{1}{2}}(\Gamma)$.

Suppose that $\frac{1}{\mu}$ is a nonzero eigenvalue of T on $H^{\frac{1}{2}}(\Gamma)$, i.e. there is a nonzero $u_\mu \in H^{\frac{1}{2}}(\Gamma)$, such $u_\mu = \mu T u_\mu$, then we have

$$E(u_\mu, v) = \mu E(Tu_\mu, v) = \mu(u_\mu, v)_\Gamma, \quad \forall v \in H^{\frac{1}{2}}(\Gamma).$$

It means that if $\frac{1}{\mu}$ is an eigenvalue of T and u_μ is the corresponding eigenfunction, then μ is an eigenvalue of the problem (8.4.23) with corresponding eigenfunction u_μ. Conversely, suppose μ is an eigenvalue of problem (8.4.23) corresponding the eigenfunction u_μ, then $\mu \neq 0$ and

$$E(u_\mu, v) = \mu(u_\mu, v)_\Gamma, \quad \forall v \in H^{\frac{1}{2}}(\Gamma). \tag{8.4.26}$$

On the other hand the definition of the operator T yields

$$E(\mu T u_\mu, v) = \mu(u_\mu, v)_\Gamma, \quad \forall v \in H^{\frac{1}{2}}(\Gamma). \tag{8.4.27}$$

Combing (8.4.26)–(8.4.27), we arrive at

$$E(u_\mu - \mu T u_\mu, v) = 0, \quad \forall v \in H^{\frac{1}{2}}(\Gamma),$$

then directly we have

$$T u_\mu = \frac{1}{\mu} u_\mu.$$

Therefore the eigenvalue problem (8.4.26) is equivalent to the eigenvalue problem of the compact operator T in the above sense.

We now return to the discrete problem (8.4.17). Similarly let $\mu_h = 1 + \lambda_h$, then the problem (8.4.17) can be rewritten as

$$\begin{cases} \text{Find non-zero } (u_h, p_h) \in V_h \text{ and number } \mu_h, \text{ such that} \\ B(u_h, v_h) + C(p_h, v_h) + (u_h, v_h)_\Gamma = \mu_h (u_h, v_h)_\Gamma, \quad \forall v_h \in S_h, & \text{(8.4.28a)} \\ -C(q_h, u_h) + A(q_h, p_h) = 0, \quad \forall q_h \in M_h. & \text{(8.4.28b)} \end{cases}$$

For given $u \in H^{\frac{1}{2}}(\Gamma)$, consider the following variational problem

$$\begin{cases} \text{Find } p_h \in M_h, \quad \text{such that} \\ A(q_h, p_h) = C(q_h, u), \quad \forall q_h \in M_h. \end{cases} \tag{8.4.29}$$

The problem (8.4.29) yields a unique solution $p_h \in M_h$.

Let $\mathcal{P}_h u = p_h$, then we obtain an operator $\mathcal{P}_h$,

$$\mathcal{P}_h : \quad H^{s+1}(\Gamma) \to M_h, \quad s \geqslant -\frac{1}{2},$$

and

$$\|\mathcal{P}_h u\|_{-\frac{1}{2},\Gamma} \leqslant C \|u\|_{\frac{1}{2},\Gamma}. \tag{8.4.30}$$

Obviously $\mathcal{P}_h u$ is the discrete approximation of $\mathcal{P} u$ in subspace M_h and the following error estimates hold

$$\|\mathcal{P} u - \mathcal{P}_h u\|_{-\frac{1}{2},\Gamma} \leqslant C h^s \|\mathcal{P} u\|_{s-\frac{1}{2}}. \tag{8.4.31}$$

We note that

$$A(\mathcal{P}_h u, \mathcal{P}_h v) = C(\mathcal{P}_h u, v) = C(p_h, v), \quad \forall v \in H^{\frac{1}{2}}(\Gamma). \tag{8.4.32}$$

Therefore the discrete eigenvalue problem (8.4.28) can be rewritten as the following equivalent form

$$\begin{cases} \text{Find non-zero } u_h \in S_h \text{ and number } \mu_h, \text{ such that} \\ E_h(u_h, v_h) = \mu_h (u_h, v_h)_\Gamma, \quad \forall v_h \in S_h, \end{cases} \tag{8.4.33}$$

with

$$E_h(u, v) = B(u, v) + A(\mathcal{P}_h u, \mathcal{P}_h v) + (u, v)_\Gamma. \tag{8.4.34}$$

For the bilinear form $E_h(u, v)$, directly we have

Lemma 8.4.3 *Bilinear form $E_h(u, v)$ on $H^{\frac{1}{2}}(\Gamma) \times H^{\frac{1}{2}}(\Gamma)$ is bounded and coercive. There exist two positive constants independent of h, such that*

$$|E_h(u, v)| \leqslant M\|u\|_{\frac{1}{2},\Gamma}\|v\|_{\frac{1}{2},\Gamma}, \quad \forall u, v \in H^{\frac{1}{2}}(\Gamma), \tag{8.4.35}$$

$$E_h(v, v) \geqslant \alpha\|v\|^2_{\frac{1}{2},\Gamma}, \quad \forall v \in H^{\frac{1}{2}}(\Gamma). \tag{8.4.36}$$

For given $g \in H^{-\frac{1}{2}}(\Gamma)$, we consider the following variational problem:

$$\begin{cases} \text{Find } u_h \in S_h, \text{ such that} \\ E_h(u_h, v) = (g, v)_\Gamma, \quad \forall v \in S_h. \end{cases} \tag{8.4.37}$$

Problem (8.4.37) exists a unique solution $u_h \in S_h$. Let $T_h g = u_h$, then we obtain an operator

$$\begin{aligned} &T_h : H^{-\frac{1}{2}}(\Gamma) \to S_h \subset H^{\frac{1}{2}}(\Gamma), \\ &\|T_h g\|_{\frac{1}{2},\Gamma} \leqslant C\|g\|_{-\frac{1}{2},\Gamma}. \end{aligned} \tag{8.4.38}$$

Suppose that (u_h, μ_h) is a solution of discrete eigenvalue prolbem (8.4.33). The definition of operator T yields

$$E_h(T_h g, v) = (g, v)_\Gamma, \quad \forall v \in S_h. \tag{8.4.39}$$

Taking $g = \mu_h u_h$ in (8.4.39), we have

$$E_h(\mu_h T_h u_h, v) = \mu_h(u_h, v)_\Gamma = E_h(u_h, v), \quad \forall v \in S_h,$$

which yields

$$E_h(u_h - \mu_h T_h u_h, v) = 0, \quad \forall v \in S_h,$$

and

$$u_h = \mu_h T_h u_h. \tag{8.4.40}$$

It means that $1/\mu_h$ is an eigenvalue of operator T_h corresponding to eigenfunction u_h. Conversely, suppose that $1/\mu_h$ is an eigenvalue of operator T_h corresponding to eigenfunction $u_h \in S_h$, then μ_h is an eigenvalue of problem (8.4.33) corresponding to eigenfunction $u_h \in S_h$. It means that the discrete eigenvalue problem (8.4.33) is equivalent to the eigenvalue problem of the operator T_h in the above sense. Therefore, we need only to estimate the error of eigenvalue of the compact operators T and T_h. At first we estimate the error of $T - T_h$.

Lemma 8.4.4 *For any given $g \in H^{-\frac{1}{2}}(\Gamma)$, we have*

$$E_h((T - T_h)g, v) = -A((\mathcal{P} - \mathcal{P}_h)Tg, \mathcal{P}v), \quad \forall v \in S_h. \tag{8.4.41}$$

Proof The definitions of T and T_h yield

$$\begin{aligned} E(Tg, v) &= (g, v)_\Gamma, \quad \forall v \in S_h, \\ E_h(T_h g, v) &= (g, v)_\Gamma, \quad \forall v \in S_h. \end{aligned}$$

Furthermore

$$\begin{aligned} 0 &= E(Tg, v) - E_h(T_h g, v) \\ &= E_h(Tg - T_h g, v) + A(\mathcal{P}Tg, \mathcal{P}v) - A(\mathcal{P}_h Tg, \mathcal{P}_h v) \\ &= E_h((T - T_h)g, v) + A((\mathcal{P} - \mathcal{P}_h)Tg, \mathcal{P}v) + \\ & A(\mathcal{P}_h Tg, (\mathcal{P} - \mathcal{P}_h)v), \quad \forall v \in S_h. \end{aligned} \tag{8.4.42}$$

On the other hand, the definition of the operators $\mathcal{P}$ and $\mathcal{P}_h$ arrive at

$$\begin{aligned} A(q, \mathcal{P}v) &= C(q, v), \quad \forall q \in M_h, \ v \in S_h, \\ A(q, \mathcal{P}_h v) &= C(q, v), \quad \forall q \in M_h, \ v \in S_h. \end{aligned}$$

Namely

$$A(q, (\mathcal{P} - \mathcal{P}_h)v) = 0, \quad \forall q \in M_h, \ v \in S_h. \tag{8.4.43}$$

Substituting (8.4.43) into (8.4.24), the Lemma 8.4.4 is proved completely. □

Lemma 8.4.5 *For any $g \in H^{-\frac{1}{2}+t}(\Gamma)(0 \leqslant t \leqslant 1)$, there exists a constant C, such that*

$$\|Tg - T_h g\|_{\frac{1}{2},\Gamma} \leqslant Ch^t \|g\|_{-\frac{1}{2}+t,\,\Gamma}. \tag{8.4.44}$$

Proof The Lemmas 8.4.3 and 8.4.4 yield

$$\begin{aligned} \|Tg - T_h g\|^2_{\frac{1}{2},\Gamma} &\leqslant \frac{1}{\alpha} E_h((T - T_h)g, (T - T_h)g) \\ &= \frac{1}{\alpha} \{E_h((T - T_h)g, Tg - \chi) + E_h((T - T_h)g, \chi - T_h g)\} \\ &= \frac{1}{\alpha} \{E_h((T - T_h)g, Tg - \chi) - \\ & A((\mathcal{P} - \mathcal{P}_h)Tg, \mathcal{P}(\chi - T_h g))\}, \quad \forall \chi \in S_h. \end{aligned}$$

The last equality is form Lemma 8.4.4.

Moreover we have

$$|E_h((T - T_h)g, Tg - \chi)| \leqslant M\|(T - T_h)g\|_{\frac{1}{2},\Gamma} \|Tg - \chi\|_{\frac{1}{2},\Gamma},$$

and

$$|A((\mathcal{P} - \mathcal{P}_h)Tg, \mathcal{P}(\chi - T_h g))|$$

$$\begin{aligned}&\leqslant |A((\mathcal{P}-\mathcal{P}_h)Tg,\mathcal{P}(T-T_h)g)|+|A((\mathcal{P}-\mathcal{P}_h)Tg,\mathcal{P}(Tg-\chi))|\\&\leqslant M\|(\mathcal{P}-\mathcal{P}_h)Tg\|_{-\frac{1}{2},\Gamma}\left\{\|\mathcal{P}(T-T_h)g\|_{-\frac{1}{2},\Gamma}+\|\mathcal{P}(Tg-\chi)\|_{-\frac{1}{2},\Gamma}\right\}.\end{aligned}$$

On the other hand, the estimate (8.4.21), (8.4.30)–(8.4.31) yield

$$\begin{aligned}\|(\mathcal{P}-\mathcal{P}_h)Tg\|_{-\frac{1}{2},\Gamma}&\leqslant Ch^t\|\mathcal{P}Tg\|_{-\frac{1}{2}+t,\Gamma}\leqslant Ch^t\|g\|_{-\frac{1}{2}+t,\Gamma},\\\|\mathcal{P}(Tg-\chi)\|_{-\frac{1}{2},\Gamma}&\leqslant C\|Tg-\chi\|_{\frac{1}{2},\Gamma},\quad\forall\chi\in S_h,\\\|\mathcal{P}(T-T_h)g\|_{-\frac{1}{2},\Gamma}&\leqslant C\|(T-T_h)g\|_{\frac{1}{2},\Gamma}.\end{aligned}$$

Therefore we obtain

$$\begin{aligned}\|Tg-T_hg\|^2_{\frac{1}{2},\Gamma}\leqslant C\Big\{&\|(T-T_h)g\|_{\frac{1}{2},\Gamma}\big[\|Tg-\chi\|_{\frac{1}{2},\Gamma}+h^t\|g\|_{-\frac{1}{2}+t,\Gamma}\big]+\\&h^t\|g\|_{-\frac{1}{2}+t,\Gamma}\|Tg-\chi\|_{\frac{1}{2},\Gamma}\Big\},\quad\forall\chi\in S_h.\end{aligned}$$

By the estimate

$$\inf_{\chi\in S_h}\|Tg-\chi\|_{\frac{1}{2},\Gamma}\leqslant Ch^t\|Tg\|_{\frac{1}{2}+t,\Gamma}\leqslant Ch^t\|g\|_{-\frac{1}{2}+t,\Gamma}.$$

The proof of Lemma 8.4.5 is completed. □

Lemma 8.4.6 *For any* $g\in H^{-\frac{1}{2}+t}(\Gamma)$, $\psi\in H^{-\frac{1}{2}+s}(\Gamma)(0\leqslant s,t\leqslant 1)$, *the following inequality holds*

$$|((T-T_h)g,\psi)_\Gamma|\leqslant Ch^{t+s}\|g\|_{-\frac{1}{2}+t}\|\psi\|_{-\frac{1}{2}+s}. \tag{8.4.45}$$

Proof The definitions of the operators T and T_h yield

$$\begin{aligned}(Tg,\psi)_\Gamma&=E(Tg,T\psi),\\(T_hg,\psi)_\Gamma&=E(T_hg,T\psi).\end{aligned}$$

Therefore we obtain

$$\begin{aligned}&((T-T_h)g,\psi)_\Gamma=E((T-T_h)g,T\psi)\\&=E_h((T-T_h)g,T\psi)+A(\mathcal{P}(T-T_h)g,\mathcal{P}T\psi)-A(\mathcal{P}_h(T-T_h)g,\mathcal{P}_hT\psi)\\&=E_h((T-T_h)g,T\psi)+A(\mathcal{P}(T-T_h)g,(\mathcal{P}-\mathcal{P}_h)T\psi)\\&=E_h((T-T_h)g,(T-T_h)\psi)+A(\mathcal{P}(T-T_h)g,(\mathcal{P}-\mathcal{P}_h)T\psi)-\\&\quad A((\mathcal{P}-\mathcal{P}_h)Tg,\mathcal{P}T_h\psi)\\&=E_h((T-T_h)g,(T-T_h)\psi)+A(\mathcal{P}(T-T_h)g,(\mathcal{P}-\mathcal{P}_h)T\psi)+\\&\quad A((\mathcal{P}-\mathcal{P}_h)Tg,\mathcal{P}(T-T_h)\psi)-A((\mathcal{P}-\mathcal{P}_h)Tg,\mathcal{P}T\psi)\\&=E_h((T-T_h)g,(T-T_h)\psi)+A(\mathcal{P}(T-T_h)g,(\mathcal{P}-\mathcal{P}_h)T\psi)+\end{aligned}$$

$$A((\mathcal{P}-\mathcal{P}_h)Tg,\mathcal{P}(T-T_h)\psi)-A((\mathcal{P}-\mathcal{P}_h)Tg,(\mathcal{P}-\mathcal{P}_h)T\psi).$$

The Lemma 8.4.3 and the estimate (8.4.31), (8.4.44) yield

$$\begin{aligned}|E_h((T-T_h)g,(T-T_h)\psi)| &\leqslant M\|(T-T_h)g\|_{\frac{1}{2},\Gamma}\|(T-T_h)\psi\|_{\frac{1}{2},\Gamma}\\ &\leqslant Ch^{s+t}\|g\|_{-\frac{1}{2}+t,\Gamma}\|\psi\|_{-\frac{1}{2}+s,\Gamma}.\end{aligned}$$

Furthermore by the boundness of the operator $\mathcal{P}$, $\mathcal{P}_h$, T, T_h, the bilinear form $A(q,p)$ and the estimate (8.4.31), (8.4.44) directly we obtain

$$\begin{aligned}|A(\mathcal{P}(T-T_h)g,(\mathcal{P}-\mathcal{P}_h)T\psi)| &\leqslant Ch^{t+s}\|g\|_{-\frac{1}{2}+t,\Gamma}\|\psi\|_{-\frac{1}{2}+s,\Gamma},\\ |A((\mathcal{P}-\mathcal{P}_h)Tg,\mathcal{P}(T-T_h)\psi)| &\leqslant Ch^{t+s}\|g\|_{-\frac{1}{2}+t,\Gamma}\|\psi\|_{-\frac{1}{2}+s,\Gamma},\\ |A((\mathcal{P}-\mathcal{P}_h)Tg,(\mathcal{P}-\mathcal{P}_h)T\psi)| &\leqslant Ch^{t+s}\|g\|_{-\frac{1}{2}+t,\Gamma}\|\psi\|_{-\frac{1}{2}+s,\Gamma}.\end{aligned}$$

Therefore the estimate (8.4.45) follows immediately. □

Let

$$\|T-T_h\|_{-\alpha,\beta}=\sup_{\psi\in H^\beta(\Gamma)}\sup_{g\in H^\alpha(\Gamma)}\frac{((T-T_h)g,\psi)_\Gamma}{\|g\|_{\alpha,\Gamma}\|\psi\|_{\beta,\Gamma}},\quad \forall\alpha,\beta\geqslant 0.$$

In the estimate (8.4.45) taking $t=\frac{1}{2}$, $s=\frac{1}{2}$, then we have

$$\|T-T_h\|_{0,0}\leqslant Ch.$$

Similarly taking $t=\dfrac{1}{2}$, $s=1$; $t=1$, $s=\frac{1}{2}$ and $t=1$, $s=1$ in (8.4.45) we obtain

$$\begin{aligned}\|T-T_h\|_{0,\frac{1}{2}} &\leqslant Ch^{\frac{3}{2}},\\ \|T-T_h\|_{-\frac{1}{2},0} &\leqslant Ch^{\frac{3}{2}},\\ \|T-T_h\|_{-\frac{1}{2},\frac{1}{2}} &\leqslant Ch^2.\end{aligned}$$

The well known eigenvalue convergence results (Bramble & Osborn, 1972) yield that if $\nu^1,\nu^2,\ldots$ are the eigenvalues of operator T ordered by decreasing magnitude, taking account of algebraic multiplicities, then for each mesh size h, then there is an ordering of the eigenvalues of the approximate operator T_h, $\nu_h^1,\nu_h^2,\ldots$ again counted according to algebraic multiplicities by decreasing magnitude, such that

(i)

$$\lim_{h\to 0}\nu_h^j=\nu^j,\quad \text{for each } j=1,2,\ldots. \tag{8.4.46}$$

(ii) Suppose that ν is an eigenvalue of the compact operators T with algebraic multiplicity m, then there are m eigenvalues $\nu_h^1,\nu_h^2,\ldots,\nu_h^m$ of the operator T_h, such that

$$\lim_{h\to 0} \nu_h^j = \nu, \quad \text{for} \quad j = 1, 2, \ldots, m.$$

Furthermore the following error estimate holds

$$\left| \nu - \frac{1}{m} \sum\nolimits_{j=1}^{m} \nu_h^j \right| \leqslant Ch^2. \tag{8.4.47}$$

Therefore for the Steklov eigenvalue problem (8.4.1) and its approximate problem (8.4.17), the following results hold

Theorem 8.4.1 *(i) If $\lambda^1, \lambda^2, \ldots$ are the eigenvalues of the Steklov eigenvalue problem (8.4.1) ordered by increasing magnitude, taking account of algebraic multiplicities, then for each h there is an ordering (again by increasing magnitude and taking account of algebraic multiplicities) $\lambda_h^1, \lambda_h^2, \ldots$,such that*

$$\lim_{h\to 0} \lambda_h^j = \lambda^j, \quad \textit{for} \ \ j = 1, 2, \ldots. \tag{8.4.48}$$

(ii) Suppose that λ is an eigenvalue of the Steklov eigenvalue problem (8.4.1) with algebraic multiplicity m, then there are m eigenvalues $\lambda_h^1, \lambda_h^2, \ldots, \lambda_h^m$ of the approximate problem (8.4.17) such that

$$\lim_{h\to 0} \lambda_h^j = \lambda, \quad \textit{for} \ j = 1, 2, \ldots, m,$$

and

$$|\lambda - \tilde{\lambda}_h| \leqslant Ch^2,$$

with

$$\tilde{\lambda}_h = \left(\frac{1}{m} \sum\nolimits_{j=1}^{m} \frac{1}{\lambda_h^j + 1} \right)^{-1} - 1.$$

The main content of this chapter comes from references Han and Guan (1992), Han, Guan, and He (1994a), Tang et al. (1998).

References

Adams, R. A., & Fournier, J. J. (2003). *Sobolev spaces*. Academic Press.

Andreev, A. B., & Todorov, T. D. (2004). Isoparametric finite-element approximation of a steklov eigenvalue problem. *IMA Journal of Numerical Analysis,24*(2), 309–322.

Armentano, M. G., & Padra, C. (2008). A posteriori error estimates for the Steklov eigenvalue problem. *Applied Numerical Mathematics,58*(5), 593–601.

Atkinson, K. E., & Chandler, G. (1990). Boundary integral equation methods for solving Laplaces equation with nonlinear boundary conditions: The smooth boundary case. *Mathematics of Computation,55*(192), 451–472.
Babuska I. & Aziz A. K. (1972). *Survey lectures on the mathematical foundations of the finite element method: The mathematical foundations of the finite element method with applications to partial differential equations*. Academic Press.
Bergman, S., & Schiffer, M. (2005). *Kernel functions and elliptic differential equations in mathematical physics*. Dover Publications.
Bramble, J. H., & Osborn, J. (1972). Approximation of steklov eigenvalues of non-selfadjoint second order elliptic operators. In *The mathematical foundations of the finite element method with applications to partial differential equations* (pp. 387–408). Elsevier.
Conca, C., Planchard, J., & Vanninathan, M. (1995). *Fluids and periodic structures* (Vol. 38). Wiley.
Courant, R., & Hilbert, D. (2008). *Methods of mathematical physics: Partial differential equations*. Wiley.
Duvaut, G., & Lions, J. L. (1976). Les inéquations en mécanique et en physique. *Travaux et Recherches Mathématiques, 21*.
Friedman, A. (1982). *Variational principle and free boundary problems*. Wiley.
Glowinski, R. (2013). *Numerical methods for nonlinear variational problems*. Springer Science & Business Media.
Glowinski, R., Lions, J.-L., & Tremolieres, R. (1981). *Numerical analysis of variational inequalities*. Amsterdam.
Goeleven, D., & Motreanu, D. (2003). *Variational and hemivariational inequalities-theory, methods and applications: Volume II: Unilateral problems* (Vol. 70). Springer Science & Business Media.
Goeleven, D., Motreanu, D., Dumont, Y., & Rochdi, M. (2004). *Variational and hemivariational inequalities: Theory, methods, and applications* (Vol. 1). Springer.
Han H. (1994a). A boundary element approximation of a Signorini problem with friction obeying coulomb law. *Journal of Computational Mathematics,* 147–162.
Han H. (1995). A boundary element method for a nonlinear boundary value problem in steady-state heat transfer. *The Journal of Integral Equations and Applications,* 17–35.
Han, H. (1988). Boundary integro-differential equations of elliptic boundary value problems and their numerical solutions. *Scientia Sinica, Series A—Mathematical, Physical, Astronomical and Technical Sciences* (Vol. 31, pp. 1153—1165).
Han, H. (1990). A direct boundary element method for Signorini problems. *Mathematics of Computation,55*(191), 115–128.
Han, H. (1991). A boundary element method for Signorini problems in three dimensions. *Numerische Mathematik,60*(1), 63–75.
Han, H. (1992). A boundary element method for Signorini problem in linear elasticity. *Numerische Mathematik,1*(1), 992.
Han, H. (1994b). The boundary element method for solving variational inequalities. *Contemporary Mathematics,163*, 55–55.
Han, H. (1994c). A boundary element procedure for the Signorini problem in three-dimensional elasticity. *Numerische Mathematik,3*, 104–117.
Han, H., & Guan, Z. (1992). An analysis of the boundary element approximation of Steklov eigenvalue problem. In *Proceedings of the 2nd Conference on Numerical Methods for Partial Differential Equations*, Tianjin, China (pp. 35—51).
Han, H., Guan, Z., & He, B. (1991). Boundary element approximation of Steklov eigenvalue problem, theory and applications of boundary element methods. In *Proceedings of 4th China–Japan Symposium on BEM*, Beijing, China (pp. 9—15).
Han, H., Guan, Z., & He, B. (1994a). Boundary element approximation of Steklov eigenvalue problem. *Journal of Chinese Universities Applied Mathematics Seris A,9*, 231–238.
Han, H., Guan, Z., & Yu, C. (1994b). The canonical boundary element analysis of a model for an electropaint process the canonical boundary element approximation of a Signorini problem. *Journal of Chinese Universities Applied Mathematics Seires A, 3*, 101.

Huang, J., & Lü, T. (2004). The mechanical quadrature methods and their extrapolation for solving BIE of Steklov eigenvalue problems. *Journal of Computational Mathematics,* 719–726.

Kikuchi, N., & Oden, J. T. (1988). *Contact problems in elasticity: a study of variational inequalities and finite element methods*. SIAM.

Kinderlehrer, D. (1981). Remarks about Signorini's problem in linear elasticity. *Annali della Scuola Normale Superiore di Pisa-Classe di Scienze,8*(4), 605–645.

Li, M., Lin, Q., & Zhang, S. (2010). Extrapolation and superconvergence of the Steklov eigenvalue problem. *Advances in Computational Mathematics, 33*(1), 25–44.

Li, Q., Lin, Q., & Xie, H. (2013). Nonconforming finite element approximations of the Steklov eigenvalue problem and its lower bound approximations. *Applications of Mathematics, 58*(2), 129–151.

Mora, D., Rivera, G., & Rodríguez, R. (2015). A virtual element method for the Steklov eigenvalue problem. *Mathematical Models and Methods in Applied Sciences, 25*(8), 1421–1445.

Nedelec, J. (1982). Integral equations with non integrable kernels. *Integral Equations and Operator Theory, 5*, 562–572.

Ruotsalainen, K., & Wendland, W. (1988). On the boundary element method for some nonlinear boundary value problems. *Numerische Mathematik, 53*, 299–314.

Signorini, A. (1933). Sopra alcune questioni di elastostatica. *Atti della societa Italiana per il progresso delle scienze, 21*(2), 143–148.

Sparrow, E. M. (2018). *Radiation heat transfer*. Routledge.

Tang, W., Guan, Z., & Han, H. (1998). Boundary element approximation of Steklov eigenvalue problem for Helmholtz equation. *Journal of Computational Mathematics,* 165–178.

Xie, H. (2014). A type of multilevel method for the Steklov eigenvalue problem. *IMA Journal of Numerical Analysis, 34*(2), 592–608.

Yang, Y., Li, Q., & Li, S. (2009). Nonconforming finite element approximations of the Ssteklov eigenvalue problem. *Applied Numerical Mathematics, 59*(10), 2388–2401.

Chapter 9
Coercive and Symmetrical Coupling Methods of Finite Element Method and Boundary Element Method

The finite element method (FEM) and boundary element methods (BEM) are two of the most widely used and the most effective computational methods in science and engineering computation. Both of them have their distinctive advantages and restrictions. The finite element method is based on the variational formula of the given problem, partition of the domain and piecewise interpolation of functions in corresponding space (Argyris & Kelsey, 1960; Courant, 1943; Hrennikoff, 1941; Turner et al., 1956). It has a solid mathematical foundation. The finite element method has very strong adaption for problems. It is suitable for solving partial differential equations with variable coefficients in complex domains and various nonlinear problems, but it can not directly solve problem on unbounded domain. The boundary element method is based on the boundary reduction of the given problem, i.e. the given problem can be reduced to equivalent boundary integral (integro-differential) equation (equations) (Brebbia, 1978; Hess & Smith, 1967; Jaswon, 1963; Massonnet, 1965; Symm, 1963). One of its main advantages is to reduce the dimension of the given problem by one dimension, thus the scale and amount of computation of the problem are greatly reduced. In particular, it is very efficient in solving problem on unbounded domain. Since the analytical expression of the fundamental solution can be found only for linear partial differential equations with constant coefficients directly. It is a real limitation. The respective limitations of the two important computational methods have been recognized since the 1970s. The scientists and engineers hope to couple them to adopt one method keeping the advantages of FEM and BEM and avoiding the limitations of them. Many of them devoted themselves to this research (Johnson et al., 1980; Wendland, 1986; Zienkiewicz et al., 1977). However, in the earliest coupling methods, only one integral boundary equation was used, and the coerciveness and symmetry of the original problem were destroyed. Han (1987, 1990), Costabel (1987), Costabel and Stephan (1990) introduced the symmetrical coupling of FEM and BEM using two boundary integral equations for given coercive and symmetrical problems. In the paper by Han (1987, 1990), the equivalent coercive coupling of FEM and BEM was also given. A remarkable advantage of these coupling methods

H. Han and D. Yin, *Mathematical Foundation of the Boundary Integro-Differential Equation Method*, https://doi.org/10.1007/978-981-95-1088-7_9

is symmetric and its equivalent formula preserves the coercive property. Obviously, the symmetrical formula yields the symmetrical algebraic system after discretization and the optimal error estimate of the numerical solution can be obtained by the coercive formula directly. The part content of this chapter comes from Han (1987, 1990) directly.

9.1 Exterior Dirichlet Problem of Poisson's Equation (I)

Consider the following exterior Dirichlet problem of Poisson's Equation:

$$\begin{cases} -\Delta u(x) = f(x), \ \forall x \in \Omega_e & (9.1.1a) \\ u(x)\Big|_{\Gamma} = 0, & (9.1.1b) \\ u(x) = c + \mathcal{O}(|x|^{1-n}) \text{ as } |x| \to \infty, & (9.1.1c) \end{cases}$$

where Ω_e is the exterior domain with boundary Γ defined in Chap. 3, $f(x)$ is a given function on Ω_e with compact support, c is an unknown constant.

9.1.1 *The Symmetric and Coercive Coupling Formula of Problem (9.1.1)*

Introduce an artificial boundary $\Gamma_A \in \Omega_e$, such that Γ_A divides Ω_e into a bounded domain Ω_1, an unbounded domain Ω_2 and $\text{Supp}\{f(x)\} \in \Omega_1$ (see Fig. 9.1 for $n = 2$).

On the unbounded domain Ω_2, $u(x)$ satisfies

Fig. 9.1 Interface

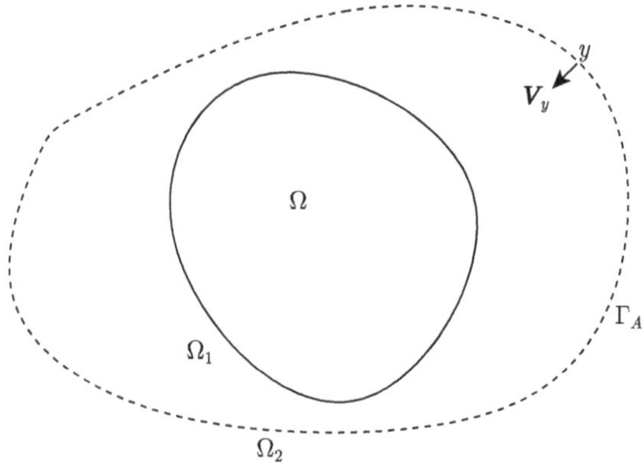

$$\begin{cases} -\Delta u(x) = 0, & \forall x \in \Omega_2 \\ u(x) = c + \mathcal{O}(|x|^{1-n}), & \text{as } |x| \to \infty. \end{cases}$$

Let

$$v(x) = u(x) - c, \quad \forall x \in \bar{\Omega}_2, \tag{9.1.2}$$

then $v(x)$ satisfies

$$\begin{cases} -\Delta v(x) = 0, & \forall x \in \Omega_2, \\ v(x) = \mathcal{O}(|x|^{1-n}), & \text{as } |x| \to \infty. \end{cases}$$

Using Green's formula on $\bar{\Omega}_2$, it yields

$$v(x) = -\int_{\Gamma_A} \frac{\partial E_n(x,y)}{\partial \boldsymbol{\nu}_y} v(y) \mathrm{d}s_y + \int_{\Gamma_A} E_n(x,y) \frac{\partial v(y)}{\partial \boldsymbol{\nu}_y} \mathrm{d}s_y \tag{9.1.3}$$

where $E_n(x, y)$ is the fundamental solution of Laplace operator given by (3.3.3), $\boldsymbol{\nu}_y$ denotes the outward unit norm vector at $y \in \Gamma_A = \partial\Omega_2$.

$$\begin{aligned} u(x) = & -\int_{\Gamma_A} \frac{\partial E_n(x,y)}{\partial \boldsymbol{\nu}_y} u(y) \mathrm{d}s_y + \int_{\Gamma_A} E_n(x,y) \frac{\partial u(y)}{\partial \boldsymbol{\nu}_y} \mathrm{d}s_y + \\ & c \int_{\Gamma_A} \frac{\partial E_n(x,y)}{\partial \boldsymbol{\nu}_y} \mathrm{d}s_y + c \\ = & -\int_{\Gamma_A} \frac{\partial E_n(x,y)}{\partial \boldsymbol{\nu}_y} u(y) \mathrm{d}s_y + \int_{\Gamma_A} E_n(x,y) \frac{\partial u(y)}{\partial \boldsymbol{\nu}_y} \mathrm{d}s_y + c, \ \forall x \in \Omega_2. \end{aligned} \tag{9.1.4}$$

The boundary condition (9.1.1c) and

$$-\Delta u(x) = 0, \ \forall x \in \Omega_e,$$

yield that

$$\int_{\Gamma_A} \frac{\partial u(x)}{\partial \boldsymbol{\nu}_x} \mathrm{d}s_x = 0. \tag{9.1.5}$$

Let

$$p(x) = \left.\frac{\partial u(x)}{\partial \boldsymbol{\nu}_x}\right|_{\Gamma_A}. \tag{9.1.6}$$

By the properties of single layer and double layer potentials, the Formula (9.1.4) yields the following relationship between $u(x)\Big|_{\Gamma_A}$ and $p(x) = \frac{\partial u(x)}{\partial \boldsymbol{\nu}_x}\Big|_{\Gamma_A}$:

$$\frac{1}{2} u(x) = -\int_{\Gamma_A} \frac{\partial E_n(x,y)}{\partial \boldsymbol{\nu}_y} u(y) \mathrm{d}s_y + \int_{\Gamma_A} E_n(x,y) p(y) \mathrm{d}s_y + c, \ \forall x \in \Gamma_A. \tag{9.1.7}$$

Furthermore using the properties of the derivatives of the single layer and double layer potential we obtain

$$\left.\frac{\partial u(x)}{\partial \boldsymbol{\nu}_x}\right|_{\Gamma_A} = \frac{1}{2}p(x) - \int_{\Gamma_A} \frac{\partial^2 E_n(x, y)}{\partial \boldsymbol{\nu}_x \partial \boldsymbol{\nu}_y} u(y)\mathrm{d}s_y + \int_{\Gamma_A} \frac{\partial E_n(x, y)}{\partial \boldsymbol{\nu}_x} p(y)\mathrm{d}s_y, \ \forall x \in \Gamma_A. \tag{9.1.8}$$

Now the problem (9.1.1) can be reduced on the bounded domain Ω_1 equivalently:

$$\begin{cases} -\Delta u(x) = f(x), \ \forall x \in \Omega_1, & (9.1.9a) \\ u(x)\big|_{\Gamma} = 0, & (9.1.9b) \\ \frac{1}{2}u(x) = -\int_{\Gamma_A} \frac{\partial E_n(x, y)}{\partial \boldsymbol{\nu}_y} u(y)\mathrm{d}s_y + \\ \int_{\Gamma_A} E_n(x, y)p(y)\mathrm{d}s_y + c, \ \forall x \in \Gamma_A, & (9.1.9c) \\ \left.\frac{\partial u(x)}{\partial \boldsymbol{\nu}_x}\right|_{\Gamma_A} = \frac{1}{2}p(x) - \int_{\Gamma_A} \frac{\partial^2 E_n(x, y)}{\partial \boldsymbol{\nu}_x \partial \boldsymbol{\nu}_y} u(y)\mathrm{d}s_y + \\ \int_{\Gamma_A} \frac{\partial E_n(x, y)}{\partial \boldsymbol{\nu}_x} p(y)\mathrm{d}s_y, \ \forall x \in \Gamma_A. & (9.1.9d) \end{cases}$$

Let

$$\begin{aligned} H_*^1(\Omega_1) &= \{v \in H^1(\Omega_1) \text{ and } v\big|_{\Gamma} = 0\}, \\ H_*^{-\frac{1}{2}}(\Gamma_A) &= \{q \in H^{-\frac{1}{2}}(\Gamma_A) \text{ and } \int_{\Gamma_A} q(x)\mathrm{d}s_x = 0\}, \\ \mathbb{V}^* &= H_*^1(\Omega_1) \times H_*^{-\frac{1}{2}}(\Gamma_A). \end{aligned}$$

Multiplying equation (9.1.9a) by the function $v \in H_*^1(\Omega_1)$ and integrating over Ω_1 it arrives at

$$a_1(u, v) + \int_{\Gamma_A} \frac{\partial u(x)}{\partial \boldsymbol{\nu}_x} v(x)\mathrm{d}s_x = \langle f, v\rangle_{\Omega_1}, \ \forall v \in H_*^1(\Omega_1), \tag{9.1.10}$$

with

$$a_1(u, v) \quad \int_{\Omega_1} \nabla u(x) \cdot \nabla v(x)\mathrm{d}x, \tag{9.1.11}$$

$$\langle f, v\rangle_{\Omega_1} = \int_{\Omega_1} f(x)v(x)\mathrm{d}x. \tag{9.1.12}$$

Substituting (9.1.8) into (9.1.10), the variational equality (9.1.10) can be rewritten as follows:

$$a_1(u,v)+\frac{1}{2}\int_{\Gamma_A}p(x)v(x)\mathrm{d}s_x-\int_{\Gamma_A}\int_{\Gamma_A}\frac{\partial^2 E_n(x,y)}{\partial \boldsymbol{v}_x\partial \boldsymbol{v}_y}u(y)v(x)\mathrm{d}s_y\mathrm{d}s_x+$$
$$\int_{\Gamma_A}\int_{\Gamma_A}\frac{\partial E_n(x,y)}{\partial \boldsymbol{v}_x}p(y)v(x)\mathrm{d}s_y\mathrm{d}s_x=\langle f,v\rangle_{\Omega_1},\ \forall v\in H_*^1(\Omega_1). \quad (9.1.13)$$

The equality (9.1.7) is equivalent to

$$\frac{1}{2}\int_{\Gamma_A}u(x)q(x)\mathrm{d}s_x+\int_{\Gamma_A}\int_{\Gamma_A}\frac{\partial E_n(x,y)}{\partial \boldsymbol{v}_y}u(y)q(x)\mathrm{d}s_y\mathrm{d}s_x-$$
$$\int_{\Gamma_A}\int_{\Gamma_A}E_n(x,y)p(y)q(x)\mathrm{d}s_y\mathrm{d}s_x=0,\ \forall q\in H_*^{-\frac{1}{2}}(\Gamma_A). \quad (9.1.14)$$

Combining the variational equalities (9.1.13) and (9.1.14), we obtain the symmetry and the coercive coupling formulations of the given problem (9.1.1):

$$\begin{cases}\text{Find}(u,p)\in H_*^1(\Omega_1)\times H_*^{-\frac{1}{2}}(\Gamma_A),\text{ such that}\\ a_1(u,v)+B(u,v)+C(p,v)=\langle f,v\rangle_{\Omega_1},\quad \forall v\in H_*^1(\Omega_1), & (9.1.15a)\\ C(q,u)-A(q,p)=0,\quad \forall q\in H_*^{-\frac{1}{2}}(\Gamma_A). & (9.1.15b)\end{cases}$$

$$\begin{cases}\text{Find}(u,p)\in H_*^1(\Omega_1)\times H_*^{-\frac{1}{2}}(\Gamma_A),\text{ such that}\\ a_1(u,v)+B(u,v)+C(p,v)=\langle f,v\rangle_{\Omega_1},\quad \forall v\in H_*^1(\Omega_1), & (9.1.16a)\\ -C(q,u)+A(q,p)=0,\quad \forall q\in H_*^{-\frac{1}{2}}(\Gamma_A). & (9.1.16b)\end{cases}$$

with

$$A(q,p)=\int_{\Gamma_A}\int_{\Gamma_A}E_n(x,y)p(y)q(x)\mathrm{d}s_y\mathrm{d}s_x, \quad (9.1.17)$$

$$\begin{aligned}B(u,v)&=-\int_{\Gamma_A}\int_{\Gamma_A}\frac{\partial^2 E_n(x,y)}{\partial \boldsymbol{v}_x\partial \boldsymbol{v}_y}u(y)v(x)\mathrm{d}s_y\mathrm{d}s_x\\ &=\sum_{1\leqslant k<j\leqslant n}\int_{\Gamma_A}\int_{\Gamma_A}E_n(x,y)\left(\frac{\partial u(y)}{\partial \boldsymbol{\tau}_y^{k,j}}\right)\left(\frac{\partial v(x)}{\partial \boldsymbol{\tau}_x^{k,j}}\right)\mathrm{d}s_y\mathrm{d}s_x, \quad (9.1.18)\end{aligned}$$

$$C(p,v)=\frac{1}{2}\int_{\Gamma_A}p(x)v(x)\mathrm{d}s_x+\int_{\Gamma_A}\int_{\Gamma_A}\frac{\partial E_n(x,y)}{\partial \boldsymbol{v}_x}p(y)v(x)\mathrm{d}s_y\mathrm{d}s_x. \quad (9.1.19)$$

The bilinear forms $A(q,p)$, $B(u,v)$ and $C(p,v)$ were defined by (3.4.16), (3.4.17) and (8.1.14) on the boundary Γ. On the space $\mathbb{V}^*=H_*^1(\Omega_1)\times H_*^{-\frac{1}{2}}(\Gamma_A)$ define the following two bilinear forms, let

$$D_S(u, p; v, q) = a_1(u, v) + B(u, v) + C(p, v) + \\ C(q, u) - A(q, p), \ \forall (u, p), (v, q) \in \mathbb{V}^*, \quad (9.1.20)$$

$$D_C(u, p; v, q) = a_1(u, v) + B(u, v) + C(p, v) - \\ C(q, u) + A(q, p). \ \forall (u, p), (v, q) \in \mathbb{V}^*. \quad (9.1.21)$$

Obviously bilinear form $D_S(u, p; v, q)$ is symmetric on $\mathbb{V}^* \times \mathbb{V}^*$, but $D_C(u, p; v, q)$ is not.

The problem (9.1.15) can be rewritten as

$$\begin{cases} \text{Find } (u, p) \in \mathbb{V}^*, \ \text{such that} \\ D_S(u, p; v, q) = \langle f, v \rangle_{\Omega_1}, \ \forall (v, q) \in \mathbb{V}^*. \end{cases} \quad (9.1.22)$$

Similarly the problem (9.1.16) is equivalent to

$$\begin{cases} \text{Find } (u, p) \in \mathbb{V}^*, \ \text{such that} \\ D_C(u, p; v, q) = \langle f, v \rangle_{\Omega_1}, \ \forall (v, q) \in \mathbb{V}^*. \end{cases} \quad (9.1.23)$$

The problems (9.1.22) and (9.1.23) are equivalent. The problem (9.1.22) is the symmetric coupling formula of finite element method and boundary element method for the given problem (9.1.1). The problem (9.1.23) is the coercive coupling formula of finite element method and boundary element method for the given problem (9.1.1).

Lemma 9.1.1 *(i) $D_S(u, p; v, q)$ and $D_C(u, p; v, q)$ are bounded bilinear forms on $\mathbb{V}^* \times \mathbb{V}^*$, namely, there exists a positive constant M, such that*

$$\begin{cases} |D_S(u, p; v, q)| \leqslant M \|(u, p)\|_{\mathbb{V}^*} \|(v, q)\|_{\mathbb{V}^*}, \\ |D_C(u, p; v, q)| \leqslant M \|(u, p)\|_{\mathbb{V}^*} \|(v, q)\|_{\mathbb{V}^*}, \ \forall (u, p), (v, q) \in \mathbb{V}^*. \end{cases} \quad (9.1.24)$$

(ii) $D_C(u, p; v, q)$ is coercive on $\mathbb{V}^* \times \mathbb{V}^*$, namely, there exists a positive constant β, such that

$$D_C(u, p; u, p) \geqslant \beta \|(u, p)\|^2_{\mathbb{V}^*}, \ \forall (u, p) \in \mathbb{V}^*. \quad (9.1.25)$$

Proof It is known that bilinear form $a_1(u, v)$ is bounded and coercive on $H^1_*(\Omega_1) \times H^1_*(\Omega_1)$ (Brenner et al., 2008), there exists two constant $M_1 > 0$ and $\beta_1 > 0$ such that

$$|a_1(u, v)| \leqslant M_1 \|u\|_{1,\Omega_1} \|v\|_{1,\Omega_1}, \ \forall u, v \in H^1_*(\Omega_1), \quad (9.1.26)$$

$$a_1(u, u) \geqslant \beta_1 \|u\|^2_{1,\Omega_1}, \ \forall u \in H^1_*(\Omega_1). \quad (9.1.27)$$

By Lemma 3.4.1 we have that $A(p,q)$ is bounded and coercive on $H_*^{-\frac{1}{2}}(\Gamma_A) \times H_*^{-\frac{1}{2}}(\Gamma_A)$, $B(u,v)$ is bounded and non-negative on $H^{\frac{1}{2}}(\Gamma_A) \times H^{\frac{1}{2}}(\Gamma_A)$, namely there exist two positive constant M_2 and β_2, such that

$$|A(p,q)| \leqslant M_2 \|p\|_{-\frac{1}{2},\Gamma_A} \|q\|_{-\frac{1}{2},\Gamma_A}, \ \forall p,q \in H_*^{-\frac{1}{2}}(\Gamma_A), \tag{9.1.28}$$

$$|B(u,v)| \leqslant M_2 \|u\|_{\frac{1}{2},\Gamma_A} \|v\|_{\frac{1}{2},\Gamma_A}, \ \forall u,v \in H^{\frac{1}{2}}(\Gamma_A), \tag{9.1.29}$$

$$A(p,p) \geqslant \beta_2 \|p\|_{-\frac{1}{2},\Gamma_A}, \ \forall p \in H_*^{-\frac{1}{2}}(\Gamma_A), \tag{9.1.30}$$

$$B(u,u) \geqslant 0, \ \forall u \in H^{\frac{1}{2}}(\Gamma_A). \tag{9.1.31}$$

Combining the inequalities (9.1.26)–(9.1.31), the proof of this lemma is completed.

The Lemma 9.1.1 and Lax-Milgram theorem (Brenner et al., 2008) yield the following Theorem directly:

Theorem 9.1.1 *Suppose that $f \in H^{-1}(\Omega_1)$, then the variational problem (9.1.23) and (9.1.22) exists a unique solution $(u,p) \in \mathbb{V}^*$ and*

$$\|(u,p)\|_{V^*} \leqslant \frac{M}{\beta} \|f\|_{-1,\Omega_1}. \tag{9.1.32}$$

9.1.2 The Numerical Solutions of Problem (9.1.1) Based on the Symmetric and Coercive Coupling Formula

We discuss the numerical solutions of problems (9.1.22) and (9.1.23). Suppose that W_h and M_h are two finite dimensional subspaces of $H_*^1(\Omega_1)$ and $H_*^{-\frac{1}{2}}(\Gamma_A)$, W_h and M_h are two regular finite element spaces, which satisfy the following approximation properties:

$$\inf_{v_h \in W_h} \|u - v_h\|_{1,\Omega_1} \leqslant Ch^t |u|_{1+t,\,\Omega_1}, \ \forall u \in H_*^1(\Omega_1) \cap H^{1+t}(\Omega_1), \tag{9.1.33}$$

$$\inf_{q_h \in M_h} \|p - q_h\|_{-\frac{1}{2},\Gamma_A} \leqslant Ch^t |p|_{-\frac{1}{2}+t,\,\Gamma_A}, \ \forall p \in H_*^{-\frac{1}{2}}(\Gamma_A) \cap H^{-\frac{1}{2}+t}(\Gamma_A),$$
$$\text{with } 0 < t \leqslant 1. \tag{9.1.34}$$

Let

$$\mathbb{V}_h^* = W_h \times M_h. \tag{9.1.35}$$

Obviously $\mathbb{V}_h^*$ is a subspace of $\mathbb{V}^*$.

The discrete problem (9.1.22) is given by

$$\begin{cases} \text{Find } (u_h, p_h) \in \mathbb{V}_h^*, \quad \text{such that} \\ D_S(u_h, p_h; v_h, q_h) = \langle f, v_h \rangle_{\Omega_1}, \quad \forall (v_h, q_h) \in \mathbb{V}_h^*. \end{cases} \tag{9.1.36}$$

Similarly by the problem (9.1.23) we obtain

$$\begin{cases} \text{Find } (u_h, p_h) \in \mathbb{V}_h^*, \quad \text{such that} \\ D_C(u_h, p_h; v_h, q_h) = \langle f, v_h \rangle_{\Omega_1}, \quad \forall (v_h, q_h) \in \mathbb{V}_h^*. \end{cases} \tag{9.1.37}$$

Certainly the problem (9.1.36) and (9.1.37) are equivalent.

The resulting linear algebraic system based on (9.1.36) is symmetric, while its equivalent formulation (9.1.37) possesses coercive property. This property yields that the problem (9.1.37) exists a unique solution $(u_h, p_h) \in \mathbb{V}_h^*$, and which can be directly applied to obtain the optimal error estimate of the numerical solutions. We have

Theorem 9.1.2 *(i) The problem (9.1.37) and (9.1.36) exists a unique solution* $(u_h, p_h) \in \mathbb{V}_h^*$.

(ii) suppose that the solution (u, p) *of problem (9.1.23) satisfies* $u \in H_*^1(\Omega_1) \cap H^{1+t}(\Omega_1)$ *and* $p \in H_*^{-\frac{1}{2}}(\Gamma_A) \cap H^{-\frac{1}{2}+t}(\Gamma_A)$ *with* $0 < t \leqslant 1$. *Then the following error estimate holds:*

$$\begin{aligned} \|(u-u_h, p-p_h)\|_{\mathbb{V}_h^*} &\leqslant \frac{M}{\beta} \inf_{(v_h,q_h)\in\mathbb{V}_h^*} \|(u-u_h, p-q_h)\|_{\mathbb{V}_h^*} \\ &\leqslant Ch^t \left\{ |u|_{1+t,\,\Omega_1} + |p|_{-\frac{1}{2}+t,\,\Gamma_A} \right\}. \end{aligned} \tag{9.1.38}$$

9.2 Exterior Dirichlet Problem of Poisson Equation (II)

In this section we consider second exterior Dirichlet problem of Poisson equation:

$$-\begin{cases} \Delta u(x) = f(x), \ \forall x \in \Omega_e, & (9.2.1\text{a}) \\ u(x)\big|_\Gamma = 0, & (9.2.1\text{b}) \\ u(x) = \begin{Bmatrix} C \log|x|, \ n = 2 \\ C|x|^{2-n}, \ \ n = 3 \end{Bmatrix} + \mathcal{O}(|x|^{1-n}), \ \text{ as } |x| \to \infty. & (9.2.1\text{c}) \end{cases}$$

The exterior domain Ω_e and function $f(x)$ are given in Sect. 9.1, as the same in which the artificial boundary $\Gamma_A \in \Omega_e$ divides Ω_e into a bounded part Ω_1 and unbounded part Ω_2 such that $\text{Supp}\{f(x)\} \in \Omega_1$. By the Theorem 3.4.6 we know that in the case $n = 2$, if the boundary Γ is a degenerate scale, the problem (9.2.1) is improperly posed. Therefore in the case $n = 2$, suppose that the boundary Γ is not a degenerate scale, namely the $\boldsymbol{\gamma}_\Gamma$-index of boundary Γ is not zero.

On the unbounded domain Ω_e, $u(x)$, the solution of problem (9.2.1), satisfies:

$$\begin{cases} -\Delta u = 0, \ \forall x \in \Omega_2, \\ u(x) = \begin{Bmatrix} C \log |x|, & n = 2 \\ C|x^{2-n}|, & n = 3 \end{Bmatrix} + \mathcal{O}(|x|^{1-n}), \ \text{as } |x| \to \infty. \end{cases}$$

Let

$$w(x) = u(x) + \int_{\Gamma_A} \frac{\partial E_n(x,y)}{\partial \boldsymbol{\nu}_y} u(y) ds_y, \ \forall x \in \Omega_2, \tag{9.2.2}$$

with $\boldsymbol{\nu}_y$, the outward unit norm vector at $y \in \Gamma_A = \partial\Omega_2$, then $w(x)$ satisfies

$$\begin{cases} -\Delta w(x) = 0, \ \forall x \in \Omega_2, & (9.2.3a) \\ w(x)\big|_{\Gamma_A} = \dfrac{1}{2} u(x) + \displaystyle\int_{\Gamma_A} \frac{\partial E_n(x,y)}{\partial \boldsymbol{\nu}_y} u(y) ds_y, \ \forall x \in \Gamma_A, & (9.2.3b) \\ w(x) = \begin{Bmatrix} C \log |x|, & n = 2 \\ C|x^{2-n}|, & n = 3 \end{Bmatrix} + \mathcal{O}(|x|^{1-n}), \ \text{as } |x| \to \infty. & (9.2.3c) \end{cases}$$

The problem (9.2.3) is an exterior Dirichlet problem of Laplace equation. The Theorem 3.4.6 yields

Theorem 9.2.1 *Assume that $u\big|_{\Gamma_A} \in H^{\frac{1}{2}}(\Gamma_A)$ and in the case $n = 2$, Γ_A is not a degenerate scale, then the problem (9.2.3) exists a unique solution*

$$w(x) = \int_{\Gamma_A} E_n(x,y)\rho(y) ds_y, \ \forall x \in \Omega_2, \tag{9.2.4}$$

where $\rho \in H^{-\frac{1}{2}}(\Gamma_A)$.

Substituting (9.2.4) into (9.2.2) we obtain

$$u(x) = -\int_{\Gamma_A} \frac{\partial E_n(x,y)}{\partial \boldsymbol{\nu}_y} u(y) ds_y + \int_{\Gamma_A} E_n(x,y)\rho(y) ds_y, \ \forall x \in \Omega_2. \tag{9.2.5}$$

On the boundary Γ_A, by (9.2.5) we obtain

$$\frac{1}{2} u(x) = -\int_{\Gamma_A} \frac{\partial E_n(x,y)}{\partial \boldsymbol{\nu}_y} u(y) ds_y + \int_{\Gamma_A} E_n(x,y)\rho(y) ds_y, \ \forall x \in \Gamma_A,$$

$$\frac{\partial u(x)}{\partial \boldsymbol{\nu}_x}\Big|_{\Gamma_A} = \frac{1}{2}\rho(x) - \int_{\Gamma_A} \frac{\partial^2 E_n(x,y)}{\partial \boldsymbol{\nu}_x \partial \boldsymbol{\nu}_y} u(y) ds_y + \int_{\Gamma_A} \frac{\partial E_n(x,y)}{\partial \boldsymbol{\nu}_x} \rho(y) ds_y, \ \forall x \in \Gamma_A.$$

Now the problem (9.2.1) can be reduced to the bounded domain Ω_2 with the auxiliary function $\rho(x)$.

$$\begin{cases} -\Delta u(x) = f(x), \ \forall x \in \Omega_1, & (9.2.6a) \\ u(x)\big|_{\Gamma} = 0, & (9.2.6b) \\ \frac{1}{2}u(x) = -\int_{\Gamma_A} \frac{\partial E_n(x,y)}{\partial \boldsymbol{v}_y} u(y) \mathrm{d}s_y + \\ \quad \int_{\Gamma_A} E_n(x,y)\rho(y)\mathrm{d}s_y, \ \forall x \in \Gamma_A, & (9.2.6c) \\ \frac{\partial u(x)}{\partial \boldsymbol{v}_x}\Big|_{\Gamma_A} = \frac{1}{2}\rho(x) - \int_{\Gamma_A} \frac{\partial^2 E_n(x,y)}{\partial \boldsymbol{v}_x \partial \boldsymbol{v}_y} u(y)\mathrm{d}s_y + \\ \quad \int_{\Gamma_A} \frac{\partial E_n(x,y)}{\partial \boldsymbol{v}_x}\rho(y)\mathrm{d}s_y, \ \forall x \in \Gamma_A. & (9.2.6d) \end{cases}$$

For any $v \in H^1_*(\Omega_1)$, $v(x)$ multiplying (9.2.6a) and integrating by parts on Ω_1, it arrives at

$$\int_{\Omega_1} \nabla u \cdot \nabla v \mathrm{d}x + \int_{\Gamma_A} \frac{\partial u(x)}{\partial \boldsymbol{v}_x} v(x)\mathrm{d}s_x = \int_{\Omega_1} f(x)v(x)\mathrm{d}x, \ \forall v \in H^1_*(\Omega_1). \quad (9.2.7)$$

Substituting (9.2.6d) into (9.2.7) and combining the variational form of the equality (9.2.6c), it yields the following variational problem, which is equivalent to the problem (9.2.6).

$$\begin{cases} \text{Find } (u,\rho) \in H^1_*(\Omega_1) \times H^{-\frac{1}{2}}(\Gamma_A), \ \text{ such that} \\ \int_{\Omega_1} \nabla u \cdot \nabla v \mathrm{d}x + \frac{1}{2}\int_{\Gamma_A} \rho(x)v(x)\mathrm{d}x - \\ \quad \int_{\Gamma_A}\int_{\Gamma_A} \frac{\partial^2 E_n(x,y)}{\partial \boldsymbol{v}_x \partial \boldsymbol{v}_y} u(y)v(x)\mathrm{d}s_y\mathrm{d}s_x + \\ \int_{\Gamma_A}\int_{\Gamma_A} \frac{\partial E_n(x,y)}{\partial \boldsymbol{v}_x}\rho(y)v(x)\mathrm{d}s_y\mathrm{d}s_x \\ = \int_{\Omega_1} f(x)v(x)\mathrm{d}x, \ \forall v \in H^1_*(\Omega_1), & (9.2.8a) \\ -\frac{1}{2}\int_{\Gamma_A} u(x)\mu(x)\mathrm{d}s_x - \int_{\Gamma_A}\int_{\Gamma_A} \frac{\partial E_n(x,y)}{\partial \boldsymbol{v}_y} u(y)\mu(x)\mathrm{d}s_y\mathrm{d}s_x + \\ \int_{\Gamma_A}\int_{\Gamma_A} E_n(x,y)\rho(y)\mu(x)\mathrm{d}s_y\mathrm{d}s_x = 0, \ \forall \mu \in H^{-\frac{1}{2}}(\Gamma_A). & (9.2.8b) \end{cases}$$

By the notations $a_1(u, v)$, $\langle f, v\rangle_{\Omega_1}$, $A(u, v)$, $B(u, v)$ and $C(\rho, v)$ used in the Sect. 9.1.1, the problem (9.2.8) can be rewritten as

$$\begin{cases} \text{Find } (u, \rho) \in H_*^1(\Omega_1) \times H^{-\frac{1}{2}}(\Gamma_A), \text{ such that} \\ a_1(u, v) + B(u, v) + C(\rho, v) = \langle f, v \rangle_{\Omega_1}, \ \forall v \in H_*^1(\Omega_1), & (9.2.9a) \\ -C(\mu, u) + A(\mu, \rho) = 0, \ \forall \mu \in H^{-\frac{1}{2}}(\Gamma_A), & (9.2.9b) \end{cases}$$

and its equivalent symmetric form

$$\begin{cases} \text{Find } (u, \rho) \in H_*^1(\Omega_1) \times H^{-\frac{1}{2}}(\Gamma_A), \quad \text{such that} \\ a_1(u, v) + B(u, v) + C(\rho, v) = \langle f, v \rangle_{\Omega_1}, \ \forall v \in H_*^1(\Omega_1), & (9.2.10a) \\ C(\mu, u) - A(\mu, \rho) = 0, \ \forall \mu \in H^{-\frac{1}{2}}(\Gamma_A). & (9.2.10b) \end{cases}$$

Let

$$\mathbb{V} = H_*^1(\Omega_1) \times H^{-\frac{1}{2}}(\Gamma_A),$$

with

$$\|(u, \rho)\|_{\mathbb{V}} = \left\{ \|u\|_{1,\Omega_1}^2 + \|\rho\|_{-\frac{1}{2},\Gamma_A}^2 \right\}^{\frac{1}{2}}, \ \forall (u, \rho) \in \mathbb{V}.$$

Obviously $D_C(u, \rho; v, \mu)$ and $D_S(u, \rho; v, \mu)$ defined by (9.1.20) and (9.1.21) are also well defined bounded bilinear form on $\mathbb{V} \times \mathbb{V}$. Then problems (9.2.9) and (9.2.10) can be rewritten as

$$\begin{cases} \text{Find } (u, \rho) \in \mathbb{V}, \quad \text{such that} \\ D_C(u, \rho; v, \mu) = \langle f, v \rangle_{\Omega_1}, \ \forall (v, \mu) \in \mathbb{V}. \end{cases} \tag{9.2.11}$$

and

$$\begin{cases} \text{Find } (u, \rho) \in \mathbb{V}, \quad \text{such that} \\ D_S(u, \rho; v, \mu) = \langle f, v \rangle_{\Omega_1}, \ \forall (v, \mu) \in \mathbb{V}. \end{cases} \tag{9.2.12}$$

By the existence and uniqueness of problem (9.2.1) and the equivalence of problem (9.2.1) with problem (9.2.11) (problem (9.2.12)) provided that Γ and Γ_A are not degenerate scales in the case $n = 2$. We obtain

Theorem 9.2.2 *Suppose that in the case $n = 2$, Γ and Γ_A are not degenerate scales and the given function $f(x)$ satisfies* Supp $\{f(x)\} \in \Omega_1$ *and $f \in H^{-1}(\Omega_1)$. Then the problem (9.2.11) and the problem (9.2.12) exists a unique solution $(u, \rho) \in \mathbb{V}$.*

We have known that the bilinear form $D_C(u, \rho; v, \mu)$ is coercive on $\mathbb{V}^* \times \mathbb{V}^*$, which yields the optimal error estimates of the numerical solution of the problem (9.1.23) (and the problem (9.1.22)). If $D_C(u, \rho; v, \mu)$ is coercive on $\mathbb{V} \times \mathbb{V}$, similarly the optimal error estimate of numerical solution based on finite element method of problem (9.2.11) (and problem (9.2.12)) follows directly. In the following we focus on the coerciveness of the bilinear form $D_C(u, \rho; v, \mu)$ on $\mathbb{V} \times \mathbb{V}$. In the case $n = 3$, we know that bilinear form $A(\rho, \mu)$ is coercive on $H^{-\frac{1}{2}}(\Gamma_A) \times H^{-\frac{1}{2}}(\Gamma_A)$ provided by the Theorem 3.4.2. It yields that bilinear form $D_C(u,\rho;v,\mu)$ is coercive on $\mathbb{V}\times\mathbb{V}$ directly.

In the case $n=2$, let γ_A denote the γ-index of the boundary Γ_A by the Definition 3.4.2, Then we know that there exists $\rho_\Gamma\in H_*^{-\frac{1}{2}}(\Gamma_A)$, such that

$$\int_{\Gamma_A}E_2(x,y)\mathrm{d}s_y=\int_{\Gamma_A}E_2(x,y)\rho_\Gamma(y)\mathrm{d}s_y+\gamma_A,\ \forall x\in\Gamma_A. \tag{9.2.13}$$

For any $\rho\in H^{-\frac{1}{2}}(\Gamma_A)$, let

$$\rho(x)=\rho_*(x)+\alpha_*,\quad \text{with}\quad \alpha_*=\frac{1}{\text{meas}\{\Gamma_A\}}\int_{\Gamma_A}\rho(x)\mathrm{d}s_x. \tag{9.2.14}$$

Obviously $\rho_*\in H_*^{-\frac{1}{2}}(\Gamma_A)$.

$$\begin{aligned}&\int_{\Gamma_A}E_2(x,y)\rho(y)\mathrm{d}s_y\\&=\int_{\Gamma_A}E_2(x,y)\rho_*(y)\mathrm{d}s_y+\alpha_*\int_{\Gamma_A}E_2(x,y)\mathrm{d}s_y\\&=\int_{\Gamma_A}E_2(x,y)(\rho_*(y)+\alpha_*\rho_\Gamma(y))\mathrm{d}s_y+\alpha_*\gamma_A,\ \forall x\in\Gamma_A.\end{aligned} \tag{9.2.15}$$

The last equality is from equality (9.2.13).

Furthermore

$$\begin{aligned}A(\rho,\rho)&=\int_{\Gamma_A}\int_{\Gamma_A}E_2(x,y)\rho(y)\rho(x)\mathrm{d}s_x\mathrm{d}s_y\\&=\int_{\Gamma_A}\int_{\Gamma_A}E_2(x,y)(\rho_*(y)+\alpha_*\rho_\Gamma(y))(\rho_*(x)+\alpha_*\rho_\Gamma(x))\mathrm{d}s_y\mathrm{d}s_x+\\&\quad\text{meas}(\Gamma_A)\alpha_*^2\gamma_A.\end{aligned}$$

Again the last equality is from equality (9.2.13), and

$$\begin{aligned}A(\rho,\rho)=A(\rho_*+\alpha_*\rho_\Gamma,\rho_*+\alpha_*\rho_\Gamma)+\text{meas}(\Gamma_A)\alpha_*^2\gamma_A,\\ \text{with }(\rho_*+\alpha_*\rho_\Gamma)\in H_*^{-\frac{1}{2}}(\Gamma_A).\end{aligned} \tag{9.2.16}$$

(i) In the case $\gamma_A>0$. Since $(\rho_*(x)+\alpha_*\rho_\Gamma(x))\in H_*^{-\frac{1}{2}}(\Gamma_A)$ and $A(\rho,\rho)$ is coercive on $H_*^{-\frac{1}{2}}(\Gamma_A)$. We know that there exists a positive constant $\beta_1>0$, such that

$$A(\rho,\rho)\geqslant\beta_1\|\rho_*+\alpha_*\rho_\Gamma\|^2_{-\frac{1}{2},\Gamma_A}+\text{meas}\{\Gamma_A\}\alpha_*^2\gamma_A,\ \forall\rho\in H^{-\frac{1}{2}}(\Gamma_A). \tag{9.2.17}$$

Moreover form inequality (9.2.17) we can find a constant $\beta_2 > 0$, such that

$$A(\rho,\rho) \geqslant \beta_2(\|\rho_*\|_{-\frac{1}{2},\Gamma_A} + |\alpha_*|)^2, \ \forall \rho \in H^{-\frac{1}{2}}(\Gamma_A). \tag{9.2.18}$$

$\|\rho_*\|_{-\frac{1}{2},\Gamma_A} + |\alpha_*|$ is a equivalent norm of $H^{-\frac{1}{2}}(\Gamma_A)$, namely there is a constant $\beta_3 > 0$, such that

$$\|\rho_*\|_{-\frac{1}{2},\Gamma_A} + |\alpha_*| \geqslant \beta_3\|\rho\|_{-\frac{1}{2},\Gamma_A}, \ \forall \rho \in H^{-\frac{1}{2}}(\Gamma_A). \tag{9.2.19}$$

Combining inequalities (9.2.18) and (9.2.19) we have

$$A(\rho,\rho) \geqslant \beta\|\rho\|^2_{-\frac{1}{2},\Gamma_A}, \ \ \forall \rho \in H^{-\frac{1}{2}}(\Gamma_A), \ \text{with } \beta = \beta_2\beta_3^2 > 0. \tag{9.2.20}$$

It means that bilinear form $A(\rho,\rho)$ is coercive on $H^{-\frac{1}{2}}(\Gamma_A) \times H^{-\frac{1}{2}}(\Gamma_A)$ provided $\gamma_A > 0$.

(ii) In the case $\gamma_A = 0$, namely Γ_A is degenerate scale. The equality (9.2.16) yields

$$A(\rho,\rho) = A(\rho_* + \alpha_*\rho_\Gamma, \rho_* + \alpha_*\rho_\Gamma), \ \forall \rho \in H^{-\frac{1}{2}}(\Gamma_A).$$

Taking $\rho = -\rho_\Gamma + 1 \in H^{-\frac{1}{2}}(\Gamma_A)$, and $\rho(x)$ is not the zero element, the equality (9.2.13) with $\gamma_A = 0$ yields that

$$A(\rho,\rho) = 0.$$

It means that in the case $\gamma_A = 0$, $A(\rho,\mu)$ is not coercive on $H^{-\frac{1}{2}}(\Gamma_A) \times H^{-\frac{1}{2}}(\Gamma_A)$.

(iii) In the case $\gamma_A < 0$, taking

$$\rho = -\rho_\Gamma + 1 \in H^{-\frac{1}{2}}(\Gamma_A). \tag{9.2.21}$$

Substituting (9.2.16) it arrives at

$$A(\rho,\rho) = \text{meas}(\Gamma_A)\gamma_A < 0.$$

Obviously $A(\rho,\mu)$ is not coercive on $H^{-\frac{1}{2}}(\Gamma_A) \times H^{-\frac{1}{2}}(\Gamma_A)$.

Synthesizing the above results, we obtain directly.

Theorem 9.2.3 *In the case $n = 3$, the bilinear form $D_C(u,\rho; v,\mu)$ is coercive on $\mathbb{V} \times \mathbb{V}$.*

In the case $n = 2$,

(i) if $\gamma_A > 0$, the bilinear form $D_C(u, \rho; v, \mu)$ is coercive on $\mathbb{V} \times \mathbb{V}$,
(ii) if $\gamma_A \leqslant 0$, the bilinear form $D_C(u, \rho; v, \mu)$ is not coercive.

In the case $D_C(u, \rho; v, \mu)$ is coercive on $\mathbb{V} \times \mathbb{V}$, the error estimates of the numerical solutions based on the finite element method of problem (9.2.11) (and problem (9.2.12)) are directly obtained through the similar discussion in Sect. 9.1.2, it omitted here.

9.3 An Exterior Displacement Problem of Nonhomogene Ous Navier System

In this section we consider the following exterior displacement problem of nonhomogeneous Navier system:

$$\begin{cases} -\nabla \cdot \boldsymbol{\sigma}(\boldsymbol{u}(x)) = \boldsymbol{f}(x), \ \forall x \in \Omega_e, & (9.3.1a) \\ \boldsymbol{u}(x)\big|_{\Gamma} = 0, & (9.3.1b) \\ \boldsymbol{u}(x) = \sum_{j=1}^{n} m_j \boldsymbol{\omega}_j(x) + \mathcal{O}(|x|^{1-n}), \quad \text{as} \quad |x| \to +\infty, & (9.3.1c) \\ \int_{\Gamma_R} \left(\boldsymbol{\sigma}(\boldsymbol{u}(x))\boldsymbol{\nu}_R(x)\right)^{\mathrm{T}} \boldsymbol{\omega}_j(x) \mathrm{d}s_x = 0, \ \text{for} \ R \geqslant R_0 & \\ \text{and} \ \Gamma_{R_0} \in \Omega_2, \ j = 1, 2 \ldots, n, & (9.3.1d) \end{cases}$$

where $\boldsymbol{f}(x) = (f_1(x), f_2(x), \ldots, f_n(x))^{\mathrm{T}}$ is given and $\mathrm{Supp}\{\boldsymbol{f}(x)\} \in \Omega_1$, $\{m_1, m_2, \ldots, m_n\}$ are unknown constants.

9.3.1 *The Coercive and Symmetrical Variational Formulation of Problem (9.3.1) on Bounded Domain Ω_1*

Obviously on the unbounded domain Ω_2, $\boldsymbol{u}(x)$, the solution of problem (9.3.1), satisfies

$$\begin{cases} -\nabla \cdot \boldsymbol{\sigma}(\boldsymbol{u}(x)) = 0, \ \forall x \in \Omega_2, & (9.3.2a) \\ \boldsymbol{u}(x) = \sum_{j=1}^{n} m_j \boldsymbol{\omega}_j(x) + \mathcal{O}(|x|^{1-n}), \quad \text{as} \quad |x| \to +\infty, & (9.3.2b) \\ \int_{\Gamma_R} \left(\boldsymbol{\sigma}(\boldsymbol{u}(x))\boldsymbol{\nu}_R(x)\right)^{\mathrm{T}} \boldsymbol{\omega}_j(x) \mathrm{d}s_x = 0, & \\ \text{for} \ R \geqslant R_0 \ \text{and} \ \Gamma_{R_0} \in \Omega_2, \ j = 1, 2 \ldots, n. & (9.3.2c) \end{cases}$$

Let

$$\boldsymbol{v}(x)=\boldsymbol{u}(x)-\sum_{j=1}^{n}m_j\boldsymbol{\omega}_j(x),\ \forall x\in\Omega_2, \tag{9.3.3}$$

then $\boldsymbol{v}(x)$ satisfies

$$\begin{cases}-\nabla\cdot\boldsymbol{\sigma}(\boldsymbol{v}(x))=0, & \forall x\in\Omega_2,\\ \boldsymbol{v}(x)=\mathcal{O}(|x|^{1-n}), & |x|\to+\infty.\end{cases}$$

The Betti-Somigliam representation formula of $\boldsymbol{v}(x)$ on Ω_2 yields

$$\boldsymbol{v}(x)=-\int_{\Gamma_A}\Big(\mathbb{T}(D_y,\boldsymbol{\nu}_y)\mathbb{E}_n(x,y)\Big)^{\mathrm{T}}\boldsymbol{v}(y)\mathrm{d}s_y+\\ \int_{\Gamma_A}\mathbb{E}_n(x,y)\mathbb{T}(D_y,\boldsymbol{\nu}_y)\boldsymbol{v}(y)\mathrm{d}s_y,\ \forall x\in\Omega_2, \tag{9.3.4}$$

with the fundamental solution $\mathbb{E}_n(x,y)$ of Navier operator given by (6.1.2) and $\boldsymbol{\nu}_y$ denotes the outward unit norm vector at $y\in\Gamma_A=\partial\Omega_2$. Substituting (9.3.3) into (9.3.4), it arrives at

$$\boldsymbol{u}(x)=-\int_{\Gamma_A}\Big(\mathbb{T}(D_y,\boldsymbol{\nu}_y)\mathbb{E}_n(x,y)\Big)^{\mathrm{T}}\boldsymbol{u}(y)\mathrm{d}s_y+\int_{\Gamma_A}\mathbb{E}_n(x,y)\mathbb{T}(D_y,\boldsymbol{\nu}_y)\boldsymbol{u}(y)\mathrm{d}s_y+\\ \sum_{j=1}^{n}m_j\int_{\Gamma_A}\Big(\mathbb{T}(D_y,\boldsymbol{\nu}_y)\mathbb{E}_n(x,y)\Big)^{\mathrm{T}}\boldsymbol{\omega}_j(y)\mathrm{d}s_y+\sum_{j=1}^{n}m_j\boldsymbol{\omega}_j(x),\ \forall x\in\Omega_2. \tag{9.3.5}$$

The progerties of the double layer potential of Navier system yields

$$\int_{\Gamma_A}\Big(\mathbb{T}(D_y,\boldsymbol{\nu}_y)\mathbb{E}_n(x,y)\Big)^{\mathrm{T}}\boldsymbol{\omega}_j(y)\mathrm{d}s_y=0,\ \forall x\in\Omega_2,\ j=1,2,\ldots,n. \tag{9.3.6}$$

Directly we obtain

$$\boldsymbol{u}(x)=-\int_{\Gamma_A}\Big(\mathbb{T}(D_y,\boldsymbol{\nu}_y)\mathbb{E}_n(x,y)\Big)^{\mathrm{T}}\boldsymbol{u}(y)\mathrm{d}s_y+\int_{\Gamma_A}\mathbb{E}_n(x,y)\mathbb{T}(D_y,\boldsymbol{\nu}_y)\boldsymbol{u}(y)\mathrm{d}s_y+\\ \sum_{j=1}^{n}m_j\boldsymbol{\omega}_j(x),\ \forall x\in\Omega_2. \tag{9.3.7}$$

Let

$$\boldsymbol{p}(x)=\mathbb{T}(D_x,\boldsymbol{\nu}_x)\boldsymbol{u}(x)\Big|_{\Gamma_A}, \tag{9.3.8}$$

then we have

$$\int_{\Gamma_A}\boldsymbol{p}(x)^{\mathrm{T}}\boldsymbol{\omega}_j(x)\mathrm{d}s_x=0,\quad j=1,2,\ldots,n. \tag{9.3.9}$$

From Formulas (9.3.7) and (9.3.8) on the boundary Γ_A, we obtain

$$\frac{1}{2}\boldsymbol{u}(x) = -\int_{\Gamma_A}\Big(\mathbb{T}(D_y,\boldsymbol{\nu}_y)\mathbb{E}_n(x,y)\Big)^{\mathrm{T}}\boldsymbol{u}(y)\mathrm{d}s_y + \int_{\Gamma_A}\mathbb{E}_n(x,y)\boldsymbol{p}(y)\mathrm{d}s_y + \sum_{j=1}^{n} m_j\boldsymbol{\omega}_j(x),\ \forall x\in\Gamma_A, \tag{9.3.10}$$

$$\frac{1}{2}\boldsymbol{p}(x) = -\int_{\Gamma_A}\mathbb{T}(D_x,\boldsymbol{\nu}_x)\Big(\mathbb{T}(D_y,\boldsymbol{\nu}_y)\mathbb{E}_n(x,y)\Big)^{\mathrm{T}}\boldsymbol{u}(y)\mathrm{d}s_y + \int_{\Gamma_A}\mathbb{T}(D_x,\boldsymbol{\nu}_x)\mathbb{E}_n(x,y)\boldsymbol{p}(y)\mathrm{d}s_y,\ \forall x\in\Gamma_A. \tag{9.3.11}$$

It is ready to reduce the exterior problem (9.3.1) on the bounded domain Ω_1 equivalently.

$$\begin{cases} -\nabla\cdot\boldsymbol{\sigma}(\boldsymbol{u}(x)) = \boldsymbol{f}(x),\ \forall x\in\Omega_1, & \text{(9.3.12a)}\\ \boldsymbol{u}(x)\Big|_{\Gamma} = 0, & \text{(9.3.12b)}\\ \frac{1}{2}\boldsymbol{u}(x) = -\int_{\Gamma_A}\Big(\mathbb{T}(D_y,\boldsymbol{\nu}_y)\mathbb{E}_n(x,y)\Big)^{\mathrm{T}}\boldsymbol{u}(y)\mathrm{d}s_y + \\ \qquad \int_{\Gamma_A}\mathbb{E}_n(x,y)\boldsymbol{p}(y)\mathrm{d}s_y + \sum_{j=1}^{n} m_j\boldsymbol{\omega}_j(x),\ \forall x\in\Gamma_A, & \text{(9.3.12c)}\\ \frac{1}{2}\boldsymbol{p}(x) = -\int_{\Gamma_A}\mathbb{T}(D_x,\boldsymbol{\nu}_x)\Big(\mathbb{T}(D_y,\boldsymbol{\nu}_y)\mathbb{E}_n(x,y)\Big)^{\mathrm{T}}\boldsymbol{u}(y)\mathrm{d}s_y + \\ \qquad \int_{\Gamma_A}\mathbb{T}(D_x,\boldsymbol{\nu}_x)\mathbb{E}_n(x,y)\boldsymbol{p}(y)\mathrm{d}s_y,\ \forall x\in\Gamma_A, & \text{(9.3.12d)}\\ \int_{\Gamma_A}\boldsymbol{p}(x)^{\mathrm{T}}\boldsymbol{\omega}_j(x)\mathrm{d}s_x = 0, \quad j=1,2,\ldots,n. & \text{(9.3.12e)} \end{cases}$$

Let

$$\mathbb{H}^1_*(\Omega_1) = \left\{\boldsymbol{v}\Big|\boldsymbol{v}\in\mathbb{H}^1(\Omega_1)\ \text{and}\ \boldsymbol{v}\Big|_{\Gamma}=0\right\},$$

$$\mathbb{H}^{-\frac{1}{2}}_*(\Gamma_A) = \left\{\boldsymbol{q}\Big|\boldsymbol{q}\in\mathbb{H}^{-\frac{1}{2}}(\Gamma_A)\ \text{and}\ \int_{\Gamma_A}\boldsymbol{p}(x)^{\mathrm{T}}\boldsymbol{\omega}_j(x)\mathrm{d}s_x = 0,\ j=1,2,\ldots,n\right\},$$

$$\mathbb{V}^* = \mathbb{H}^1_*(\Omega_1)\times\mathbb{H}^{-\frac{1}{2}}_*(\Gamma_A).$$

The unknowns in problem (9.3.12) are unknown vector-value function $\boldsymbol{u}(x)$ on $\bar{\Omega}_1$, unknown vector-value function $\boldsymbol{p}(x)$ on Γ_A and unknown constants $\{m_1, m_2, \ldots, m_n\}$.

An application of the Betti-Somigliam on Ω_1 for $\boldsymbol{u}(x)$ satisfied equation (9.3.12a) and $\boldsymbol{v}\in\mathbb{H}^1_*(\Omega_1)$ yields

$$a(\boldsymbol{u}, \boldsymbol{v}) + \int_{\Gamma_A} \boldsymbol{p}^{\mathrm{T}} \boldsymbol{v} \mathrm{d}s_x = \int_{\Omega_1} \boldsymbol{f}^{\mathrm{T}} \boldsymbol{v} \mathrm{d}s_x, \ \forall \boldsymbol{v} \in \mathbb{H}^1_*(\Omega_1), \tag{9.3.13}$$

with the bilinear form $a(\boldsymbol{u}, \boldsymbol{v})$ defined by (8.3.1). Inserting (9.3.12d) into (9.3.13) and coupling the variational formula of Eq. (9.3.12c), we obtain the equivalent variational problem of (9.3.12):

$$\begin{cases} \text{Find}\Big(\boldsymbol{u}, \boldsymbol{p}\Big) \in \mathbb{V}^*, \quad \text{such that} \\ a(\boldsymbol{u}, \boldsymbol{v}) - \\ \int_{\Gamma_A}\int_{\Gamma_A} \boldsymbol{v}(x)^{\mathrm{T}} \mathbb{T}(D_x, \boldsymbol{\nu}_x)(\mathbb{T}(D_y, \boldsymbol{\nu}_y)\mathbb{E}_n(x, y))^{\mathrm{T}} \boldsymbol{u}(y) \mathrm{d}s_y \mathrm{d}s_x + \\ \frac{1}{2}\int_{\Gamma_A} \boldsymbol{p}(x)^{\mathrm{T}} \boldsymbol{v}(x) \mathrm{d}s_x + \\ \int_{\Gamma_A}\int_{\Gamma_A} \boldsymbol{p}(y)^{\mathrm{T}} (\mathbb{T}(D_x, \boldsymbol{\nu}_x)\mathbb{E}_n(x, y))^{\mathrm{T}} \boldsymbol{v}(x) \mathrm{d}s_y \mathrm{d}s_x \\ = \int_{\Omega_1} \boldsymbol{f}(x)^{\mathrm{T}} \boldsymbol{v}(x) \mathrm{d}s, \ \forall \boldsymbol{v} \in \mathbb{H}^1_*(\Omega), & (9.3.14\mathrm{a}) \\ -\frac{1}{2}\int_{\Gamma_A} \boldsymbol{q}(x)^{\mathrm{T}} \boldsymbol{u}(x) \mathrm{d}s_x - \\ \int_{\Gamma_A}\int_{\Gamma_A} \boldsymbol{q}(x)^{\mathrm{T}} (\mathbb{T}(D_y, \boldsymbol{\nu}_y)\mathbb{E}_n(x, y))^{\mathrm{T}} \boldsymbol{u}(x) \mathrm{d}s_y \mathrm{d}s_x + \\ \int_{\Gamma_A}\int_{\Gamma_A} \boldsymbol{q}(x)^{\mathrm{T}} \mathbb{E}_n(x, y) \boldsymbol{p}(x) \mathrm{d}s_y \mathrm{d}s_x = 0, \ \forall \boldsymbol{q} \in \mathbb{H}_*^{-\frac{1}{2}}(\Gamma_A). & (9.3.14\mathrm{b}) \end{cases}$$

$\mathbb{A}(\boldsymbol{p}, \boldsymbol{q})$, $\mathbb{B}(\boldsymbol{u}, \boldsymbol{v})$ defined by (6.3.10), (6.3.11) and $\mathbb{C}(\boldsymbol{p}, \boldsymbol{v})$ defined by (8.3.10). On the boundary Γ_A they are

$$\mathbb{A}(\boldsymbol{p}, \boldsymbol{q}) = \int_{\Gamma_A}\int_{\Gamma_A} \boldsymbol{p}(x)^{\mathrm{T}} \mathbb{E}_n(x, y) \boldsymbol{q}(y) \mathrm{d}s_y \mathrm{d}s_x, \ \forall \boldsymbol{p}, \boldsymbol{q} \in \mathbb{H}^{-\frac{1}{2}}(\Gamma_A),$$

$$\begin{aligned} \mathbb{B}(\boldsymbol{u}, \boldsymbol{v}) &= -\int_{\Gamma_A}\int_{\Gamma_A} \boldsymbol{v}(x)^{\mathrm{T}} \mathbb{T}(D_x, \boldsymbol{\nu}_x)(\mathbb{T}(D_y, \boldsymbol{\nu}_y)\mathbb{E}_n(x, y))^{\mathrm{T}} \boldsymbol{u}(y) \mathrm{d}s_y \mathrm{d}s_x \\ &= \frac{1}{2(n-1)\pi} \sum_{1 \leqslant k \leqslant j \leqslant n} \int_{\Gamma_A}\int_{\Gamma_A} \gamma(x, y) \frac{\partial \boldsymbol{u}(x)}{\partial \boldsymbol{\tau}_y^{k,j}} \cdot \frac{\partial \boldsymbol{v}(x)}{\partial \boldsymbol{\tau}_x^{k,j}} \mathrm{d}s_y \mathrm{d}s_x + \end{aligned}$$

$$\int_{\Gamma_A}\int_{\Gamma_A} (\mathbb{U}(D_y, \boldsymbol{\nu}_y)\boldsymbol{u}(y))^{\mathrm{T}} (4\mu\mathbb{E}_n(x, y) -$$

$$\frac{\mu}{(n-1)\pi} \gamma(x, y) I) \mathbb{U}(D_x, \boldsymbol{\nu}_x) \boldsymbol{v}(x) \mathrm{d}s_y \mathrm{d}s_x +$$

$$\frac{\mu}{2(n-1)\pi} \sum_{i,j,k=1}^{n} \int_{\Gamma_A}\int_{\Gamma_A} U_{k,i}(D_y, \boldsymbol{\nu}_y) u_j(y) \gamma(x, y) U_{k,j}(D_x, \boldsymbol{\nu}_x) v_i(x) \mathrm{d}s_y \mathrm{d}s_x,$$

$$\forall \boldsymbol{u}, \boldsymbol{v} \in \mathbb{H}^{\frac{1}{2}}(\Gamma_A),$$

$$\mathbb{C}(\boldsymbol{p}, \boldsymbol{v}) = \frac{1}{2} \int_{\Gamma_A} \boldsymbol{p}(x)^{\mathrm{T}} \boldsymbol{v}(x) \mathrm{d}s_x + \int_{\Gamma_A} \int_{\Gamma_A} \boldsymbol{p}(y)^{\mathrm{T}} (\mathbb{T}(D_x, \boldsymbol{\nu}_x) \mathbb{E}_n(x, y))^{\mathrm{T}} \boldsymbol{v}(x) \mathrm{d}s_y \mathrm{d}s_x.$$

Then the variational problem (9.3.14) can be rewritten as

$$\begin{cases} \text{Find } (\boldsymbol{u}, \boldsymbol{p}) \in \mathbb{V}^*, \quad \text{such that} \\ a(\boldsymbol{u}, \boldsymbol{v}) + \mathbb{B}(\boldsymbol{u}, \boldsymbol{v}) + \mathbb{C}(\boldsymbol{p}, \boldsymbol{v}) = \int_{\Omega_1} \boldsymbol{f}(x)^{\mathrm{T}} \boldsymbol{v}(x) \mathrm{d}x, \ \forall \boldsymbol{v} \in \mathbb{H}_*^{\frac{1}{2}}(\Gamma_A), & (9.3.15a) \\ -\mathbb{C}(\boldsymbol{q}, \boldsymbol{u}) + \mathbb{A}(\boldsymbol{q}, \boldsymbol{u}) = 0, \ \forall \boldsymbol{q} \in \mathbb{H}_*^{-\frac{1}{2}}(\Gamma_A). & (9.3.15b) \end{cases}$$

The equivalent symmetric form of problem (9.3.15) is given by

$$\begin{cases} \text{Find } (\boldsymbol{u}, \boldsymbol{p}) \in \mathbb{V}^*, \quad \text{such that} \\ a(\boldsymbol{u}, \boldsymbol{v}) + \mathbb{B}(\boldsymbol{u}, \boldsymbol{v}) + \mathbb{C}(\boldsymbol{p}, \boldsymbol{v}) = \int_{\Omega_1} \boldsymbol{f}(x)^{\mathrm{T}} \boldsymbol{v}(x) \mathrm{d}x, \ \forall \boldsymbol{v} \in \mathbb{H}_*^{\frac{1}{2}}(\Gamma_A), & (9.3.16a) \\ \mathbb{C}(\boldsymbol{q}, \boldsymbol{u}) - \mathbb{A}(\boldsymbol{q}, \boldsymbol{p}) = 0, \ \forall \boldsymbol{q} \in \mathbb{H}_*^{-\frac{1}{2}}(\Gamma_A). & (9.3.16b) \end{cases}$$

Let

$$\mathbb{D}_C(\boldsymbol{u}, \boldsymbol{p}; \boldsymbol{v}, \boldsymbol{q}) = a(\boldsymbol{u}, \boldsymbol{v}) + \mathbb{B}(\boldsymbol{u}, \boldsymbol{v}) + \mathbb{C}(\boldsymbol{p}, \boldsymbol{v}) - \mathbb{C}(\boldsymbol{q}, \boldsymbol{u}) + \mathbb{A}(\boldsymbol{q}, \boldsymbol{p}), \tag{9.3.17}$$

$$\mathbb{D}_S(\boldsymbol{u}, \boldsymbol{p}; \boldsymbol{v}, \boldsymbol{q}) = a(\boldsymbol{u}, \boldsymbol{v}) + \mathbb{B}(\boldsymbol{u}, \boldsymbol{v}) + \mathbb{C}(\boldsymbol{p}, \boldsymbol{v}) + \mathbb{C}(\boldsymbol{q}, \boldsymbol{u}) - \mathbb{A}(\boldsymbol{q}, \boldsymbol{p}). \tag{9.3.18}$$

Furthermore the problem (9.3.15) and (9.3.16) can be rewritten as the following

$$\begin{cases} \text{Find } (\boldsymbol{u}, \boldsymbol{p}) \in \mathbb{V}^*, \quad \text{such that} \\ \mathbb{D}_C(\boldsymbol{u}, \boldsymbol{p}; \boldsymbol{v}, \boldsymbol{q}) = \int_{\Omega_1} \boldsymbol{f}(x)^{\mathrm{T}} \boldsymbol{v}(x) \mathrm{d}x, \ \forall (\boldsymbol{v}, \boldsymbol{q}) \in \mathbb{V}^*. \end{cases} \tag{9.3.19}$$

$$\begin{cases} \text{Find } (\boldsymbol{u}, \boldsymbol{p}) \in \mathbb{V}^*, \quad \text{such that} \\ \mathbb{D}_S(\boldsymbol{u}, \boldsymbol{p}; \boldsymbol{v}, \boldsymbol{q}) = \int_{\Omega_1} \boldsymbol{f}(x)^{\mathrm{T}} \boldsymbol{v}(x) \mathrm{d}x, \ \forall (\boldsymbol{v}, \boldsymbol{q}) \in \mathbb{V}^*. \end{cases} \tag{9.3.20}$$

The problems (9.3.19) and (9.3.20) are equivalent. Obviously the bilinear form $\mathbb{D}_S(\boldsymbol{u}, \boldsymbol{p}; \boldsymbol{v}, \boldsymbol{q})$ on $\mathbb{V}^* \times \mathbb{V}^*$ is symmetric, but it is not coercive. The bilinear form $\mathbb{D}_C(\boldsymbol{u}, \boldsymbol{p}; \boldsymbol{v}, \boldsymbol{q})$ on $\mathbb{V}^* \times \mathbb{V}^*$ is not symmetric, the coerciveness of $\mathbb{D}_C(\boldsymbol{u}, \boldsymbol{p}; \boldsymbol{v}, \boldsymbol{q})$ is given in the following lemma:

Lemma 9.3.1 *(i) The bilinear form $\mathbb{D}_C(\boldsymbol{u}, \boldsymbol{p}; \boldsymbol{v}, \boldsymbol{q})$ is bounded on $\mathbb{V}^* \times \mathbb{V}^*$, namely there exists a positive constant M, such that*

$$\left|\mathbb{D}_C(\boldsymbol{u}, \boldsymbol{p}; \boldsymbol{v}, \boldsymbol{q})\right| \leqslant M\|(\boldsymbol{u}, \boldsymbol{p})\|_{\mathbb{V}^*}\|(\boldsymbol{v}, \boldsymbol{q})\|_{\mathbb{V}^*}, \ \forall(\boldsymbol{u}, \boldsymbol{p}), \ (\boldsymbol{v}, \boldsymbol{q}) \in \mathbb{V}^*. \tag{9.3.21}$$

(ii) The bilinear form $\mathbb{D}_C(\boldsymbol{u}, \boldsymbol{p}; \boldsymbol{v}, \boldsymbol{q})$ is coercive on $\mathbb{V}^* \times \mathbb{V}^*$, in other words there is a positive constant β, such that

$$\mathbb{D}_C(\boldsymbol{u}, \boldsymbol{p}; \boldsymbol{v}, \boldsymbol{q}) \geqslant \beta\|(\boldsymbol{v}, \boldsymbol{q})\|^2_{\mathbb{V}^*}, \ \forall(\boldsymbol{v}, \boldsymbol{q}) \in \mathbb{V}^*. \tag{9.3.22}$$

In the case $n = 2$, the proof of Lemma 9.3.1 was given in the paper (Han, 1987, 1990). In the case $n = 3$, the proof is provided directly by the coerciveness of $a(\boldsymbol{u}, \boldsymbol{v})$ on $\mathbb{H}^{\frac{1}{2}}_*(\Omega_1) \times \mathbb{H}^{\frac{1}{2}}_*(\Omega_1)$, $\mathbb{A}(\boldsymbol{p}, \boldsymbol{q})$ on $\mathbb{H}^{-\frac{1}{2}}_*(\Gamma_A) \times \mathbb{H}^{-\frac{1}{2}}_*(\Gamma_A)$ and $\mathbb{B}(\boldsymbol{u}, \boldsymbol{v})$ is non-negative on $\mathbb{H}^{\frac{1}{2}}(\Gamma_A) \times \mathbb{H}^{\frac{1}{2}}(\Gamma_A)$.

An application of Lax-Milgram Theorem it arrives at

Theorem 9.3.1 *For given $\boldsymbol{f} \in \mathbb{H}^{-1}(\Omega_1)$, the problem (9.3.19) exists a unique solution $(\boldsymbol{u}, \boldsymbol{p})$ and*

$$\|(\boldsymbol{u}, \boldsymbol{p})\|_{\mathbb{V}^*} \leqslant \frac{M}{\beta}\|\boldsymbol{f}\|_{\mathbb{H}^{-1}(\Omega_1)}. \tag{9.3.23}$$

Furthermore the problem (9.3.20) has a unique solution, which is the unique solution of problem (9.3.19).

9.3.2 *The Discrete Approximation of Problems (9.3.19) and (9.3.20)*

Consider the numerical solution of problems (9.3.19) and (9.3.20). Let $\mathbb{W}_h$ and $\mathbb{M}_h$ be two finite element spaces of $\mathbb{H}^{\frac{1}{2}}_*(\Omega_1)$ and $\mathbb{H}^{-\frac{1}{2}}_*(\Gamma_A)$, which satisfy

$$\inf_{\boldsymbol{v}_h \in \mathbb{W}_h} \|\boldsymbol{u} - \boldsymbol{v}_h\|_{\mathbb{H}^1(\Omega_1)} \leqslant Ch^t|\boldsymbol{u}|_{\mathbb{H}^{1+t}(\Omega_1)}, \ \forall \boldsymbol{u} \in \mathbb{H}^1_*(\Omega_1) \cap \mathbb{H}^{1+t}(\Omega_1),$$

$$\inf_{\boldsymbol{q}_h \in \mathbb{M}_h} \|\boldsymbol{p} - \boldsymbol{q}_h\|_{\mathbb{H}^{-\frac{1}{2}}(\Gamma_A)} \leqslant Ch^t|\boldsymbol{p}|_{\mathbb{H}^{-\frac{1}{2}+t}(\Gamma_A)}, \ \forall \boldsymbol{p} \in \mathbb{H}^{-\frac{1}{2}}_*(\Gamma_A) \cap \mathbb{H}^{-\frac{1}{2}+t}(\Gamma_A),$$

with positive constant C and $0 < t \leqslant 1$.

Let $\mathbb{V}^*_h = \mathbb{W}_h \times \mathbb{M}_h$, the discrete approximation of problem (9.3.19) is given by

$$\begin{cases} \text{Find } (\boldsymbol{u}_h, \boldsymbol{p}_h) \in \mathbb{V}^*_h, \ \text{ such that} \\ \mathbb{D}_C(\boldsymbol{u}_h, \boldsymbol{p}_h; \boldsymbol{v}_h, \boldsymbol{q}_h) = \displaystyle\int_{\Omega_1} \boldsymbol{f}(x)^{\mathrm{T}}\boldsymbol{v}_h(x)\mathrm{d}x, \ \forall(\boldsymbol{v}_h, \boldsymbol{q}_h) \in \mathbb{V}^*_h. \end{cases} \tag{9.3.24}$$

and form the problem (9.3.20) we obtain

$$\begin{cases} \text{Find } (\boldsymbol{u}_h, \boldsymbol{p}_h) \in \mathbb{V}_h^*, \quad \text{such that} \\ \mathbb{D}_S(\boldsymbol{u}_h, \boldsymbol{p}_h; \boldsymbol{v}_h, \boldsymbol{q}_h) = \displaystyle\int_{\Omega_1} \boldsymbol{f}(x)^{\mathrm{T}} \boldsymbol{v}_h(x) \mathrm{d}x, \ \forall (\boldsymbol{v}_h, \boldsymbol{q}_h) \in \mathbb{V}_h^*. \end{cases} \tag{9.3.25}$$

The problems (9.3.24) and (9.3.25) are equivalent. The problem (9.3.25) yields the symmetric linear algebraic equation. Straightly we get the following theorem:

Theorem 9.3.2 *(i) The problem (9.3.24) and (9.3.25) exists a unique solution* $(\boldsymbol{u}_h, \boldsymbol{p}_h) \in \mathbb{V}_h^*$.

(ii) Assume that the solution $(\boldsymbol{u}, \boldsymbol{p})$ *of problem (9.3.19) satisfies*

$$\boldsymbol{u} \in \mathbb{H}_*^1(\Omega_1) \cap \mathbb{H}^{1+t}(\Omega_1), \quad \boldsymbol{p} \in \mathbb{H}_*^{-\frac{1}{2}}(\Gamma_A) \cap \mathbb{H}^{-\frac{1}{2}+t}(\Gamma_A) \ \textit{with} \ 0 < t \leqslant 1.$$

The following error estimate holds

$$\begin{aligned} \|(\boldsymbol{u} - \boldsymbol{u}_h, \boldsymbol{p} - \boldsymbol{p}_h)\|_{\mathbb{V}} &\leqslant \frac{M}{\beta} \inf_{(\boldsymbol{v}_h, \boldsymbol{q}_h) \in \mathbb{V}_h^*} \|(\boldsymbol{u} - \boldsymbol{v}_h, \boldsymbol{p} - \boldsymbol{q}_h)\|_{\mathbb{V}} \\ &\leqslant Ch^t \left\{ |\boldsymbol{u}|_{\mathbb{H}^{1+t}(\Omega_1)} + |\boldsymbol{p}|_{\mathbb{H}^{-\frac{1}{2}+t}(\Gamma_A)} \right\}. \end{aligned} \tag{9.3.26}$$

In this chapter, we discuss the coercive and symmetrical coupling methods of finite element method and boundary element method through three typical examples. The coupling methods of finite element method and boundary element method are widely used in scientific and engineering computation, such as elastoplastic interface problem (Costabel & Stephan, 1990), elastic-plastic problem on the unbounded domain (Wu, 1992), electromagnetic scattering problem (Hiptmair, 2003), exterior Stokes problem (Gatica et al., 2013), and various type of nonlinear problems (Carstensen & Gwinner, 1997; Gatica & Hsiao, 1992; Gatica & Wendland, 1996, 1997; Gatica et al., 2021; Meddahi et al., 2000). Finally, the interested reader can also refer to the review paper by Stephan (2018).

References

Argyris J. H., & Kelsey S. (1960). *Energy theorems and structural analysis* (Vol. 60). Springer.

Brebbia, C. A. (1978). *The boundary element method for engineers*. Wiley.

Brenner S. C., & Scott, L. R. (2008). *The mathematical theory of finite element methods* (Vol. 3). Springer.

Carstensen, C., & Gwinner, J. (1997). Fem and BEM coupling for a nonlinear transmission problem with Signorini contact. *SIAM Journal on Numerical Analysis, 34*(5), 1845–1864.

Costabel, M. (1987). *Symmetric methods for the coupling of finite elements and boundary elements (invited contribution)*. Springer.

Costabel, M., & Stephan, E. P. (1990). Coupling of finite and boundary element methods for an elastoplastic interface problem. *SIAM Journal on Numerical Analysis,27*(5), 1212–1226.

Courant, R. (1943). Variational methods for the solution of problems of equilibrium and vibrations. *Bulletin of the American Mathematical Society, 49*(1), 1–23.

Gatica, G. N., & Hsiao, G. C. (1992). On the coupled BEM and fem for a nonlinear exterior Dirichlet problem in R2. *Numerische Mathematik, 61*(1), 171–214.

Gatica, G. N., Hsiao, G. C., & Meddahi, S. (2021). Further developments on boundary-field equation methods for nonlinear transmission problems. *Journal of Mathematical Analysis and Applications, 502*(2), 125262.

Gatica, G. N., Hsiao, G. C., Meddahi, S., & Sayas, F. J. (2013). On the dual-mixed formulation for an exterior stokes problem. *ZAMM—Journal of Applied Mathematics and Mechanics/Zeitschrift für Angewandte Mathematik und Mechanik, 93*(6–7), 437–445.

Gatica, G. N., & Wendland, W. L. (1996). Coupling of mixed finite elements and boundary elements for linear and nonlinear elliptic problems. *Applicable Analysis, 63*(1–2), 39–75.

Gatica, G. N., & Wendland, W. L. (1997). Coupling of mixed finite elements and boundary elements for a hyperelastic interface problem. *SIAM Journal on Numerical Analysis, 34*(6), 2335–2356.

Han, H. (1987,1990). A new class of variational formulations for the coupling of finite and boundary element methods. Technical Report, Department of Applied Mathematics, Tsinghua University, 1987. *Journal of Computational Mathematics, 8*(3), 223–232.

Hess, J. L., & Smith, A. O. (1967). Calculation of potential flow about arbitrary bodies. *Progress in Aerospace Sciences, 8*, 1–138.

Hiptmair, R. (2003). Coupling of finite elements and boundary elements in electromagnetic scattering. *SIAM Journal on Numerical Analysis,41*(3), 919–944.

Hrennikoff, A. (1941). Solution of problems of elasticity by the framework method. *Journal of Applied Mechanics, 8*(4), A169–A175.

Jaswon, M. (1963). Integral equation methods in potential theory. I. *Proceedings of the Royal Society of London. Series A. Mathematical and Physical Sciences, 275*(1360), 23–32.

Johnson, C., Nédélec, J., et al. (1980). On the coupling of boundary integral and finite element methods. *Mathematics of Computation, 35*(152), 1063–1079.

Massonnet, C. (1965). Numerical use of integral procedures. *Stress Analysis*, 198–235.

Meddahi, S., González, M., & Pérez, P. (2000). On a FEM-BEM formulation for an exterior quasi-linear problem in the plane. *SIAM Journal on Numerical Analysis, 37*(6), 1820–1837.

Stephan, E. P. (2018). Coupling of boundary element methods and finite element methods. *Encyclopedia of Computational Mechanics Second Edition,* 1–40.

Symm, G. (1963). Integral equation methods in potential theory. II. *Proceedings of the Royal Society of London. Series A. Mathematical and Physical Sciences, 275*(1360), 33–46.

Turner, M. J., Clough, R. W., Martin, H. C., & Topp, L. (1956). Stiffness and deflection analysis of complex structures. *Journal of the Aeronautical Sciences, 23*(9), 805–823.

Wendland, W. (1986). On asymptotic error estimates for the combined boundary and finite element method. In *Innovative numerical methods in engineering* (pp. 55–69).

Wu, L. (1992). An elastic-plastic problem in an unbounded domain. *Acts Scientiarum Naturalium Universitatis Pekinensis, 28*(1), 26–35.

Zienkiewicz, O., Kelly, D., & Bettess, P. (1977). The coupling of the finite element method and boundary solution procedures. *International Journal for Numerical Methods in Engineering, 11*(2), 355–375.

www.ingramcontent.com/pod-product-compliance
Lightning Source LLC
Chambersburg PA
CBHW061744040626
46109CB00003BA/52
9789819510870